D1474654

FIVEFOLD SYMMETRY

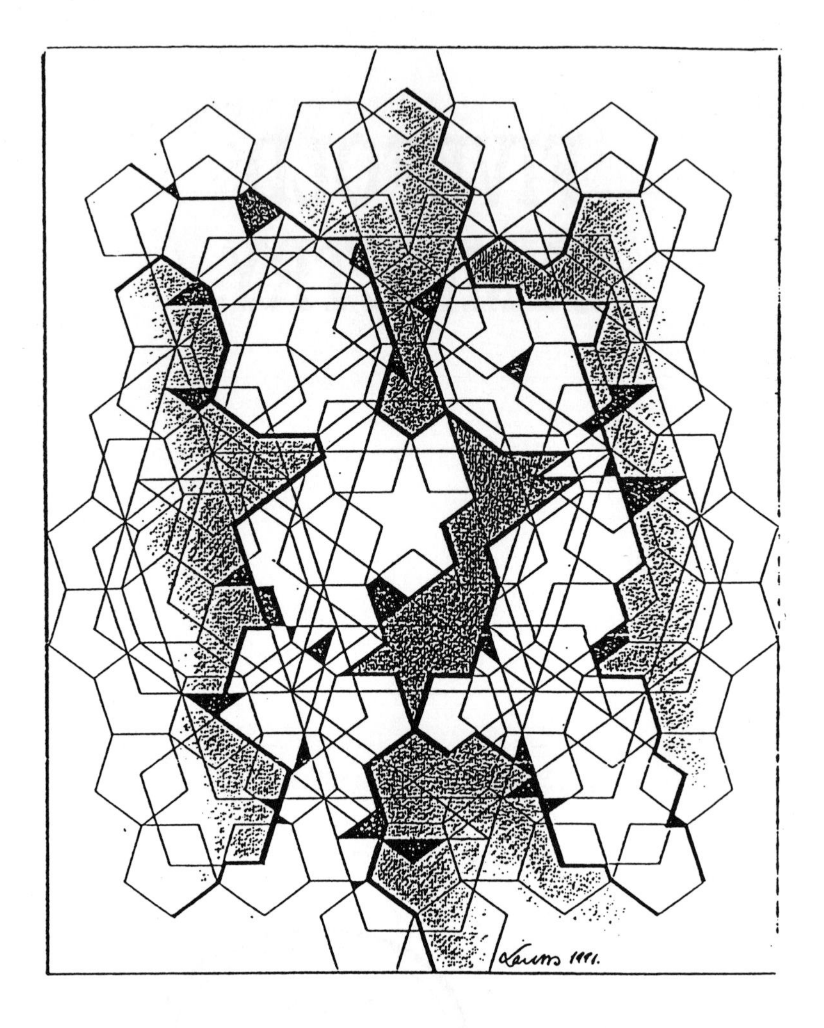

Artist's rendition of fivefold symmetry: "Pentagonal Interference I," by Ferenc Lantos, Pécs, 1991.

FIVEFOLD SYMMETRY

editor

István Hargittai

Hungarian Academy of Sciences
Budapest, Hungary

World Scientific
Singapore • New Jersey • London • Hong Kong

Published by

World Scientific Publishing Co. Pte. Ltd.
P O Box 128, Farrer Road, Singapore 9128
USA office: Suite 1B, 1060 Main Street, River Edge, NJ 07661
UK office: 73 Lynton Mead, Totteridge, London N20 8DH

Library of Congress Cataloging-in-Publication data is available.

FIVEFOLD SYMMETRY

ISBN 981-02-0600-3

Cover illustration: Carrion flower, *Stapelia Gigantea Pallida*.
Photograph taken in Honolulu, Hawaii, by I. and M. Hargittai.

Printed in Singapore by General Printing Services Pte. Ltd.

CONTENTS

CONTRIBUTORS

1. István Hargittai
Hungarian Academy of Sciences
H-1431 Budapest, Pf. 117
Hungary

2. Joe Rosen
School of Physics and Astronomy
Tel-Aviv Univ.
69978 Tel-Aviv
Israel

3. Josef Brandmüller
Faculty of Physics
LM-Univ. of Munich
Private address:
Hubertusstrasse 61
W-8035 Gauting
Germany

4. Jay Kappraff
Dept. of Mathematics
New Jersey Institute of Technology
Newark, NJ 07102
U.S.A.

5. Emil Makovicky
Institute of Mineralogy
Univ. of Copenhagen
Ostervolgade 10
DK-1350 Copenhagen K
Denmark

6. G.C. Shephard
School of Mathematics
Univ. of East Anglia
Norwich NR4 7TJ
England UK

7. Haresh Lalvani
School of Architecture
Pratt Institute
Brooklyn, NY 11205
U.S.A.

8. Douglas Dunham
Dept. of Computer Science
Univ. of Minnesota
Duluth, MN 55812
U.S.A.

9. Robert Dixon
125 Cricklade Avenue
London SW2
England UK

10. Clifford A. Pickover
IBM T. J. Watson Research Center
Yorktown Heights, NY 10598
U.S.A.

11. J. Chris Fisher
Mathematics Dept.

and

Norma Fuller
Computer Science Dept.
Univ. of Regina
Regina S4S DA2
Canada

12. I.S. Zheludev
Institute of Crystallography
Academy of Sciences of the U.S.S.R.
Leninskii pr. 59
117333 Moscow
U.S.S.R.

13. Gábor Gévay
Educational Technology Center
JATE Univ.
Szeged, Boldogasszony sgt 2.
H-6722, Hungary

14. Stephen C. Baer
Zomeworks Corporation
P.O. Box 25805
Albuquerque, NM 87125
U.S.A.

15. David Booth
Green Meadows Waldorf School
Hungry Hollow Road
Spring Valley, NY 10977
U.S.A.

16. Szaniszló Bérczi
Dept. of General Technics
Eötvös Univ.
Rákóczi út 5
Budapest, H-1088
Hungary

and

László Papp
Honvéd Square Calvinist Church
Szeged, Hungary

17. Paulus Gerdes
Faculty of Mathematics and Physics
Higher Pedagogical Institute
P.O. Box 915, Maputo
Mozambique

18. G.M. Fleurent
Abdijstraat 1
B-3271 Averbode
Belgium

19. Arthur Loeb
Dept. of Visual and Environmental Studies
Carpenter Center for the Visual Arts

and

Wasma'a K. Chorbachi
Dept. of Near Eastern Languages and Civilization
Harvard Univ.
Cambridge, MA 02138
U.S.A.

20. Lima de Freitas
Rua Ribeiro Sanches 24–3
1200 Lisboa
Portugal

21. Hugo F. Verheyen
Gloriantlaan 64–15B
2050 Antwerpen
Belgium

22. Koji Miyazaki
College of Liberal Arts
Kyoto Univ.
Yoshida, Sakyo-ku
Kyoto 606
Japan

23. Nenad Trinajstić
The Rugjer Bošković Institute
41001 Zagreb
P.O. Box 1016
Republic of Croatia

24. L.A. Cummings
School of Architecture
Univ. of Waterloo
Waterloo, Ontario N2L 3G1
Canada

25. Akhlesh Lakhtakia
Dept. of Engineering Science and Mechanics
227 Hammond Building
Pennsylvania State Univ.
University Park,
PA 16802–1484
U.S.A.

26. Donald W. Crowe
Dept. of Mathematics
Univ. of Wisconsin
Madison, WI 53706
U.S.A.

27. R.A. Dunlap
Dept. of Physics
Dalhousie Univ.
Halifax, Nova Scotia B3H 3J5
Canada

28. Roger V. Jean
Univ. of Québec
300 avenue des Ursulines
Rimouski, Québec G5L 3A1
Canada

29. Magdolna Hargittai
Hungarian Academy of Sciences
H-1431 Budapest, Pf. 117
Hungary

PREFACE

This book is about fivefold symmetry, and it may appear strange that we are devoting a whole book to this particular symmetry. Before embarking on this project, I asked several colleagues about such an idea. I remember one of the responses especially vividly. "Fivefold symmetry is truly an important subject to compile a book about," this colleague said, "almost as important as, for example, twofold, threefold, fourfold, or sixfold symmetry." Of course, separate volumes could be put together about each of these symmetries, and yet, I didn't feel any urge to do so. Judging by the responses of the potential contributors, whether scientists or humanists or artists, the original suggestion concerning fivefold symmetry has proved legitimate. Nevertheless, I feel that I owe the reader an explanation.

Fivefold symmetry is relatively common in flowers, fruits, molecules, logos, buildings, that is, generally speaking, in our world. It does not strike us as something very peculiar, even though it may indeed not be as common as the other symmetries mentioned above. There is one area, however, in which fivefold symmetry is a forbidden symmetry, and that is in the world of crystals. Three-dimensional bodies with fivefold symmetry cannot be packed with full utilization of the available space, and fivefoldedness is excluded from the possible operations when crystal structures are built. A simpler version of this problem is the impossibility to cover a surface without gaps or overlaps with same-size regular pentagons.

Pentagonal dodecahedron as growth morphology in quasicrystalline Al-Cu-Ru, synthesized by slow cooling from a melt. Specimen: Professor S. Politis, K.F.K. Karlsruhe. Scanning electron micrograph magnification 2200 times. Photograph courtesy of Professor Hans-Ude Nissen, ETH, Zürich, 1991.

The impossibility of fivefold symmetry in crystals used to be commonly accepted until 1984. In that year, a report appeared by Dan Shechtman and co-workers about an experiment in which an aluminium alloy was produced from melt by rapid solidification, and the electron diffraction patterns produced from such a sample showed characteristics of fivefold symmetry. There are few such seemingly perfect domains in science as the symmetry considerations and classifications of crystallography. Thus, it is no wonder that the discovery of Shechtman et al. caused a minirevolution in crystallography and solid state physics. Even *The New York Times* ran an article about the New Matter.

I am not a crystallographer; my immediate interest is in the structure of free molecules, that is, of molecules that are not embedded in a crystal structure. The isolated molecules may, and many do, have fivefold symmetry. I was not particularly interested in the absence of fivefoldedness in crystal structures, just accepted it as a fact, maybe even with slight amusement that the seemingly perfect crystals do lack something. However, by the time I heard about the discovery of Shechtman et al., I was more interested in it than I would have ordinarily been, that is, if I had not been personally exposed to this problem by Alan L. Mackay.

Through my interest in symmetry I was fortunate to meet with Alan Mackay during the 1981 Ottawa Crystallography Congress. In 1982 we invited him for a visit to Budapest, where he gave a lecture on Fivefold Symmetry. It was, in fact, a two-part lecture, given on two evenings. The first part was entitled Generalized Crystallography and the second, Fivefold Symmetry. By prior agreement with the lecturer, we invited a broadly interdisciplinary audience. The lectures were a great success, generating much discussion and having a strong impact. I count my increased interest in fivefold symmetry as coming from those lectures. In a way, in my mind, the Shechtman discovery somehow organically followed Mackay's lectures.

I don't know how much serendipity there was in Shechtman's discovery. To me, however, at least in hindsight, this discovery was very well prepared. Tiling the plane with figures of fivefold symmetry has been a problem for centuries, including Kepler's notable attempts. Then came the discovery of the Penrose pattern of aperiodic tilings in 1974. Mackay himself has attempted to extend this tiling into three dimensions, and even communicated a simulated diffraction pattern in 1982. The discovery was in the air, so to speak, which doesn't diminish a bit the value and significance of the discovery itself. In a way I feel that science has acted at its best by virtually predicting a heretofore unknown phenomenon, and not just unknown, but something that has been found and proven impossible many times over.

There was also a recent revolutionary result in chemistry, related to fivefold symmetry. Harry Kroto and co-workers discovered a superstable all-carbon C_{60} molecule in 1985 in which the 60 carbon atoms lie at the vertices of a truncated icosahedron. The molecule was appropriately named *buckminsterfullerene*, as Buckminster Fuller's ideas have stimulated the search for such structures. In 1990, Donald Huffman and colleagues synthesized buckminsterfullerene in recoverable quantities. Several research groups are working now on establishing the properties of this and related substances. It has been suggested that buckminsterfullerene is a third form of carbon in addition to diamond and graphite.

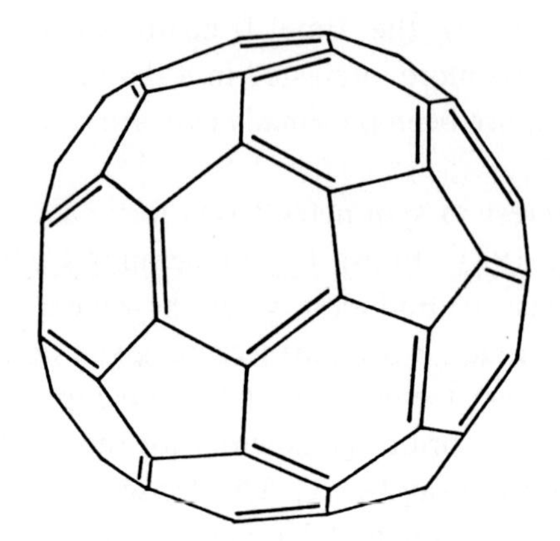

The proposed buckminsterfullerene structure from H.W. Kroto in *Symmetry 2: Unifying Human Understanding* (ed. I. Hargittai). Pergamon Press, Oxford, 1989, p. 418.

A whole new chemistry seems to be in the making, possibly with even implications for astronomy, as buckminsterfullerene or its derivatives may be important in interstellar matter.[a]

Fivefold symmetry is, of course, important in many areas beyond crystals and molecules. This volume is the bringing together of authors and ideas from the most diverse fields. May it also be a small offering to bringing closer together what C.P. Snow called "the two cultures."

The organizational work for this book commenced during the academic year 1988/89 which my wife and I spent at the University of Connecticut at Storrs. The actual editorial work was shared between my wife and myself during the Fall of 1989, which we spent at the University of Hawaii in Honolulu. We feel that this project owes a lot to Hawaii. The environment, the atmosphere, and the genuine interest and valuable assistance of our colleagues at the Chemistry Department of the University of Hawaii, all have contributed considerably to the success of our work.

Budapest, January, 1991

István Hargittai

[a]Quasicrystals and buckminsterfullerene are the subject of another recent book: *Quasicrystals, Networks, and Molecules of Fivefold Symmetry* (I. Hargittai, ed.). VCH, New York (1990).

FIVEFOLD SYMMETRY: A LESSER SYMMETRY

Joe Rosen

1. The "Missing Symmetry" Mystery

I suppose I could be considered reasonably sophisticated in matters of symmetry. Thus, I know quite well that a regular hexagon is more symmetric than a regular pentagon and that the latter, in turn, is more symmetric than a square. It is simply counting the "symmetry operations," or "symmetry elements"; the more the figure has, the more symmetric it is. The square is symmetric under rotations about its center by multiples of 90°, giving four rotation symmetry operations (including, as is customary, the trivial operation of rotation by 0°, the "identity operation"). Similarly, the regular pentagon possesses five rotation symmetry operations (rotations by multiples of 72°), and the regular hexagon has six (rotations by multiples of 60°) (see Fig. 1).

The regular polygons additionally possess reflection symmetry with respect to certain reflection lines through their centers. These are lines of "bilateral symmetry," along which the figures might be folded so that one half becomes superposed on the other. The square has four such reflection symmetry operations, the regular pentagon has five, and the regular hexagon has six (see Fig. 2). So whichever way we look at the situation — rotationally, reflectionally, or both combined — the regular hexagon is quantitatively more symmetric than the regular pentagon, which is more symmetric than the square, 6:5:4 or 12:10:8.

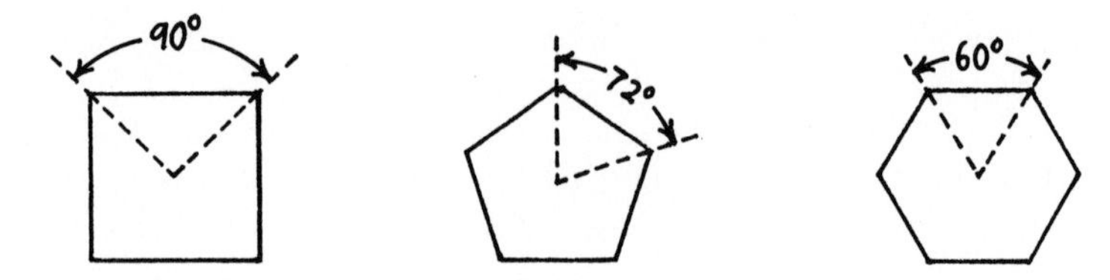

Fig. 1 A square, a regular pentagon, and a regular hexagon are symmetric under rotations about their centers by the four, five, and six multiples of 90°, 72°, and 60°, respectively.

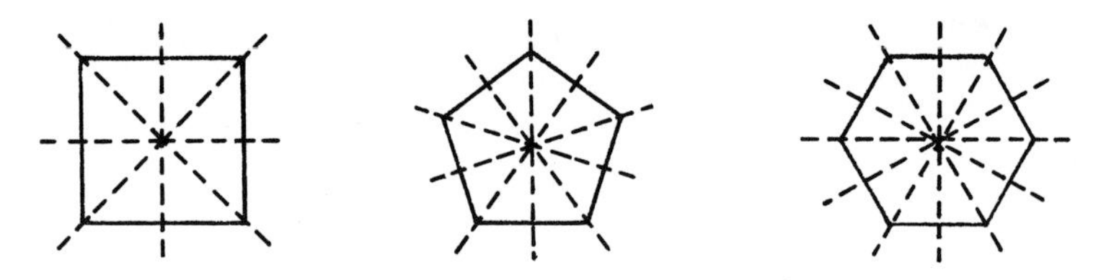

Fig. 2 The four, five, and six reflection lines (lines of bilateral symmetry) for a square, regular pentagon, and regular hexagon, respectively.

Yet I cannot help feeling that the pentagon looks less symmetric than the other two! And no amount of recounting symmetry rotations and reflections can offer objective support for that; it always comes out 6:5:4 or 12:10:8 with the pentagon in the middle. So the perceived lesser symmetry of the pentagon must be subjective.

2. A Questionnaire

It is subjective, but is it only me? To check this point I prepared a questionnaire, the original being in Hebrew and an English version appearing as Fig. 3, designed for people with no symmetry sophistication. The subjects were informed that this is a psychological survey of symmetry perception (which it is) and that there are no right or wrong answers. The subjects were requested to overlook minor irregularities in the drawings and to reply to questions 1–4 without first looking at the following questions (the questionnaire was folded back to help), indicating their first impressions. The subjects were requested to reply to the last questions only after finishing questions 1–4. If their thinking about the last questions made them change their minds about the first four

At first glance which of these figures seems to you
1. the most symmetric?______________
2. the least symmetric?______________

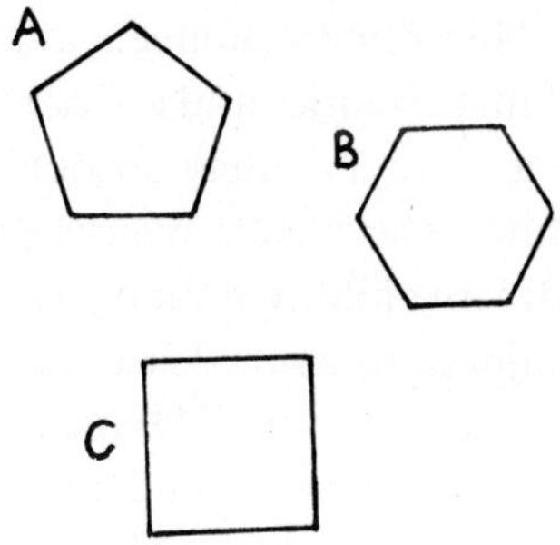

A B C

At first glance which of these figures seems to you
3. the most symmetric?______________
4. the least symmetric?______________

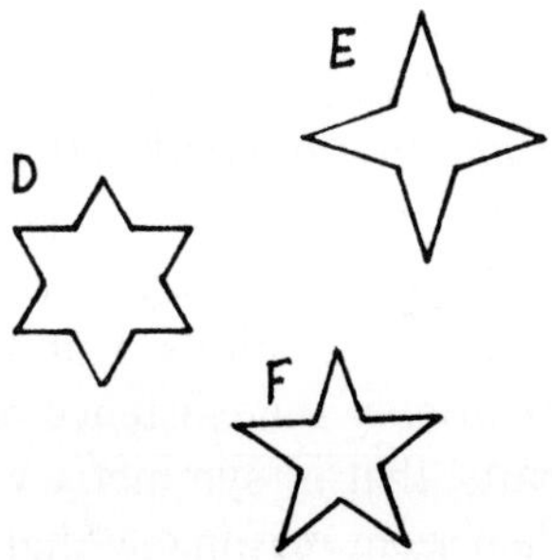

E D F

------------------------------ (folded back) ------------------------------

5. Attempt to explain briefly (in a few words) the concept of symmetry.______________

6. Attempt to explain briefly your reply to question 2 ______________

______________ and your reply to question 4. ______________

Fig. 3 English version of the original questionnaire.

questions, they were requested not to change their replies, but to indicate the fact of their mind change in their replies to the last questions.

Thirty one volunteers among my students, first-year university students of mathematics and college students of education in physics/mathematics, who were definitely unsophisticated in symmetry, answered the questionnaire. The results overwhelmingly support my own feeling. The figures with fivefold symmetry (A and F in Fig. 3) were declared by a large majority to be the least symmetric among the figures presented. The exact percentages were 77% for A and 74% for F. (It might be of interest that the 17 mathematics students were more sure of themselves than the 14 education students. The results from the former were 100% for A and 88% for F, while those from the latter were 71% for A and 57% for F, with the difference between the two groups stemming essentially from the fact that four of the education students did not reply at all!) Thus convinced that *pentasymmetrophobia* (I am sure someone can come up with a better term!) is not my own private perversion, I believe it is worth looking into.

But before doing that, it is very interesting to see what was given in reply to questions 5 and 6. The explanation of symmetry was almost invariably the existence of a line dividing the figure into two congruent parts, that is, symmetry was identified with bilateral symmetry. It would then seem reasonable that the more directions of bilateral symmetry (i.e., the more lines of reflection symmetry) a figure has, the more symmetric it should be considered. Yet almost none of the subjects changed their minds about their replies to questions 1–4. In question 6, the explanations of the first impressions as given in the replies to questions 1–4 ran as follows: the fivefold symmetric figures have less lines of bilateral symmetry than the others or they have none at all or they just look less symmetric. The last is no explanation, but only a restatement. The explanations involving the number of lines of bilateral symmetry are, of course, objectively wrong. Still, they are indicative of more difficulty for nonsophisticates in finding such lines in fivefold symmetry (and presumably in oddfold symmetry) than in four- and sixfold (and presumably in evenfold) symmetry.

3. Detective Work

Returning now to fivefold symmetry as a lesser symmetry, since the tried and true quantitative measure of symmetry does not correlate with our perception, let us look for qualitative differences between fivefold symmetry (and presumably also oddfold symmetry), on the one hand, and four- and sixfold (and presumably also evenfold) symmetry on the other. The most obvious is just the property of evenness or oddness of the foldness. Although this property correlates perfectly, it seems too abstract to be the direct cause of our effect, and I would prefer some concrete derivative property more closely related to the symmetry elements themselves.

Two such properties present themselves. One is the existence or nonexistence of an inversion center, which follows from, respectively, the evenness or oddness of the symmetry. Existence of an inversion center means that the figure is "balanced" with respect to its center, that along any diameter, the parts on one side of the center are in congruence with those on the other side (see Fig. 4). This is the same as twofold symmetry,

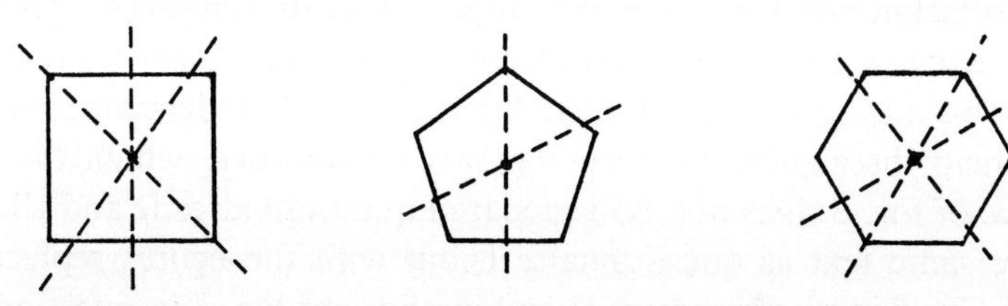

Fig. 4 The square and regular hexagon each possess an inversion center. Note the congruence of parts on both sides of every diameter. Note the lack of such congruence for the regular pentagon, which has no inversion center.

symmetry under 180° rotation, which is an element of evenfold symmetry while excluded by oddfold symmetry.

Another possibility is the existence or nonexistence of pairs of orthogonal reflection lines, suggested by Nathan Rosen, which also follows from, respectively, the evenness or oddness of the symmetry (see Fig. 5). With suitable orientation of the figure, the existence of such orthogonal pairs implies the simultaneous appearance of both a "vertical" and a "horizontal" reflection line, and we are very used to seeing such pairs of lines, though not necessarily reflection lines, in our environment (such as trees and horizon, buildings).

So which is it? Do we give extra weight to the balance about inversion centers, considering this symmetry to be of greater significance than that of just another rotation symmetry element, or does symmetry involving orthogonal reflection lines gain extra value for this fact, perhaps due to

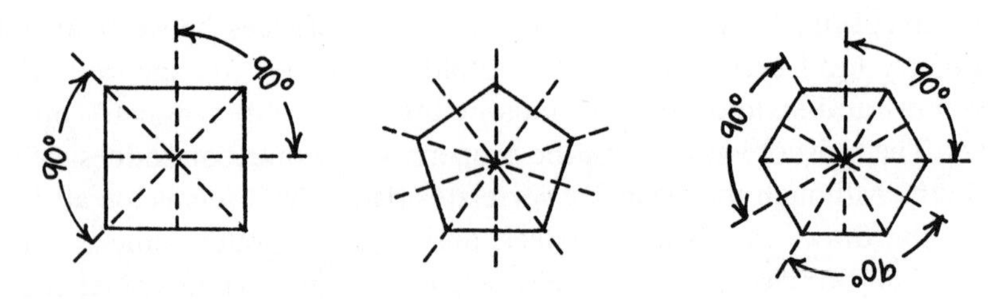

Fig. 5 Two and three pairs of orthogonal reflection lines for the square and regular hexagon, respectively. There are none for the regular pentagon.

our possible conditioning or innate propensity for orthogonal, especially vertical and horizontal, lines? Or is it both together?

4. More Questionnaires

Now rotation symmetry does not imply reflection symmetry. An *n*-fold symmetric figure needs have no reflection lines. Thus, a survey involving four-, five-, and sixfold symmetric figures with no reflection symmetry should help distinguish between the two possibilities, which the questionnaire of Fig. 3 does not. So I prepared questionnaires II and III, both with the same text as questionnaire I, but with the figures replaced by those of Fig. 6 in questionnaire II and by those of Fig. 7 in questionnaire

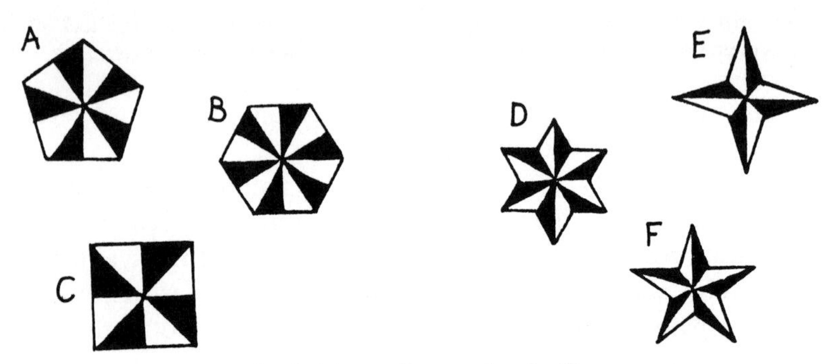

Fig. 6 Figures for questionaire II.

III. The figures of II are basically the original figures of I, modified by the addition of reflection asymmetric shading to give them chirality (handedness), thus removing the reflection symmetry while maintaining the original rotation symmetry. Nevertheless, the outlines of these figures are still reflection symmetric. The figures in questionnaire III are also based

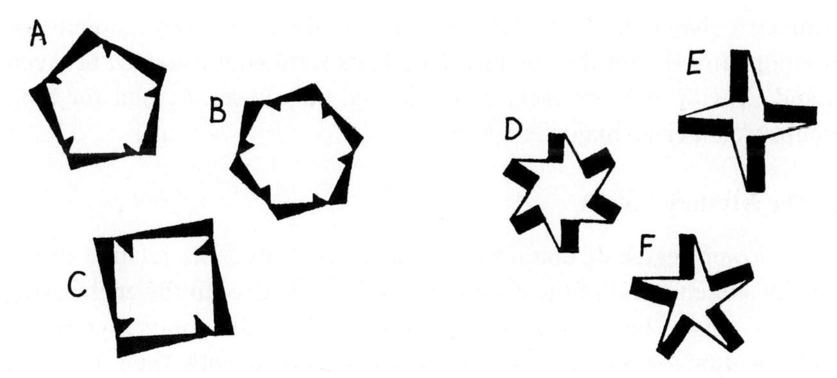

Fig. 7 Figures for questionaire III.

on the original figures, but are distorted to remove any vestige of reflection symmetry. The different modified questionnaires were presented to different subjects[a] who were also different from those of the first questionnaire.

Before examining the results of surveys II and III, my own impression of the modified figures, for whatever it is worth at this stage, remains as before, that the fivefold symmetric figures consistently look less symmetric than both their evenfold partners, whether they retain reflection symmetry only in their outline or lack reflection symmetry altogether.

Questionnaire II was presented to 34 first-year university physics and biology students. The results were similar to those of questionnaire I, but with some reduction in the majorities. The percentages were 68% for A and 65% for F. (Again there was a difference between the two groups, with the 16 biology students more nearly unanimous. Their results were 75% for A and 69% for F, while those from the 18 physics students were 61% for each of A and F.) The runners-up were the fourfold symmetric C (26%) and E (24%).

Questionnaire III was presented to 36 first-year university chemistry and engineering students. There was a further reduction in the majorities, with A receiving 56% and F 61%. The only significant runner-up was C, which got 25%. There was, however, a very interesting difference between the two groups as to their choice between A and C as the less symmetric figure. The 18 engineering students overwhelmingly considered A less

[a]I thank my colleagues Zvi Mazeh, Yona Oren, Eli Piasetzki, and Akiva Yaniv, as well as their students, for their kind cooperation.

symmetric than C by 72 to 11%, very much like the biology students for questionnaire II. But the chemistry students were *evenly divided* between A and C, with 39% for each. Nevertheless, the averaged result for both groups was a clear majority for A.

5. The Mystery Solved

The lower degree of unanimity of the physics students relative to the biology students and of the chemistry students relative to the engineering students might be due to the possession of rudimentary symmetry sophistication by some of the students in accord with their fields of interest, physics and chemistry. Yet, on the whole, all groups of subjects were symmetry nonsophisticates.

My own impression of the modified figures, then, is well supported by the surveys. The conclusion of all this is, it seems to me, that the presence or absence of an inversion center (which is the same as twofold rotation symmetry) is of special importance in the unsophisticated evaluation of symmetry or its lack. That stands out in the result for figures D, E, and F of questionnaire III, which was overwhelmingly in favor of F as the least symmetric figure. These figures are quite different from anything possessing reflection symmetry, and it is very hard to assign them reflection lines, even to a rough approximation. Still, the trend of the decreasing degree of unanimity from questionnaire I to II to III seems to indicate that the presence or absence of orthogonal reflection lines is of some importance.

Thus it is that fivefold symmetry is a lesser symmetry, lesser in that it is generally *perceived* as lesser than fourfold symmetry, while technically it is of higher degree than the latter. The mechanism for that seems to be mainly the lack of an inversion center in oddfold symmetry and the presence of one in evenfold symmetry. When reflection symmetry is also present, even if only to an approximation, the lack of orthogonal reflection lines in oddfold symmetry and their presence in evenfold symmetry seem to carry some weight in the perception mechanism, but to a considerably lower degree than the lack or presence of an inversion center.

The surveys upon which my conclusions are based were not designed or analyzed in anywhere near a statistically or psychologically professional manner. Yet they are certainly indicative, I think, and sufficiently so as to justify drawing the conclusions I have drawn, even if those conclusions should really be held as tentative, pending the availability of data of more professional quality.

FIVEFOLD SYMMETRY IN MATHEMATICS, PHYSICS, CHEMISTRY, BIOLOGY, AND BEYOND

Josef Brandmüller

Dedicated to Prof. Dr. Hans Wondratschek, Karlsruhe

1. The Point Groups with Fivefold Rotation Axes

Seven pentagonal point groups are needed to describe the symmetry of molecules with fivefold rotation axes.[1,2] Molecules with regular icosahedral structures are also known, such as $B_{12}H_{12}^{--}$ [3,4] and the dodecahedran $C_{20}H_{20}$.[5] These are described by the two icosahedral point groups 235 and $(2/m)\bar{3}\,\bar{5}$.[1] Mackay[6] has presented arguments that a close noncrystallographic packing of equal spheres is possible and that this arrangement has the symmetry of an icosahedral group. Analogously to a work of Kepler,[7] he then presented "De Nive Quinquangula"[8] and showed that it is possible also to have infinite nonperiodic patterns with partial structures containing fivefold symmetry axes. Mackay[9] then generalized the planar Penrose pattern to three dimensions. Finally, Shechtman et al.[10] made the decisive discovery of the so-called quasi-crystal. The extensive literature on the subject is summarized elsewhere.[11] The first X-ray single crystal study was carried out in 1989.[12]

The character tables (e.g., Wilson et al.[13]) of the point groups with fivefold rotation axes, both the non-Abelian pentagonal and the icosahedral point groups, contain the Fibonacci number τ, in which

$$\tau = \frac{\sqrt{5} + 1}{2} \tag{1}$$

2. The Fibonacci Number and the Golden Section

When the alternate corners of a regular pentagon are joined (Fig. 1), a pentagram is obtained. It follows on geometrical grounds that the angle $\alpha = 36°$. For cos α the following equation is obtained

$$4 \cos^2 \alpha - 2 \cos \alpha + 1 = 0, \tag{2}$$

where $\cos \alpha = (\sqrt{5} + 1)/4$. This is a remarkable result, since this value is related to the well-known Fibonacci number from the theory of numbers (Eq. (1)), by

$$\cos 36° = \tau/2. \tag{3}$$

From this, the cosine of the angle for a fivefold rotation comes to $\cos(2\pi/5) = (\tau - 1)/2$. The distance d between two alternate corners is obtained from the cosine rule, $d^2 = 2c^2(1 - \cos 3 \cdot 36°)$, from which it follows that $d = \tau c$. The ratio of the lengths thus equals the Fibonacci number τ. It has thus been shown that the fivefoldness has something to do with the Golden section, for the condition for the Golden section is

$$\tau / c = c / (\tau - c). \tag{4}$$

If c is now normalized to 1, the following equation results (cf. Eq. (2)),

$$\tau^2 - \tau - 1 = 0, \tag{5}$$

with the positive root $\tau = (\sqrt{5} + 1)/2$.

In a regular pentagon, the connecting lines of alternate corners are divided in the ratio of the Golden section.

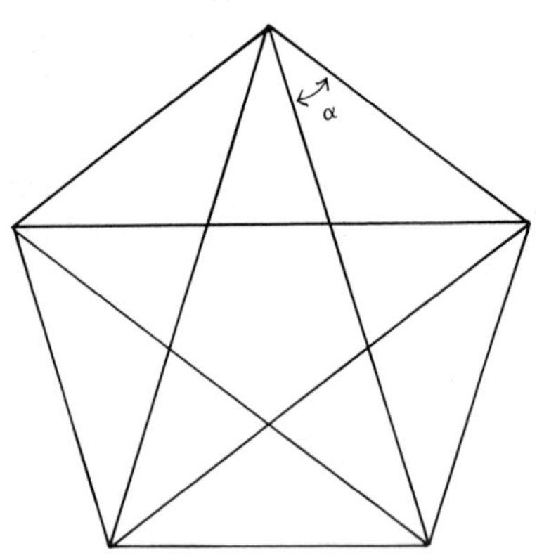

Fig. 1.

Gauss found that a regular n-agon can be constructed with a compass and ruler if and only if the odd prime factors of n differ from each other by the so-called Fermatic prime numbers[15,16].

$$p_k = 2^{2^k} + 1.$$

For $k = 1$, the Fermatic prime number $p_1 = 5$, and thus a regular pentagon can be constructed.

The Fibonacci number appears in yet another quite different connection. Leonardo di Pisa[15,17] found a sequence of whole numbers that follows the following rule: start with 0 and 1, and each subsequent number is the sum of the two previous:

$$0, 1, 1, 2, 3, 5, 8, 13, 21, 34, 55, 89, 144, \text{etc.} \tag{6}$$

This is described as a "recurrent number sequence."[18,19] If one now takes the ratio of each number with the one immediately previous, the values on the left side of Table 1 are produced. The right side of this table shows the ratio of a given number in the series with that immediately following. The limiting value on the left column is the Fibonacci number τ, which Kepler had already noticed,[20] but that was first proved in the eighteenth century by R. Simon.[15] The limiting value on the right side is its reciprocal

$$1 / \tau = \tau - 1. \tag{7}$$

The Fibonacci number τ, whose designation τ comes from the Greek word for cut, ἡ τομή, is also occasionally called Φ,[14,21] and is known to 45 places after the decimal:

$$\tau = \Phi = 1.61803\ 39887\ 49894\ 84820\ 45868\ 34365\ 63811\ 77203\ 09180\ldots$$

Fibonacci, alias Filius Bonaccio or Leonardo di Pisa (ca. 1180–1250), was self-educated and came from the merchant class. During his extensive travels he learned Hindu–Arabic arithmetic and introduced Arabic numerals to Europe in his arithmetic book "liber abacci" in 1202, which was revised in 1228. This provided the premises for the further development of algebra. He was a member of the circle of scholars round Emperor Frederick II (1212–1250).[22]

Table 1. The Fibonacci Number τ as a Limit from Ratios of the Numbers of the Fibonacci Series[a]

$\frac{1}{0} = \infty$	$\frac{0}{1} = 0$
$\frac{1}{1} = 1$	$\frac{1}{1} = 1$
$\frac{2}{1} = 2$	$\frac{1}{2} = 0.5$
$\frac{3}{2} = 1.5$	$\frac{2}{3} = 0.6666666667$
$\frac{5}{3} = 1.6666666667$	$\frac{3}{5} = 0.6$
$\frac{8}{5} = 1.6$	$\frac{5}{8} = 0.625$
$\frac{13}{8} = 1.625$	$\frac{8}{13} = 0.6153846154$
$\frac{21}{13} = 1.6153846154$	$\frac{13}{21} = 0.619047619$
$\frac{34}{21} = 1.619047619$	$\frac{21}{34} = 0.6176470588$
$\frac{55}{34} = 1.6176470588$	$\frac{34}{55} = 0.6181818182$
$\frac{89}{55} = 1.6181818182$	$\frac{55}{89} = 1.6179775281$
$\frac{144}{89} = 1.6179775281$	$\frac{89}{144} = 0.6180555556$
$\frac{233}{144} = 1.6180555556$	$\frac{144}{233} = 0.6180257511$
$\frac{377}{233} = 1.6180257511$	$\frac{233}{377} = 0.6180371353$
$\frac{\sqrt{5}+1}{2} = 1.6180339887 = \tau$	$\frac{\sqrt{5}-1}{2} = 0.6180339887 = \tau - 1 = \frac{1}{\tau}.$

[a]0, 1, 1, 2, 3, 5, 8, 13, 21, 34, 55, 89, 144, 233, 377, . . .

The Fibonacci number obeys several other interesting relationships in number theory.[14] One is

$$\tau^{n} = \tau^{n-1} + \tau^{n-2} \tag{8}$$

which holds also for negative values of *n*. Finally, τ can be described by a chain fraction:

$$\tau = 1 + \cfrac{1}{1 + \cfrac{1}{1 + \cfrac{1}{1 + \cfrac{1}{1 + \cfrac{1}{1 + \ldots}}}}} \tag{9}$$

or by the root:

$$\tau = \sqrt{1 + \sqrt{1 + \sqrt{1 + \sqrt{1 + \sqrt{1 + \sqrt{1 + \ldots}}}}}} \tag{10}$$

The power series

$$1, \tau, \tau^2, \tau^3, \tau^4, \ldots, \tau^n \tag{11}$$

has a special property: it is both additive and geometric. Each term equals the sum of the two previous (cf. Eq. (8)). In the power series

$$1, \frac{1}{\tau}, \frac{1}{\tau^2}, \frac{1}{\tau^3}, \frac{1}{\tau^4}, \ldots, \frac{1}{\tau^n} \tag{12}$$

each term equals the sum of two following (cf. Eq.(8)). Ghyka[14] emphasized that the property of this series to be simultaneously additive and geometric characterizes it straightaway, and is one of the reasons for its role in the growth of living organisms, especially plants.

Perhaps it is interesting in this connection that Kepler[20] treated the Fibonacci number in his "Harmonices mundi libri quinque," and found a connection with nonmathematical areas. Kepler started out from another remarkable property of the Fibonacci series (6):

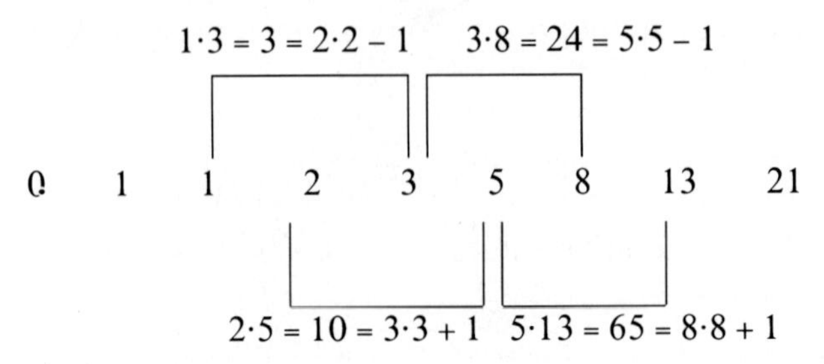

Kepler wrote the following: *"Das Rechteck aus 1 und 3 erzeugt ein Weibchen, denn es fehlt ihm eine Einheit gegenüber dem Quadrat von 2. Das Rechteck aus 2 und 5 erzeugt ein Männchen, denn es überschreitet das Quadrat von 3 um eine Einheit. Das Rechteck aus 3 und 8 erzeugt ein Weibchen, denn es fehlt ihm eine Einheit gegenüber dem Quadrat von 5. Aus 5 und 13 entsteht wieder ein Männchen, im Hinblick auf das Quadrat von 8; aus 8 und 21 ein Weibchen im Hinblick auf das Quadrat von 13 usw. bis ins Unendliche. Das ist also die Natur der Teilung, die bei der Herstellung des Fünfecks Verwendung findet. Nun hat aber Gott der Schöpfer die Gesetze der Zeugung jener Teilung entsprechend gestaltet und zwar entsprechend der . . . Proportion der Zahlenglieder die Fortpflanzungsverhältnisse bei den Pflanzen, die . . . den Samen in sich selber tragen, während die Verwendung je zweier Zahlenproportionen . . . der Verbindung von Mann und Frau entspricht. Ist es also verwunderlich, dass die Abkömmlinge des Fünfecks, die Durterz 4/5 und die Mollterz 5/6 die Seelen, die Ebenbilder Gottes, in Stimmungen versetzen, wie sie beim Zeugungsakt auftreten?"* (A tentative translation: "The rectangle from 1 and 3 produces a female, for it is less by unity than the square of 2. The rectangle from 2 and 5 generates a male, for it exceeds the square of 3 by unity. The rectangle from 3 and 8 produces a female, for it is less by 1 than the square of 5. From 5 and 13 a male is obtained again, in respect to the square of 8; . . . and so on, to infinity. That is also the nature of the division which is used in producing a pentagon. But now God the creator has formed the laws of reproduction corresponding to that division, that is according to the proportion of the terms the reproduction of the plants, . . . while the use of two number-proportions each . . . corresponds to the inosculation of man and wife. Is it therefore surprising, that the descendants of the pentagon, the major third 4/5 and the minor third 5/6, put God's images in moods which occur in the act of procreation?")

3. Fivefold Symmetry in Physics

In order for the point group symmetry of a crystal to be compatible with its translation symmetry in three-dimensional space, all point symmetry operations must satisfy a condition. Only those rotations and rotation-inversions are possible about an angle that satisfies

$$2 \cos \phi \in \{-2, -1, 0, 1, 2\}. \tag{13}$$

This is satisfied only for the rotation angles $\phi = 0$, 60, 90, 120, 180 and 360°. The "degree" (n) of a rotation is defined by $\phi = 2\pi / n$. The requirement of compatibility between translation and point symmetry of a crystal thus reduces the number of rotations in three dimensions to the values $n = 1, 2, 3, 4$, and 6. The 32 classical crystallographic point groups are made up from these. In higher dimensional spaces, on the other hand, other degrees are compatible with translation symmetry.[23] In this way, a Patterson analysis of the decagonal phase of the Al–Mn system was carried out in five-dimensional space, in which the structures possess translational symmetry (superspace group P 10_5mc).[12]

Most physical properties of crystals can be described with tensors. Consequently, the irreducible tensors of point groups with fivefold rotation axes were worked out for general tensors up to the fourth rank, including all irreducible representations.[1] From these, the selection rules for static and dynamic properties can be worked out.[24] The Raman and hyper-Raman tensors for all point groups with fivefold rotation axes were reproduced in Brandmüller and Claus.[2] Hargittai and Hargittai[4] mention that very small gold particles do not have the usual surface centred cubic gold grid, but icosahedral shells. The most stable configurations contain 55 or 147 gold atoms. Janssen[25] also treated the crystallography of quasi-crystals. In particular, the connection with incommensurable structures was demonstrated.[26] Quasi-crystals are always an up-to-date subject. In Urban and Nissen,[27] a scanning electron microscopic picture that showed fivefold symmetry especially well was reproduced.

4. Fivefold Symmetry in Chemistry

Group and representation theory produces the selection rules for the properties, especially for the spectroscopic behaviour of molecules. At present, there are known molecular examples, albeit only a few, of molecules in 6 of the 7 pentagonal point groups. Examples are the pentamethyl cyclopentadienyl radical $(CH_3)_5C_5$ for the point group $5 = C_5$, pentamethyldifluoridine iodine $(CH_3)_5F_2I$ for $\frac{5}{m} = C_{5h}$, ferrocene $(C_5H_5)_2Fe$ (prismatic form) for $\frac{5}{m}\frac{2}{m} = D_{5h}$, bis(cyclopentadienyl) iron(II) $(C_5H_5)_2Fe$ for $52 = D_5$, tridecahydroundecaborane $(B_{11}H_{13})^{--}$ for $5m = C_{5v}$, and ferrocene $(C_5H_5)_2Fe$ (antiprismatic form) for the point group $\bar{5}\,2m = D_{5d}$. Apparently, the group $\bar{5} = S_{10}$ does not seem to appear in nature.[1] The dodecahedrane $C_{20}H_{20}$ has the structure of a pentagon-dodecahedron. It is remarkable how this same structure provides evidence of a Celtic culture in Avenches, the Roman Aventicum in Vaud, Switzerland. Vainshtein[28] writes that the simplest polyoma virus has an icosahedral structure with symmetry 235.[29] The virus forms a cage in which RNA is contained.[4]

5. Fivefold Symmetry in Living Organisms

In the living world, fivefoldness appears to play a much bigger role than in the inanimate.[14,30] Plants with a fivefold petal arrangement occur relatively frequently.[31] The angle γ of the Golden section is important in spiral growth of plants. γ is defined by the relation

$$\frac{360°}{360° - \gamma} = \tau, \text{ from which it follows } \gamma = 360°(2 - \tau) = 137.51°.$$

Dixon[32] points out that the stem axes of two successive leaves are 137.5° apart (Fig. 2). In that way, no one leaf is found directly over another. There is always open space for each successive leaf, which is important in photosynthesis.[4] Eigen and Winkler[33] write, "Symmetry appears in

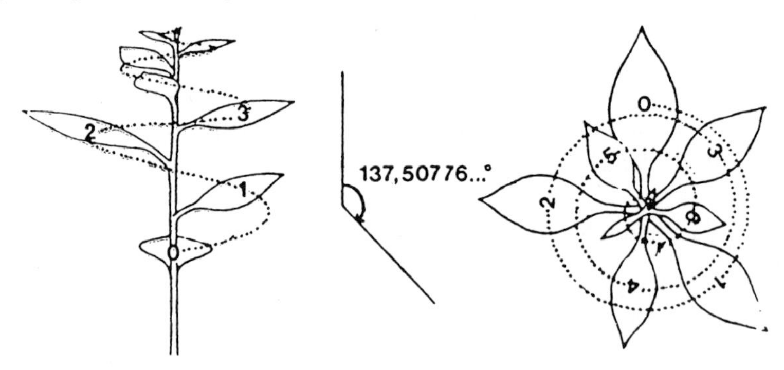

Fig. 2.

nature as the result of an evolutionary process; it is in no way the cause. Symmetry must be produced through a selective advantage, for otherwise it neither can be maintained nor win out in the interaction between mutation and selection. In this way life plays with symmetry, in much the same way a composer plays with rhythm and harmony." Dixon carries out a computer simulation of the growth of inflorescence of a composite flower (Fig. 3). The angle between successive petals is each time 137.5°.

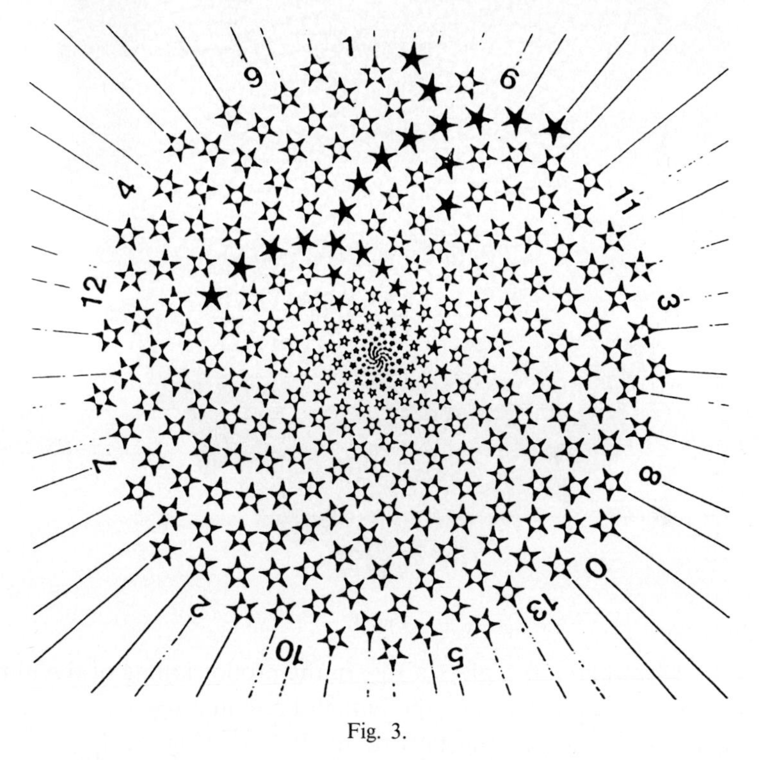

Fig. 3.

The computer calculates and shows a growth spiral. A large sunflower (Fig. 4) shows in fact such a spiral arrangement of the petals.[30] Wolf and Wolff[34] give further examples of fivefoldness of flower and shoots.

Fivefoldness is also observed in the animal world. Ernst Häckel[35] gives examples from the radiolaries which have icosahedral and pentagon-dodecahedral symmetry $(2/m)\bar{3}\ \bar{5}$. Starfish engulfing scalop, following the outer phenotype, have the point symmetry of 5m. The starfish *Ophiotrix capillaris* has the same symmetry.[31]

Fig. 4.

Fivefold symmetry also plays a role in the proportioning of the human body. Leonardo da Vinci drew the human body in a regular pentagon in an illustration for the architect Vitruvius Pollio[28] (Fig. 5). Le Corbusier showed that the proportioning of the human body is determined in various ways by the ratio of the Golden section.[34,36] Kappraff[28] demonstrated the role of the Fibonacci number in the representation of Venus in the painting "The Birth of Venus" by Botticelli (1444–1510).

6. Fivefold Symmetry in Architecture, Representative Art, and Music

In Buddhist architecture, there are centrosymmetric shrines, an example of which is the Borodubar Stupa in Java, which consists of eight

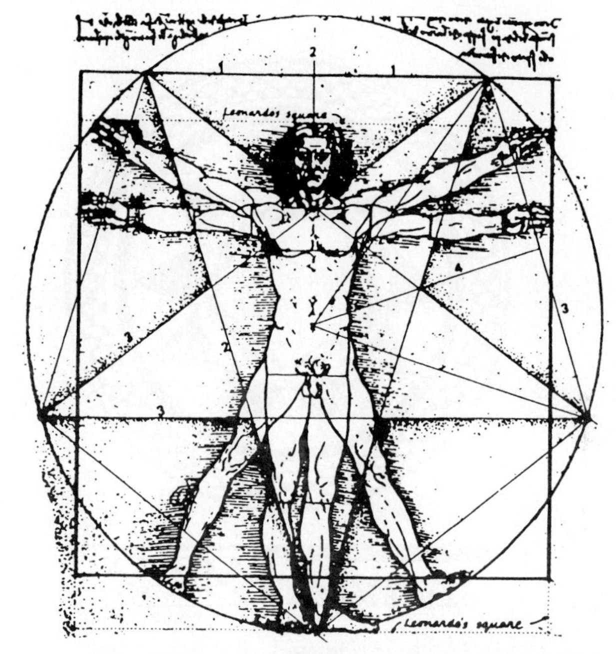

Fig. 5.

climbing balustrades with flat reliefs and shrines in niches.[15] The five lower terraces are square, while the upper three are round. Mainzer pointed out that 3, 5, and 8 are numbers in the Fibonacci series. The ratio of the Golden section is also found in Chinese and Japanese pagodas.[37] The number ratio of the Golden section, τ, occurs frequently in an approximate ratio of 8:5 in Western churches of the twelfth and thirteenth centuries, for example in Chartres.[15] The medieval castles, fortresses, and citadels in Antwerp, Caprarola, Ferrara, Modena, Parma, Paris (Issy-les-Moulineaux and Mont Valérien) and Viterbo, as well as the plague hospital in Ancona have a ground-plan involving fivefold symmetry. A recent (1940) building with the same symmetry is the Pentagon in Arlington, Virginia.

Fivefoldness has been used in various ways in representative art. Von Wickede[38] mentions a vessel (Fig. 6) (see Mallowan, Ref. 39). This may be one of the oldest portrayals of a pentagon. The vessel comes from Tell-Halaf Ceramics, which indicates a prehistoric period in the Near East that began in the sixth millennium B.C.E. Fivefold symmetry is also

Fig. 6.

found in Minoan and Mycenean Seals[40] (Fig. 7). Much later, an ornamental bronze disk of alemannic origin from the seventh century was found in a grave in Schaan (Liechtenstein), on which styled dragon heads are arranged in strict fivefold symmetry. The so-called Altar of Verdun in the lower Austrian Abbey Klosterneuburg near Vienna, completed around the year 1181 by Nicholas of Verdun, contains an enamel panel depicting Noah's Ark.[41] The noble animals (lion, horse, boar, deer, hare, and ox) appear in the regular pentagon, the next lowest in the upper level, while the lower animals are placed on the lower level. A work of Piero della Francesca (1420–1492) is to be found in the National Gallery at Perugia. A geometric analysis of the picture has shown that the Golden section was used by the artist in constructing the design.[42] Luca Pacioli (1445–1514) was a pupil of Piero della Francesca. A renaissance picture[43] shows him together with a pupil at a mathematics lesson. A pentagonal dodecahedron is found as illustrative material in the picture. The painting "Nature Morte Vivante" of Salvador Dalí in the Dalí Museum in St Petersburg, Florida, is based on the geometric arrangement of so-called $\sqrt{\tau}$ rectangle.[14,51]

Fig. 7.

In 1963, Escher[44] designed a metal biscuit tin as a present for the 25th anniversary of the Dutch firm Verblifa. This has strict icosahedral symmetry 235. Mussels and starfish are arranged such that any symmetry plane is missing. I have already mentioned that the simplest virus has the same symmetry. In addition, another picture by Escher, which is entitled "Reptiles" and shows a cycle, contains the pentagon dodecahedron.[45] The Singapore dollar of 1987 is embossed with fivefold symmetry. An Italian stamp, "La Natura e Poesia" of 1989, shows a styled tree with a fivefold top.

Fivefoldness is also found in music. Lendvai[36] treated dualism and synthesis in the music of Béla Bartók (1881–1945). As an example, Lendvai analysed a sonata for two pianos and percussion. He found that, "down to the minutest details," the number sequence of the Fibonacci series and the number ratio of the Golden section are the guiding principles. Lendvai wrote: "This music wants to be an expression of universal laws." Hargittai and Hargittai[4] write: "It is not known, and perhaps never will be, however, whether he consciously applied symmetry or was simply led intuitively to the Fibonacci numbers and the Golden section so often present in his music." A "Fibonacciana" was broadcast in 1985/86 on the Bayerischer Rundfunk; this was a piece of music which showed fivefoldness in various ways as a composition principle.

7. Fivefold Symmetry in Mythology, Philosophy and Religion

A pentagram shows five times over the letter , which goes back to Egyptian and Cretan hieroglyphics, possibly of Oriental origin. The pentagram served as a secret sign for the Pythagoreans (sixth century B.C.E.); they made mathematical harmony the central theme of their philosophy. In the Middle Ages, the German word *Drudenfuß* was also used; this comes from the middle high German *trûte* meaning monster or goblin. A pentagram looked like the pigeontoed feet of a witch or sorceress; it served as a means of protection against evil spirits. It had magic meaning in astrology and alchemy as well. In Goethe's *Faust*, Mephistopheles was exorcised with a pentagram (see Appendix). It may perhaps be a remnant of the idea of the pentagram as protection against evil spirits that a quarter of all national flags, for very different social structures, has a pentagram as an emblem. It is also painted on American tanks, and the Soviet star is five pointed as well.

Of the five Platonic bodies, two (namely the icosahedron and the pentagon dodecahedron) have fivefold rotational axes. The point symmetry of both of these bodies is the complete icosahedral group $(2/m)\,\bar{3}\,\bar{5}$. In the Platonic tradition, regular bodies are regarded as elements of matter and of the Universe.[15] The icosahedron symbolises the element water, and the pentagon dodecahedron is regarded as a portrayal of the whole universe. Kepler[20] very meticulously investigated the stereometry of both of these bodies. In his *Mysterium Cosmographicum*,[46] he connected the orbits of the six planets known at the time to a certain order of five regular bodies boxed inside each other (Fig. 8). A dodecahedron was ascribed to the Earth's orbit and was embraced by the sphere of Mars. An icosahedron was placed inside the Earth's orbit, and the sphere inside contained the orbit of Venus, etc.: "In the fivefoldness of regular bodies, geometry, in the sixfoldness of the planets, the Solar system . . . that is the secret of the world." Kepler had this inspiration in Graz on June 19, 1595. Weyl[30] writes about this: "We still share his belief in the mathematical harmony of the universe. This belief has withstood the test of constantly changing experience. But we no longer look for this harmony in static forms such as regular bodies, but in dynamic principles." Heisenberg[47] took up these thoughts 350 years later: "Modern physics stays in fact very near these ideas, but somewhat different mathematical forms have superseded geometrical forms. . . . For the modern atomic physicist the *causa formalis* is in the foreground."

Although throughout the Middle Ages the ancient idea of warding off evil spirits with a pentagram was still observed, a new meaning appeared. Among evil spirits was the devil. Thus fivefoldness is found in church windows in Bamberg Cathedral (Fig. 9, point group 5m).[34] The north window of the Altenberg Cathedral near Cologne dates from around 1300; it contains a depiction of a pentagram in two places.[48] The meaning and symbolism of numbers with respect to geometrical figures were apparently generally familiar in the Middle Ages. In Christian areas, fivefoldness was given a new meaning as a symbol of Christ crucified with five wounds.[28] In this connection, Doczi pointed to the crucifix of Brunelleschi (1377–1446) in the church of Santa Maria Novella in Florence.[28] He showed that a pentagram can be seen as a geometrical

Fig. 8.

Fig. 9.

principle underlying this cross.[49] The great sepulchre of Sir Thomas Gorges and his wife in Salisbury Cathedral, erected in the first half of the seventeenth century, is capped with an icosahedron and a pentagon-dodecahedron.

Even in modern times, fivefoldness appears in Christian themes in art. In 1955, Salvador Dalí painted the Last Supper inside a pentagondodecahedron, which symbolized the universe as a whole in Greek philosophy. Dalí had adopted the theme of five regular bodies in his so-called "Lanterns Dalí's."

The construction of the main facade of Milan cathedral is based on a $\sqrt{3}$ = 1,732 symmetry, which is fairly near in value to the Golden section τ = 1,618. In 1392, when he was asked for his expert advice on the continuation of the famous cathedral, the Gothic master builder Jean Vignot said to the city fathers of Milan, *"Ars sine Scientia nihil."*

Paul Feyerabend,[50] professor of the philosophy of science at ETH Zürich, and a pupil of Sir K.R. Popper, produced the reverse, as it were, of Vignot's statement, "Knowledge without the assistance of art is impossible." Thus perhaps in Latin translation, *Scientia sine Arte nihil.* This is a justification for natural scientists to become informed about matters outside of their fields.

Acknowledgements

I thank G.-C. Fini, University of Bologna, for several suggestions, M.C.L. Gerry, University of British Columbia, Vancouver, Canada, and

H.W. Schrötter, LM-University of Munich, for the translation into English, and F. Aussenegg, University of Graz, for his hospitality at the Institute for Experimental Physics (1988–1990).

Appendix

From *Faust: A Tragedy*, part one, by J.W. von Goethe, translated by J.S. Blackie, W. Blackwood (Edinburgh), and T. Cadell (Strand, London), 1834.

Mephistopheles:
Let me speak plain! there is a small affair,
That, without your assistance, bars my way,
Thin goblin-foot upon the threshold there—
Faust:
The pentagram stands in your way!
Ha! tell me then, thou son of hell,
If this be such a powerful spell
To keep thee in; why kept it not thee out?
What could have cheated such a powerful spirit?
Mephistopheles:
That is not hard to say, 'tis not well drawn, look near it;
The farthest corner, that which is turn'd out
Toward the door, is left a little open.
Faust:
Sufficient for a poddle-dog to hop in!
Here Fortune hit the nail upon the head;
Thus were the devil Faustus' prisoner made!
Chance is not always blind, a people say.
Mephistopheles:
The thoughtless cur saw nothing in its way.
But now the matter looks more serious;
The devil cannot move out of the house.
Mephistopheles:
But now this threshold's charm disenchant,
The tooth of a rat is all I want;
Nor need I make a lengthen'd conjuration,
I hear one scraping there in preparation.
The lord of the rats and of the mice,

Of the flies, and frogs, and bugs, and lice!
Commands you with your teeth's good saw,
The threshold of this door to gnaw;
Forth come, and there begin to file
Where he lets fall this drop of oil.
Ha! there he jumps! that angle there,
With thy sharp teeth I bid thee tear,
Which jutting forward, sad disaster,
Unwilling prisoner keeps thy master.
Briskly let the work go on,
One bite more and the task is done!
Now, Faust, until we meet again, dream on!

References

1. J. Brandmüller and R. Claus, The irreducible tensors of the point groups with fivefold rotational axes, *Croatica Chem. Acta* **61**, (1988) 267–300.
2. J. Brandmüller and R. Claus, Raman and hyper Raman tensors for groups with fivefold rotation axes, *Indian J. of Pure & Appl. Physics* **26** (1988) 60–67.
3. D.C. Harris and M.D. Bertolucci, *Symmetry and Spectroscopy.* Oxford University Press, New York (1978).
4. I. Hargittai and M. Hargittai, *Symmetry Through the Eyes of a Chemist.* VCH, Weinheim (1986).
5. R.J. Ternansky, D.W. Balogh, and L.A. Paquette, Dodecahedrane, *J. Am. Chem. Soc.* **104** (1982) 4503–4504.
6. A.L. Mackay, A dense non-crystallographic packing of equal spheres, *Acta Crystallogr.* **15** (1962) 916–918.
7. J. Kepler, *De Nive Sexangula*, Gottfried Tampach, Frankfurt (1611).
8. A.L. Mackay, De Nive Quinquangula: On the pentagonal snowflake, *Sov. Phys. Crystallogr.* **26** (1981) 517–522.
9. A.L. Mackay, Crystallography and the Penrose patterns, *Physica*, **114A** (1982) 609–613.
10. D. Shechtman, I. Blech, D. Gratias, and J.W. Cahn, Metallic phase with long-ranged orientational order and no translational symmetry, *Phys. Rev. Lett.* **53** (1984) 1951–1953.
11. P.J. Steinhardt and S. Ostlund, *The Physics of Quasi-Crystals*, World Scientific, Singapore (1987).
12. W. Steurer, Fünfdimensionale Pattersonanalyse der dekagonalen Phase des Sytems Al-Mn [Fivedimensional Patterson-analysis of the decagonal phase of the system A1-Mn], *Z. Kristallog.* **186** (1989) 284–285.
13. E.B. Wilson, Jr., J.C. Decius, and P.C. Cross, *Molecular Vibrations.* McGraw Hill, New York (1955).
14. M. Ghyka, *The Geometry of Art and Life*, Dover, New York (1977).

15. Kl. Mainzer, *Symmetrien der Natur, ein Handbuch zur Natur-und Wissenschaftsphilosophie* [Symmetries in nature, a handbook on philosophy of nature and science]. W de Gruyter, Berlin (1988).
16. J. Tropfke, *Geschichte der Elementarmathematik, Bd. 4, Ebene Geometrie* [History of elementary mathematics, Vol. 4, plane geometry], S. 257ff. Berlin (1923).
17. Leonardo di Pisa, *Scritti di Leonardo Pisano, Bd. 1, Il Libro Abbaci* [Manuscripts of Leonardo Pisano, Vol I, II (Abbacus' book)] (Liber Abaci 1202/1228), B. Boncomagni, ed., Rome (1857).
18. D.R. Hofstadter, *Gödel, Escher, Bach, an Eternal Golden Braid*, Basic Books, New York (1979); *Ein endlos geflochtenes Band*, Ernst Klett, Stuttgart (1985).
19. M. Kohmoto, Electronic states of quasiperiodic systems: Fibonacci and Penrose lattices, *Int. J. Modern Physics B* **1** (1987) 31–49.
20. J. Kepler, *Harmonices Mundi Libri Quinque* [Five books on the harmony of the world], Joannes Plancus, Lincii Austria, 1619; translated and commented by M. Caspar, *Weltharmonik* [Harmony of the world], R. Oldenbourg-Verlag, Munich (1978).
21. R.C. West, *Handbook of Chemistry and Physics.* CRC Press, Cleveland (1975).
22. H. Götze, *Castel del Monte, Gestalt und Symbol der Architektur Friedrichs II* [Castel del Monte, shape and symbol of the architecture of Frederic II], Prestel-Verlag, Munich (1984).
23. H. Brown, R. Bülow, J. Neubüser, H. Wondratschek, and H. Zassenhaus, *Crystallographic Groups of Four-Dimensional Space.* Wiley (Interscience), New York (1978).
24. J. Brandmüller and F.X. Winter, Influence of symmetry on the static and dynamic properties of crystals, calculation of the sets of the Cartesian irreducible tensors for the crystallographic point groups, *Z. Kristallogr.* **172** (1985) 191.
25. T. Janssen, Crystallography of quasi-crystals, *Acta Crystallogr.* **A42** (1986) 261.
26. T. Janssen, from incommensurate to quasi-crystals, *Proc. 2nd Int. Conf. Phonon Physics*, Budapest. J. Kollar et al., eds. World Scientific, Singapore (1985) 260.
27. K. Urban and H.-U. Nissen, Quasikristalle, noch immer ein aktuelles Thema [Quasi-crystals, still an up-to-date topic]. *Physikal. Blätter* **44** (1988) 144.
28. I. Hargittai, *Symmetry: Unifying Human Understanding*, Pergamon, New York (1986).
29. K.W. Adolph, D.L.D. Caspar, C.J. Hollingshed, E.E. Lattman, W.C. Phillips, and W.T. Murakami, Polyoma vivion and capsid crystal structures, *Science* **203** (1979) 1117.
30. H. Weyl, *Symmetrie*, Birkhäuser, Basel (1955).
31. A.V. Shubnikov and V.A. Koptsik, *Symmetry in Science and Art*, Plenum, New York (1974).
32. R. Dixon, The mathematical daisy, *New Scientist* **92** (1981) 792. See also *Naturwissenschaft. Rundschau* **35** (1982) 461.

33. M. Eigen and R. Winkler, *Das Spiel, Naturgesetze steuern den Zufall* [The game, laws of nature govern random], Piper, Munich (1975).
34. K.L. Wolf and R. Wolff, *Symmetrie*, Böhlau, Münster, (1956).
35. E. Häckel, *Kunstformen der Natur* [Forms of art in nature], Bibliographisches Institut, Leipzig (1899–1904).
36. G. Kepes, *Modul, Proportionen, Symmetrie, Rhythmus,* La Connaissance, Brussels (1969).
37. G. Doczi, *The Power of Limits, Proportional Harmonies in Nature, Art, and Architecture*, Shambhala, Boston (1985).
38. A. von Wickede, Die Ornamentik der Tell Halaf Keramik, ein Beitrag zu ihrer Typologie [The ornaments of Tell Halaf ceramics; a contribution to its typology], *Acta Praehistorica Archaeologica* **18** (1986) 7–32.
39. M.E.L. Mallowan, Excavations at Brak and Chagar Bazar, *Iraq* **9** (1947) 1–259.
40. I. Pini, in *Corpus der Minoischen und Mykenischen Siegel* [Handbook of the minioc and mycenic seals], edited by von F. Matz and I. Pini, Akademie der Wissenschaften und der Literatur, Mainz, Mann, Berlin (1970).
41. H. Buschhausen, *Der Verduner Altar* [The alter by Nicolo from Verdun], Edition Tusch, Vienna (1981).
42. O. Del Buono and P. de Vecchi, *L'Opera Completa di Piero della Francesca* [The complete opus of Piero della Francesca], Rizzoli Editore, Milan (1981).
43. H. Magazine, R.M. Ketchum, and J.H. Plumb, *Il Rinascimento* [The renaissance], Feltrinelli Editore, Milan (1961).
44. J.L. Locher, *Leben und Werk M.C. Escher* [Life and work of M.C. Escher], Rheingauer, Eltville (1984).
45. B. Ernst, *Der Zauberspiegel des M.C. Escher* [The magic mirror of M.C. Escher], Deutscher Taschenbuch, Munich (1982).
46. J. Kepler, J., *Mysterium Cosmographicum* [The secret of the world], Editio altera (1621), Das Weltgeheimnis, translated and introduced by M. Caspar, Augsburg (1923).
47. W. Heisenberg, *Vortrag auf der Nobelpreisträgertagung in Lindau* [Talk on the Nobel Prize laureates conference in Lindau] (1953).
48. B. Lymant, *Die Mittelalterlichen Glasmalereien der ehemaligen Zisterzienserkirche Altenberg* [The medieval glass paintings of the former cistercian church Altenberg], Altenberger Domverein, Bergisch Gladbach (1979).
49. P. Cowen, *Die Rosenfenster der gotischen Kathedralen* [The rosette windows of the gothic cathedrals], Herder, Freiburg (1979).
50. P. Feyerabend and Chr. Thomas, *Wissenschaft und Tradition* [Science and tradition], Verlag der Fachvereine Zürich, Zurich (1983).
51. R. Morse, *Salvador Dali ... Catalog of a Collection, Ninety-Three Oils 1917–1970*, Salvador Dali Museum, Beachwood, OH.

THE RELATIONSHIP BETWEEN MATHEMATICS AND MYSTICISM OF THE GOLDEN MEAN THROUGH HISTORY

Jay Kappraff

"Number is the bond of the eternal continuance of things"
Philolaus

1. Introduction

A study of the golden mean requires us to take serious account of ancient history. Most books venture only to the edge of the philosophy and mathematics in ancient Greece. Plato, Aristotle, and Pythagoras are considered the progenitors of modern mathematics and science. Yet there is evidence that Pythagoras was primarily a medium through which even more ancient esoteric knowledge was conveyed from the cultures of Egypt, India, and the neolithic people that inhabited the British Isles and France between one and four millenia before the birth of Pythagoras. Since the works of Plato constitute much of what has been explicitly handed down to us concerning the workings of the ancient mind, the works of Plato should rather be thought of as a kind of Rosetta stone of ancient knowledge than its origin.

I am reminded of Thomas Mann's book, *Joseph and his Brothers,* in which time beyond recorded history is portrayed as a spiral. We are never sure at any moment whether Mann is referring to *the* Joseph or the countless Josephs that came before him. If we look back in time, we find that much of what is attributed to Pythagoras and later, Plato, was known to far more ancient peoples. For example, the archeologist, Alexander Thom, has shown convincing evidence that the neolithic people of the British Isles erected stone monoliths by elaborate systems of geometry involving circles, ellipses, and ovals in order to mark celestial phenomena including the complex cycle of the moon.[1] Fig. 1 shows a set of spherical models constructed out of hard stone of the five Platonic solids created by these people a millenium before Plato. Besides being elaborately complex sculptures, these solids show that their creators were remarkably sophisticated people who mastered the art of stereometry undoubtably for the purpose of developing the means to study the complexities of the heavens.

Pythagoras is also generally credited with developing the rudiments of the musical scale. Yet, Ernest McClain has shown that the most ancient cultures of India used the musical scale as a way to describe their own cosmology and calendars.[2] One of the colorful legends that has been conveyed to us from ancient times has to do with the star pentagon which served as the sacred symbol of the cult society that surrounded Pythagoras, in which the penalty was death to anyone who revealed its secrets. As we shall see, this symbol was a supreme monument to the golden mean and could just as well have symbolized the Indian culture to which McClain refers. It also may have symbolized the many ways in which the golden mean and fivefold symmetry reveal themselves in the natural world.

One thing that we can say for sure about the thought processes of ancient man is that they differed markedly from our own. In ancient times, man saw himself as an intermediary between the demonic powers of the Earth which threatened his existence and the beatific powers of the Gods to which he owed his existence. He sought to find order in the chaos of the forces of nature by studying the heavens, but never attemped to spring out of the natural order to gain control over it. Thus the ancient Mayans could develop an extraordinarily sophisticated calendric system while never developing their technology even to the point of inventing the wheel. Knowledge was viewed as a connected whole, and all attempts were resisted to study nature by reducing it to theories and models. Until Aristotle introduced observation and measurement as the only way to arrive at truth, it appears as though reality was seen as something that transcended measurement or verbal description best described by numbers, music, and poetry.

R.A. Schwaller di Lubicz[3] feels that the strange combination of myth, legend, and symbol conveyed by ancient writings were the only way

Fig. 1. A full set of Scottish Neolithic "Platonic solids" a millenium before Plato's time. From *Time Stands Still*, by Keith Critchlow (redrawn by Bruce Brattstrom from a photo by Graham Challifour).

information about workings of the universe could be conveyed without reducing its *true* meaning. According to di Lubicz the ancient Egyptians felt that:

> Measure was an expression of Knowledge; that is to say that measure has for them a universal meaning linking the things of here below with things Above and not solely an immediate practical meaning — quantity is unstable: only function has a value durable enough to serve as a basis [for description]. Thus the Egyptians' unit of measurement was always variable — measure and proportions were adapted to the purpose and the symbolic meaning of the idea to be expressed. [For example] the cubit will not necessarily be the same from one temple to another, since these temples are in different places and their purposes are different.

One of the theses of this paper will be that certain scientific and mathematical descriptions associated with the modern theories of chaos and the growth of plants cannot be understood through direct measurement, and can only be studied through numerical structures inherent in the phenomena.

According to Plato, the nature of things and the structure of the universe lay in the study of music, astronomy, geometry, and numbers. To a great extent, this essay will show how the golden mean and its related fivefold symmetry satisfies these prescripts.

2. Fibonacci Numbers and the Golden Mean

In 1209, Leonardo of Pisa, otherwise known as Fibonacci, introduced the series: 1 1 2 3 5 8 13 . . . as the solution to a famous problem concerning the propogation of populations of rabbits through successive generations.[4] We refer to this so-called Fibonacci series as the *F*-series. It has many interesting properties, some of which will be explored in this article. Anne Tyng[5] has looked at a "chain of linked form units" corresponding to the *F*-series as a model for neuron chains. For example, by studying Fig. 2, you will see that the *F*-series results in hierarchical patterns of series within series. At each level, a "link" acts either as an element in an ongoing chain or as one of the initiator of a new chain. Similar hierarchical arrangements result in countless other patterns.

The ratio of adjacent elements of the *F*-series approaches the irrational number $\phi = (1 + \sqrt{5})/2$ in a limiting sense. The number, ϕ, is known as the golden mean and is the solution to the algebraic equation,

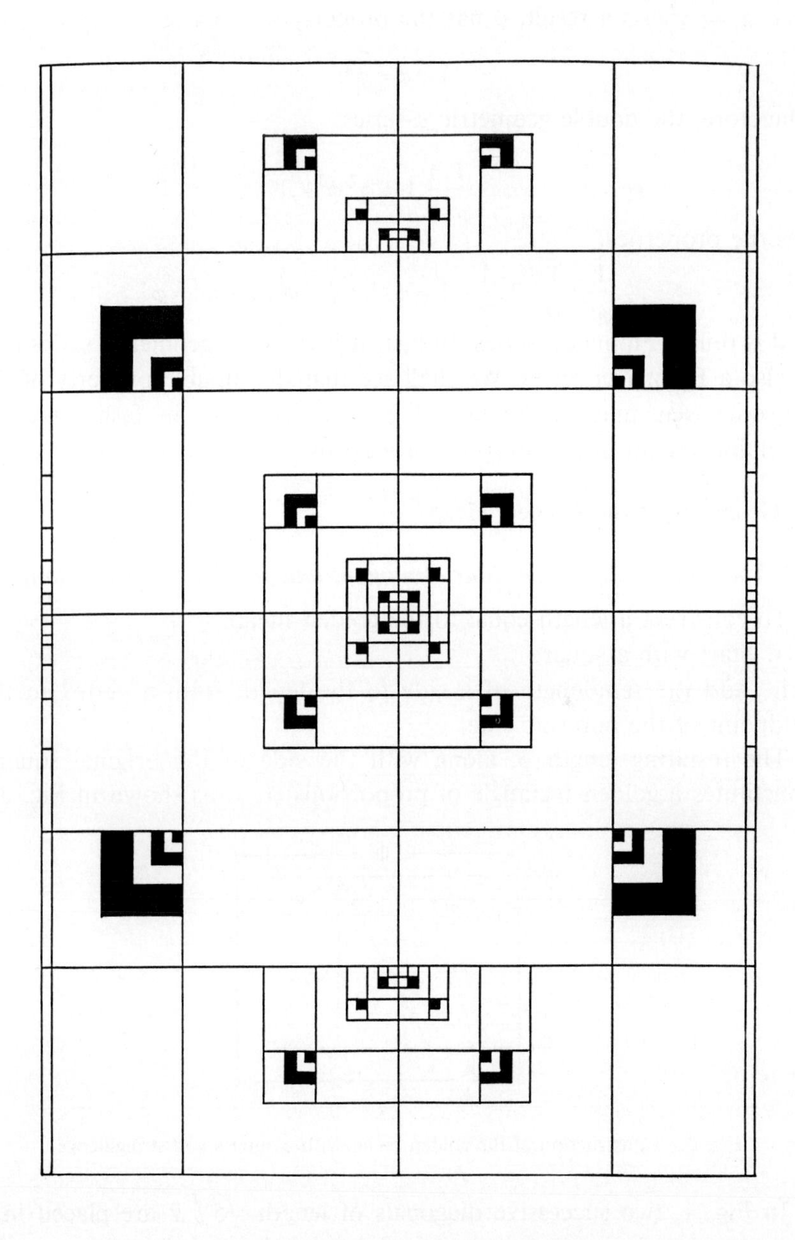

Fig. 2. Fourth stage in the development of a pattern of Fibonacci hierarchies. From *The Mathematics of Design Class of Jay Kappraff*, by Brian Getts.

$1 + x = x^2$. As a result, ϕ has the property:

$$1 + \phi = \phi^2. \tag{1}$$

Therefore, the double geometric ϕ-series:

$$\ldots \frac{1}{\phi^2} \frac{1}{\phi} \; 1 \; \phi \; \phi^2 \; \phi^3 \ldots \tag{2}$$

has the properties,

$$\ldots \frac{1}{\phi^2} + \frac{1}{\phi} = 1, \; \frac{1}{\phi} + 1 = \phi, \; 1 + \phi = \phi^2, \; \phi + \phi^2 = \phi^3 \tag{3}$$

and is thus a Fibonacci series. In fact, it is the only geometric series that is also a Fibonacci series. We shall see that this double property of the ϕ -series was made great use of by Le Corbusier in fashioning his Modulor system of architectural proportion.

3. Geometry of the Golden Mean

3.1 *The Golden Rectangle, Star Pentagon, and the 3,4,5-Right Triangle*

To construct a length equal to the golden mean:

i. Start with a square

ii. Add the semilength of a side to the length from a vertex to the midpoint of the opposite side.

The resulting length ϕ, along with the side of the original square, constitutes a golden rectangle of proportions $\phi : 1$, as shown in Fig. 3.

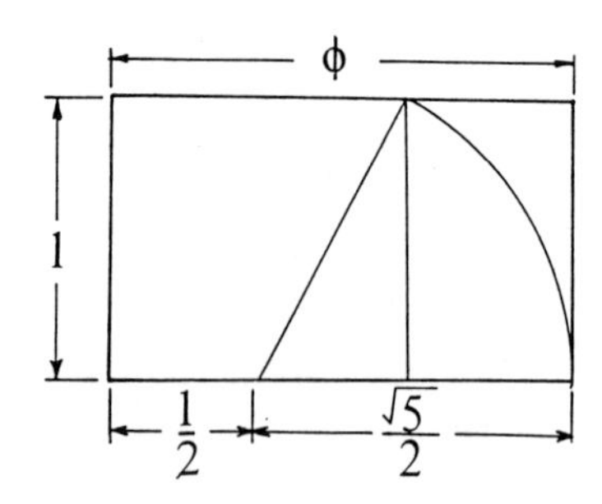

Fig. 3. Construction of the golden mean with compass and straightedge.

In Fig. 4, two successive diagonals of length $\sqrt{5} / 2$ are placed in a square to give rise, surprisingly, to a 3,4,5-right triangle.[6] In this way, the 3,4,5-triangle can be seen to be related to the golden mean. The

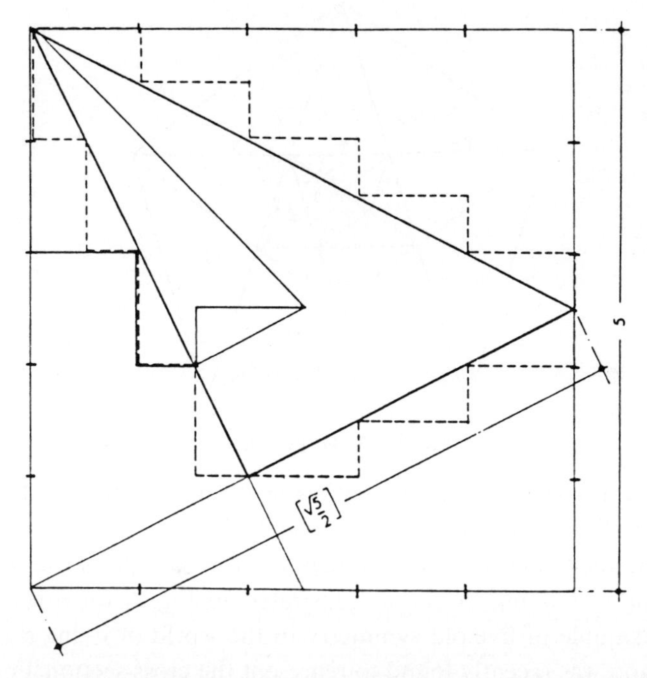

Fig. 4. A relationship between the 3,4,5-right triangle and the golden mean (from Ref. 6).

3,4,5-triangle appears to have great significance in ancient cosmology. It is known as the "Egyptian triangle" since it had sacred significance to the ancient Egyptians. H.F. Verheyen (see this volume) has shown how it is connected with the structure of the Great Pyramid of Cheops. Based on the data and methods of Thom, Keith Critchlow has shown how the pattern of a Druid temple at Inverness was related to the 3,4,5-triangle. In Sec. 8, we shall see how the musical scale can be constructed from the integers 3,4,5, and in Sec. 9, we shall learn about John Michell's[8] cosmological model based on a 3,4,5-right triangle that conforms to many sacred structures.

According to legend, the society of Pythagoras used the star pentagon shown in Fig. 5 as their sacred symbol. The pentagon is certainly an ideal way to commemorate the golden mean since the ratio of diameter to side of a pentagon is $\phi : 1$, and any pair of diagonals cut each other in the golden section in addition to the many other ways in which this figure

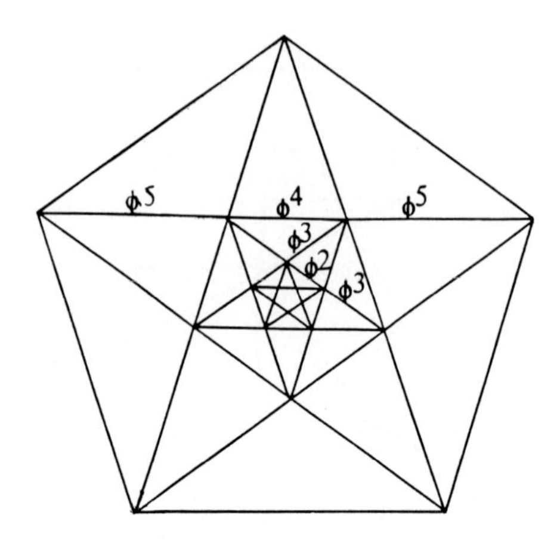

Fig. 5. A star pentagon.

embodies the golden mean. The decagon in also related to the golden mean since the ratio of radius to side is $\phi : 1$.

It is well known that fivefold symmetry of the pentagon arises naturally in the world of living things, as illustrated in Fig. 6 for a starfish. As another example of fivefold symmetry in the world of living things, the star decagon was recently found to represent the cross-sectional profile of the DNA molecule, shown in Fig. 7, where each vertex of the star represents one of the ten bases of DNA situated along one repeating segment of the double helix. In addition, the ratio of the length of one of the periodic lengths of the ten bases of B-DNA, the most common form of DNA, to the diameter of the star decagon measures 34 : 20, which is within experimental error to the golden mean ratio.

It has been well documented in Refs. 4, 9, 10, 11, and 12 that logarithmic spiral forms are found in the patterns of growth of the shells of sea animals, such as the Nautilus, and the horns of horned animals, and that these spirals are derived from the geometry of geometric series often with a common ratio related to such as in the ϕ series. These researchers have also determined that the proportions of the human body can be related to the golden mean. In fact, in his book, *The Temple of Man,* Schwaller Di Lubicz [3] shows evidence that the Egyptian Temple of Luxor recreates the image of man not only in its exterior form, but also the position of the rooms and hallways reproduce the inner cavities and organs of a man with great detail. Much of his analyses were based on the

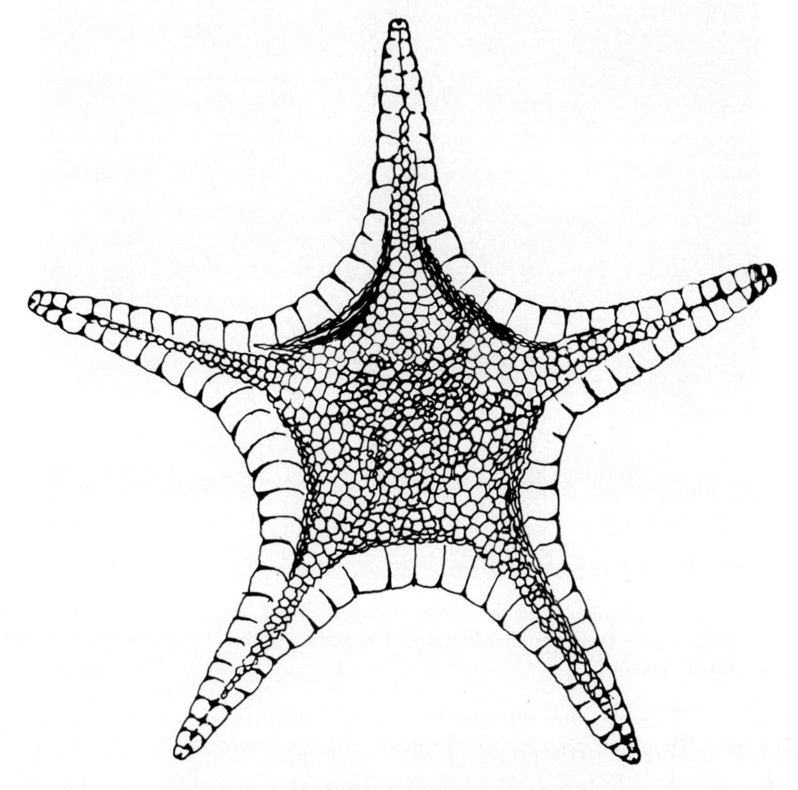

Fig. 6. A starfish. Drawn by Bruce Brattstrom.

methods of Jay Hambidge, which in turn were based on the golden mean proportions inherent in the human body and face.

3.2 *The Platonic Solids*

In Timaeus,[13] four of the five Platonic solids, shown in Fig. 8, were related to the four elements: earth — cube, air — icosahedron, fire — tetrahedron, and water — octahedron. The fifth solid, the dodecahedron, represented the entire cosmos (or perhaps the ether). The mathematics of the Platonic solids was studied in Euclid's *Elements — Book XII*. In this book, Euclid shows how the two Platonic solids with fivefold symmetry, the icosahedron and the dodecahedron, are related to the others through the golden mean. One striking example of this is shown in Fig. 9, where

Fig. 7. A cross-sectional view of DNA showing a star decagon. (Photo by Robert Langridge of the Computer Graphics Laboratory of the University of California in San Francisco.)

the 12 vertices of the icosahedron are determined by the corners of through mutually orthogonal golden rectangles, which envelop a cube. Verheyen (this volume) has shown that the structure of the Great Pyramid is closely related to the icosahedron. Fig. 10 shows a striking example of how the structure of the HTLV-2 virus has the form of an icosahedron.

4. Number Theory Properties of the Golden Mean

4.1 *Continued Fractions*

Perhaps the most profound property of the golden mean is that, in a sense we now define, it is the "most irrational" number in the number system. There are infinitely many irrational numbers, actually an "uncountable" infinity of them, as mathematicians say. Each irrational can be approximated arbitrarily close by rational numbers (numbers expressible as the ratio of integers P/Q where $Q \neq 0$). However, the

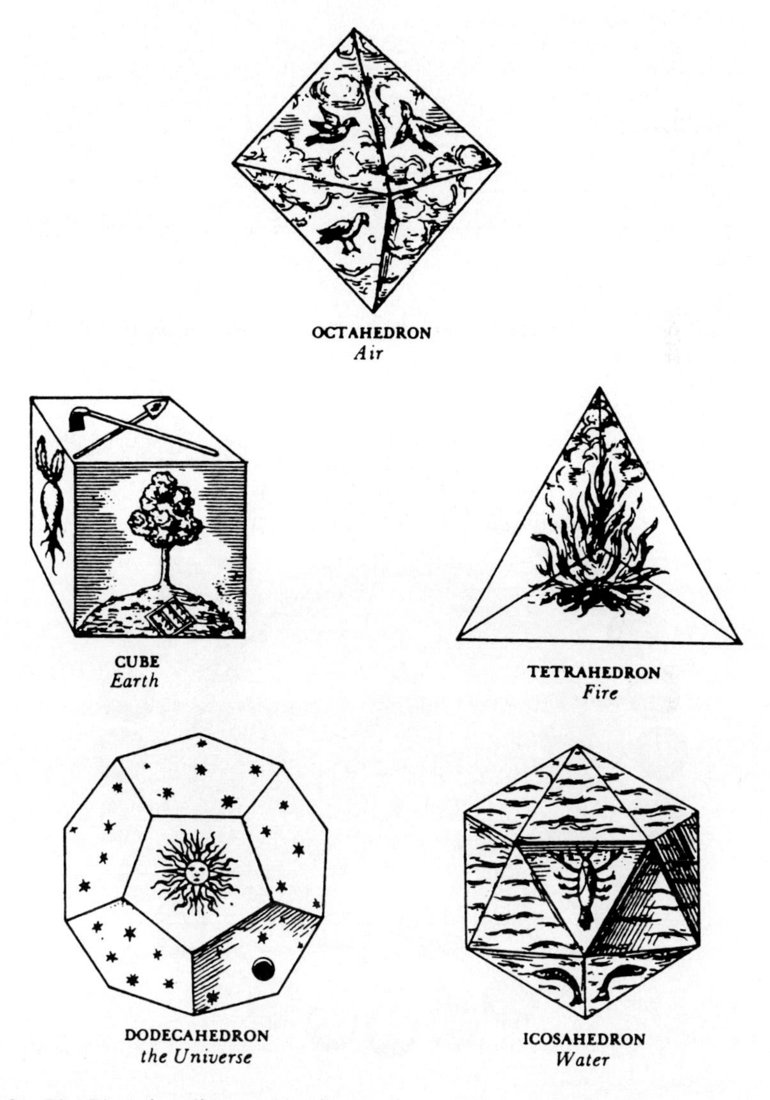

Fig. 8. The Platonic solids representing the four classical elements of the ancient world with the Dodecahedron as the cosmos.

expansion of an irrational number as an infinite compound fraction known as a continued fraction[14] yields the best rational approximants to

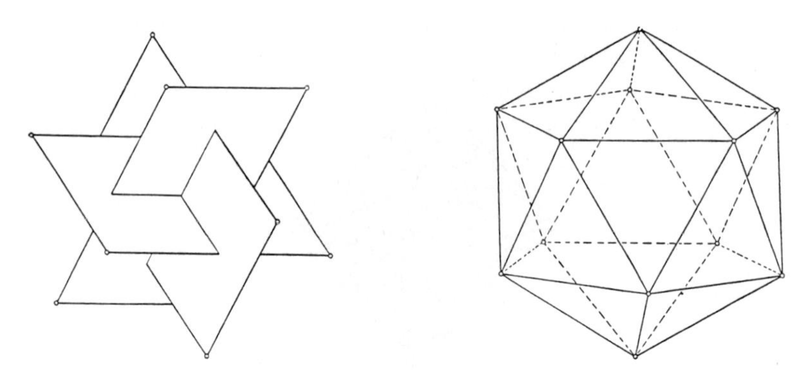

Fig. 9. The vertices of three interlocking golden rectangles lie at the vertices of an icosahedron.

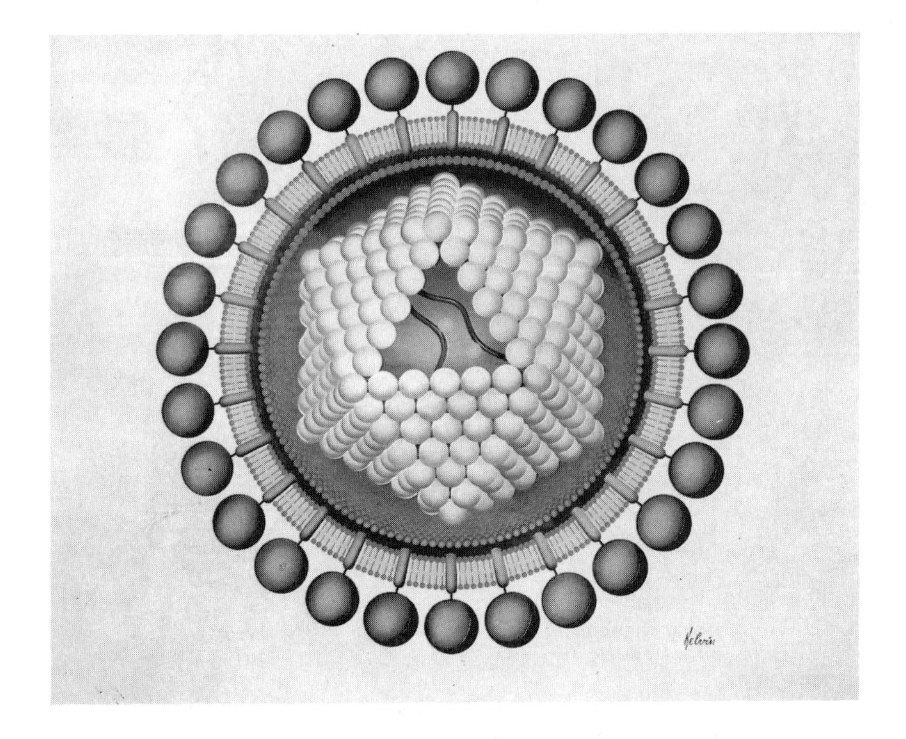

Fig. 10. The icosahedral structure of the HTLV — 2 virus (AIDS Virus). From "The First Human Retrovirus," by Robert C. Gallo. Illustration by George V. Kelvin, Science Graphics. © 1986 by *Scientific American, Inc.* All rights reserved.

the irrational, i.e., there exist no closer approximations to the irrational with denominators greater than the ones obtained by truncating the continued fraction at successive stages. Let's see how this works.

In general, it can be shown that the rational approximants of an irrational number α, where,

$$\alpha = a_0 + \cfrac{1}{a_1 + \cfrac{1}{a_2 + \cfrac{1}{a_3 + \cfrac{1}{\ddots}}}} \tag{4}$$

given in shorthand as $\alpha = [a_0; a_1, a_2, a_3, \ldots]$ are,

$$P_1/Q_1, \quad P_2/Q_2, \quad P_3/Q_3, \ldots$$

For example, the number ϕ has all ones down the diagonal, i.e.,

$$\phi = [1; 1\ 1\ 1\ \ldots]$$

and the rational approximants of ϕ are the ratio of adjacent terms of the F-series:

$$1/1, \quad 2/1, \quad 3/2, \quad 5/3, \quad 8/5, \ \ldots \tag{5}$$

A measure of their approximation is given by Ref. 14

$$\frac{1}{Q_k^2(a_{k+1}+2)} \leq |\alpha - P_k / Q_k| \leq \frac{1}{Q_k^2 a_{k+1}}\ . \tag{6}$$

As the result of this inequality, it is clear that the approximants of irrational numbers with large values of a_k admit good approximations, whereas the approximants of ϕ are the worst approximations of any irrational, since all values of a_k equal to 1. It is for this reason that ϕ can be considered the "most irrational" number. This will turn out in Sec. 7 to be one of the keys to understanding the onset of chaos in dynamical systems.

It can also be shown that the approximants oscillate on either side of ϕ on the number line, as indicated by Eq. (5), as they approach ϕ. This property will also manifest itself in the growth of plants, as we shall see in Sec. 6.

4.2 *A Golden Mean "Decimal System"*

Another number theoretical property of ϕ is the fact that the ϕ-series works much like the number system base two. Any real number can be represented uniquely as a sum of non-consecutive numbers from the ϕ-series (see Eq. 2), i.e.,

$$\alpha = \sum_{n=1}^{\infty} \phi^n \varsigma_n(\alpha), \tag{7}$$

where,

$$\varsigma_n(\alpha) = 0 \text{ or } 1 \text{ and } \varsigma_n(\alpha)\,\varsigma_{n+1}(\alpha) = 0.$$

In a similar fashion, any integer J can be expressed uniquely as a finite series of numbers from the F-series as follows,

$$J = \sum_{n=1}^{\infty} F_n \varsigma_n(J) \tag{8}$$

where, $\varsigma_n(J) = 0$ or 1 and $\varsigma_n(J)\varsigma_{n+1}(J) = 0$.

Furthermore, the first number in this decomposition is obtained by extracting the largest number of the F-series less than the given number. The second number is the largest number from the F-series less than the remainder, and so on. For example,

$$32 = 21 + 8 + 3.$$

Although the "decimal systems" given by Eqs. (7) and (8) may seem unnatural at first glance, there may be applications for which these systems are the most natural ones, just as the irrational number e is the "natural" base of logarithms. For example, we shall see in Sec. 5.2 that this ability to approximate arbitrary lengths from the ϕ-series makes this series suitable for use as a scale of architectural proportions as Le Corbusier accomplished for the Modulor. In Sec. 7, we shall see that Eq. (8) is most natural for organizing the boundaries of trajectories in the chaotic regime.

4.3 *The Golden Mean and Optimal Spacing*

The following spacing theorem[15] appears to lie at the basis of why the golden mean arises naturally in the growth of plants and other biological organisms, to be discussed in Sec. 6.

Theorem 1 : Let x be any irrational number. When the points $[x]_f$, $[2x]_f$, $[3x]_f$, ..., $[nx]_f$ are placed on the line segment [0, 1], the $n+1$ resulting line segments have at most three different lengths. Moreover, $[(n+1)x]_f$ will fall into one of the largest existing segments ($[\ ]_f$ means "fractional part of").

It turns out that segments of various lengths are created and destroyed in a first-in-first-out manner. Of course, some irrational numbers are better that others at spacing intervals evenly. For example, an irrational that is near 0 or 1 will start out with many small intervals and one large one. The two numbers $1/\phi$ and $1/\phi^2$ lead to the "most uniformly distributed" sequence among all numbers between 0 and 1.[15] These numbers section the largest interval into the golden mean ratio $1:\phi$. In Fig. 11a, this theorem is illustrated for a sequence of points, $[n1\ /\ \phi]_f$ for

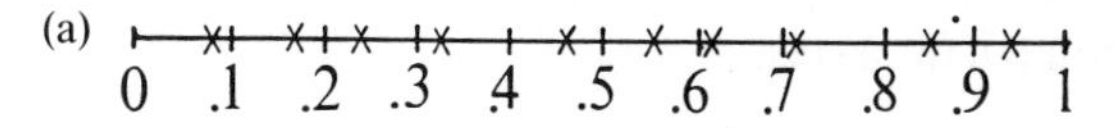

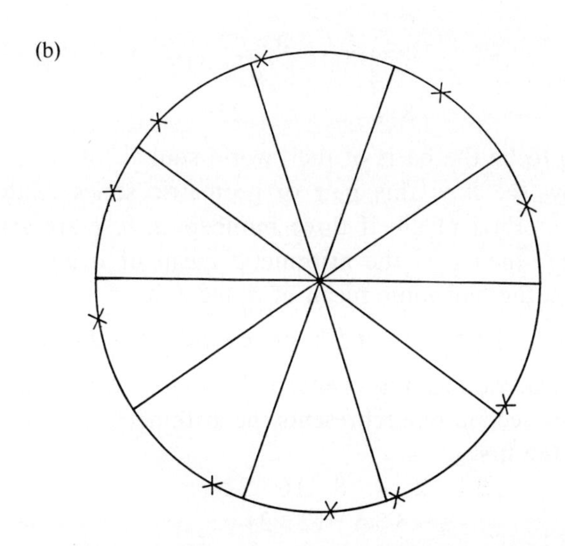

Fig. 11. (a) The points n 1/mod 1 for n = 1,2,... 10 are evenly spaced on the unit interval; (b) the points n 1/mod 2 are evenly spaced on the circumference of a circle.

$n = 1$ to 10. This is equivalent to placing the points $2\pi n / \phi$ mod 2π for $n = 1$ to 10, around the periphery of a circle as shown in Fig. 11b. This placement of points around a circle is more relevant to the study of plant growth, as we shall see.

5. Proportion, the Musical Scale, and the Golden Mean

5.1 *The Musical Scale*

In Greek mythology, the number 2 was the first feminine number and represented the first stage of creation, the split into the mutually dependent opposites of positive-negative, hot-cold, moist-dry. The number 3 was the first masculine number and represented the second stage of creation, the productive union of negative and positive which follows the separation and refinement of these opposite elements.

The two geometrical series:

1 2 4 8 ... and 1 3 9 27 ...

based on the prime numbers 2 and 3, were arranged by Plato in *Timaeus* into a lambda, Λ, configuration:

$$\begin{array}{ccccccc} & & & 1 & & & \\ & & 2 & & 3 & & \\ & 4 & & & & 9 & \\ 8 & & & & & & 27 \end{array}$$

and considered to be the basis of the "world soul."[13]

We shall now see how this pair of geometric series relates to the musical scale.[4,16] First of all, if three numbers a, b, c are arranged in increasing order, then c is the arithmetic mean of a and b if $c = (a + b) / 2$; c is the harmonic mean of a and b if

$$(c - a) / a = (b - c) / b \quad \text{or} \quad c = 2ab / (a + b); \tag{9}$$

and c is the geometric mean of a and b if $c = \sqrt{ab}$. We arrange the two series so that the second one represents the arithmetic means of pairs of numbers from the first:

$$\begin{array}{ccccccccccccc} 1 & & 2 & & 4 & & 8 & & 16 & & 32 & & \dots \\ & & & 3 & & 6 & & 12 & & 24 & & \dots & \end{array}$$

Notice that while each element of the second series is the arithmetic mean of the two numbers that brace it in the upper series, each number of the upper series is the harmonic mean of the pair of numbers that brace it from below. Also, each series cuts the other in the ratio 3 : 2 and 4 : 3. This may be continued again and again to form endless geometric series in the ratio of 2 : 1 from left to right, 3 : 2 along the left leaning diagonal, and 4 : 3 along the right leaning diagonal involving integers only:

$$\begin{array}{cccccccc} 1 & 2 & 4 & 8 & 16 & 32 & \ldots & \\ & 3 & 6 & 12 & 24 & \ldots & & \\ & & & 9 & 18 & 36 & 72 & \ldots \\ & & & & & 27 & \ldots & \end{array}$$

Thus Plato's lambda is formed by the boundary of these geometric series.

The ancient seven-tone scale of Pythagoras was based on string lengths involving the integers 1, 2, 3, 4 which made up the tetraktys. If the fundamental length of a string of unit length sounds the tone c , then a bridge placed at the 3/4 point at the string gives rise to the fourth tone of the scale, F , also known as the musical fourth; the bridge placed at the 2/3 point results in the fifth tone of the scale, G , the musical fifth; the bridge placed at the midpoint, i.e., the 1/2 point results in the tone one octave above the fundamental, C , as shown in Fig. 12. Therefore, Plato's lambda relates directly to the musical scale.

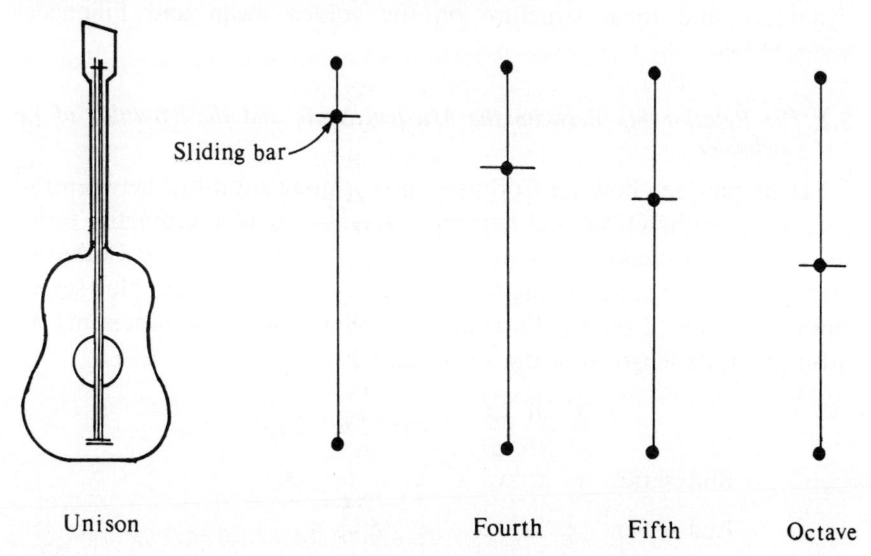

Fig. 12. An illustration of the meaning of a musical unison, fourth, fifth, and octave.

From another perspective, the arithmetic and harmonic means can be seen to have a reciprocal relationship to each other with respect to a given interval. For example, the diagram below shows the arithmetic and harmonic means placed in the interval of an octave:

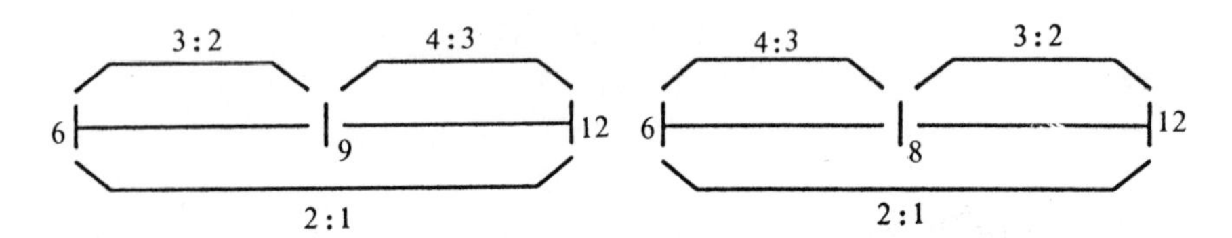

Notice that the musical fourth and fifth are related to each other by reflection in the octave, i.e., $6:9 = 8:12$ and $6:8 = 9:12$. More will be said about the musical scale in Sec. 8.

Thus, the musical scale forms the basis of a tightly woven system of proportionality in which a few ratios are replicated at different magnitudes. It is for this reason that the neoplatonists of the Renaissance, most notably Leon Battista Alberti and Palladio, fashioned the proportions of their great architectural masterpieces from the proportions of the musical scale.[17] Bartók is also known to have based his music from its rhythm, dynamics, and tonal structure on the golden mean and Fibonacci series.[18,4]

5.2 *The Relationship Between the Musical Scale and the Modulor of Le Corbusier*

Let us now see how Le Corbusier used this relationship between the geometric, arithmetical, and harmonic means of double geometric series to create his *Modulor* series of architectural proportions.[4] Le Corbusier created a double scale of lengths which he called the red and blue series. Both series are ϕ-series, illustrated as follows, where d represents an arbitrary scale length (not drawn to scale),

$$\begin{array}{lccccccc} & \frac{2d}{\phi^3} & \frac{2d}{\phi^2} & \frac{2d}{\phi} & 2d & 2d\phi & 2d\phi^2 & 2d\phi^3 \\ \text{Blue series:} & x & x & x & x & x & x & x \\ \text{Red series:} & x & x & x & x & x & x & \\ & \frac{d}{\phi} & d & d\phi & d\phi^2 & d\phi^3 & d\phi^4 & \end{array}$$

Each length of the red series is the arithmetic mean of successive lengths of the blue series. Therefore, according to the musical analogy, each point of the red series is the harmonic mean of the two elements of the blue series that brace it. It is easy to show, using Eq. (9), that the harmonic mean of the interval between successive blue elements divides the interval by the golden section.

Le Corbusier felt that the *Modulor* would serve as a useful system of architectural proportion because of the relationship between the golden mean and human scale,[4] and even symbolized his system by a "Modulor man" (see Brandmüller, this volume). He used this scale to proportion rooms of a house and the facades of buildings by creating rectangular modules with length and/or width from the red and blue series. The fact that according to Eq. (8), an arbitrary length can be approximated to within any preset tolerance by a combination of elements from the red and blue series makes these series capable of tiling any rectangular area with arbitrarily close precision.

Le Corbusier mentions in his book, *Modulor*,[20] that inspiration for creating this system of proportions came from his study of the arrangement of florets on the surface of plants. We shall now turn to the way in which the golden mean governs the growth of plants.

6. The Golden Mean and Plant Growth

As a young man, Le Corbusier studied the elaborate spiral patterns of stalks, or "paristichies," as they are called, on the surface of pine cones, sunflowers, pineapples, and other plants.[15] This led him to make certain observations about plant growth that have been known to botanists for over a century. The stalks, or florets, of a plant lie along two nearly orthogonal intersecting spirals, one clockwise and the other counterclockwise. The numbers of counterclockwise and clockwise spirals on the surface of the plants are generally successive numbers from the F-series. These successive numbers are called the *phyllotaxis* numbers of the plant. For example, there are 55 clockwise and 89 counterclockwise spirals lying on the surface of some sunflowers; thus sunflowers are said to have 55,89–phyllotaxis. On the other hand, pineapples are examples of 5,8–phyllotaxis (although, since 13 counterclockwise spirals are also evident on the surface of a pineapple, it is sometimes referred to as 5,8,13–phyllotaxis.

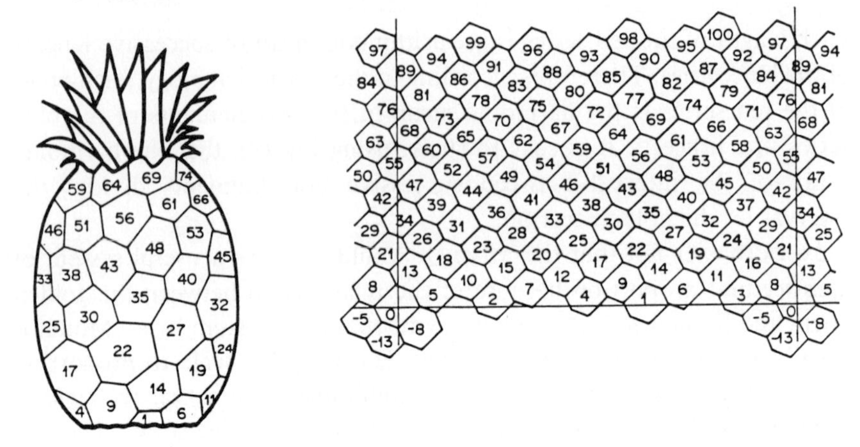

Fig. 13. Subdivision of a pineapple into a two-dimensional lattice. (From *Introduction to Geometry* by H.S.M. Coxeter, published by John Wiley.)

In Fig. 13, H.S.M. Coxeter[21] considers the pineapple transformed to a semi-infinite cylinder which has been opened up to form a period rectangle (the left and right sides of the rectangle are identified). Notice the three families of spirals. Also notice that the stalks are labeled chronologically according to the order in which they have appeared in the growth process. Each stalk occurs at an angle $\lambda = 2\pi / \phi^2$ radians or 137.5 degrees displaced from the preceding one where λ is called the divergence angle. Also notice that a series of stalks alternate on either side of the initial stalk numbered by Q_k from the Fibonacci series, very much as in Fig. 13. Each of these stalks also occur after the entire series has rotated about the cylinder by a number of times P_k , also given by the Fibonacci series. For example, in Fig. 13, the 13th stalk occurs after 8 revolutions around the stem of the pineapple. This follows from Eq. (7), where the Q_kth stalk occurs after P_k revolutions about the cylinder, i.e.,

$$\lambda Q_k - P_k \approx 0 \bmod 2\pi.$$

There remains the question as to why divergence angles are related to the golden mean. The answer to this question appears to be that golden mean divergence angles ensure that successive stalks are inserted at positions on the surface of the plant "where they have the most room." Theorem 1 of Sec. 4.3 makes this precise. If the center of gravity of each stalk is projected onto the base of the period rectangle then the next stalk divides the largest of the three intervals predicted by the theorem in the golden section (just as in the red and blue series). Any other divergence angle would place stalks too near the radial directions of other stalks, and therefore make the stalks less than optimally spaced; however, the divergence angle $2\pi / \phi^2$ leads to the most uniformly distributed set of stalks.[15]

In Fig. 14, we show the results of computer-generated pictures of plant growth due to N. Rivier.[22] Fig. 14a shows a growth process with divergence angle $\lambda = 13/21$, a close Fibonacci approximation to $1/\phi$ where ϕ is the golden mean. Contrast its spider web appearance with the plant-like appearance of Fig. 14b, which has a divergence angle of $1/\phi$. Thus, even close rational approximations to $1/\phi$ are not enough to properly space the florets of a plant, and so the morphology of plants is extremely sensitive to measurements. This goes against our common experience, in which there can be no difference between rationals and irrationals, since measurements are always in error and there is always a rational number with measurable distance to any irrational. The dynamics of oscillating systems will be discussed in the next section, and they too will be shown to be governed by numbers beyond the capability of measurement.

7. The Golden Mean and the Dynamics of Chaos

Just as the growth of plants can be looked at in terms of optimal placement of points around the periphery of a circle, so can the motion of any oscillating system. The fundamentals of dynamical systems have been described by the physicist Leo Kadinoff[23] as follows:

> In general, in a mapping problem one investigates the properties of a sequence of points $z_0, z_1, z_2, \ldots$ each point generated from the last by the application of a defined function, R:
>
> $$z_{j+1} = R(z_j).$$
>
> These problems serve as simple models of dynamical behavior, in which one can think of the z_j as a description of the state of the system at a time $t_j = j_t$. One is particularly interested in universal or generic properties of the set z_j — that is, properties which do not depend in detail upon the form of R. Any such robust property has a chance of being important for the behavior of the more complex dynamical systems manifested in the physical world.

One set of universal behavior concerns maps in which z is a real number

(a)

(b)

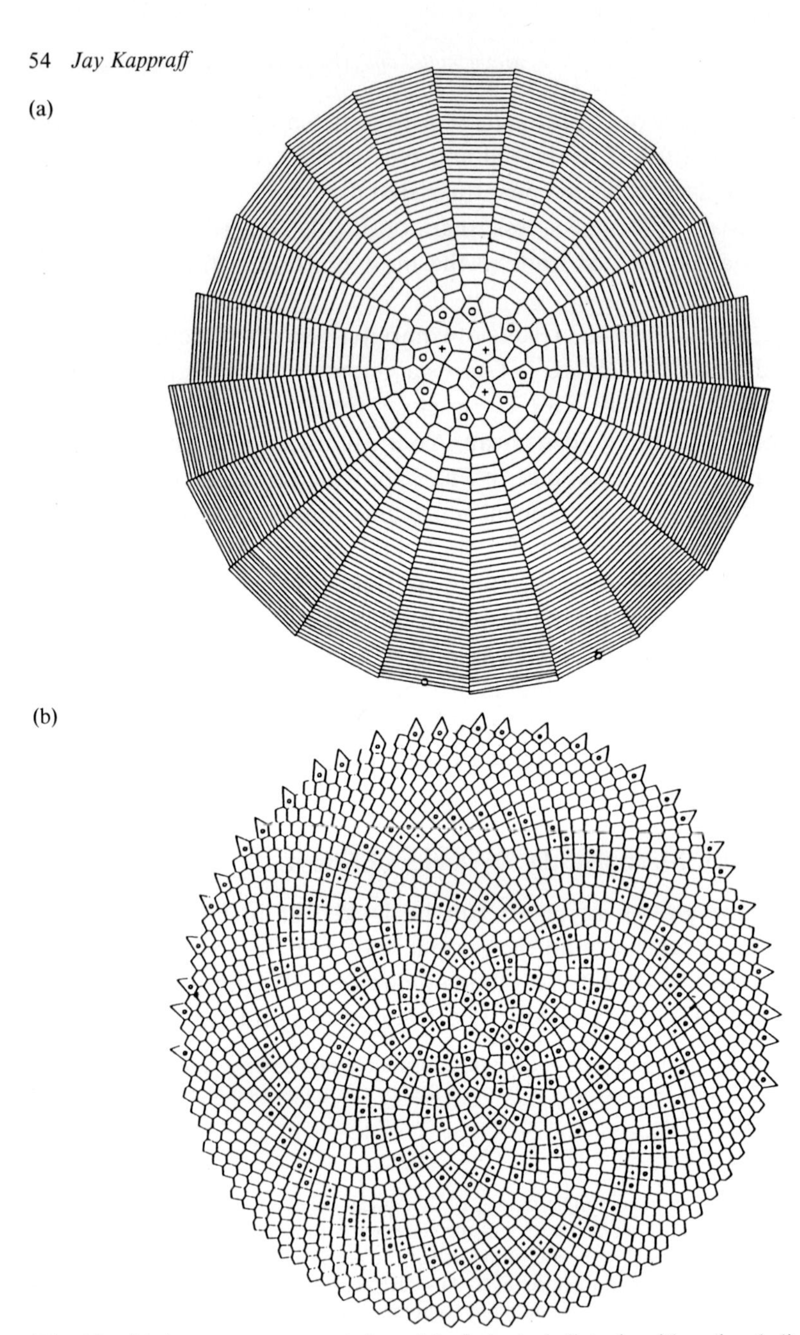

Fig. 14. (a) A computer generated model of plant phyllotaxis with rational divergence angle, = 13/21. Note the spider web appearance; (b) irrational divergence angle $1/\phi$. Note the daisy like appearance. (Courtesy of N. Rivier)

and R is a map that obeys a kind of periodicity condition,

$$R(z+1) = 1 + R(z).$$

For example,

$$R(z) = z - k / 2\pi \sin 2\pi z. \tag{10}$$

where the parameter k is roughly related to the energy of the system. In this model, the oscillation of points around a circle, i.e., mod 2π, are represented by their movement through the interval [0, 1], i.e., mod 1 just as in the discussion of plant growth in the last section.

The trajectories produced by these maps are characterized by a winding number,

$$w(z_0) = \lim_{j \to \infty} z_j / j.$$

which describes the average number of revolutions traveled per step. When w is rational, one can describe the system as commensurable. For example, if $w = p / q$, then the orbit repeats itself after q iterations after cycling p times through the interval [0, 1]. Irrational winding numbers correspond to trajectories that never repeat themselves although, as we have shown in Sec. 4.1, they can be best approximated by rationals using continued fractions.

One can talk about the stability of these orbits. Perturb the initial point of an orbit with a rational winding number ever so slightly, and the resulting orbit is no longer periodic and may differ significantly from the original. Such orbits are said to be unstable. What about the stability of irrational orbits? Their stability depends on the value of the parameter k in Eq. (10). If $k < 1$, then the trajectories, $\{z_j\}$, with irrational winding numbers fill up the entire interval as $j \to \infty$. However, depending on whether the orbit with an irrational winding number has sufficiently close neighboring orbits with rational winding numbers, the trajectories will be more or less stable. The orbits with sufficiently close neighboring orbits with rational winding numbers destabilize first as k is increased.[24]

When $k = 1$, all orbits with irrational winding numbers have been destabilized, and further increase in k results in chaos. In the chaotic regime, the nature of the orbit depends on the *exact* position of the starting point, in that the smallest perturbation of the starting point may result in a totally different kind of orbit. Also, the orbit may no longer spread out to fill the entire circle, but instead bunch into a set of narrow disconnected regions of, as mathematicians say, "total measure zero." Kadinoff has found by using Eq. (7) that these intervals are directly related to a system of hierarchies within hierarchies of Fibonacci numbers.[23]

It is now easy to see why the golden mean is the key to understanding the structure of ordered and chaotic orbits. Since, as we showed in Sec. 4.1, the golden mean is the "most irrational number," the orbit with the golden mean winding number has the most distant neighboring rational orbits of any winding number and is therefore the last orbit to be destabilized as the energy of the system, k, is increased. At $k = 1$ the golden mean orbit is destabilized, after which chaos ensues.

The theme of order and chaos has only recently emerged as an intense area of mathematical and scientific research. In fact, it is thought by many mathematicians and scientists that the ordered results of science and mathematics upon which most of the modern theories of science have been built may correspond to only a narrow range of observed phenomena. In fact, the themes of order and chaos have been central to many myths throughout all ages. In the next section we shall explore one source of these myths.

8. Ancient Cosmology and the Musical Scale

In his book, *The Myth of Invariance*, Ernest McClain[2] analyzes the mysterious mathematical and geometric details that abound in India's oldest sacred book, the *Rig Veda*. This book develops, through the language of poetry, a system to deal with the twin problems characterizing the movement of the sun and moon on the one hand and the construction of the musical scale on the other. Although this book does not refer directly to the golden mean, it has its basis in fivefoldness (although not fivefold symmetry).

According to McClain,

> Its hymns link Sun and Moon and all creation to incestuous couplings within a pantheon of deities in which sons create their own mothers and all are counted. The universe emerges as a victory of gods over demonic forces which can be defeated but never eradicated.

McClain feels that the *Rig Veda* was attempting to come to practical terms with the observed fact that the twelve lunar months are not quite coherent with the solar year. In fact, a full moon occurring on a given day of our calendar will not repeat on this day until 19 years have elapsed, i.e., the moon moves along a 19-year cycle.

According to Alexander Thom, this 19-year cycle was also built into structure of the stone monoliths erected by the neolithic people of the British Isles and France. This complex relationship between solar and lunar cycles posed great difficulty to the formulation of a rational calendar necessary for survival. The central geometric image of the *Rig Veda* is the "single-wheeled chariot of the Sun, harmonizing moon months with solar years and the signs of the zodiac":

> Formed with twelve spoke, by length of
> time unweakened rolls around the
> heaven this wheel of during order.
> Twelve are the fellies, and the wheel is single.

The poetry of the *Rig Veda* combines imagery of the Gods and cosmos together in such a way that it can be read as pertaining to the musical scale. Why the musical scale? Well, the musical scale also has twelve semitones, and the twelve spokes which locate the tones could be thought of as the rays of the Vedic Sun-gods, as shown in Fig. 15. The musical scale is also well known to have problems with respect to the commensurablity of its tones. Let's see why.

In the form that we know it, the scale is divided into twelve tones dividing the ratio of the octave, 2:1, into twelve equal ratios. In other words, if they were arranged evenly around a circle, the twelve tones would have frequencies of $2^{n/12}$ where $n = 1$ to 12. This is known as the well-tempered scale. As a scale, it has an important drawback. The ratios inherent in the well-tempered scale are all irrational numbers, however, pleasant sounding tones are found to be the ratio of small whole numbers. The best solution to the problem of representing tones as the ratio of small whole numbers is the Just scale, often attributed to the Roman astronomer Ptolemy. If a unit length of a string is taken to be the tone, D, then eleven of the twelve half-tones of this scale correspond to the following fractions of the string;

D	E^b	E	F	F	$G^\#$	A	B	B^b	C	$C^\#$	D
1	$\frac{15}{16}$	$\frac{9}{10}$	$\frac{5}{6}$	$\frac{4}{5}$	$\frac{3}{4}$	$\frac{2}{3}$	$\frac{5}{8}$	$\frac{3}{5}$	$\frac{5}{9}$	$\frac{8}{15}$	$\frac{1}{2}$

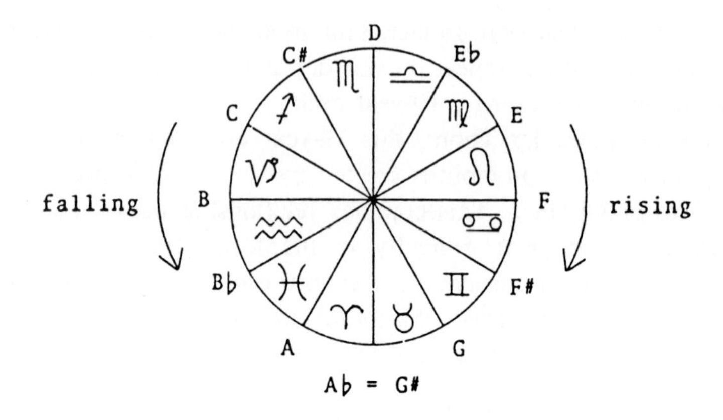

Maṇḍala of The Single-Wheeled Chariot of the Sun

This hypothetical "tonal zodiac" shows how a twelve-spoked maṇḍala harmonizes music and astronomy at an abstract geometrical level. In ancient times neither the constellations nor the intervals of the chromatic scale divided the cycle equally.

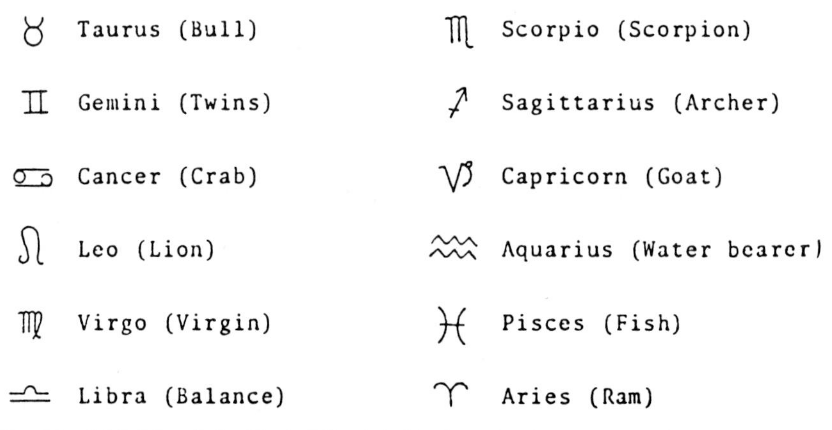

Fig. 15. Mandala of the Single-Wheeled Chariot of the Sun. This hypothetical "tonal zodiac" shows how a twelve-spoked mandala harmonizes music and astronomy at an abstract geometrical level. In ancient times neither the constellations nor the intervals of the chromatic scale divided the cycle equally. (From *The Myth of Invariance*, by E.G. McClain, published by Harper and Row.)

Observe that the numerators and denominators are all divisible by the primes 2, 3 and 5. Also notice that the half-tone corresponding to $G^{\#}$ or equivalently A^{b} is missing. It cannot be represented as the ratio of small whole numbers, but must be assigned the irrational number, $\sqrt{2}$.

This throws off the entire scale if one wishes to change the fundamental note from D to some other tone, yet preserve the ratios between successive tones and have a scale with only a finite number of tones. The *Rig Veda* is a search for a solution to this problem.

We shall not be able to explore this problem in depth in this short article, however McClain considers the eleven tones to be arranged on the Mandala of Fig. 16, where the falling pitches are the reflection in the

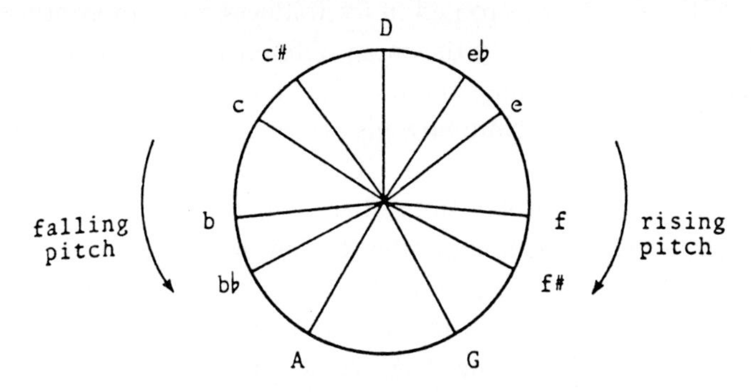

Fig. 16. The Hindu-Greek diatonic scale and its reciprocal. (From *The Myth of Invariance,* by E.G. McClain, published by Nicolas Hays, Inc.)

octave of the rising pitches (see Sec. 5.1). Next, McClain notices that the eleven tones are organized in two symmetric groups of five about the reference tone. In Plato's Atlantis myth, these eleven elements represent "Poseiden and his five pairs of twin sons." In the *Rig Veda* , Usas in her role as universal "bride" is offered the following wedding prayer:

> Vouchsafe to her ten sons, and make
> her husband the eleventh man.

Here Usas represents the female number 2 or the octave which results in no new tone when multiplying the fundamental tone. The other tones of the scale are generated by the male numbers 3 and 5.

The *Rig Veda* is also rich in the imagery of order and chaos. In Fig. 16, the undifferentiated circle represents "the whole undifferentiated primordial chaos ruled by the dragon Vrtra. The ratio 1 : 2 belongs to mother, and every point in the circle which we can define by an integer belongs to God. The model for all Existence — hence of everything which can be named or numbered — is Indra. The conflict between Indra and Vrtra can never end; it is the conflict between the field of rational numbers and the continuum of real numbers. Integers which introduce new "cuts" in the tone-mandala demonstrate "Indra power" over Vrtra; Vrtra is "cut to pieces" in every battle with the Gods, but his death would be their own. "Without Vrtra there would be no Indra, not even the Gods for he is their container." McClain carries this allegory to far reaches and connects it with a great deal of our own cultural history.

Much of the *Rig Veda* is rooted in fivefoldness most appropriate to the theme of this paper. Among the many references to fivefoldness, Indra the dancer is the Lord of men and he rules "the fivefold race of those who dwell upon the Earth." "Note how his horses are harnessed: Sixfold they bear him or by fives are harnessed." The *Rig Veda* also refers directly to four sets of five tone progressions that span the octave. Two of these "pentatonic scales" encompass the eleven tones referred to above. McClain shows how these pentatonic scales are constructed from the ratio 3:4 in conjunction with the number 5 (reminiscent of the 3,4,5-Egyptian triangle). The Rig Vedic creation hymn unifies the pentatonic and diatonic constructions:

> Upon this five-spoked wheel revolving ever
> all living creatures rest and are dependent.
> Its axle, heavy-laden, is not heated: the
> nave from ancient time remains unbroken.
> The wheel revolves, unwasting, with its felly:
> Ten draw it, yoked to the far-stretching car-pole.

The *Rig Veda* shows the degree to which the ancient mind was aware of the importance of the subtle interplay between rational and irrational numbers, something that modern mathematics and science are only beginning to appreciate. In the next section, we shall explore the ways in which number was uppermost in the minds of ancient cultures as they attempted to recreate the order of the heavens from their Earthly abodes.

9. The City of Revelations: The Cosmology of John Michell

John Michell has created a system that describes solar and lunar cycles and geometry while at the same time replicating the geometrical patterns of certain sacred structures such as Stonehenge, the mythical city referred to in Revelations 21, the plan of the allegorical city in Plato's Laws, and the layout of St. Joseph's chapel at Glastonbury. The system is well described in Michell's book, *The Dimensions of Paradise.*[8]

The system is derived from a 3,4,5-right triangle, and this triangle in turn is related to the golden mean as we showed in Fig. 4. The 3,4,5-right triangle can be related to the cosmos by considering a knotted rope of 12 equal circular segments representing the constellations of the zodiac. This mandala pattern can be cut open to a straight line and then wrapped into a 3,4,5-right triangle, as shown in Fig. 17.

Beginning with a 3,4,5-right triangle, Michell constructs a large square of side 11 units with four small squares of side 3 units symmetrically placed around it like the one shown in Fig. 18.

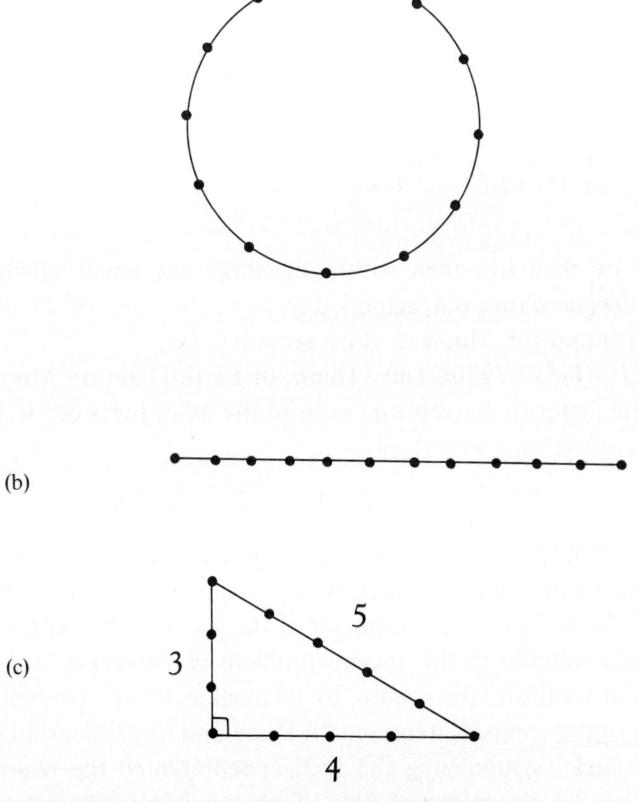

Fig. 17. (a) Representation of the twelve seasons of the Zodiac by a knotted rope; (b) the rope is cut open to a straight line; (c) the line is bent into a 3,4,5-right triangle.

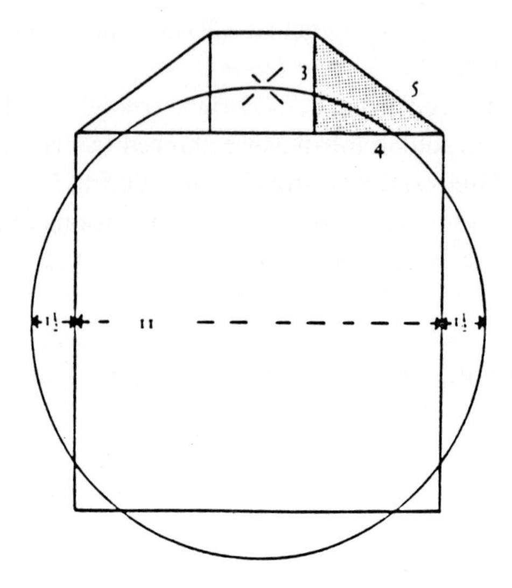

Fig. 18. The construction of a squared circle beginning with a 3,4,5-right triangle. Multiplied by 720 these base numbers are raised to the dimensions to the New Jerusalem diagram. (From *The Dimensions of Paradise*, by John Michell, published by Harper and Row.)

Circles are then inscribed within the large and small squares; their diameters D and d turn out, remarkably, to be the ratios of the diameters of the Earth and the Moon to close accuracy, i.e.,

$$D/d = 11/3 = 7920/2160 = \text{Diam. of Earth/Diam. of Moon}$$

The scale factor to convert one ratio to the other turns out to be equal to

$$720 = (3 + 4 + 5)\ (3 \times 4 \times 5) = 12 \times 60,$$

again reflecting the strange power of number. Furthermore, if another circle is placed through the center of the moon circles, its circumference, using 22/7 for π, equals the perimeter of the square. This represented an approximate solution to the ancient problem of "squaring" the circle.

In ancient tradition, the square, by its axial geometry symbolizing the directions of the compass, represented Earth and the dimension of space while the circle, symbolizing the zodiac, represented the realm of the heavens and the dimension of time. Thus, ancient mathematics, architecture, astronomy, cosmology, and music were all entwined with myth and sacred scriptures to form an holistic view of the cosmos. An attempt was made to bring Heaven down to Earth and replicate it at all scales and to synchronize space and time.

Fig. 19 shows Michell's New Jerusalem diagram with 28 lunar circles representing the 28 days of the lunar cycle placed around the Earth circle arranged so that three circles are to the North, three to the South, three to the East and three to the West, as described in Revelation 21 as the

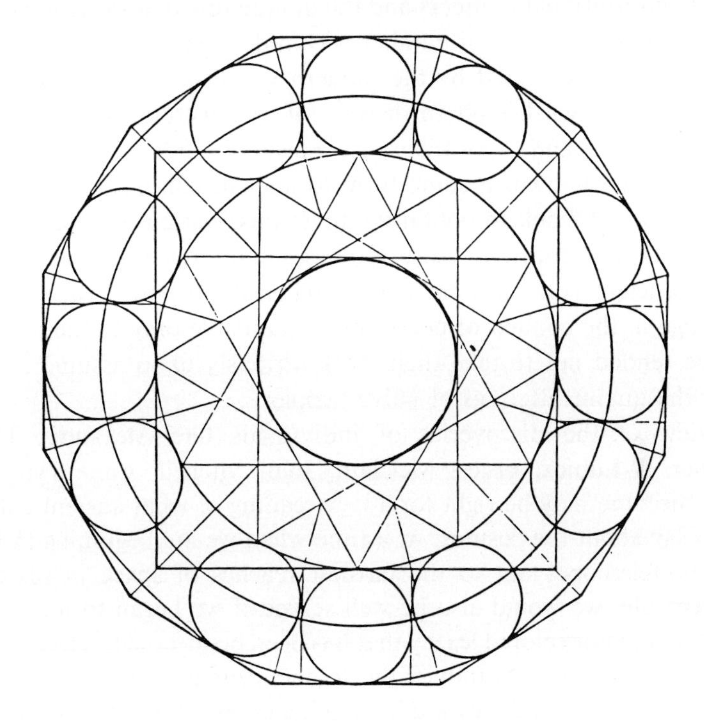

Fig. 19. The New Jerusalem diagram of ancient cosmology also providing the groundplans of Stonehenge and the Glastonbury sanctuary. Dimensions:
12 small circles, each diameter 21.80 ft.
circle through center of small circles, diameter 100.8 ft.
circle within square, diameter 79.20 ft.
As a scheme of cosmology the diagram is measured in units of 100 miles instead of feet. Thus, the diameter of the circle within the square becomes 7920 miles, the diameter of the earth, and the diameter of the small circles 2160 miles, equal to the moon's diameter. (From *The Dimensions of Paradise*, by John Michell, published by Harper and Row.)

structure of the mythical city. Much more can be said of this diagram, but not in this short article.

10. Conclusion

I have tried to show that the golden mean and its inherent fivefold symmetry have played an important role in both ancient and modern mathematical and scientific consciousness. The different natures of rational and irrational numbers and the unique role that the golden mean plays as the most irrational number was understood by ancient cultures and is being rediscovered by the modern world of science and mathematics. In this regard, modern theories of architectural proportionality, plant phyllotaxis, and chaos theory have begun to rediscover ideas which may have been known in some form to ancient cultures. Much of this information is difficult to obtain by direct observation or measurement, which alters the delicate physical processes and reduces the phenomena. Because ancient cultures were either preliterate or coded their esoteric knowledge in the context of poetry understandable only to the initiated, we have tended not to take their work seriously or to assume it to be merely the quaint utterings of naive peoples.

In view of the discoveries of individuals like Alexander Thom, Schwaller di Lubicz, Ernest McClain, John Michell, Anne Tyng, and others, the time may be right for a new reading of what ancient cultures have to say about the cosmos. At a time when we are beginning to focus our radio-telescopes out to the farthest reaches of space in search of intelligent life, we would also be well served if we began to focus more sharply on the unexplored legacy that has been bequeathed to us from our most ancient cultures. A study of the golden mean and fivefold symmetry should play a crucial role toward linking the knowledge of the ancient and modern worlds.

References

1. K. Critchlow, *Time Stands Still.* St. Martin's Press, New York (1982).
2. E.G. McClain, *The Myth of Invariance.* Nicolas Hays, Inc., York Beach, Maine (1976).
3. R.A. Schwaller di Lubicz, *The Temple in Man.* Robert and Deborah Lawlor, trans. Inner Traditions Int., New York (1977).
4. J. Kappraff, *Connections: The Geometric Bridge Between Art and Science.* McGraw Hill, New York (1991).

5. A.G. Tyng, *Simultaneous Randomness and Order: The Fibonacci Divine Proportion as a Universal Forming Principle.* Dissertation. University Microfilms Int., Ann Arbor, Michigan (1975).
6. R. Lawlor, *Sacred Geometry.* Crossroad, New York (1982).
7. H.F. Verheyen, see this volume.
8. J. Michell, *The Dimensions of Paradise.* Harper and Row, San Francisco (1987).
9. T.A. Cook, *The Curves of Life.* Dover, New York (1967).
10. J. Hambridge, *Dynamic Symmetry.* Dover, New York (1967).
11. M. Ghyka, *The Geometry of Art and Life.* Dover, New York (1978).
12. G. Doczi, *Power of Limits.* Shambala Publ., Denver (1981).
13. Plato, *Timaeus.* Desmond Lee, Penguin Books (1977).
14. A.I. Khinchin, *Continued Fractions.* Univ. of Chicago Press, Chicago (1989).
15. C. Marzec and J. Kappraff, Properties of maximal spacing on a circle relating to phyllotaxis and to the golden mean, *J. Theor. Biol.* **103** (1983) 201–226.
16. H.S.M. Coxeter, Music and mathematics, *Math. Teacher.* (1968) 312–320.
17. R. Wittkower, *Architectural Principles in the Age of Humanism.* Originally published by Tiranti, London (1962). W.W. Norton and Co., New York (1971).
18. E. Lendvai, Duality and synthesis in the music of Bela Bartok, in *Module, Proportion, Symmetry, Rhythm*, Gyorgy Kepes, ed. George Braziller, New York (1966).
19. J. Brandmüller, see this volume.
20. Le Corbusier, *The Modulor.* MIT Press (1968).
21. H.S.M. Coxeter, Golden mean phylolotaxis and Wythoff's game, *Scripta Mathematica.* Vol. XIX, No. 2–3 (1953).
22. N. Rivier, J. Occeli, and A. Lissowski, Structure of Benard convection cells, phyllotaxas and crystallograph in cylindrical symmetry, *J. Physique.* **45** (1984) 49–63.
23. L.P. Kadinoff, *Supercritical Behavior of an Ordered Trajectory.* James Franck and Enrico Fermi Institute Research Paper (1985).
24. M. Tabor, *Chaos and Integrability in Hamiltonian Systems.* Wiley (1988).

800-YEAR-OLD PENTAGONAL TILING FROM MARĀGHA, IRAN, AND THE NEW VARIETIES OF APERIODIC TILING IT INSPIRED

Emil Makovicky

1. Introduction

Aperiodic tiling with pentagonal geometry, discovered by Penrose,[1,2] has been, in its different versions, the object of intensive study by numerous mathematicians and crystallographers. The present discovery of a similar, 800-year-old tiling from (post) Saljuq Iran therefore represents a matter of considerable interest. Besides giving a surprising insight into the skills of ancient geometric artists, it also reveals some new aspects of Penrose tiling and leads toward further generalizations. This discovery results from an extensive search through literature for complex historic patterns based on pentagonal geometry.

2. Setting

The historical town of Marāgha is situated in western Iran, close to the shores of Lake Urmiya. One of the chief monuments of Marāgha is the Blue Tomb, Gunbad-i-Quābūd, which was built in A.D. 1196–1197, a quarter of a century before the Mongol invasion.[3,4] The tomb is often ascribed erroneously to the mother of the first Il Khanid ruler, but its true age is given by the portal inscription, which gives a date 60 years before the coming of Il Khans.[5] In 1259, this town became the seat of the first real astronomical observatory in the world, although it is impossible to say whether this was a reflection of a high local scholastic tradition.

Fig. 1. The Blue Tomb, Gunbad-i-Quābūd (built A.D. 1196–1197) in Marāgha, Western Iran. (Reproduced from Ref. 3 with permission.)

The Blue Tomb (Fig. 1) represents a polygonal tower, each side of which is crowned by a niche with a pointed arch. The name of the tomb comes from the masterly combination of unglazed, buff surfaces and ornamental ribs interlaced with turquoise blue ribbons and inscriptions. The unglazed, and the turquoise-glazed curvilinear nets sometimes represent liberally combined dual nets.

3. The Marāgha Pattern

3.1 *Pattern Description*

The principal object of this study is a grand polygonal ornamental net (Figs. 2a and b), which envelops the entire shaft of the building below the niches, without interruption at the corner pilasters. It was worked out in strong brick ribs; all its polygons are filled by small-scale curvilinear turquoise nets radiating in most instances from centrally placed small pentagons.

The large-scale net is composed of the following elements (throughout, letters (a)–(g) correspond to parts (a)–(g) of Fig. 3): (a) regular pentagons; (b) complex decagons, hereafter called *butterflies*, with convex angles of 72° and reentrant angles of 108°; (c) deltoids ("kites") *and* a pair of partly overlapping pentagons that always form together a *rhomb* with "deltoid-marked" corners of 72° and unmarked corners of 108°; and (d) occasional nested pentagons with five spokes.

The combination rules are simple: only straight-line segments of the net intersect (at 72°), whereas all line breaks (of 108°or 144°) are outside these intersections. Polygons of the same kind do not share edges. Butterfly wings terminate in pentagons and are surrounded either by four additional pentagons or by an additional *cis* pair of pentagons and a *cis* pair of rhombs (each straddling the long diagonal). These groupings represent an elongated hexagon and a regular pentagon, respectively, both with a butterfly and both possessing mirror symmetry, (e) and (f). The nested pentagons (d) appear to be inserted into a finished pattern and replace the large pentagon (f) without any adjustment problems. If they are replaced by (f), some of the surrounding large "unfinished" pentagons will become complete (f) versions whereas others will remain incomplete, with three additional pentagons and one rhomb attached to the butterfly polygon (g). The vertical edges of Fig. 2a as well as the edges and the central vertical rib of Fig. 2b represent the corner ribs of the building and mirror planes of the pattern. The height of Figs. 2a and 2b represents the entire pattern executed on the building, reconstructed painstakingly from photographs by Godard[3,4] of the damaged complex. The vertical ribs (pilasters), especially, have suffered greatly from the vicissitudes of time. At one end of each rib is a decagon (though without corresponding internal symmetry), from the edges of which rays of rhombs and butterflies radiate vertically and diagonally, 36° apart from one another.

Fig. 2a (above) Line reconstruction of the principal ornamental net from one side of the tomb tower and from the adjacent corner pilasters (shaded) of the tomb.

Fig. 2b (next page) Line reconstruction of the principal ornamental net from the shaft of the Blue Tomb at Marāgha. Two sides of the polygonal building are depicted, with the corner pilasters represented by the vertical sequences at the edges and in the center of the field.

Fig. 2b

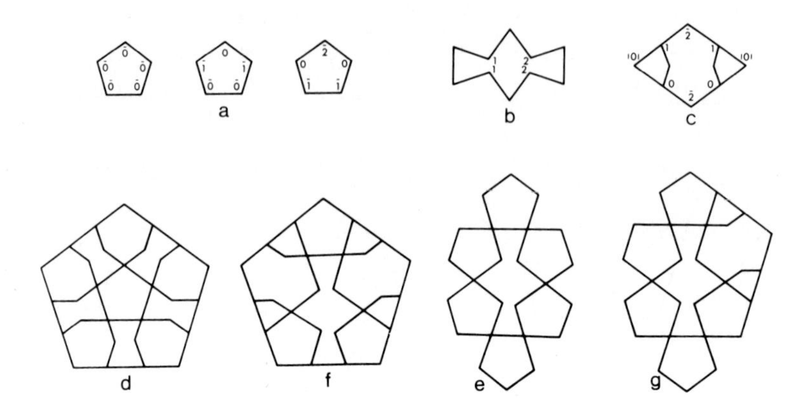

Fig.3. Principal tiles and tile patches from the pattern in Fig. 2b: (a) pentagons, (b) butterfly tiles, (c) rhombs, (d) nested pentagons, (e) composite compressed hexagons, (f) complete composite pentagons, and (g) "unfinished pentagons." (a), (b), (c), (f) and (g) constitute the elements of M1 tile set. Penrose-like *corner* markings have been applied to the elements (a)–(c) for the M1 tiling.

One of them, at 36° from the vertical, reaches the other decagon at the opposite end of the next corner of the building. The periodicities of diagonals and of the vertical ribs do not tally (a band pattern *p2gm*), but this can be rectified by altering the ratio of rhombs and butterflies in the vertical rays. This correction will produce a two-dimensionally periodic, centered pattern with both horizontal and vertical mirror planes. However, the former are not obeyed by the internal organization of the tile patches with the decagonal outlines that are positioned at the intersections of 36° diagonals and verticals (Fig. 2b).

The entire pattern is too complex to be understood at a glance. It requires long contemplation, and almost appears to be designed by a mathematician rather than an artist. Its badly damaged lowermost portions can be safely reconstructed because of the good state of preservation of the corresponding uppermost portions. However, a new photograph of the tomb in Ref. 6, which shows details of an almost undamaged side of the building, has revealed that in a small part of the bottom portions of the pattern the artist gained the upper hand over the mathematician. The tenfold stars, which can be traced in the polygonal net on both sides of the partly overlapping nested pentagons (d) at the bases of corner pilasters (e.g., in the left-hand lower corner of Fig. 2a) were emptied of their original polygonal contents and were filled by fivefold "rosettes." Eye-attracting rosettes of this kind are common in Islamic wall ornaments, but those used here (only once per each side of the building) are completely foreign to the rest of the pattern, and will therefore be omitted from further consideration (Fig. 2b). The Marāgha pattern represents the peak of Saljuq architectural traditions, just before the destruction inflicted by the Mongols.

3.2 *Penrose Tiles?!*

At first glance, the similarity of the Marāgha pattern to a cartwheel Penrose tiling, somewhat adapted to the periodicity requirements of the polygonal building, became inescapably apparent. The rays ought to correspond to Conway worms (see, e.g., Ref. 7) in Penrose tiling and the two kinds of rosettes, as well as other elements, should have their counterparts as well.

After some searching, the analogy became straightforward. We take the oldest set of Penrose tiling, which is composed of pentagons, narrow rhombs, stars, and half-stars, shown in Fig. 4. It will be henceforth denoted as P1.[7] Then we mark centers of sides of all pentagons in this tiling. We find that the Marāgha tiles represent a kind of *interstitial net* to P1, interconnecting the just constructed midpoints (Fig. 5). Marāgha pentagons are inscribed in the Penrose ones; Penrose rhombs become butterflies in this process, whereas Penrose stars and the surrounding pentagons taken together correspond to the nested pentagons (d) from Marāgha. The transformation just described can be followed comparing Figs. 4 and 5. The entire series of Figs. 4–7 is based on the same cutout of P1 tiling. Therefore, the relationships of different tilings can be easily traced by copying them onto transparencies and superimposing any combinations of these.

Above we have considered the nested pentagons (d) to be a secondary element to the other ones and have shown that they can be substituted by the complex pentagon (f), which contains an asymmetrically placed butterfly, two rhombs, and three pentagons. This substitution can happen along any of the five arms of a Penrose star (Fig. 4); that is, this star and the adjacent pentagons (or their equivalent, the element (d)) represent an

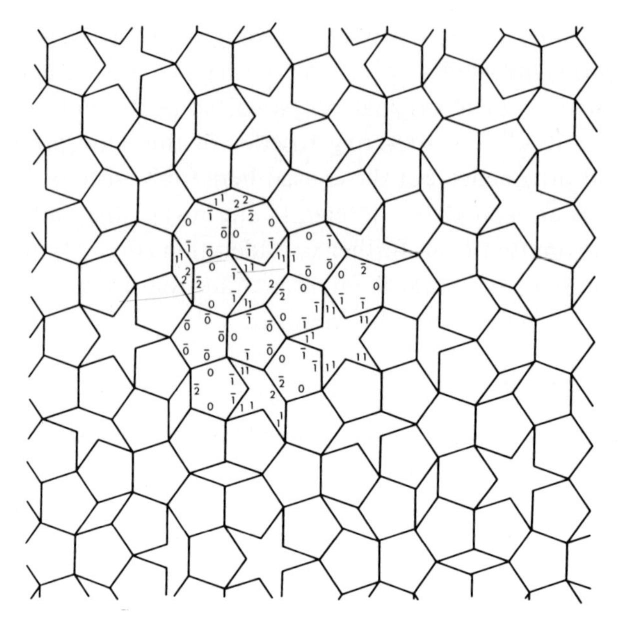

Fig. 4. A representative area of Penrose tiling P1. (Redrawn from Ref. 7 with minor alterations.)

averaged motif corresponding to five complex pentagons (f) rotated 72° from one another, each of them with only a single mirror plane as an element of internal symmetry. Because of Penrose matching requirements,[7] this star is surrounded by half-stars or lozenges, which will cause rhombs (c) of the Marāgha tiling transformed by the (d) → (f) substitution to be always situated between two butterflies (Fig. 6).

The same principles apply to the half-stars. Here only two butterfly positions exist, rotated 144° against each other and the Marāgha rhomb (c) will lie on their "starry" side, situated between this and an adjacent butterfly (cf. Figs. 4–6).

4. Novel Tilings

4.1 *The M1 Tiling and Its Relationship to the Classical Penrose Pattern*

We have just seen that P1 can be completely transformed into a new tiling with no "ornamental" fivefold elements (Fig. 6). The optional elements (d) that bring back the high averaged local symmetry have been

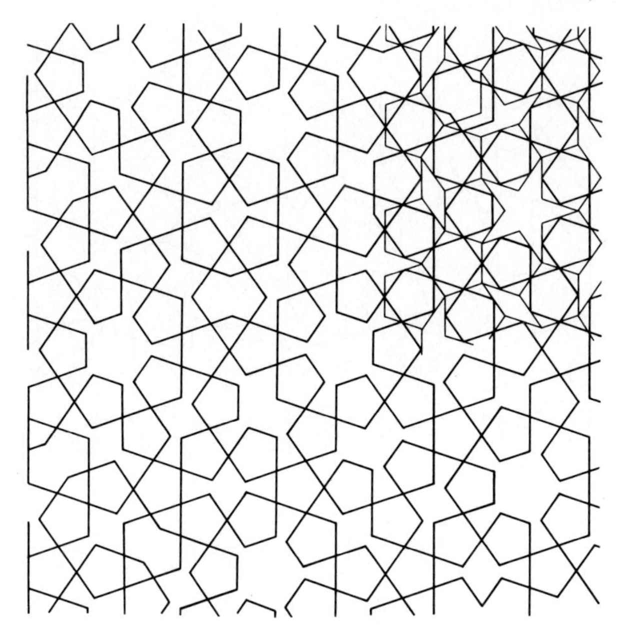

Fig. 5. The same area of aperiodic pentagonal tiling, executed using selected tiles and relationships from the Marāgha pattern. Elements (a), (b), (c) and (d) (for Penrose stars and their surroundings) from Fig. 3 were used. Half-stars were partly reproduced by incomplete elements (d), partly as incomplete elements (f). The underlying P1 tiling is partly shown.

eliminated. This tiling of undisputable dynamism (Fig. 6) will be called hereafter M1 in allusion to the origin of its tiles.

Treated summarily, the *classical Penrose tilings* (i.e., P1 and P2: darts and kites[7]) represent *averaged patterns*, encompassing several analogous M1 tilings that differ from each other only by the orientations of butterflies and rhombs connected to them — five distinct cases for every Penrose star and two distinct cases for every half-star. A "PM1" modification of P1 with numbered edges, analogous to M1, can be constructed (Fig. 7) by replacing all stars and half-stars by the arrow lozenges, marked in agreement with Penrose as 1-1-2-2, and by introducing a compressed hexagon with the following angles and edge markings: $\bar{2}$-108°-0-144°-1-108°-$\bar{2}$-108°-1-144°-0-108°-$\bar{2}$. The two parallel edges of this hexagon, marked as "$\bar{2}$", always join the edges marked as "2" of the lozenges. All three types of marked pentagon from P1 remain preserved. M1 and PM1 represent an extension of the concept of

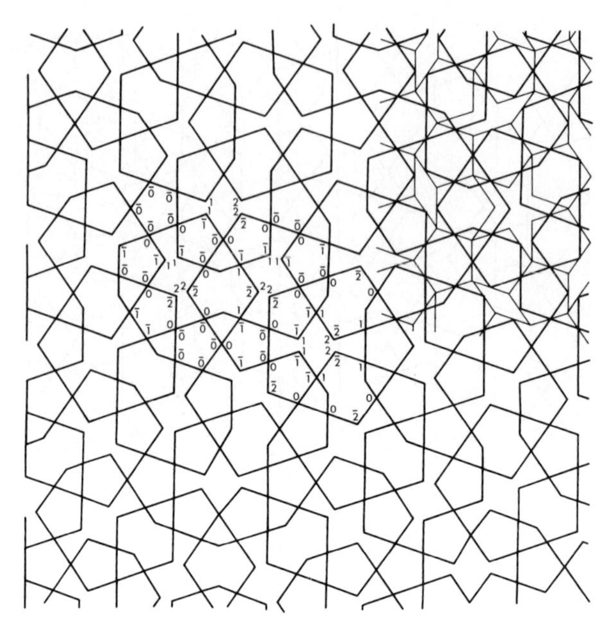

Fig. 6. The same area of aperiodic pentagonal tiling as in Fig. 4, executed in the tile set M1 with all constraints applied (see text). Underlying P1 tiling is partly shown.

pentagonal aperiodic tiling, reducing the occurrence of local fivefold symmetry to pinwheels around the "$\bar{0}$ pentagons." The M1 and PM1 tilings should therefore be of considerable interest to crystallographers: the perfect fivefold motifs, rare in crystallography, occur in them with minimal frequency.

Does the M1 (and PM1) tiling guarantee aperiodicity? Marking of the M1 set analogous to that of P1 is not easy to produce. As the corners and intersections in M1 stand for edges in P1, the markings from PM1 can be used: 1-1 and 2-2 will be applied to the pairs of close reentrant angles of butterflies, the three sets of pentagon markings of P1 will be used for *corners* of inscribed pentagons, |0| markings for the acute vertices of the rhombs, $\bar{2}$ for their obtuse vertices, and 1 and 0 for the points where the "kite" markings intersect their edges (these markings differ for the two sides situated across the long diagonal of the rhomb) (Fig. 3). Alternatively, the following verbal adjacency conditions are valid: each butterfly is bounded either by two rhombs in a *cis* arrangement and 2(+ 2 terminal) pentagons or by 1 rhomb and 3(+ 2) pentagons. Each rhomb

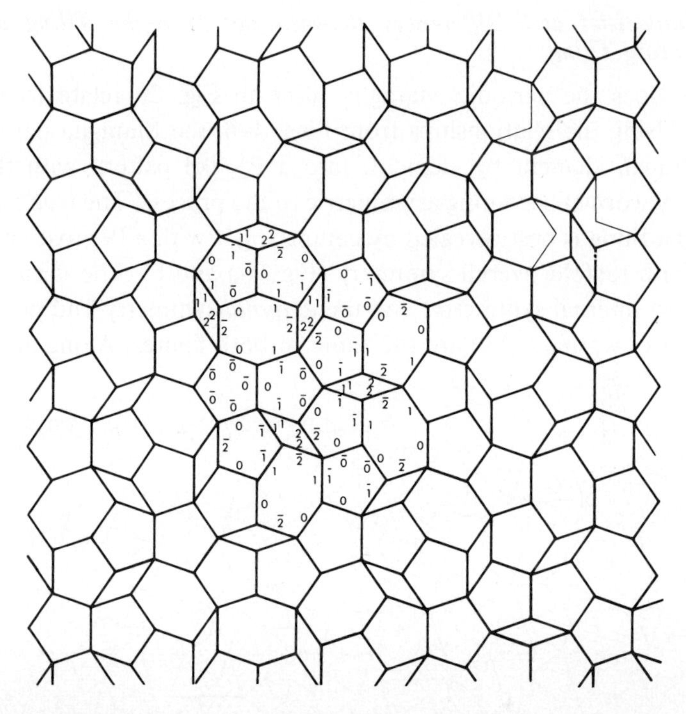

Fig. 7. The same area of aperiodic pentagonal tiling redrawn as a modified Penrose tiling PM1. The underlying P1 net is partly drawn (as thin lines, only where it differs from PM1).

has for neighbours, two butterflies (in a *cis* or *trans* arrangement), but at its acute vertices, it is accompanied by 2 + 2 pentagons. Study of P1 and PM1 suggests that these conditions ought to be equivalent to the original Penrose conditions (i.e., markings).

However, to increase the randomness of the pattern (to exclude worms of high periodicity seen in Fig. 2b) additional, more stringent conditions were introduced for M1. Rhomb–rhomb contacts as well as the rhomb-butterfly body-rhomb, etc., periodic chains are to be avoided. The latter rule concurs with the exclusion of groupings composed of a butterfly bounded by 4(+ 2) pentagons (denoted as element (e) in Fig. 3). Two butterflies can touch each other only at their distal corner; they can abut nested pentagons (d) only at a reentrant side (Fig. 6). These conditions were also translated into the PM1 pattern in Fig. 7.

4.2 *Similarities and Differences Between the Marāgha Tiling and the Penrose Tiling*

How does the periodic Marāgha tiling in Fig. 2b relate to Penrose tiling? Using the relationships from Figs. 4–6, the Marāgha pattern can be redrawn, element by element, into a P1-like pattern with the rays (Conway worms) becoming asymmetric in the process. The true nature of Marāgha tiling is best revealed by comparing it with a Penrose cartwheel tiling with tenfold overall symmetry (Figs. 8a and b). The decagon core (with low internal symmetry) and the *external* symmetry and periodicity of Conway worms (rays) are the same in both figures. Along any worm

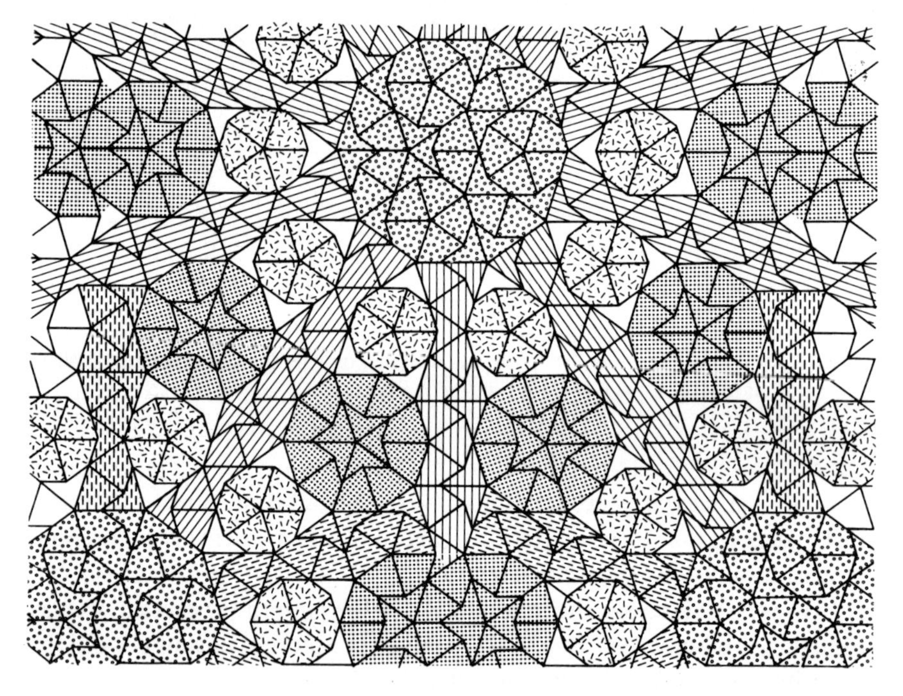

Fig. 8a. A portion of the tenfold Penrose cartwheel tiling P2 (composed of darts and kites) directly comparable with (b), the Marāgha pattern from two adjacent walls of the tomb. Shading indicates elements common to both patterns. Decagonal tile patches are granulated, primary Conway worms/rays are ruled, secondary worms/rays are indicated by dashing (not indicated after the first interruptions), the two sets of pentagonal rosettes by different forms of stippling. Widths or sizes of shaded areas are partly arbitrary to some extent. (Based on Fig. 10.5.1. c of Ref. 7.)

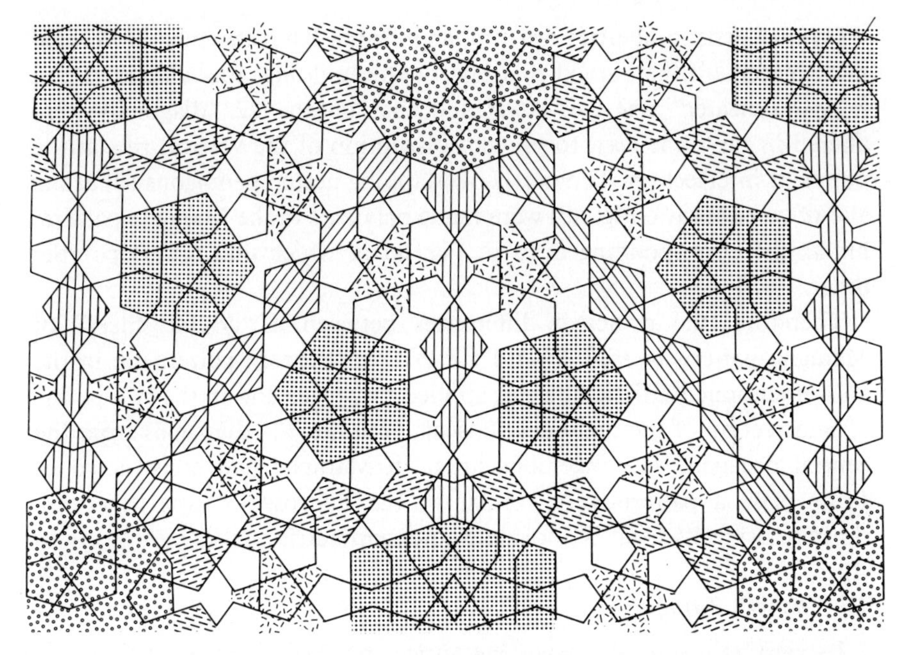

Fig. 8b

(typified by the 36° diagonal in Figs. 2b and 8) we arrive, after a sequence rhomb *r* and butterfly *b-r-r-b-r*, at the first *local repetition* of the original decagon. This decagon is repeated *together* with its radial Conway worms in ten radial directions. And now comes the difference: in a cartwheel Penrose pattern, all these secondary worms (rays) are *interrupted* whenever they intersect the primary ones (Fig. 8a). The artist of Marāgha made all *vertical* rays *primary* and all those at 72° to them *secondary* (the 36° diagonals interconnect directly with the two closest decagons and are always *primary*) (Fig. 8b). In this way, a *basis* for a two-dimensionally periodic, centered pattern was created. You may observe that the secondary worms are readily recognizable also after the above described interruptions in the cartwheel P2 pattern, but they become extremely rudimentary in the Marāgha pattern. However, this does not constitute a difference between the Marāgha and the cartwheel Penrose patterns; when the cartwheel P2 pattern is redrawn using P1 tiles, exactly the same phenomenon as in the Marāgha pattern is observed.

Further artistic liberties are similar in kind. The sequence of vertical rays, placed on the curved pilasters of the building, was altered into a regular, "neater" one: *-r-b-r-b-r-* and that on the 72° diagonals were altered in the same way, to prevent disruption of the vertical ribs at the points of intersection. The partly overlapping nested pentagons (d) at the end of vertical rays (Fig. 8b) were made balanced in the Marāgha pattern, unlike the corresponding discs in Fig. 8a in which one "overlaps" the other.

Therefore, the principal differences between a "double sheet" of Marāgha pattern in Fig. 2b and a cartwheel Penrose pattern lie in the abovementioned artistic liberties applied to worms and in the two points of intersection of the marginal verticals with the 72° diagonals from the central decagon. The ingenious artist in Marāgha had constructed in 1196–1197 a pattern very similar to the Penrose cartwheel pattern, skilfully using its local isomorphies and symmetries.

4.3 *The M2 Tiling*

Its aesthetic appeal notwithstanding, the M1 tiling is difficult to comprehend because of its complexity. Therefore, a simplified M2 tiling was derived from the M1 and PM1 tiling. Centers of all adjacent small pentagons in either tiling are connected by straight lines (passing always over 0-0 contacts). In the case of M1, the rhombs (minus kites) are considered to be two overlapping pentagons, and for each of them a center must be found. For the P1 tiling, centers must be found for both positions of all "reversible" pentagons that lie between any star and a half-star adjacent to it (i.e., for both pentagonal, overlapping halves of the new compressed hexagon in PM1). In this way, two tile types are obtained:

1. *Elongated hexagons*, which encircle butterflies in M1 or are centered on lozenges in PM1 (also on all lozenges that have been selected from stars and half-stars of P1).
2. *"Hourglass" polygons* centered on rhombs of M1 (i.e., on compressed hexagons/"reversible pentagons" of PM1/P1: Fig. 9). The entire pattern visually "decomposes" into decagons (which contain 3 hexagons + 1 hourglass), truncated decagons (2 hexagons + 1 hourglass) and pentagonal stars (1 hexagon + 2 hourglasses (Fig. 9)).

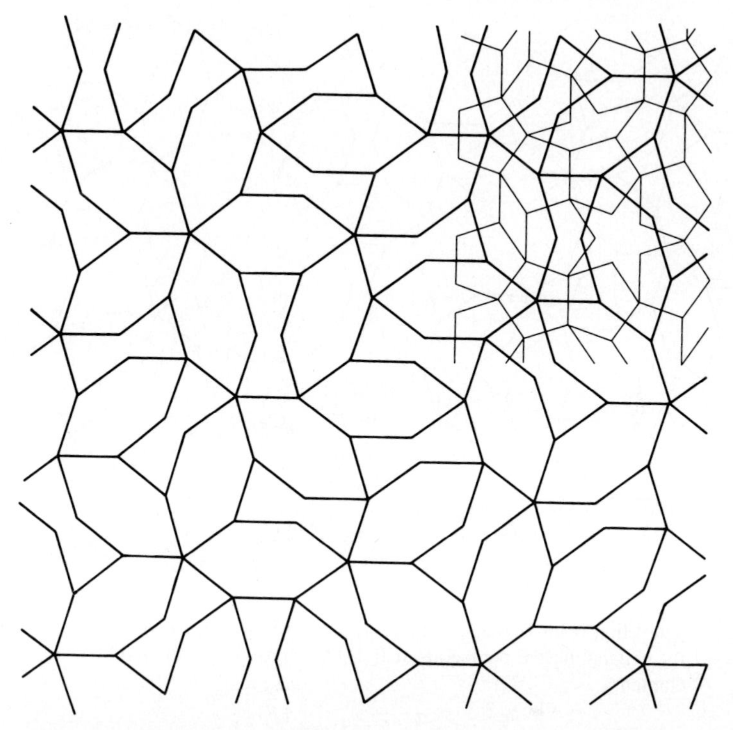

Fig. 9. The same area of aperiodic pentagonal tiling as in Fig. 4, executed in the tile set M2 with all constraints applied. Underlying P1 net is partly drawn.

To ensure aperiodicity, adjacency conditions analogous to those just given were introduced: every hexagon must have one or two adjacent, *subparallel* hourglasses in a *cis* configuration. Hourglasses that abut the hexagon by their short "bases" are *not counted* here. Every hourglass polygon must have two adjacent, subparallel hexagons in the *cis* or *trans* position. Hourglasses can be interconnected only via corners, not via sides. As in the case of M1, these conditions can always be satisfied, if necessary by rearranging the contents of one or more adjacent decagons. The last condition mentioned excludes simple Conway worms that are present in the Marāgha pattern, which has been transformed into M2 (Fig. 10). Chain-ups and worms being excluded and with no high symmetry elements present, this tiling has a random character over most of its area (Fig. 9).

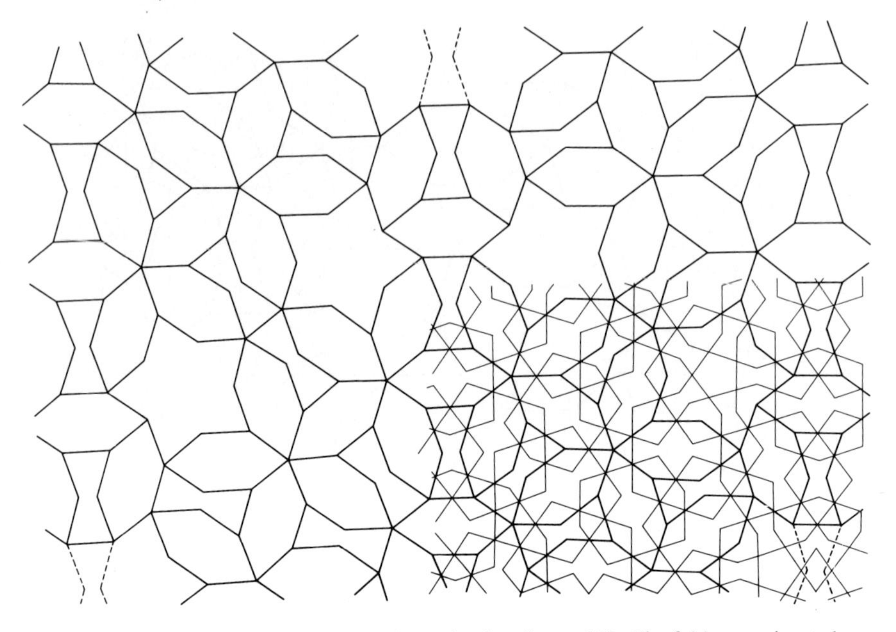

Fig. 10. The tiling from Marāgha, redrawn in the tile set M2. Fivefold stars situated on the original nested pentagons, ((d) in Fig. 3), were left unfilled by lower symmetry elements.

Both M1 and M2 tilings allow building of stripes with band group *p* 2 *gm* as does Penrose tiling P1, but they produce neither two-dimensionally periodic nor simple cartwheel patterns. Although not yet examined in this respect, they probably force fixed limiting ratios of individual tiles when they stretch to infinity. When some of the conditions have been relaxed (adjacency rules, chain-up), arbitrary random patterns without guaranteed aperiodicity can be easily obtained (Figs. 11 and 12).

5. Sources for the Marāgha Artist

The constructor of the Marāgha pattern did not work in an artistic vacuum. A survey of Islamic patterns that use pentagonal and decagonal motifs[8] shows the presence of all elements observed in the Marāgha pattern: small pentagons, butterflies, marked rhombs, nested pentagons, and kites, as well as the composite pentagons (f) and butterflies flanked by six small pentagons (Fig. 13). Some of them are still extant — for

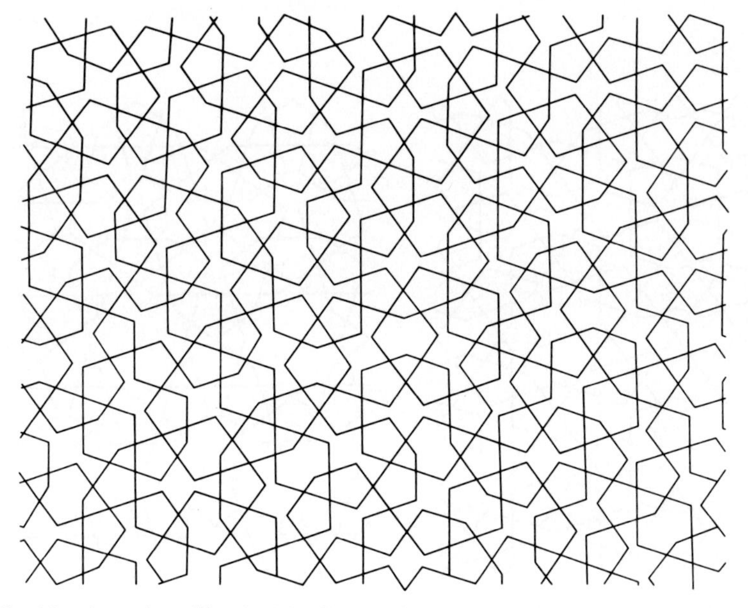

Fig. 11. A random tiling by M1 tile set with some matching conditions relaxed.

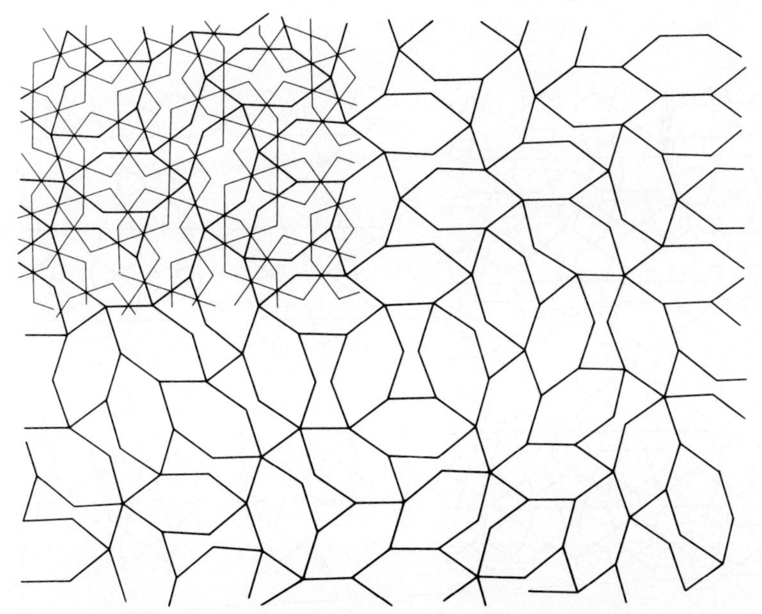

Fig. 12. The same tiling redrawn in the M2 set with relaxed matching conditions. The underlying M1 set is partly shown.

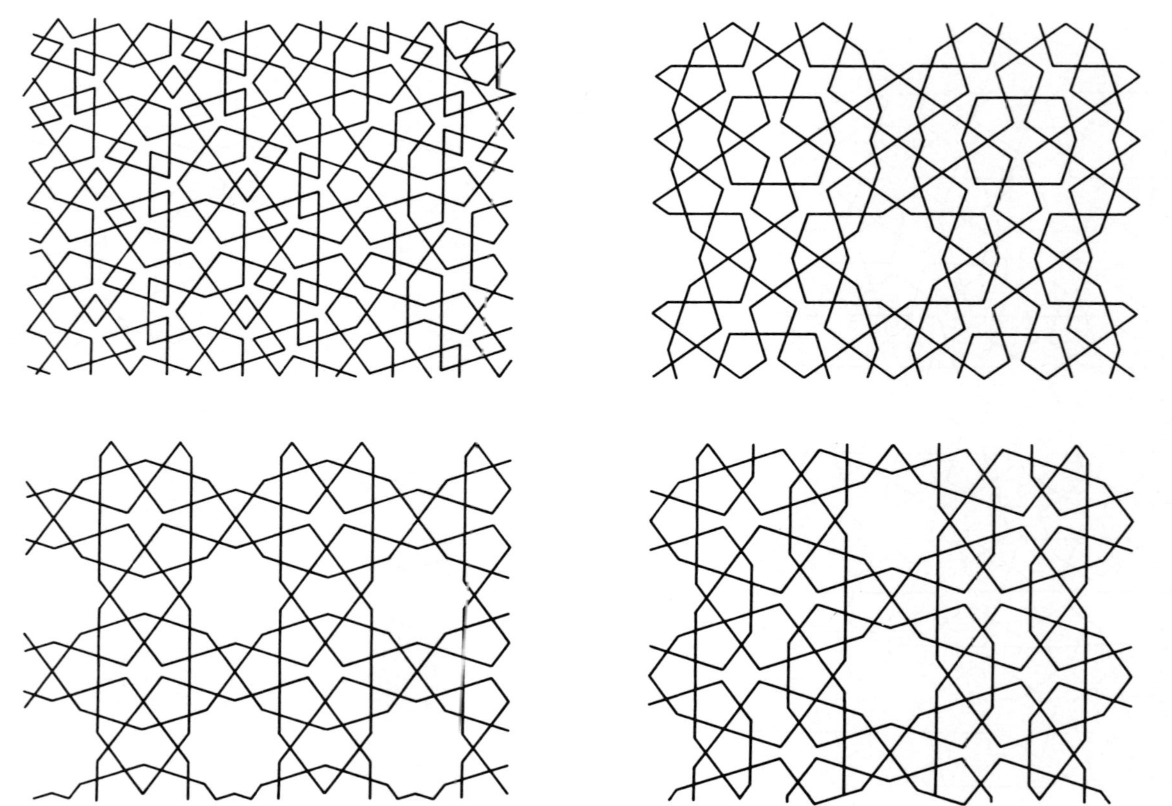

Fig. 13. Representative portions of four Islamic patterns, illustrating the generation and low key use of elements (a)–(f) from Fig. 3. (From Bourgoin, Ref. 8.)

example, in Cairo. However, in all these patterns the pentagonal elements are combined with decagonal ones and/or mirror and glide planes to produce relatively simple patterns. Therefore, their pentagonal metrics, local symmetry, and combinational possibilities are utilized poorly or not at all. It required the geometric genius of the creator of the Marāgha ornaments to free these elements from all the limitations imposed by lesser artists and construct a truly pentagonal pattern in which the decagonal symmetry was used only as the basis of a large-scale cartwheel principle.

6. Conclusions

1. The Pattern from Marāgha, Iran (1196–1197), is based on tiles that can readily be obtained by a transformation of the Penrose pattern of pentagons, stars, and lozenges. It deviates from a true cartwheel Penrose tiling only in several geometric and artistic adaptations.
2. On this basis, novel Penrose-type tilings M1, PM1, and M2 were derived in which highly symmetrical elements have been largely eliminated (Figs. 6, 7, and 9). In this respect, the tilings M1, M2, and PM1 may be of special interest to crystallographers.
3. Penrose tiling by stars, etc. (P1) as well as the tiling by darts and kites (P2) were shown to represent averaged patterns over a set of mutually very similar, underlying patterns (PM1 in Fig. 7) in which "stars" and "half-stars" are, respectively, always replaced by one of the five or two possible orientations of a lower symmetric tile patch. This way of understanding P1 and P2 may be important for crystallography as well.
4. In replacing "stars" and "half-stars," oriented overlapping pentagons (modified into rhombs in M1, hourglass polygons in M2, and compressed polygons in PM1) are created as new tiling elements.

7. Acknowledgements

The kind interest of Prof. I. Hargittai, the editor of this volume, useful reviews by Prof. A. Mackay (London), Dr. J. Bailey (Copenhagen), and an unknown reviewer, assistance of lecturer M. Krustrup (Royal Danish Academy of Arts) in the research of architectural literature, skillful draftsmanship of Mr. A. Helmbaek and Mr. A. Assour, as well as typing assistance of Mrs. M.L. Johansen, are gratefully acknowledged.

References

1. R. Penrose, The role of aesthestics in pure and applied mathematical research, *Bull. Inst. Math. Appl.* **10** (1974) 266–271.
2. R. Penrose, Pentaplexity, *Eureka* **39** (1978) 16–22.
3. A. Godard, Notes complémentaires sur les tombeaux de Marāgha, *Athār-é-Īrān*, **1** (1936) 125–160.
4. A. Godard, *The Art of Iran.* Allen & Unwin, London (1965).
5. A. Hunt and L. Harrow, *Islamic Architecture, vol. 1, Iran.* Scorpion, London (1977).
6. J.D. Hoag, *Islamic Architecture.* Faber & Faber-Electra, London (1987).
7. B. Grünbaum and G. C. Shephard, *Tilings & Patterns.* Freeman, New York (1987).
8. J. Bourgoin, *Arabic Geometrical Patterns and Design.* Dover, New York (1973).

PLANE PROJECTIONS OF REGULAR POLYTOPES WITH FIVEFOLD SYMMETRIES

G. C. Shephard

Until recently, diagrams of polyhedra and polytopes in books and research papers, were drawn laboriously by hand. This is no longer necessary — the availability of computer graphics enables diagrams to be drawn in a few seconds which previously took weeks, or even months, of painstaking work. In this paper we reproduce computer-drawn diagrams showing the result of projecting, onto various (2-dimensional) planes, the regular polytopes in four dimensions which have fivefold symmetries. More precisely the term "having fivefold symmetries" means that their symmetry groups contain rotations through angles $2k\pi/5$ ($k = 1, 2, 3$ and 4). The diagrams were drawn on a Calcomp 1051 plotter using GINO-F software on a VAX 8650 mainframe computer. The author wishes to thank the director and staff of the Computing Centre at the University of East Anglia for their help in carrying out this project.

There are twelve regular four-dimensional polytopes with fivefold symmetries (and none in higher dimensions). In addition to the familiar 120-cell (bounded by 120 pentagonal dodecahedra) and the 600-cell (bounded by 600 regular tetrahedra) there are ten star polytopes. These are listed below by their Schläfli symbols and their names (which are either classical, or suggested by H. S. M. Coxeter and J. H. Conway).

I (i) {3, 3, 5} the 600-cell, {3, 5, 5/2} the icosahedral 120-cell, {5, 5/2, 5} the great 120-cell, and {5, 3, 5/2} the grand 120-cell,

(ii) {5/2, 5, 3} the stellated 120-cell, and {5, 5/2, 3} the great grand 120-cell,

(iii) {5/2, 3, 5} the great stellated 120-cell, {5/2, 5, 5/2} the grand stellated 120-cell, {5, 5/2, 5} the great 120-cell, and {3, 3, 5/2} the grand 600-cell.

II (iv) {5, 3, 3} the 120-cell,

(v) {5/2, 3, 3} the great grand stellated 120-cell.

All ten of the polytopes in Part I of the list have the same set of 120 vertices as the 600-cell, whilst those in Part II have the same set of 600 vertices as the 120-cell. All polytopes preceded by the same small Roman numeral ((i), (ii), . . . , (v)) have the same set of edges. Hence as we can only represent the edges and vertices (and not the 2- and 3-dimensional faces) in a figure we need only five sets of diagrams to represent all twelve polytopes. The notation used in the list is adopted in the caption to each of the diagrams. For example I(ii)a denotes the image of the stellated 120-cell (or equally of the great grand 120-cell) under orthogonal projection onto the 2-plane denoted by a, as explained below.

It is convenient to take the coordinates of the 120 vertices of the 600-cell in the form (see Ref. 1, p. 247)

A_κ (k even): $(\cos k\theta,\ \sin k\theta,\ d\cos 11k\theta,\ d\sin 11k\theta)$,
B_κ (k odd): $(b\cos k\theta,\ b\sin k\theta,\ c\cos 11k\theta,\ c\sin 11k\theta)$,
C_κ (k odd): $(c\cos k\theta,\ c\sin k\theta,\ -b\cos 11k\theta,\ -b\sin 11k\theta)$,
D_κ (k even): $(d\cos k\theta,\ d\sin k\theta,\ -\cos 11\ k\theta,\ -\sin 11k\theta)$,

where the subscripts are calculated modulo 60, $\theta = \pi/30 = 6°$, and

$$b = 2\cos 11\theta,\ c = (2\cos 7\theta)^{-1},\ \text{and}\ d = (\cos 12\theta)/(\cos 4\theta)\,.$$

If these are scaled so that the circumradius is 1, then the edge lengths are, in the three cases: (i) $1/\tau$, (ii) 1, and (iii) τ. The 600 vertices of the 120-cell can be found as the centres of the tetrahedral 3-faces of the 600-cell and if these are scaled so that the circumradius is 1, then the edge lengths are (iv) $1/\tau^2\sqrt{2}$ and (v) $\tau^2/\sqrt{2}$.

The set of edges joining the 30 vertices $A_0, A_2, \ldots, A_{58}$ of the 600-cell in order, forms what is known as a Petrie polygon (see Ref. 1, p. 223). This is a zigzag polygon made up of edges of the 600-cell such that each consecutive three (but no consecutive four) edges belong to the same (tetrahedral) 3-face of the polytope.

The 2-plane a (which joins the origin to the points (1, 0, 0, 0) and (0, 1, 0, 0)) has the property that the image of the Petrie polygon A_0, $A_2, \ldots, A_{58}$ of the 600-cell under orthogonal projection onto the plane a is the external regular 30-gon of Figure I(i)a. The external regular 30-gon of Figure II(iv)a is a Petrie polygon of the 120-cell. If we denote by a′ the 2-plane absolutely orthogonal to a, then images under orthogonal projection of each of the five kinds of polytopes onto the plane a′ are identical with those onto the 2-plane a. In particular, if we project the 600-cell onto a′, the external regular 30-gon is the image of the Petrie polygon given in 13.61 of Ref. 1, p. 248.

These projections are aesthetically attractive; other 2-planes also yield interesting diagrams. Out of the many possibilities we have chosen the 2-plane b (which joins the origin to the points (1, 0, 1, 0) and (0, 1, 0, 1)) and the 2-plane c (which joins the origin to (1, 0, 0, 1) and (0, 1, 1, 0)). It will be observed that the images of projections onto these planes have 10-fold and 12-fold symmetries, respectively. (Note that Figures II(v)b and II(v)c are not reproduced here. The long edge-length makes these diagrams so complicated as to be unintelligible.)

Once the computer program has been written, it is a triviality to produce diagrams showing the result of projecting any of the polytopes onto any prescribed 2-plane. Many of the figures that arise in this way are very beautiful, and it only lack of space that prevents us from reproducing them here.

Reference

1. H.S.M. Coxeter, *Regular Polytopes*, 2nd Edn. Dover, New York (1973).

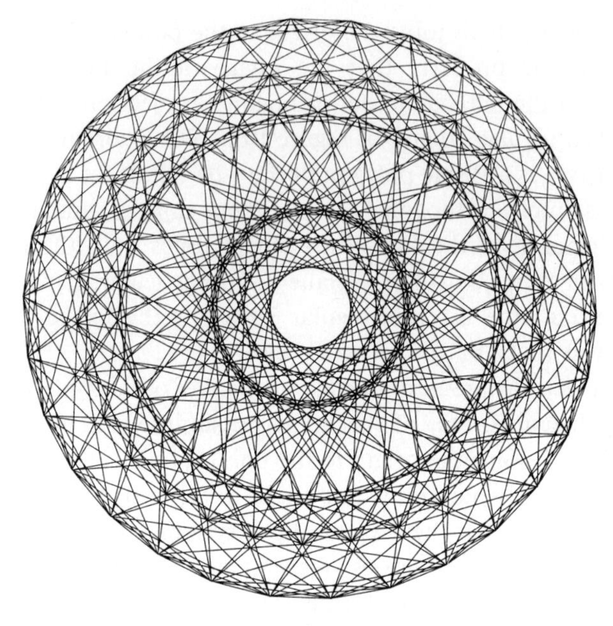

I(i)a

I(ii)a

I(iii)a

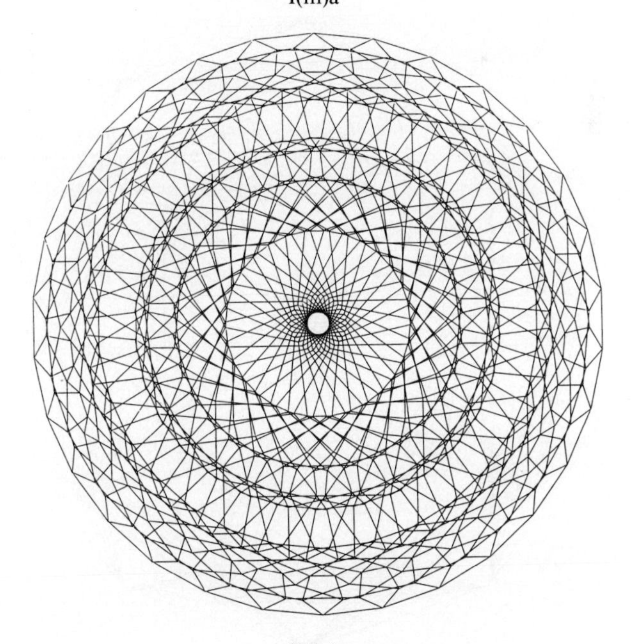

II(iv)a

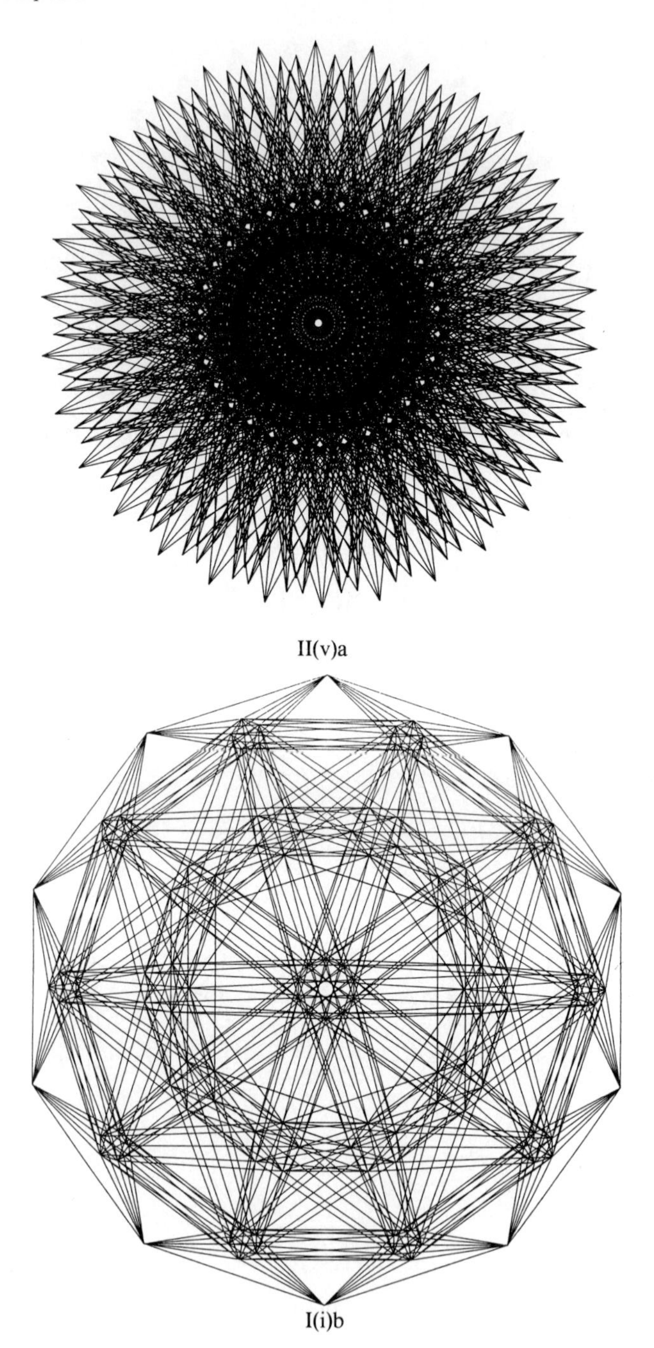

II(v)a

I(i)b

I(ii)b

I(iii)b

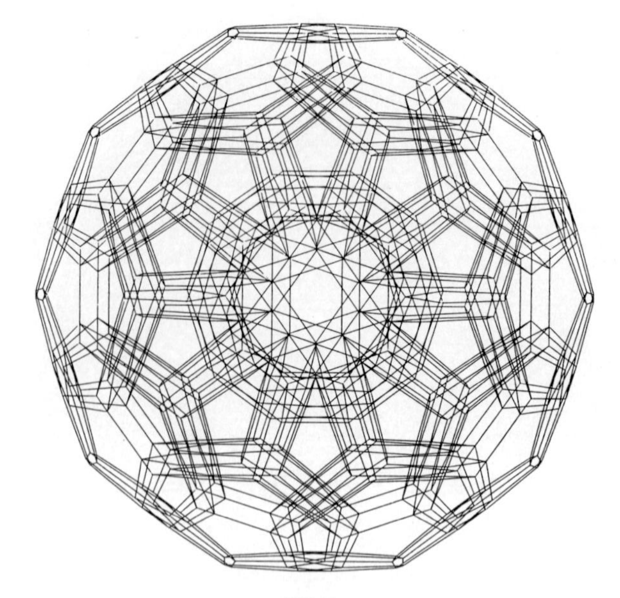

II(iv)b

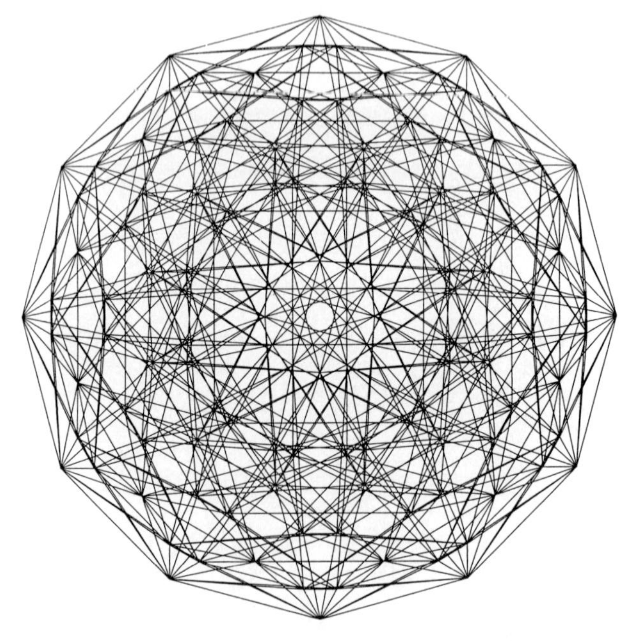

I(i)c

I(ii)c

I(iii)c

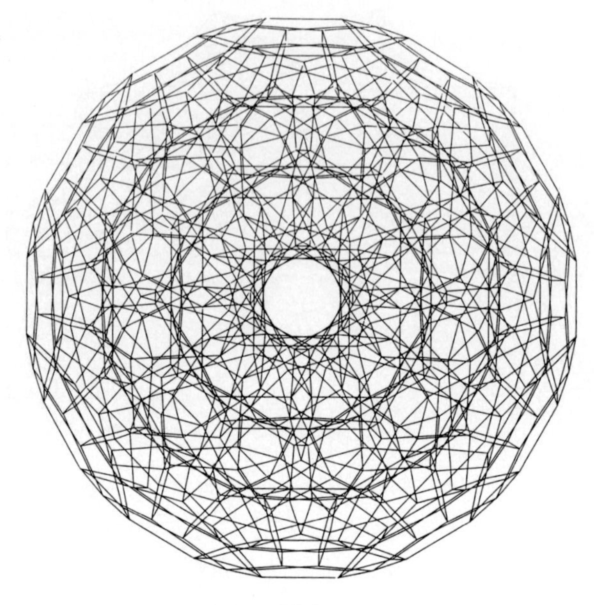

II(iv)c

CONTINUOUS TRANSFORMATIONS OF NON-PERIODIC TILINGS AND SPACE-FILLINGS

Haresh Lalvani

1. Introduction

This paper presents aspects of a morphologic system for continuous transformations of non-periodic space structures. An example of the two-dimensional non-periodic structure is provided by the well-known case of the Penrose tiling based on pentagonal symmetry.[1,2] This tiling is composed of two types of rhombii with face angles of 36° and 72° (Fig. 1). The algebraic generalisation of this type of Penrose tiling in two-dimensions was shown by de Bruijn from projection of five-dimensional space.[3] The concept of the infinite family of non-periodic rhombic tilings was independently discovered by Lalvani from the geometry of two-dimensional projections of n-cubes viewed along their n-fold axis.[4,5] The three-dimensional analog, consisting of two rhombohedra and based on icosahedral symmetry, was suggested independently by various authors in different fields (Ammann,[6] Mackay,[7] Kramer,[8] and Levine and Steinhardt[9]). In architecture and design areas, this space-filling was independently suggested by Stuart[10] and Miyazaki,[11] the latter following the earlier work on icosahedral space-filling by Baer,[12] and families of non-periodic space-fillings were suggested by Lalvani.[5]

This paper extends the method for generating non-periodic tilings and space-fillings using the vector-star method described in the author's earlier work,[5] which has proceeded in parallel with developments in the

Fig 1. The Penrose tiling, $P = n = i = 5$, with a portion of six overlapping decagons highlighted

field of "quasi-crystal." This method uses a star of n vectors, termed an n-star, as a generator for the tilings and space-fillings. In addition, these generators can be i-stars derived from n-stars, where $i \leq n$. The tilings and space-fillings have their edges parallel to the n vector directions and are corresponding projections from n-dimensional Euclidean space. The organisation of the vector-stars into families and continuous star transformations within these families has also been suggested in Ref. 5. This paper proposes a system for continuous transformations of non-periodic tilings and space-fillings and presents a new class of structures which are here termed *hyper tilings* and *hyper space-fillings.*

2. The Transforming Vector-Star

Vector-star transformations, which in turn determine the transformations of their corresponding non-periodic tilings and space-fillings, are

obtained by transforming three variables: the symmetry **P**, **Q** of the star, the dimension n from which the stars are projected into two- and three-dimensions, and the number of vectors i. The value of n can be any integer and the geometry of the star can be regular or irregular. In the two-dimensional cases, the vector-stars are *planar* and have a symmetry **P**, where P is any number; **Q** is a degenerate axis and is always 1.
In the three-dimensional cases, the stars are *spatial* and have four classes of symmetries. The first is an infinite class derived from the prisms of symmetry **P, 2**, where P is any number, and Q is always 2. The remaining three belong to the symmetry **P, 3** which is restricted to values of $P = 3$, 4 and 5, and where Q is always 3. The latter three correspond to the *tetrahedral*, *octahedral* and *icosahedral* families, respectively. All spatial symmetries have an additional dependent axis of symmetry R which is 2 in all four families. The values of $P > 5$ and $Q > 3$ provide examples of symmetries in non-Euclidean (hyperbolic) space. Non-periodic structures from non-Euclidean stars remain to be explored.
Corresponding to all planar and spatial stars, higher-dimensional vector-stars, or *hyper-stars*, exist. In n-dimensional space, the planar hyper-stars have a hyper-symmetry $(\mathbf{P})_n$ and the spatial hyper-stars have a hyper-symmetry $(\mathbf{P}, \mathbf{Q})_n$. When projected in two- or three-dimensions, the hyper-stars are deformed variants of stars derived from the various families of symmetries.

3. Families of Non-Periodic Tilings

Non-periodic tilings, like their generating stars, can be characterised by three variables, $(\mathbf{P})$ or $(\mathbf{P})_n$, n and i, and can be notated as $(\mathbf{P})_n(n\ i)$. Since several types of i-stars are possible from the same dimension, a more specific notation i_j ($j = 1,2,3,4 \ldots\ldots$) can be used to specify i. The infinite family of rhombi from regular n-stars is first described. The well-known case of the non-periodic Penrose tiling is described next as an introduction to the tilings that follow. A new infinite class of *hyper-Penrose tilings* is described next. Two degenerate cases, one where the Penrose tiling loses its vectors, and the other where some of the vectors become collinear are described next. All tilings are shown as transformable to others in a continuum.

3.1 *The Infinite Family of Rhombi*

An infinite family of planar n-stars of symmetry P are derived from distinct directions of radial lines which join the center of a regular $2n$-sided polygon to its vertices. Each distinct pair of vectors, i.e., $i = 2$, from the n-star determine the edges of a rhombic tile and all rhombi from dimension n make up an infinite family shown in Fig. 2.[5] Each column shows the rhombi from an n-star, and the rows in a column list the types of rhombi within dimension n. The number of rhombi equal $n/2$ for even n, and $(n-1)/2$ for odd n. The types of rhombi are denoted by 2_1, 2_2, $2_3 \ldots . \, 2_j$, where 2 indicates the two vector directions from the n-star.

The angles between the vectors of the n-star are whole number

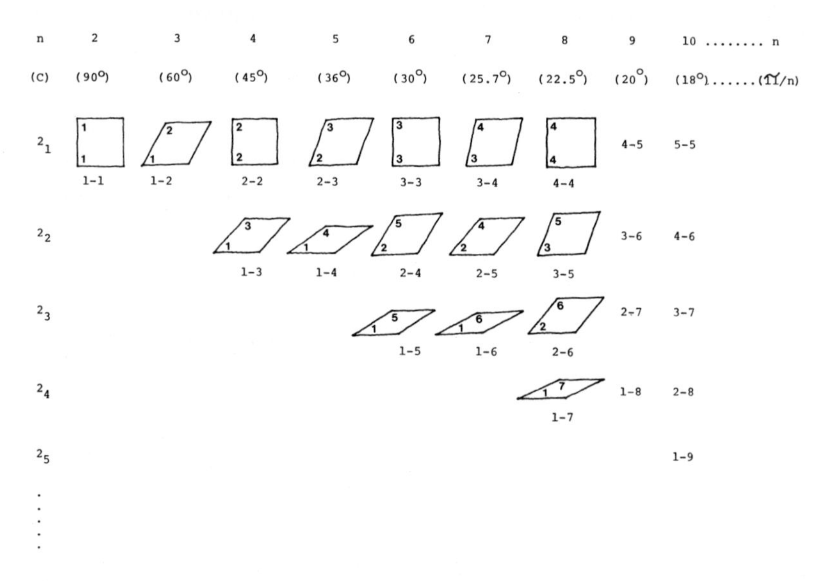

Fig 2. A table showing the infinite family of rhombi ($i = 2_1, 2_2, 2_3 \ldots . \, 2_j$) projected from dimension n.

multiples of the central angle C(= 180°/n) of the $2n$-sided polygon. In Fig. 2, each rhombus is described in terms of these whole numbers. Since a rhombus has two complementary face angles, the table denotes each by distinct whole number pairs which add up to n. For example, under the column $n = 7$, there are only three distinct pairs of numbers that equal n, namely, 1 and 6, 2 and 5, and 3 and 4. These are denoted as 1-6, 2-5, and

3-4. Each number is a multiple of the angle C (= 180°/7 or 25.714°), and each pair produces a distinct rhombus from the 14-sided regular polygon (Fig. 3). The acute face angles of the three rhombi equal 25.714° (1 × C), 51.428° (2 × C) and 77.142° (3 × C), respectively. The face angles of the other rhombi can be similarly derived from this table.

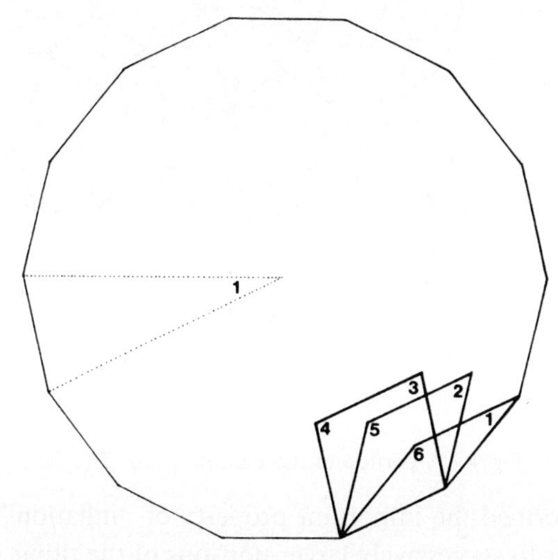

Fig 3. Derivation of the three rhombi for the $n = 7$ case from a regular 14-sided polygon. Their face angles are distinct pairs which add up to 7.

3.2 *The Penrose Tiling (5)(5,5) : P = n = i = 5*

The Penrose tiling, based on a regular pentagonal star (P = 5) is composed of two types of rhombi, 1-4 and 2-3, listed under column $n = 5$ in Fig. 2 These are the only two pairs of numbers that add up to five. It is a projecion from five dimensions ($n = 5$), it has five vectors ($i = 5$) and is noted here as (**5**)(5,5). The distinction between n and i for the same symmetry is significant as described later and the Penrose tiling is a special case where P = $n = i = 5$. The tiling has eight types of vertices,[3] and the sum of numbers at every vertex equals ten. A detail segment of this tiling consisting of six overlapping decagons, highlighted in profile in Fig. 1, is shown in Fig. 4. Numbers associated with the angles are shown at few of the vertices and a regular five-vector star, based on a regular half-decagonal star, is shown alongside. By inspection, we see that all edges of the tiling are parallel to the five directions of the star.

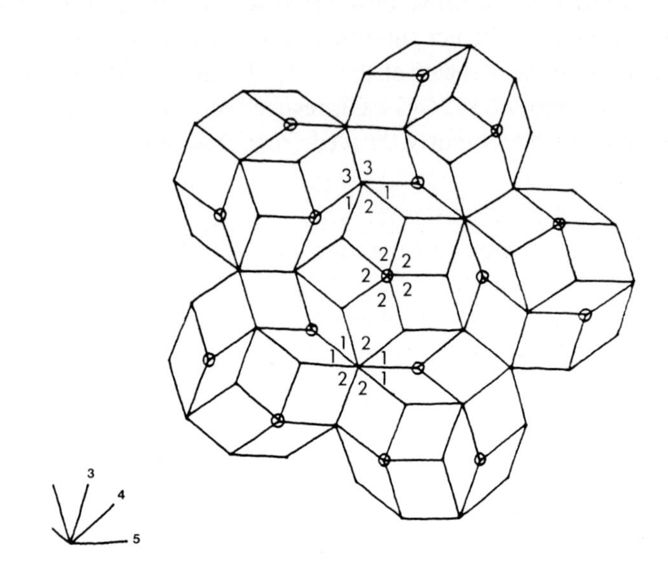

Fig 4. A portion of the Penrose tiling (**5**)(5,5).

Penrose described the important property of "inflation" by which the tiles are similar to successively larger portions of the tiling. Further, these increments are in golden ratio, an irrational number, and thus ensure the non-periodicity. The self-similarity also proves that the tiling fills the plane. A recursive plane-filling procedure using half-rhombi, also in increments of golden ratio, was suggested independently by de Bruijn and the author.[13]

The only other known case of a non-periodic tiling composed of rhombi from Fig. 2 and having the property of inflation is the Ammann tiling.[14] This is the tiling (**4**)(4,4) where $P = n = i = 4$. It is based on a four-star ($i = 4$), corresponds to $P = 4$, and is a projection from $n = 4$. The general class of non-periodic tilings $P = n = i$ with the property of inflation remains to be established for higher values of P, though all the tiles are listed in Fig. 2.

3.3 *Hyper Tilings* $(P)_n(n,i)$

The tilings $P = n = i$ can be transformed to an infinite family of other tilings $P_n(n,i)$, all of which are topologically isomorphic to these but are geometrically distinct. Such tilings, which are here termed *hyper tilings*, are derived from *hyper n-stars* and have a symmetry $(\mathbf{P})_n$ in n-space. The hyper-symmetry $(\mathbf{P})_n$ is retained in the inflated state. In their two-dimensional projection, the inflated states of hyper-tilings are composed of deformed rhombi with "kinked" edges, and the inflated hyper-stars are correspondingly kinked. Continuous transformations between the hyper-tilings can be achieved by transforming $(\mathbf{P})_n$, n or i continuously, or in gradual unit increments. All transformations can be mapped in a three-dimensional table, here termed *structure space*, where each of the three variables represents an independent axis of this space. For a fixed $(\mathbf{P})_n$, the two variables, n and i, are mapped in an infinite two-dimensional structure space (Fig. 5) and grow in unit increments along the x-axis and the y-axis, respectively. Each pair n, i represents a geometrically distinct vector star, and hence a distinct tiling. Examples are illustrated for the family of tilings $\mathbf{5}_n(n, i)$ and extend to other values of $\mathbf{P}$ by analogy.

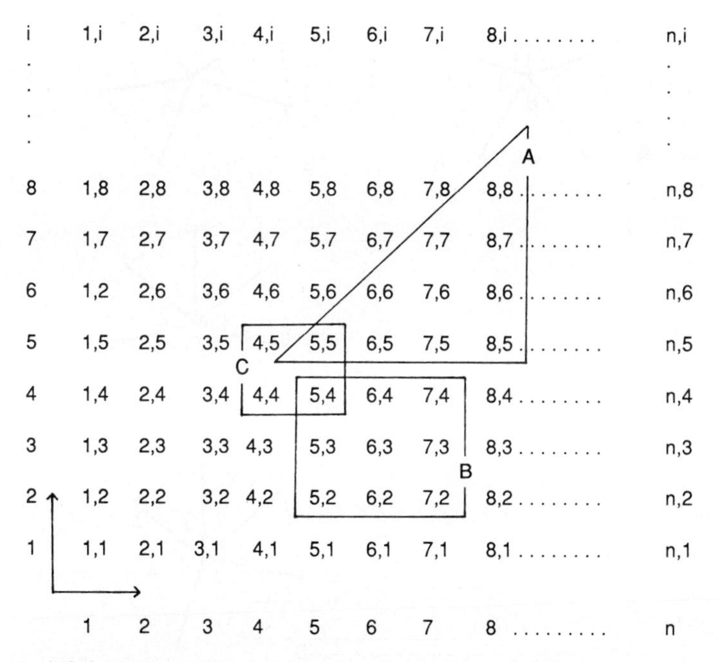

Fig 5. An infinite table of tilings n, i for any hyper-symmetry $(\mathbf{P})_n$. Portions A, B and C are highlighted and exemplified later (see Figs. 11–14).

3.3a *Hyper-Penrose Tilings* $(\mathbf{5})_n(n, i)$; n and $i \geq 5$

The hyper-tilings based on symmetry $(\mathbf{5})_n$, where n and $i \geq 5$, and bearing a one-to-one correspondence with the Penrose tiling are possible. These tilings are topologically indistinguishable from the Penrose tiling and can be termed *hyper-Penrose tilings.* Since the Penrose tiling is plane-filling by Penrose's inflation argument, all tilings topologically identical to this tiling would also be plane-filling. In these tilings, the regular pentagonal star of the Penrose tiling (Fig. 6a) changes to various hyper-pentagonal stars as shown with the three types from $n = 7$ (Fig. 6, b–d). The three hyper-stars 7,5 are based on the pentagons derived from a regular heptagon. Examples of hyper-Penrose tilings $(\mathbf{5})_n(n, i)$, derived from various hyper-stars, are shown in Figs. 7–12. Each illustration shows six overlapping decagons and its associated half-star. The vertices, edges and the rhombi bear a one-to-one correspondence to the portion of the Penrose tiling in Fig. 4. Each case is a transformation of the other.

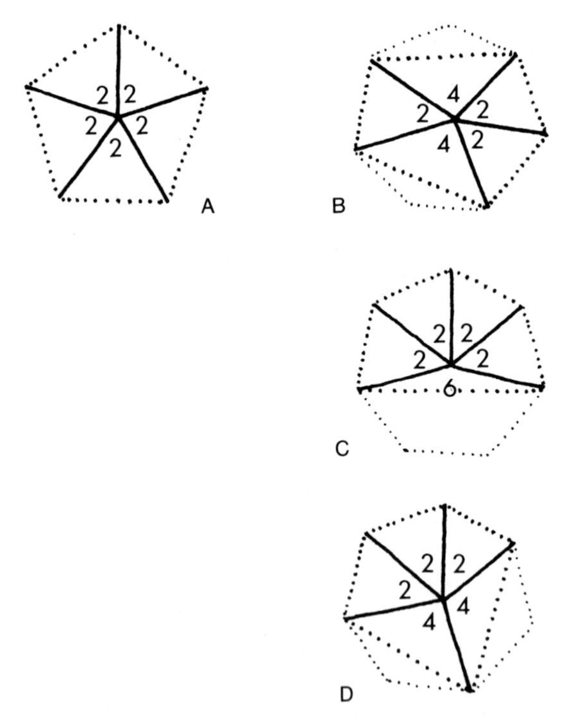

Fig. 6. A regular pentagonal star ($n = 5$) and three hyper pentagonal stars ($i = 5$) from $n = 7$.

In each, the connectivity is preserved and the eight types of vertices of the Penrose tiling retain their topology while relaxing the angles.

Figs. 7 and 8 show tilings based on hyper star $(\mathbf{5})_6$. These two tilings, $(\mathbf{5})_6(6,5)$ and $(\mathbf{5})_6(6,6)$, respectively, use the three rhombi for the $n = 6$ case shown in Fig. 2, namely 3-3, 2-4 and 1-5. The numbers, corresponding to the face angles, are shown at three vertices (compare with Fig. 4). Note that for plane-filling, the number adds to $2n$ at every vertex. In each figure, the vector star is shown alongside, and the vector directions are indicated in bold lines within the tiling. The tiling $(\mathbf{5})_6(6,5)$ has five vectors and is a "squished" version of the Penrose tiling (which is $(\mathbf{5})_5(5,5)$ in this general notation), while $(\mathbf{5})_6(6,6)$ appears to correct itself and uses 6 vector directions.

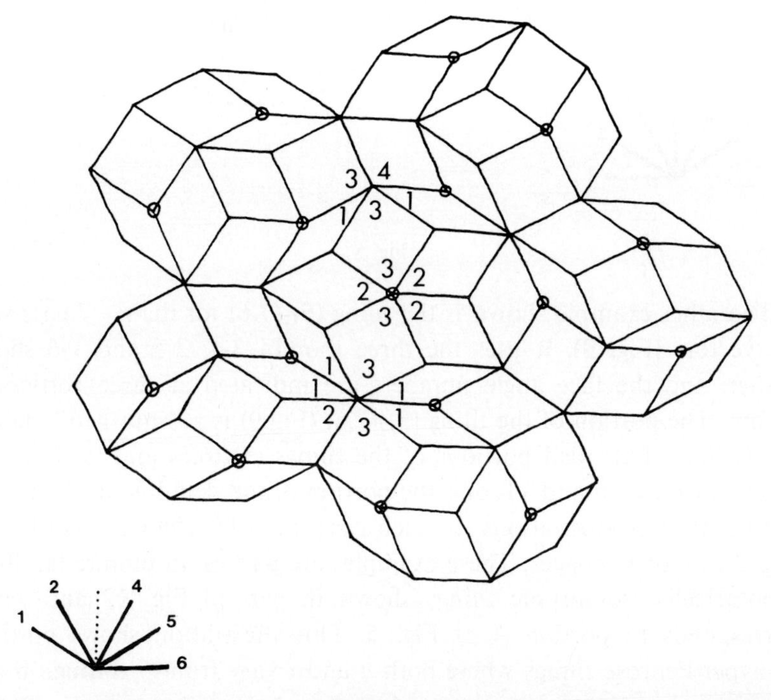

Figs. 7–9 Portions of hyper-Penrose tilings $(\mathbf{5})_6(6,5)$, $(\mathbf{5})_6(6,6)$, and $(\mathbf{5})_6(7,6)$, respectively, each based on hyper-pentagonal symmetry $(\mathbf{5})_n$. All three cases are topologically identical to each other and the Penrose tiling of Fig. 4, and are composed of rhombi from Fig. 2.

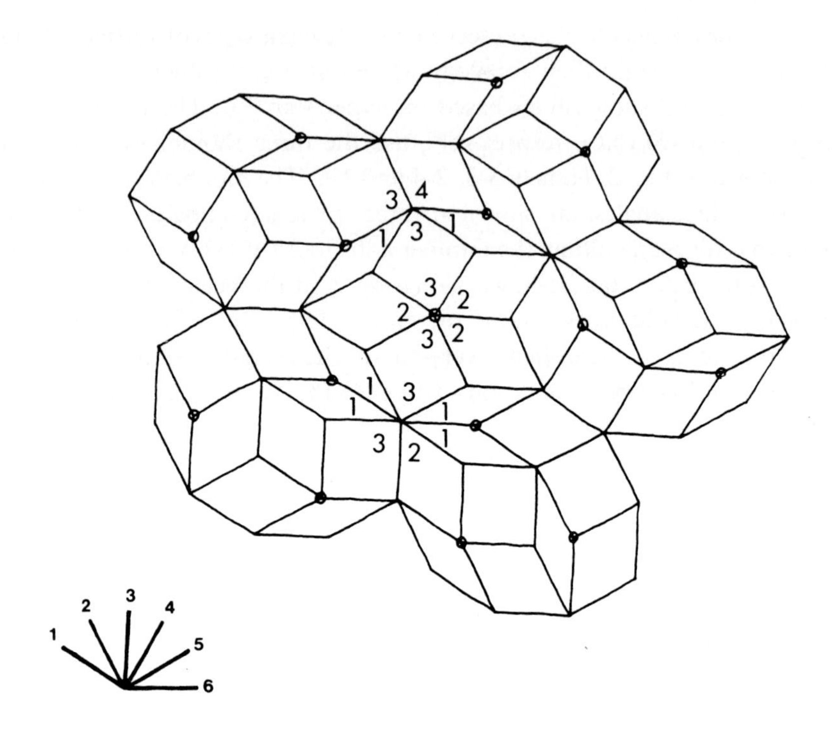

Fig. 8

The other example shown is the tiling $(\mathbf{5})_7(7{,}6)$ for the $n = 7$ case with six vectors (Fig. 9). It uses the three rhombi 3-4, 2-5 and 1-6 shown earlier, and the face angle numbers are indicated at three vertices as before. The portion of the tiling $(\mathbf{5})_7(7{,}6)$ (Fig.9) is a "squished" version of $(\mathbf{5})_6(6{,}6)$. Extended portions of the tilings $(\mathbf{5})_6(6{,}6)$ and $(\mathbf{5})_7(7{,}6)$ are shown in Figs. 10 and 11, and the portion of the detail is highlighted in profile in both illustrations. In each case, the tiles change according to Fig. 2 as n or i changes. These examples are part of an infinite family of topologically isomorphic tilings shown in part in Fig. 12 and which corresponds to portion A of Fig. 5. This illustration shows portions of hyper-Penrose tilings where both n and i vary from 5 through 8 and $n > i$.

A computer-animation showing a continuous transformation between some of these tilings was prepared, jointly with D. Sturman, earlier.[15]

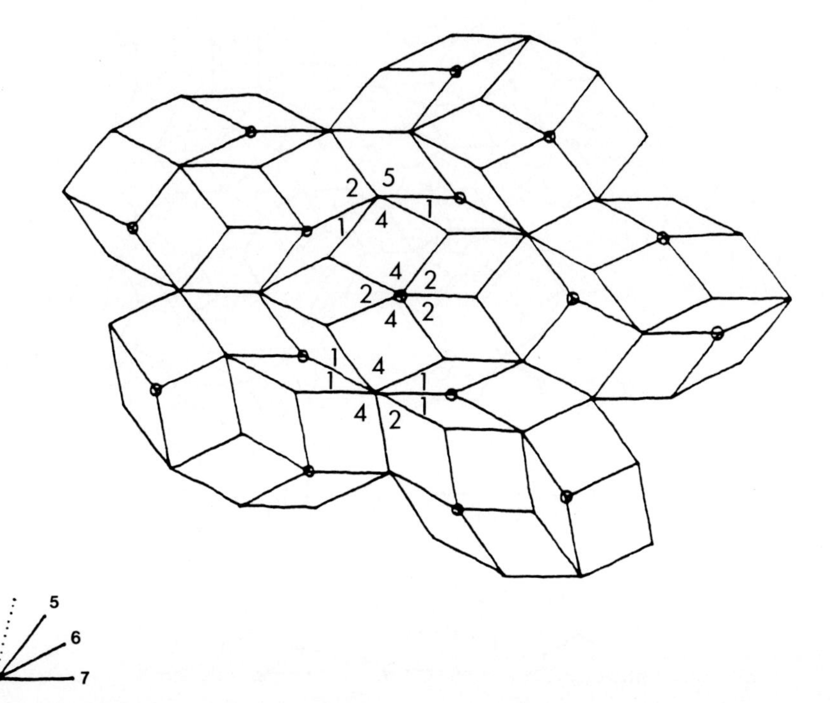

Fig. 9

Details of the plane-filling procedure have been established for the tilings $(\mathbf{5})_n(n,\ 5)$[16] and remain to be extended to other cases at the time of this writing.

3.3b *Tilings With Lost Vectors*: $P = i,\ n > i$

In cases $n > i$, where n is kept fixed and i is changed incrementally, an interesting class of degenerate tilings is obtained. In these cases, $\mathrm{P} = i$ and the symmetry $(\mathbf{P})_n$ changes with i. The transformations of this class of tiling takes place in the vertical columns in Fig. 5. The process is similar to that of adding or removing zones in the generation of zonohedra. In a sense, "non-periodic zones" are added as we increase i, and removed as we decrease i, leading to a corresponding change in symmetry. Details are shown with the portion B of Fig. 5. As the Penrose tiling $(\mathbf{5})_5(5,5)$ (not shown in the figure) loses a zone, it becomes $(\mathbf{4})_5(5,4)$, which in turn becomes $(\mathbf{3})_5(5,3)$ and then $(\mathbf{2})_5(5,2)$, the successive stages being produced by losing a single vector each time. P changes from 5 through 2. The last

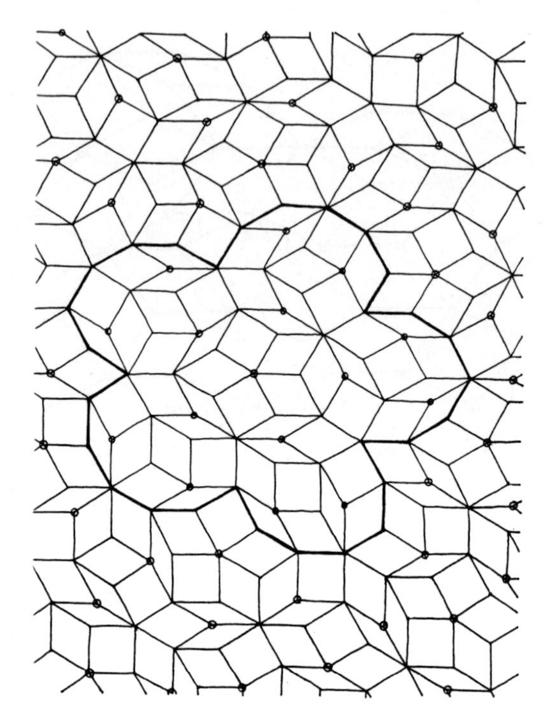

Fig. 10. The hyper-Penrose tiling $(\mathbf{5})_6(6,6)$.

three stages of these transformations are shown in the left-hand column of Fig. 13. The middle column shows corresponding transformations of $(\mathbf{5})_6(6,5)$ (also not shown in the figure) to $(\mathbf{4})_6(6,4)$, $(\mathbf{3})_6(6,3)$ and $(\mathbf{2})_6(6,2)$, and the right hand column shows the corresponding analogs for $n = 7$ case. Note that the $i = 4$ cases shown in the top row consist of overlapping octagons, the $i = 3$ cases have overlapping hexagons, and the $i = 2$ cases are rhombic tilings. It is easy to visualise how the latter cases in the horizontal rows transform to others using the lazy-tong principle. Continuous transformations of non-periodic tilings by loss of vectors was also prepared as a part of the computer-animation cited earlier.

3.3c *Tilings With Excess Vectors*: $P = n$, $n < i$

Another interesting class of degenerate tilings is where some of the vectors are redundant, i.e., $i > n$. In such cases, $\mathrm{P} = n$, and the symmetry $(\mathbf{P})_n$ changes with n. The redundant vectors become collinear with some others in a transformation process, but the tilings retain the non-

Fig. 11. The hyper-Penrose tiling $(\mathbf{5})_6(7,6)$.

periodicity and continue to be plane-filling. Details are shown with the portion C of Fig. 5 and are illustrated in Figs. 14 and 15. Fig. 14 shows a single zonogon, also notated n, i and filled with rhombic tiles. On the top right is the familiar regular decagon 5,5 of the Penrose tiling. As we go down the column, the decagon transforms to an octagon 5,4 by losing one vector direction, a process already described in the previous section. Here an intermediate stage is shown to illustrate the continuum. Proceeding leftward, as n decreases from 5 to 4, some of the angles change from 72° to 90°, while others change from 36° to 45°. In the bottom row, a regular octagon 4,4 is produced, but in the top row an "octagonal" decagon 4,5 is produced in the top left-hand corner. The intermediates further clarify the transformations, but more importantly enable the visualisation of the transformation from 4,4 to 5,5 along the diagonal. The latter suggests an important case of inter-symmetry transformations where P changes from 4 to 5 continuously.

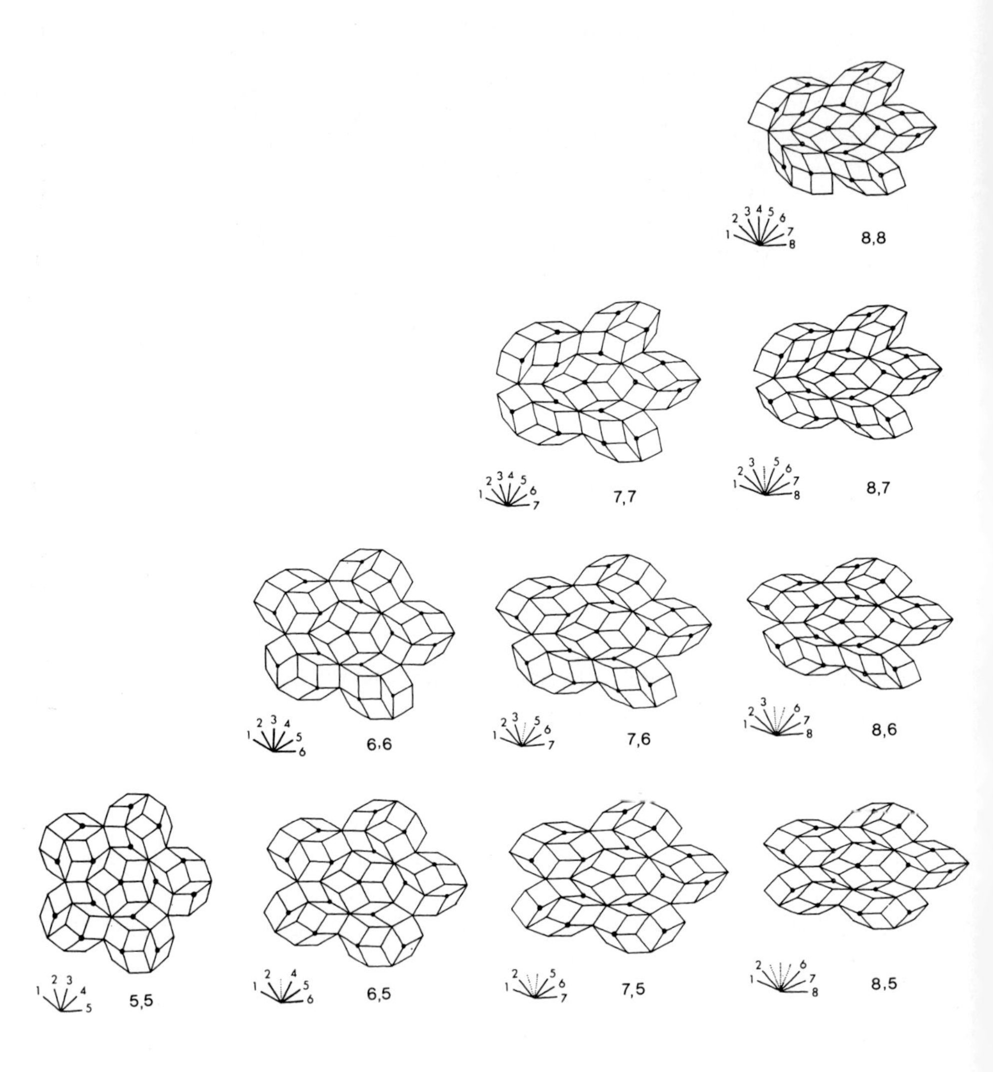

Fig. 12. A portion of the infinite family of hyper-Penrose tilings $(\mathbf{5})_n(n\ i)$ based on hyper-symmetry $(\mathbf{5})_n$ where both n and i are greater than 5.

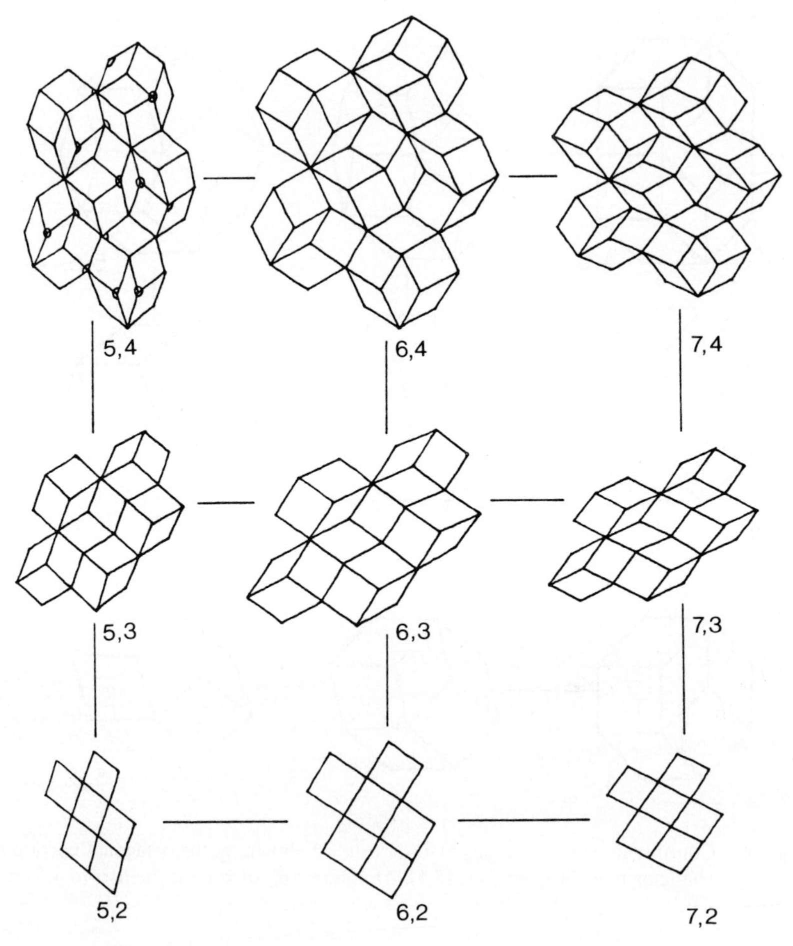

Fig. 13. A table of degenerate tilings, P = i, $n > i$, where the higher tilings have lost some vectors.

In Fig. 15, the transformations of the single zonogon are extended to portions of corresponding tilings. Note how $(\mathbf{4})_5(4,4)$ at bottom left changes to $(\mathbf{5})_5(5,5)$ at the top right through an intermediate along the diagonal. Along the vertical, as $(\mathbf{4})_5(4,4)$ changes to $(\mathbf{4})_5(4,5)$, the tiling appears to split at places. Along the horizontal, as n changes from 4 to 5, the vertical lines "open" up into rhombi. In this sense, the tiling $(\mathbf{4})_5(4,5)$ is also isomorphic to the Penrose tiling when we think of the two

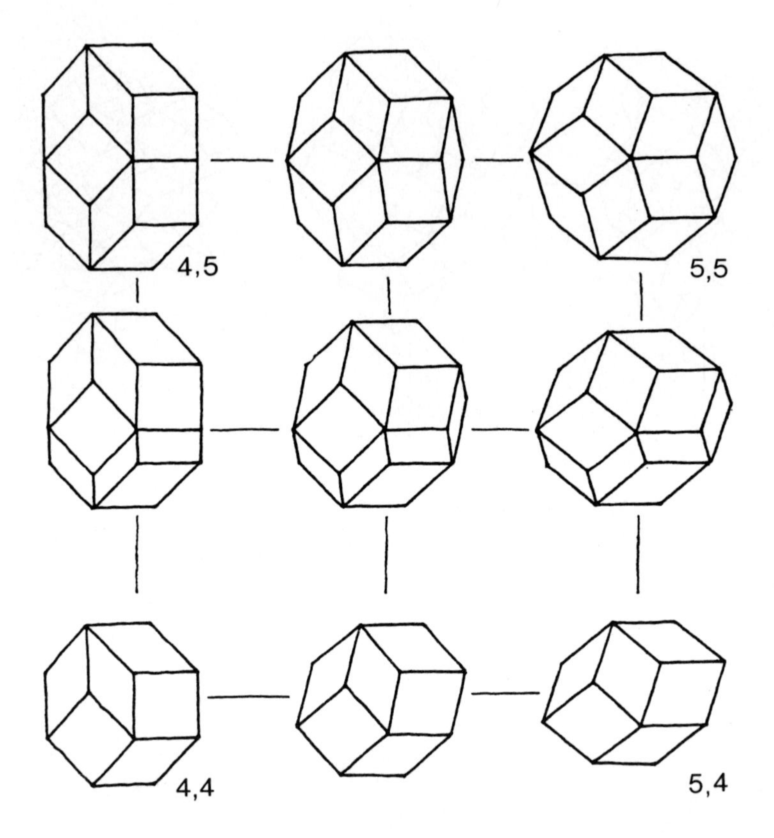

Fig. 14. Continuous transformation of $n = 4$ to $n = 5$ shown by the octagonal portion 4,4, changing to a decagonal one (5,5). An interesting surprise is the hybrid 4,5 on top left.

collinear and adjacent vectors as collapsed rhombi. Note that both $(\mathbf{4})_5(4,4)$ and $(\mathbf{4})_5(4,5)$ resemble the Ammann tiling, but are different. In these three cases, the same rhombi of 45° and 90° are rearranged differently.

4. Families of Non-Periodic Space-Fillings

The transformations described for the two-dimensional cases extend to three-dimensions in a natural way. The stars have a dimension n, number of vectors i, and the symmetry $\mathbf{P}$ changes to $\mathbf{P}$, $\mathbf{Q}$. The space-filling can

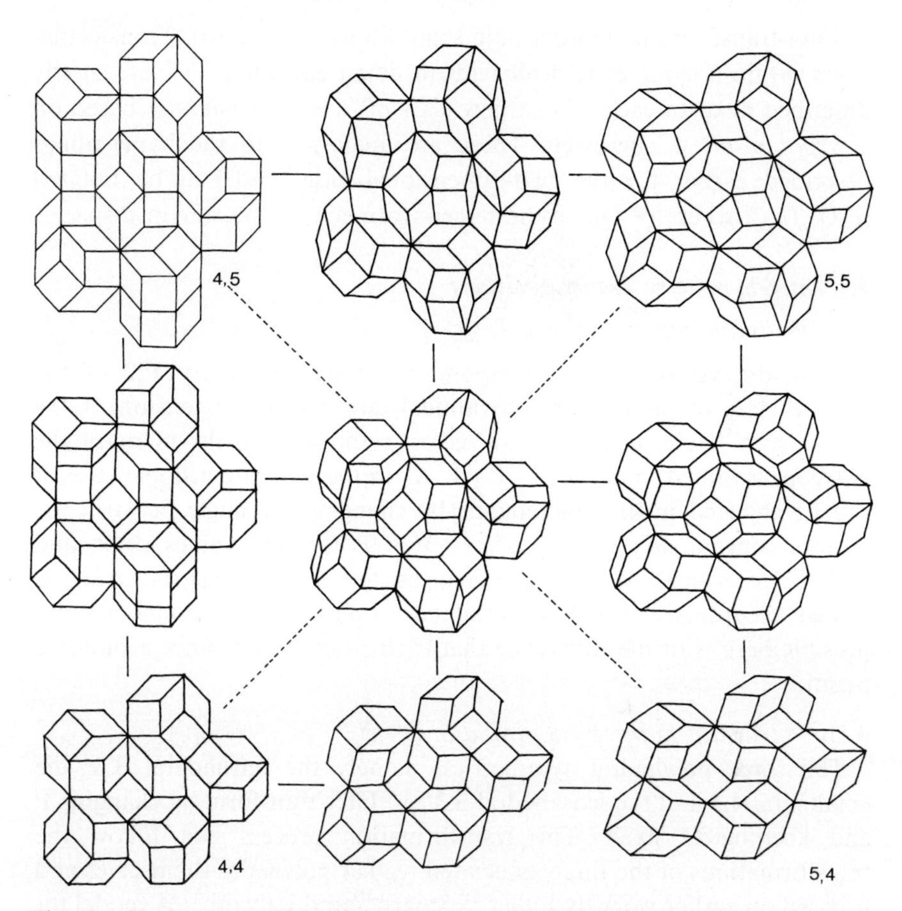

Fig. 15. Continuous transformation of the Ammann-related tiling $(\mathbf{4})_5(4,4)$ on bottom left to the Penrose tiling $(\mathbf{5})_5$ (5,5) on top right through various paths.

be notated $(\mathbf{P},\mathbf{Q})(n,i)$, where **P**, **Q** correspond to the four families already mentioned in Sec. 2. The n-star is derived by joining the vertices of various polyhedra to their centers within any symmetry family, and i-stars are derived from this as in the two-dimensional case. The symmetry and topology of the star determines n, which occurs in specific values.

The transformations *between* symmetries are described first and are based on a model for a three-dimensional continuum of polyhedral transformations. These deal with changes in **P**, **Q**. Appropriate examples of inter-transforming non-periodic space-fillings are shown. Transformations within a symmetry, dealt with in detail elsewhere,[17,18] are briefly described next. In each case, a new class of *hyper space-fillings* based on a *hyper n-star* is suggested. These are analogous to the hyper-tilings described earlier for the two-dimensional case, and can be notated $(\mathbf{P},\mathbf{Q})_n(n,i)$, where $(\mathbf{P},\mathbf{Q})_n$ is the hyper-symmetry of the star in n-space.

4.1 *Inter-Symmetry Transformations*

4.1a *Prismatic Stars: Variable P; Q* = 2

Prismatic vector stars, corresponding to the symmetry **P,2** of the infinite class of prisms (or bipyramids), are a direct extension of the planar stars. Here an additional vector, orthogonal to the plane of the polygon, is added, making N = P + 1. As in the two-dimensional case, P can be changed in unit increments by changing the angle between the vectors continuously. This angle is 360°/P for odd values of P, and 180°/P for even P. As the angle changes, the vector stars, and hence n, change accordingly. All tilings obtained in the two-dimensional case are possible here with the difference that each rhombus becomes a rhombic prism.

4.1b *Polyhedral Stars*: *P* = *3*, *4 and 5*; *Q* = *3*

The three polyhedral symmetries, namely the tetrahedral **3,3**, the octahedral **4,3** and the icosahedral **5,3** can inter-transform by changing P and keeping Q fixed. The transformation process can follow the transformations of the three associated regular polyhedra themselves and is based on earlier work by Fuller,[19] Stuart,[20] and Lalvani.[17] A model for a continuum of polyhedral transformations between the tetrahedron, octahedron and the icosahedron was described in Ref. 17 as a part of rotary transformations of regular-faced polyhedra within the tetrahedral family. This is described here in a slightly revised form.

4.1b(i) *Fundamental region*

A cube is subdivided into 24 identical fundamental regions corresponding to the tetrahedral symmetry by dissecting it with its six mirror planes which pass through the opposite pair of edges (Fig. 16). Each face of the cube is now divided into four isosceles triangles and has twofold symmetry (or a fourfold symmetry in special cases) while the threefold axes are retained. Various polyhedra are derived by placing a vertex

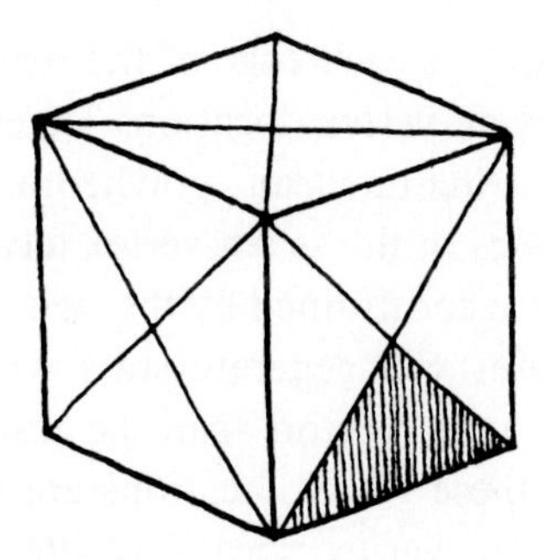

Fig. 16. Surface of a cube subdivided into 24 fundamental regions corresponding to tetrahedral symmetry.

within one fundamental region (the shaded triangle in the illustration) and repeating over every alternate region of the cube in the case of rotational symmetry transformations. Joining the vertices to each other generates a polyhedron, and joining the n non-collinear vertices to the center of the cube generates a polyhedral n-star.

Details of vertex locations within the fundamental region are shown in Fig. 17. This region has the same orientation as the shaded triangle in Fig. 16. A family of seven types of vertex locations are generated from any triangular fundamental region where one vertex is placed at each

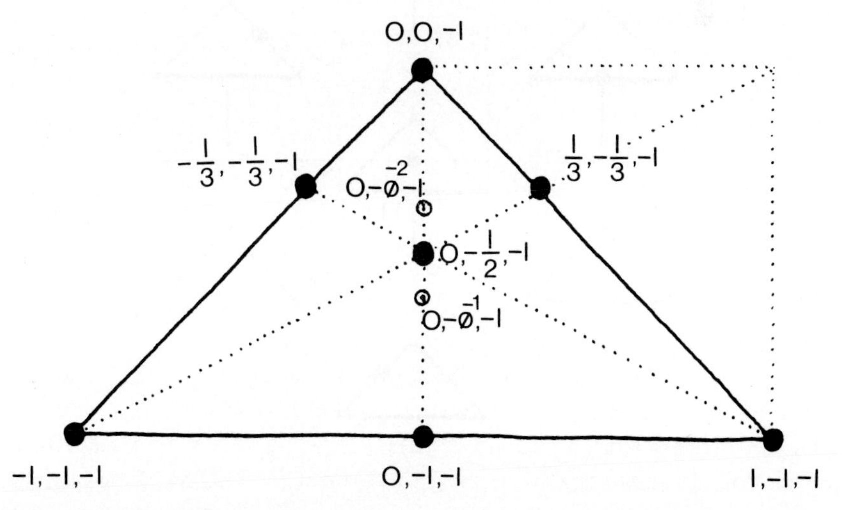

Fig. 17. Vertex locations on one fundamental region corresponding to various regular-faced polyhedra.

corner of the triangle, one at each edge of the region, and on the face of the triangle. These seven vertex locations determine the vertices of various Platonic and Archimedean polyhedra. For the tetrahedral symmetry, the coordinates of the seven vertex locations are given in Fig. 17. These coordinates are constrained by the size of the base cube of unit half-edge. This constraint will generate stars with different lengths of vectors. Alternatively, unit vectors can be used, where the vector directions specified by these vertex locations are retained.

Once the seven basic vertex locations are specified, continuous transformations between them are generated by moving a vertex to any other location continuously. The polyhedron, and the associated n-star, transform to another continuously. P changes to 3 to 4 to 5 in this process.

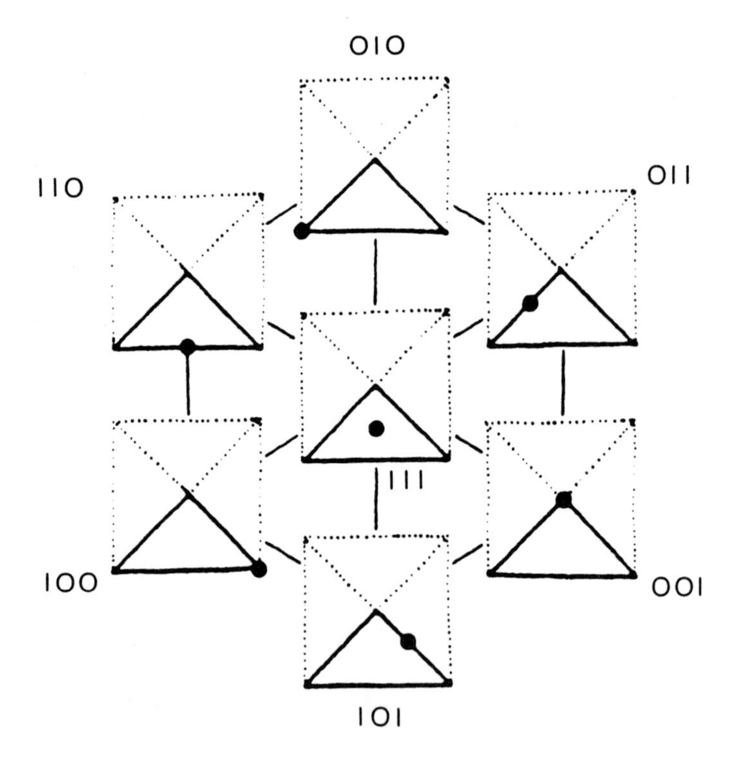

Fig. 18. Organisation of the seven basic types of vertex locations in a three-dimensional cubic structure space, seen here along its threefold axis.

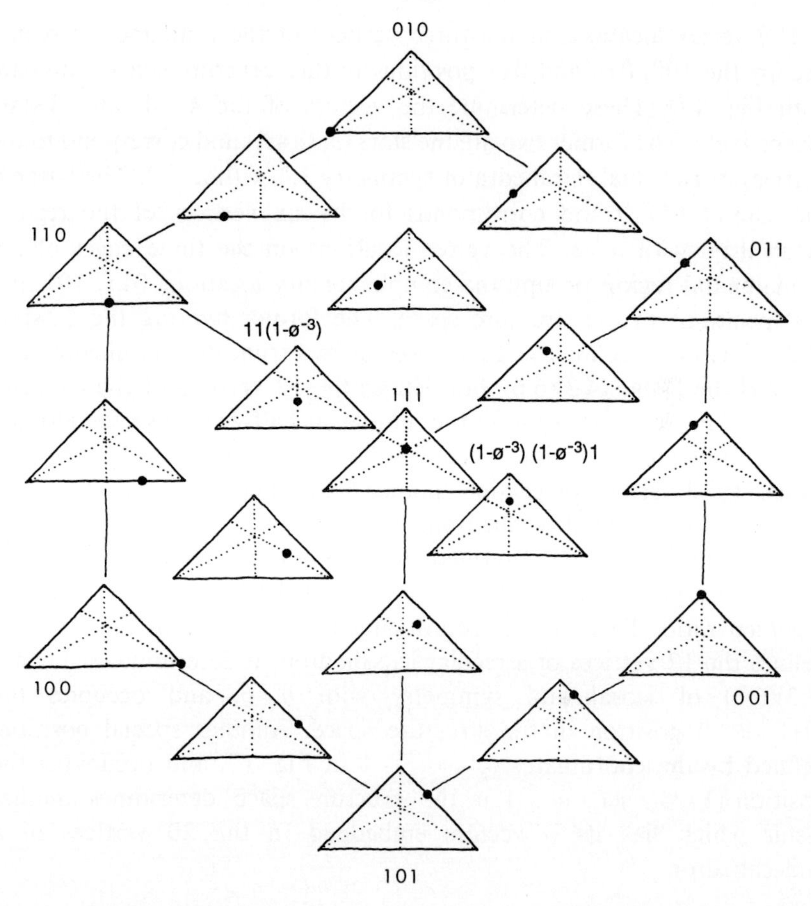

Fig. 19. Continuous transformation of a vertex location in one fundamental region within the structure space permitting transformations between a tetrahedron, octahedron and the icosahedron.

4.1b(ii) *3-cube structure space*

The seven vertex locations can be mapped on the seven vertices of a three-dimensional cubic structure space as described in Ref 17. These seven are shown in Fig. 18 where the structure space is viewed along the threefold axis showing only three of its six faces, and seven of its eight vertices. Intermediate stages in the continuum of transformations are shown in the corresponding Fig. 19. In the stars obtained from these vertex locations, *n* equals *i* and is determined by the number of vertices of the associated regular-faced polyhedra.

The vertex locations at the three vertices of the fundamental region occupy the 100, 010 and 011 positions in this structure space (compare with Fig. 17). These determine the vectors of the 4-, 4- and 3-stars respectively. The former two are the stars **(3,3)**(4,4) and correspond to the vertices of two dual tetrahedra of symmetry **3,3** with $n = 4$. The latter is the star **(4,3)**(3,3) and corresponds to the six vertices of the regular octahedron with $n = 3$. The vertex locations on the three edges of the fundamental region occupy the complementary locations 011, 101 and 100 positions in the structure space. The former two are the 12-stars **(3,3)**(12,12) and define the 12 vertices of two truncated tetrahedra. The latter is the 6-star **(4,3)**(6,6) and defines the 12 vertices of the cuboctahedron. The vertex location in the middle, at half-way along the altitude of the fundamental region, occupies the 111 position in the structure space. Its 12 vertices determine another 6-star defined by the alternating vertices of a truncated octahedron.

A special case of an intermediate is produced when the vertex is located at a point that divides the altitude of the fundamental region in a golden ratio. This has the coordinates 0, $-\phi^{-1}, -1$ in Fig. 17, and defines the 12 vertices of a regular icosahedron. It determines the 6-star **(5,3)**(6,6) of icosahedral symmetry with $n = 6$, and occupies the $11(1 - \phi^{-3})$ position in the structure space. Another special position, defined by the coordinates $0, -\phi^{-2}, -1$ in Fig. 17, and occupying the position $(1 - \phi^{-3})(1 - \phi^{-3})1$ in the structure space, determines another 6-star which has its 6 vectors embedded in the 20 vertices of a dodecahedron.

4.1b(iii) *Inter-transforming 6-stars ($i = 6$), an example*

An interesting derivative of this scheme is to think in terms of vertex locations on only three faces of the cube (Fig. 20), and where each star has six vectors. As before, each face is considered to have a local twofold symmetry, and the inter-symmetry transformations here are produced by a continually changing 6-star. The value of i remains 6 throughout, while n changes from 6 to 4 to 3 as some vectors become collinear. In this variant scheme, the 4-star in position 010 in the structure space changes to **(3,3)**(4,6); the 3-star in the position 100 changes to **(3,3)**(3,6), but retains the Miraldi angles of 70°32' for its faces; position 001 changes to the 3-star **(4,3)**(3,6); and positions 011 and 101 become 6-stars **(3,3)**(12,6) determined by the central angles of a truncated tetrahedron. The

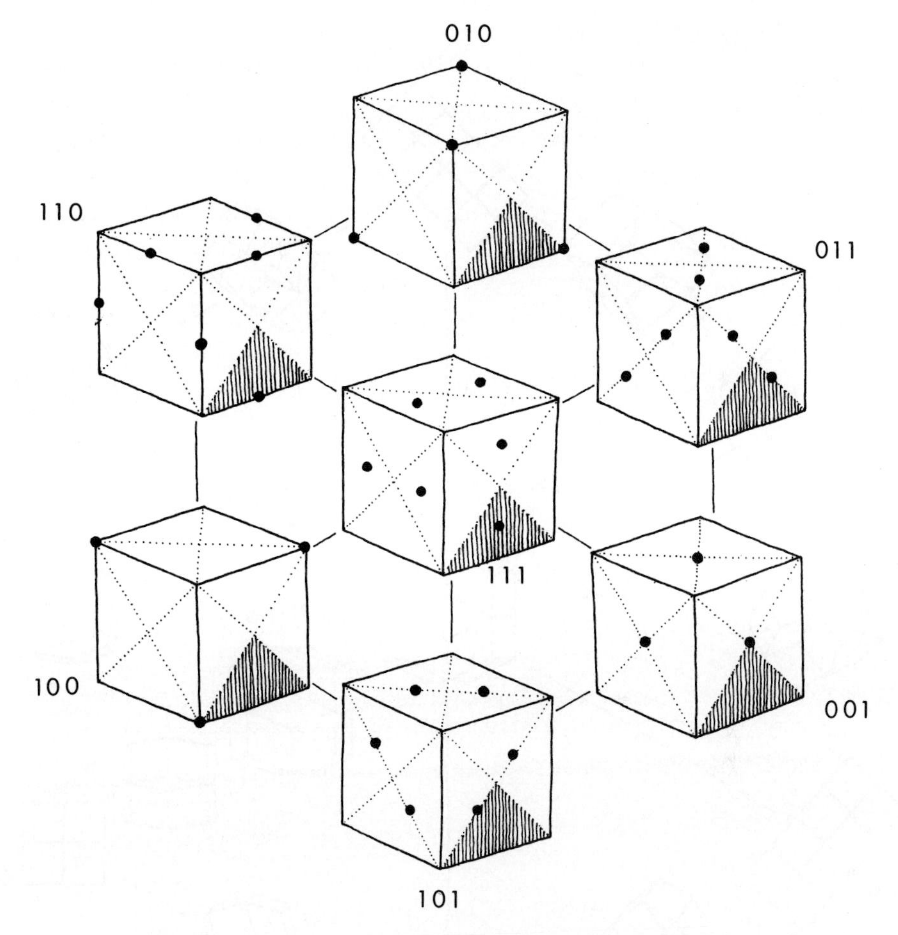

Fig. 20. Vertex locations on alternating fundamental regions on three faces of a cube. When these vertices are joined to the center of the cube, three-dimensional vector-stars are generated. The transformations between stars follows Fig. 19.

icosahedral star **(5,3)**(6,6) remains unchanged. All other intermediate positions within the structure space are various 6-stars. The 6-cells and the space-fillings based on these stars is described next as one relevant example of inter-transformations of non-periodic space-fillings.

4.1b(iii)a *Inter-transforming 6-cells*

The 6-cells based on the 6-stars of Fig. 20 are shown in Fig. 21; the intermediate stages correspond to Fig. 19. Positions 100, 010 and 001 are

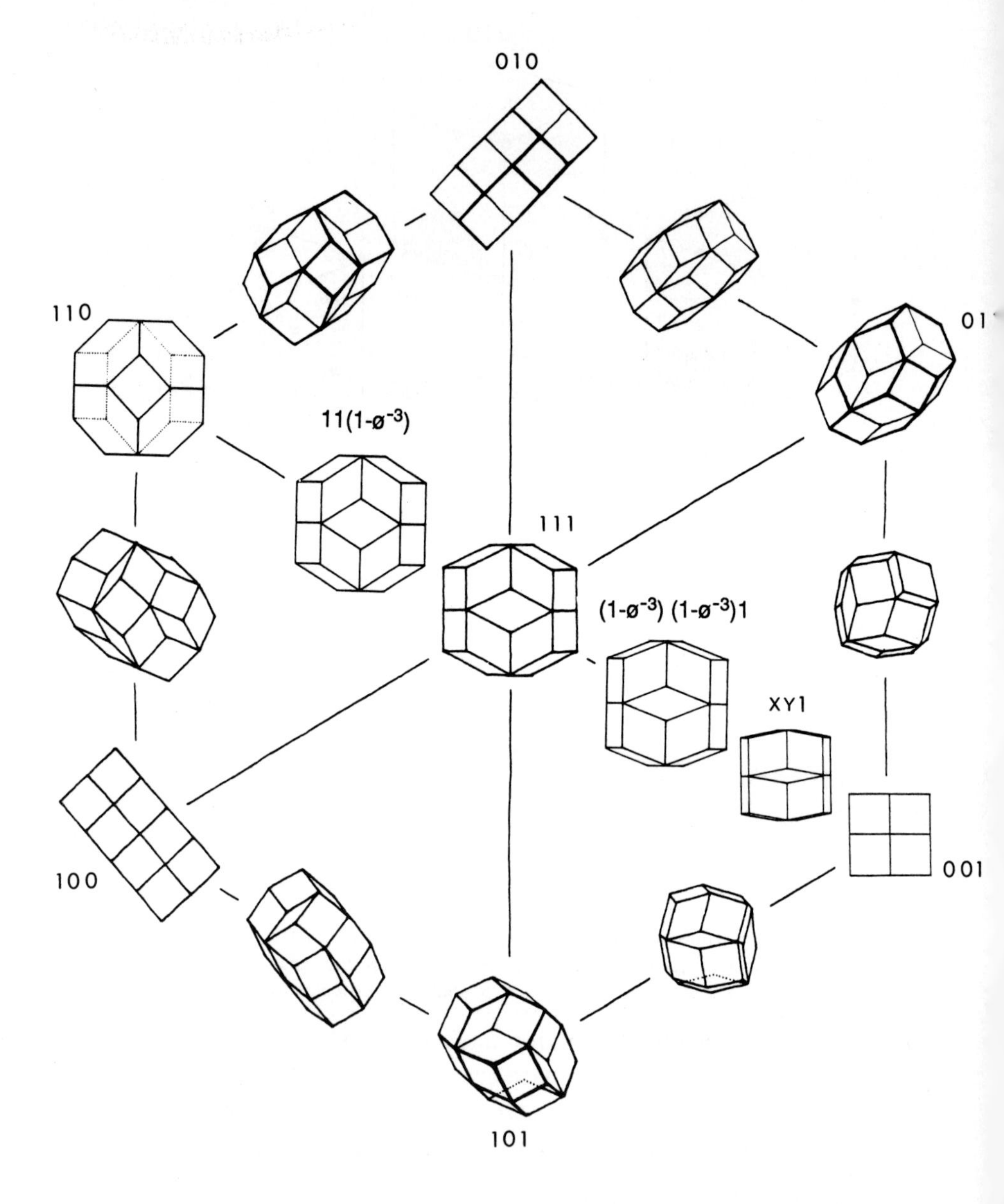

Fig. 21. Continuous transformations of a 6-cell corresponding to the vector-stars based on Fig. 19.

singularities in this continuum since n degenerates to 3 or 4 and some rhombohedra collapse in the process. In the figures, all the cells are viewed along the same twofold axis of the star. Note that the twofold axis becomes a fourfold axis in position 001 and is defined by the edges of the regular cubes. In positions 100 and 010, the squares are the Miraldi rhombi of 72° 32' viewed along the twofold axis and defined by shorter face diagonals of the Miraldi rombohedra; this view also leads to the packing of rhombic dodecahedra in position 010. The 6-cell in position 110 is the truncated octahedron composed of three types of rhomohedra,[5] including a flat one, from two rhombi of 90° and 60°. The 6-cell in position $11(1-\phi^{-3})$ is the familiar rhombic triacontahedron composed of two rhombohedra from a 63° 26' rhombus. The latter two transform to one another by the transformation of the vector star which changes the face angles from two to one, and reduces the types of rhombohedra from three to two. Note that position $(1-\phi^{-3})\,(1-\phi^{-3})1$ generates a 6-cell from the vertices of a regular pentagonal dodecahedron and is composed of four rhombohedra[21] from a Miraldi rhombus and a 41° 49' rhombus.

The continuous transformations between the truncated octahedron and the rhombic triacontahedron has been suggested independently by Mackay[22] and Lalvani.[23] The vector-star transformation between the octahedral and icosahedral symmetries was suggested earlier by Kramer.[24] The continuous transformation using a variable star has also been independently suggested by Torres et al.[25] and has come to the author's attention during the preparation of this paper.

4.1b(iii)b *Inter-transforming space-fillings* $(PQ)(n, 6)$

This scheme produces the continuous inter-transformations between the following space-fillings: the periodic space-filling corresponding to a simple cubic lattice (position 001), the periodic space-filling with Miraldi rhombohedra corresponding to a rhombodhedral lattice (position 100), the four-directional space-filling with Miraldi rhombohedra (position 010) corresponding to the bcc lattice, and a variety of six-directional space-fillings. The latter include non-periodic space-filling based on octahedral symmetry (position 110) where the vertices correspond to the fcc lattice, and the quasi-crystalline lattice corresponding to icosahedral

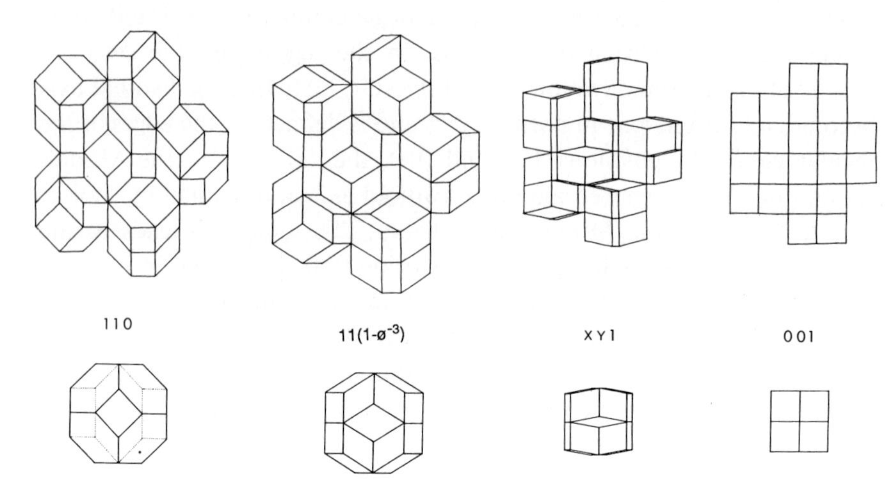

Fig. 22. A few examples of portions of space-fillings corresponding to the cells of Fig. 21. The stages shown go from the octahedral (cubic) to the icosahedral to the simple cubic lattice.

symmetry (position $11(1-\phi^{-3})$). All intermediates in the transformations are non-periodic space-fillings based on the infinite intermediate states of the transforming 6-star.

A linear path of transformations from the continuum is illustrated in Fig. 22. The successive positions 110, $11(1-\phi^{-3})$, $xy1$, and 001 are portions of space-fillings corresponding to their respective 6-cells shown in Fig. 21; $xy1$ is a general position, where x and y lie equally between 0 and 1. The extended space-filling for 110 is shown in Fig. 23. Note that this space-filling, seen here in its fourfold view, corresponds to the tiling $(\mathbf{4})_5(4,5)$ shown earlier in Fig. 15 (in the top left corner). This suggests that all tilings n, i are the n-fold views of space-fillings. Referring to Fig. 15, the tiling $(\mathbf{4})_5(4,5)$ is the fourfold view of the space-filling from the star **(4,3)**(6,6), the tiling $(\mathbf{5})_5(5,5)$ is the fivefold view of the space-filling **(5,3)**(6,6), tiling $(\mathbf{4})_5(4,4)$ is the fourfold view of the space-filling **(4,3)**(6,5), and the tiling $(\mathbf{4})_5(5,4)$ is the fivefold view of the space-filling **(5,3)**(6,5). The tilings of Fig. 15 represent the continuous transformation of the space-fillings from stars of symmetry **4,3** to **5,3** as they change their orientation continually from a fourfold view to a view along the fivefold axis of symmetry of the vector-star. Extensions to transformations

Fig. 23. A "fourfold" view of a quasi-symmetric space-filling based on octahedral symmetry. In the view shown it corresponds to the plane tiling $(\mathbf{4})_5(4,5)$ shown in Fig. 14 on top left.

between other symmetry views for all symmetries follow. Note that this provides a technique for generating tilings from space-fillings, and vice versa.

4.2 *Intra-Symmetry Transformations: Fixed P,Q*

There are two classes of continuous transformations *within* a fixed symmetry **P,Q**. The first deals with the derivation of specific values of n from the symmetry, and the second derives various values of i for a fixed n.

4.2a *Variable n*

Different values of n can be determined by the various families of polyhedra that can be derived within any symmetry. These include analogous derivation of continually transforming polyhedra for each symmetry. The transformations between the polyhedra change n, but preserve symmetry. Details of polyhedral transformations have been described in Ref. 17. The values of n from regular and semi-regular polyhedra corresponding to the four symmetry families have been dealt with elsewhere.[5] Within each family, different subdivisions of the regular and semi-regular polyhedra lead to additional families of n-stars. These subdivisions, and their transformations, have also been described elsewhere.[17] Methods of *recursive* subdivisions lead to corresponding infinite families of n-stars. A class of recursive subdivisions was described in Ref. 17 and a new recursive class was developed by Hanrahan-Lalvani.[26] Additional stars are possible from the geodesic subdivision of spheres.[5] Comparative details of n-stars from different classes of subdivisions of polyhedra, their inter-transformations and space-fillings generated by these subdivisions, remain to be studied.

4.2b *Fixed n, Variable i*

Once the n-stars are established within any symmetry, various i-stars ($i = 0,1,2,3,4,5,\ldots. n-2, n-1, n$) can be derived from directions of all distinct combinations of non-collinear vectors. For the icosahedral symmetry, all i-stars and their associated zonohedra based on $n = 6$, 10 and 15 have been described in Ref. 27, and correspond to the stars from the fivefold, threefold and twofold axes. For the octahedral symmetry, all i-stars based on $n = 3,4,6$ and 9, and their associated zonohedra, have been worked out.[28] The i-stars from the tetrahedral symmetry are embedded in the octahedral family, except for $n = 12$ case based on the truncated tetrahedron. For the prismatic symmetries, the i-stars from $P > 7$ remain to be worked out. The systematic continuous transformations between the different i-stars remains to be established.

4.3 *Hyper Space-Fillings* $(P,Q)_n(n,i)$

We now come to the last class of continuous transformations of space-fillings. These are an analog of the hyper-Penrose tilings described earlier. As an extension of the hyper-polygonal stars $(\mathbf{P})_n(n,i)$ in the two-dimensional case, *hyper-prismatic* (or hyper-pyramidal) stars $(\mathbf{P,2})_n(n,i)$ are possible. Here, an additional vector is inclined at *any* angle to the plane of the polygon. The transformations between the hyper-stars, and between their associated hyper space-fillings, follow the scheme shown earlier in Fig. 5. An interesting example based on a hyper pyramidal star is the space-filling $\mathbf{(5,3)}_6(15,6)$ based on a 6-star derived from the icosahedral 15-star. It is composed of parallel layers of close-packed rhombohedra, where each layer is planar and is the Penrose tiling in its plane view. The successive layers are shifted in the plan view, and all its edges are parallel to the twofold axes of symmetry of an icosahederon.

Corresponding to any polyhedral star, an infinite number of *hyper-polyhedral* stars are possible, each generating its own hyper space-filling. As an example, an infinite class of hyper-icosahedral stars $\mathbf{(5,3)}_6(n,6)$ corresponds to the regular icosahedral 6-star **(5,3)** (6,6). Two examples are shown in Fig. 24: (a) is the hyper 6-star $\mathbf{(5,3)}_6(13,6)$ based on the cubic symmetry and has its vectors along the two, three and fourfold axes of symmetry of a cube, (b) is the hyper 6-star $\mathbf{(5,3)}_6(31,6)$ based on vectors along the two, three and fivefold axes of icosahederal symmetry.

In space-fillings based on hyper-icosahedral stars, the hyper-triacontahedral cells overlap in three-dimensions in a manner that is analogous to the two-dimensional case. As an example, the tiling $\mathbf{(5)}_6(6,6)$ shown in Fig. 10 can be thought of as a hyper-fivefold view of the space-filling $\mathbf{(5,3)}_7(7,7)$, where an additional seventh vector is perpendicular to the plane of the paper. The hyper space-fillings transform in the

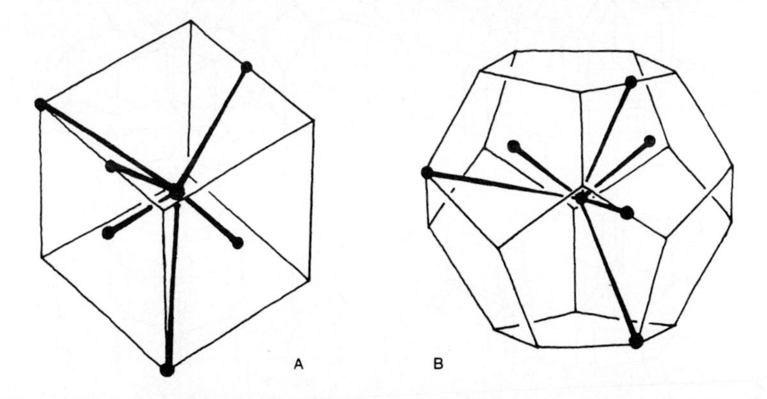

Fig. 24. Two examples of hyper-icosahedral stars with 6 vectors: (a) shows the star $\mathbf{(5,3)}_6(31,6)$ based on cubic symmetry, and (b) shows the star $\mathbf{(5,3)}_6(31,6)$ based on icosahedral symmetry.

same manner as shown Figs. 12, 13 and 15 for the two-dimensional cases. Some of the transformations are topology-preserving as before, some lose vectors, and in some others hyper-cells collapse and disappear. When projected into three-dimensions, the geometry of hyper space-fillings is a deformation of its regular counterparts, as in the case of tilings.

5. Escher-Like Metamorphosis

Non-periodic analogs of Escher's metamorphosis of periodic patterns are possible based on the variety of transformations described here. This can be achieved by first filling-in the intermediates, as shown in Fig. 14 for a single zonogon or a zonohedron, and in Fig. 15 for space-fillings or tilings, followed by using Escher's techniques for metamorphoses of patterns. One example is shown in Fig. 25, albeit *in space*. Metamorphosis, *in time*, are possible with computer animations and may be more relevant to transformations processes in nature, as in the case of crystal–quasicrystal transition.[29]

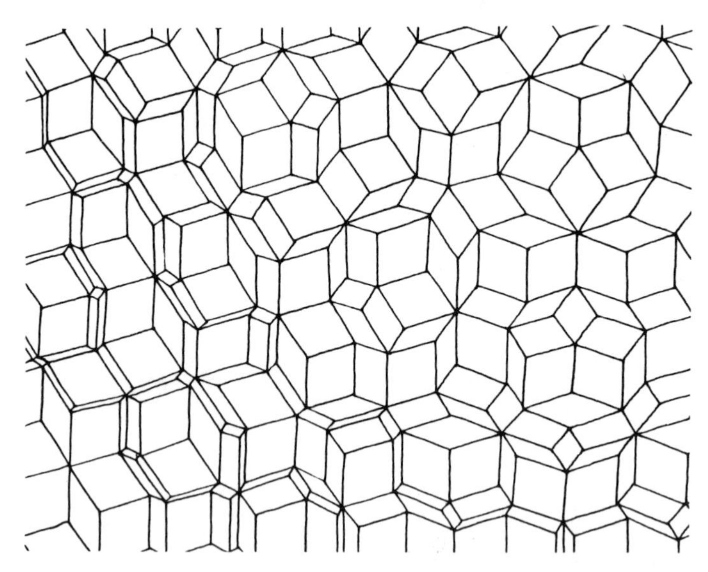

Fig. 25. An Escher-like metamorphosis of the tiling $(\mathbf{5})_5(5,5)$ on bottom left, through $(\mathbf{4})_5(5,4)$, to $(\mathbf{3})_5(5,3)$ on upper right. It is also a fivefold view of the space-filling $(\mathbf{5,3})(6,6)$ from the icosahedral star as it transforms from overlapping 6-cells to 5-cells to 4-cells to 3-cells.

Notes and References

1. M. Gardner, *Scientific American* (Jan. 1977) 110–121.
2. P. Penrose, Pentaplexity, *Math. Intelligencer* **2** (1979) 32–37.
3. N. de Bruijn, Algebraic theory of Penrose's non-periodic tilings of the plane: 1, *Proc. Koninklijke Nederlandse Akademie van Wetenschappen* **A84** (1) (1981) 39–66
4. H. Lalvani, *Generating Non-Periodic Tilings and Space-Fillings*, a two-part presentation at the conference "Geometric Problems in Crystallography," Universitat Bielefeld, Aug. 1985; see Ref. 5.
5. H. Lalvani, Non-periodic space structures, *Space Structures* **2** (1986/87) 93–108; based on Ref 4.
6. R. Ammann, cited in Ref. 2.
7. A.L. Mackay, Crystallography and the Penrose pattern, *Physica* **114A** (1982) 609–613.
8. P. Kramer and R. Neri, On periodic and non-periodic space-fillings of E^m obtained by projection, *Acta Cryst* **A40** (1984) 580–587.
9. D. Levine and P. Steinhardt, Quasicrystals: a new class of ordered structures, *Phys. Rev. Lett.* **53**(26) (1984) 1951–3.
10. D. Stuart, brought to author's attention by Dr. A. Tyng (personal communication, 1978, 1989).
11. K. Miyazaki and I. Takada, Uniform ant-hills in the world of golden isozonohedra, *Structural Topology* **4** (1980) 21.
12. S. Baer, *The Zome Primer*. Zomeworks Corp., Albuquerque, New Mexico (1970).
13. N. de Bruijn, personal communication (1986).
14. B. Grünbaum and G.C. Shephard, *Tilings and Patterns*. W.H. Freeman (1987); see Ammann's aperiodic set A5, p. 556–557.
15. H. Lalvani and D. Sturman, *Transformations of the Penrose Tiling*, a computer animation video, N.Y.I.T. and M.I.T. Media Lab. (1988).
16. H. Lalvani and D. Sturman, unpublished (1987).
17. H. Lalvani, *Multi-Dimensional Periodic Arrangements of Transforming Space Structures*, Ph.D. Thesis, Univ. of Pennsylvania, 1981 (published by University Microfilms International, Ann Arbor, Michigan, 1982); see also Ref. 18.
18. H. Lalvani, *Structures on Hyper-Structures*. Lalvani, New York (1982); based on Ref. 17.
19. R.B. Fuller, *Synergetics 1*. McMillan (1975; p. 119–215).
20. D. Stuart., Polyhedral and mosaic transformations, *Student publication of the School of Design,* North Carolina State Univ. at Raleigh (1963, p. 3–28)
21. This observation is due to P. Reissig, working with the author; models of these inter-transforming 6-cells were built earlier with P. Reissig (unpublished, 1989).
22. A. Mackay, What has the Penrose tiling to do with the icosahedral phases? Geometrical aspects of the icosahedral quasicrystal problem, *Journal of Microscopy*, **146** (3)(1987) 233–243.
23. H. Lalvani, Morphological aspects of space structures, in *Space Structures, Theory and Practice*, H. Nooshin, ed. Multi-Science (in press).
24. P. Kramer, personal communication (1986).

25. M. Torres, G. Pastor, I. Jiminez, and J. Fayos, Geometric models for continuous transitions from quasicrystals to crystals, *Phil. Mag. Lett.* **59** (4) (1989) 181–188.
26. P. Hanrahan and H. Lalvani, *Recursive Topological Transformations* (1983, unpublished).
27. R. Haase, L. Kramer, P. Kramer, and H. Lalvani, Polyhedra of three quasilattices associated with the icosahedral group, *Acta. Cryst.* **A 43** (1987) 574–587.
28. H. Lalvani, assisted by P. Reissig, unpubl., 1987–88; all drawings and models have been constructed.
29. Key frames for a continuous transformations of a 6-cube through the stages of the octahedral, icosahedral and the simple cubic lattice were prepared in 1988 (unpublished); three stills have appeared in J. Kappraff's *Connections, The Geometric Bridge Between Art and Science*, McGraw-Hill (1991) p. 377.

FIVEFOLD SYMMETRY IN HYPERBOLIC CRYSTALLOGRAPHY

Douglas Dunham

1. Introduction

The well-known "crystallographic restriction" (see Coxeter,[1] Sec. 4.5) prohibits a perfect two-dimensional "crystal" from having global fivefold rotational symmetry in the Euclidean plane, and this extends also to higher dimensional Euclidean space. Of course, people have designed patterns with local fivefold symmetry and patterns of points with global fivefold symmetry that are dense in the plane (the set of points having both coordinates of the form $r + s\sqrt{5}$ with r and s rational — see Grünbaum and Shephard,[2] p. 6). Also, crystals with local fivefold symmetry have been discovered. On the other hand, there is no restriction on the kinds of rotational symmetry that can occur among repeating patterns of the hyperbolic plane. In fact, for every $n \geq 3$ there is a hyperbolic pattern having global n-fold rotational symmetry. In particular, there is nothing special about hyperbolic patterns with fivefold symmetry, and moreover there is no other value of n (such as 5 in the Euclidean case) for which global n-fold symmetry is prohibited.

The next section discusses repeating hyperbolic patterns of the hyperbolic plane, particularly regular tessellations that provide examples of repeating hyperbolic patterns exhibiting n-fold rotational symmetry for all values of $n \geq 3$. Then we specialize to the case $n = 5$ and show some sample patterns. The following section explains the relation between

n-fold symmetry in hyperbolic, Euclidean, and spherical geometry. Also, we note that a different kind of restriction applies to repeating hyperbolic and spherical patterns, namely, they cannot be "stretched" by a similarity transformation. Finally, the fivefold symmetries that occur in the regular tessellations ("honeycombs") of higher dimensional hyperbolic space are described.

2. Repeating Patterns of the Hyperbolic Plane

A *repeating pattern* of the hyperbolic plane (or sphere or Euclidean plane) is a pattern made up of congruent copies of a basic subpattern or *motif* (which we assume to be "nice" – bounded by a simple closed curve, for example). For instance, any one of the fish of Fig. 1 would serve as a motif.

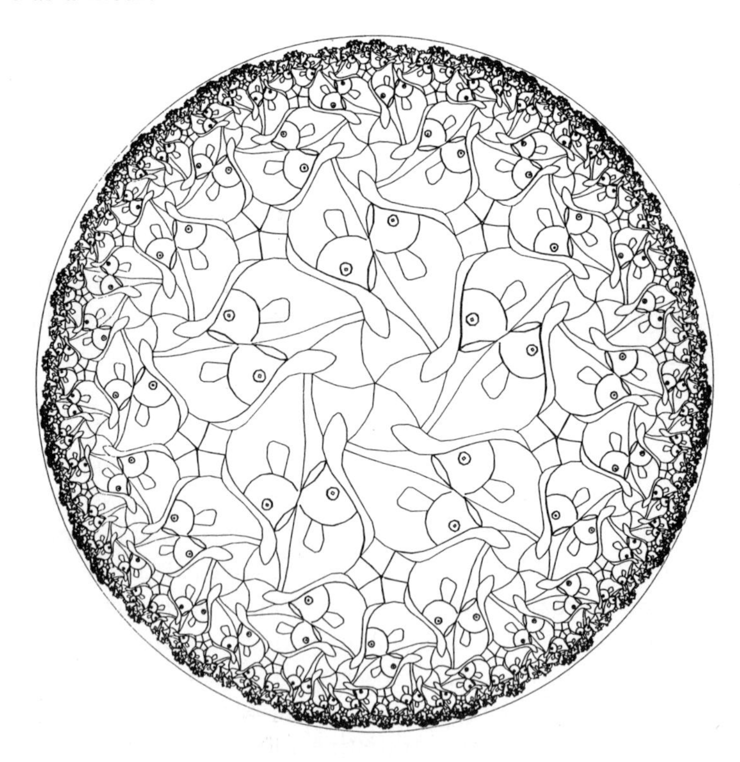

Fig. 1 A repeating hyperbolic pattern of fish exhibiting fivefold symmetry and based on a repeating Euclidean pattern of fish by the artist M.C. Escher. The symmetry group of this pattern is $[5, 5]^+$.

An important kind of repeating pattern is the *regular tessellation*, $\{p, q\}$, of the hyperbolic plane by regular p-sided polygons, or *p-gons,* meeting q at a vertex. Such tessellations are also defined for the sphere and the Euclidean plane. It is necessary that $(p - 2)(q - 2) > 4$ for the tessellation to be hyperbolic. Fig. 2 shows the tessellation $\{6, 4\}$ (solid lines) and its dual tessellation $\{4, 6\}$ (dotted lines). Any one of the small right triangles bounded by solid, dotted, and dashed lines forms a motif for either tessellation.

Since the hyperbolic plane cannot be isometrically embedded in Euclidean 3-space, the *Poincaré circle model* has been chosen to display the figures in this article. Poincaré's circle model has two useful properties: it is conformal (i.e., the hyperbolic measure of an angle is equal to its Euclidean measure), and it lies within a bounded region of the Euclidean plane, allowing an entire hyperbolic pattern to be displayed. The "points" of this model are the interior points of a *bounding circle* in the Euclidean plane. The (hyperbolic) "lines" are circular arcs orthogonal

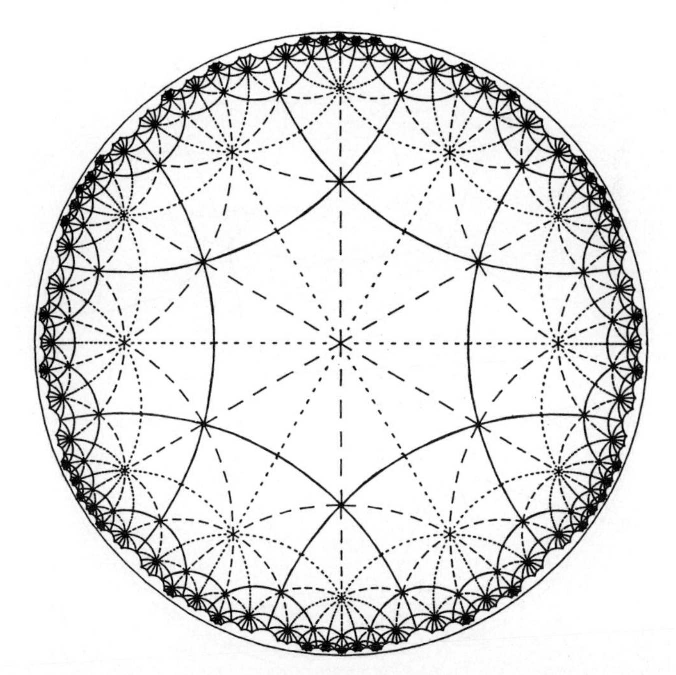

Fig. 2 The hyperbolic tessellation $\{6, 4\}$ (solid lines), its dual tessellation $\{4, 6\}$ (dotted lines), and other lines (dashed) of reflection symmetry of the two tessellations.

to the bounding circle, including diameters. All of the lines (solid, dotted, and dashed) of Fig. 2 represent hyperbolic lines. Note that equal hyperbolic distances are represented by smaller and smaller Euclidean distances as one approaches the bounding circle. Thus, the fish of Fig. 1 are all the same hyperbolic size (as are the small right triangles of Fig. 2). Also note that the center of the bounding circle plays no special role in the Poincaré circle model. For example, the dotted tessellation {4, 6} of Fig. 2 could be translated so that the center of one of the 4-gons would be at the center of the bounding circle; this is shown in Fig 3.

A *symmetry operation* or simply a *symmetry* of a repeating pattern is an isometry (hyperbolic distance-preserving transformation) of the hyperbolic plane that transforms the pattern onto itself. For example, reflections across any of the lines in Fig. 2 are symmetries of that pattern (reflections across hyperbolic lines of the Poincaré circle model are inversions in the circular arcs representing those lines [or ordinary Euclidean reflections across diameters]). Symmetries of Fig. 1 include

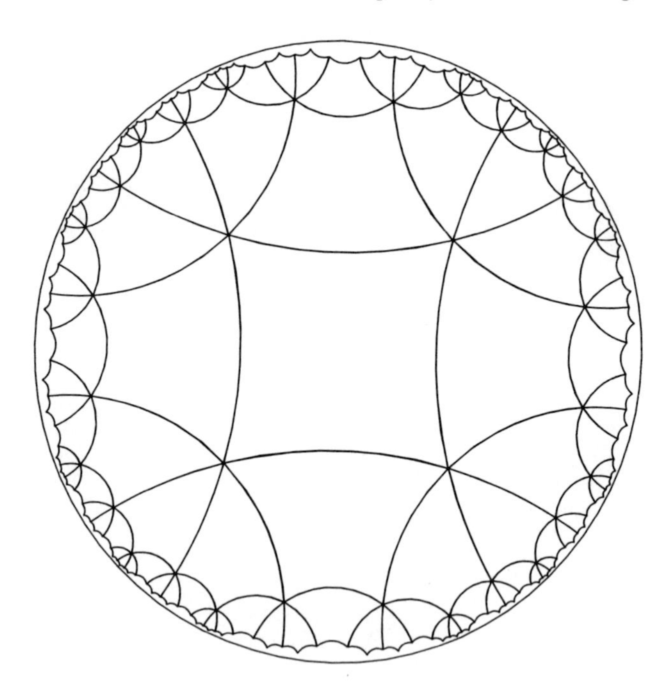

Fig. 3 The tessellation {4, 6} with a face centered within the bounding circle.

rotations by 180° about the points where lower fish jaws meet, and rotations by 72° about points where the tails or dorsal fins meet (in hyperbolic geometry, as in Euclidean geometry, the product of reflections across two intersecting lines produces a rotation about the intersection point by twice the angle of intersection).

The *symmetry group* of a pattern is the set of all symmetries of the pattern. The symmetry group of the tessellation $\{p, q\}$, denoted $[p, q]$, can be generated by reflections across the sides of a right triangle with acute angles of $180/p°$ and $180/q°$ (i.e., all symmetries in the group $[p, q]$ may be obtained by successively applying a finite number of those three reflections). Thus [6, 4] is the symmetry group of the tessellation {6, 4} formed by the solid lines of Fig. 2 — in fact [6, 4] is the symmetry group of the entire pattern of Fig. 2, and in particular of {4, 6}. In general, $\{p, q\}$ and $\{q, p\}$ have isomorphic symmetry groups (denoted either $[p, q]$ or $[q, p]$).

The orientation-preserving subgroup of $[p, q]$ is denoted $[p, q]^+$ and consists of symmetries made up of an even number of reflections. The symmetry group of Fig. 1 is $[5, 5]^+$ since we see only the left sides of the fish. Notice that the pattern of Fig. 1 exhibits fivefold rotational symmetry about both the meeting points of the tails and the meeting points of the dorsal fins.

We note that both the symmetry groups $[p, q]$ and $[p, q]^+$ contain p-fold and q-fold rotational symmetries. Consequently, for any value of $p \geq 3$, there is a repeating pattern of the hyperbolic plane with global p-fold rotational symmetry. In fact, a regular tessellation $\{p, q\}$, for any $q \geq 7$, is such a pattern. Hence, there is no "crystallographic restriction" to repeating patterns of the hyperbolic plane.

A computer program has been written to create repeating patterns of the hyperbolic plane based on the tessellations $\{p, q\}$ for any p and q satisfying $(p - 2)(q - 2)>4$ (see Dunham, Ref. 3).

3. Fivefold Symmetry in the Hyperbolic Plane

As mentioned previously, for any value of n, values of p and q can be chosen so that $\{p, q\}$ exhibits n-fold rotational symmetry. If we chose p to be 5, then $\{5, q\}$ will be hyperbolic for any value of $q \geq 4$, since $(5 - 2)(q - 2)>4$ in that case. In particular {5, 4}, shown in Fig 4, exhibits global fivefold symmetry. Dually, $\{p, 5\}$ (for $p \geq 4$) also exhibits

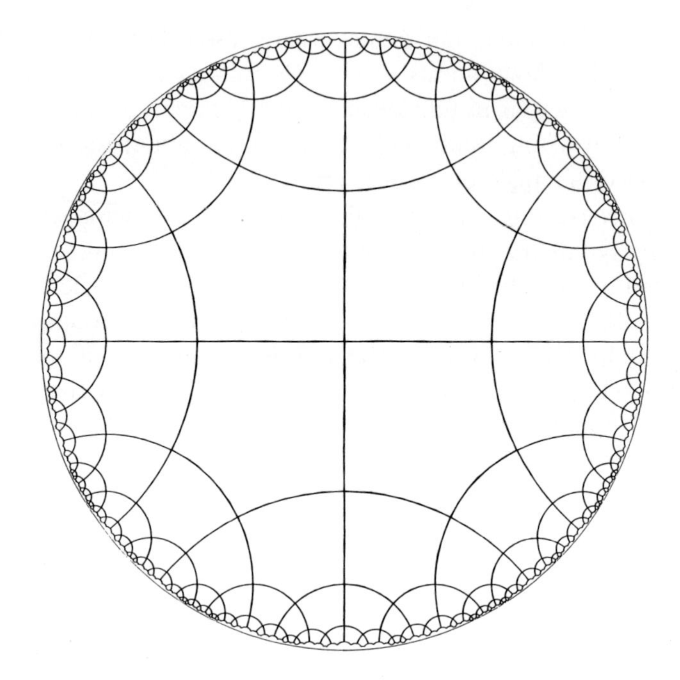

Fig. 4 The tessellation {5, 4} with a vertex centered in the bounding circle. This tessellation exhibits fivefold symmetry.

global fivefold symmetry; {4, 5} is shown in Fig. 5. Fig. 6 shows a repeating pattern based on the tessellation {4, 5}.

In fact, p and q can both be five, yielding the hyperbolic tessellation {5, 5} shown in Fig. 7. The pattern of Fig. 1 is based on the {5, 5} tessellation. As noted previously, that pattern has fivefold rotational symmetry about both the dorsal fin and the tail meeting points of fish. In fact, the pattern's symmetry group $[5, 5]^+$ can be generated by the two fivefold rotations attached to any one fish. A colored version of this pattern exhibiting fivefold five-color symmetry is shown on p. 398 and discussed on p. 248 of Dunham.[4]

4. Relation to Euclidean and Spherical Geometry

If the ">" in the relation $(p - 2)(q - 2) > 4$ is replaced by "=" or "<", one obtains tessellations of the Euclidean plane and the sphere, respectively. In the Euclidean case, the only solutions to the equation

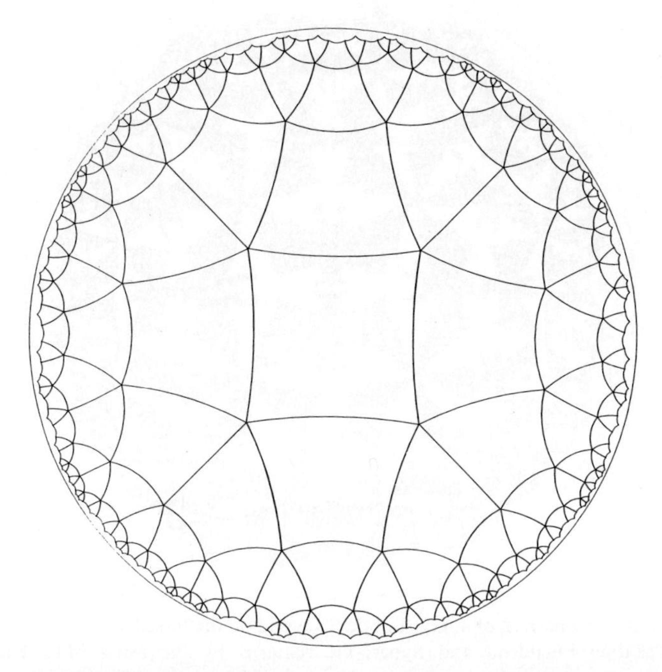

Fig. 5 The tessellation {4, 5}, dual to the tessellation of Fig. 4, with a face centered within the bounding circle.

$(p - 2)(q - 2) = 4$ are {4, 4}, {3, 6}, and {6, 3}. The corresponding symmetry groups of these tessellations are [4, 4] = $p4m$ and [3, 6] = $p6m$, which contain all 17 of the plane crystallographic groups as subgroups (see Sec. 4.6 and Table 4 of Coxeter and Moser[5]). Since neither of the groups [4, 4] = $p4m$ and [3, 6] = $p6m$ contain fivefold symmetries, no repeating Euclidean pattern (of "nice" motifs) can contain any global fivefold symmetries. Of course, these groups do not contain any n-fold symmetries for $n > 6$ either.

In the spherical case, one obtains the tessellations {3, 3}, {3, 4}, {4, 3}, {3, 5}, and {5, 3} as the only solutions (with $p \geq 3$ and $q \geq 3$) to the inequality $(p - 2)(q - 2) < 4$. These tessellations are central projections of the tetrahedron, octahedron, cube, icosahedron, and dodecahedron, respectively, onto their circumscribing spheres. Thus, the last two tessellations exhibit fivefold rotational symmetry. None of these tessellations contains any n-fold symmetries for $n \geq 6$.

Fig. 6 A repeating pattern of angels and devils based on the tessellation {4, 5} of Fig. 5 and repeating Euclidean and hyperbolic patterns by the artist M.C. Escher. The symmetry group of this pattern contains fivefold rotations about meeting points of wing tips.

We also note that there is a "metric" restriction on hyperbolic and spherical repeating patterns: they cannot be "stretched." For example, once the radius of a sphere is specified, so is the length of an edge of any particular regular spherical tessellation. Similarly, once the curvature of hyperbolic space is specified, so is the length of an edge of $\{p, q\}$. This is in contrast to the Euclidean case in which regular tessellations (and any repeating patterns, in fact) come in all sizes. To put it another way, there are no similarity transformations of the hyperbolic plane or the sphere. This is true of the higher dimensional spheres and hyperbolic spaces too.

5. Fivefold Symmetry in Higher Dimensional Hyperbolic Tessellations

A regular tessellation or regular *honeycomb* $\{p, q, r\}$ of three-dimensional hyperbolic space is a tessellation by regular cells $\{p, q\}$ having a (regular) vertex figure $\{q, r\}$. As in the two-dimensional case, this notation extends to tessellations (or honeycombs) of Euclidean 3-space and the 3-sphere. Also, just as Platonic solids may be "blown-up" onto

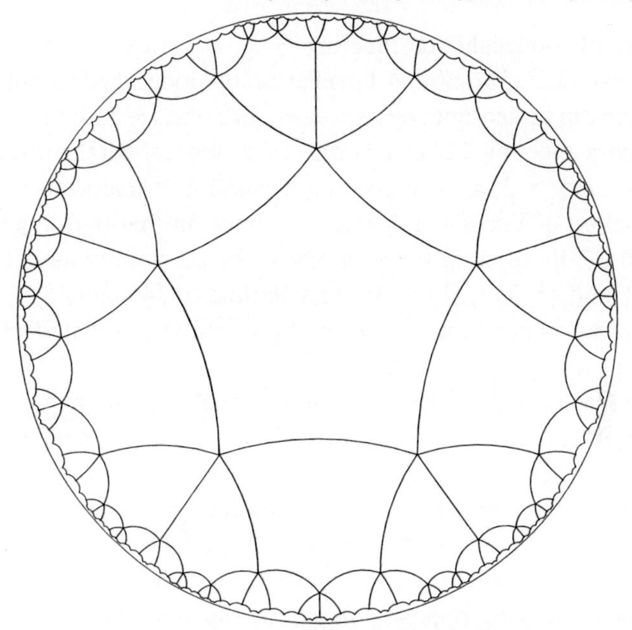

Fig. 7 The tessellation {5, 5}. This tessellation exhibits two kinds of fivefold symmetry: about the vertices and about the centers of the pentagons.

regular tessellations $\{p, q\}$ of the 2-sphere, regular 3-polytopes may be "blown-up" (centrally projected) onto regular tessellations of their circumscribing 3-sphere. In fact, these remarks also apply in higher dimensions.

In order for a regular three-dimensional honeycomb $\{p, q, r\}$ to exist, both the cells, $\{p, q\}$ and the vertex figures, $\{q, r\}$, must be finite (Platonic solids, i.e., regular tessellations of the 2-sphere), as pointed out by Coxeter (Ref. 6, Sec. 10). Consequently, there are only a finite number (11) of three-dimensional honeycombs. Again, this observation extends to higher dimensions: in a regular $(d + 1)$-dimensional honeycomb, both the cells and vertex figures correspond to regular d-dimensional honeycombs of the d-sphere. Thus, by induction, there are only a finite number of regular honeycombs in each of the dimensions three or greater.

We now examine three-dimensional hyperbolic honeycombs. It turns out that there are four combinations of cells and vertex figures (chosen from Platonic solids) that lead to regular hyperbolic honeycombs: {4, 3, 5} (a tessellation of three-dimensional hyperbolic space by "cubes" meeting 20 at a vertex in an icosahedral arrangement), its dual {5, 3, 4} (a tessellation by dodecahedra meeting 8 at a vertex in an octahedral arrangement), {5, 3, 5} (another tessellation by dodecahedra, but meeting 20 at a vertex in an icosahedral arrangement), and {3, 5, 3} (a tessellation by icosahedra meeting 12 at a vertex in a dodecahedral arrangement). The honeycomb {5, 3, 4} is interesting because it is analogous to {4, 3, 4} (the tessellation of Euclidean 3-space by cubes) in that both honeycombs are formed by dividing up their respective spaces with planes. The faces of {4, 3, 4} and {5, 3, 4} form the tessellations {4, 4} and {5, 4}, respectively, in those planes. Fig. 4 shows one of those tessellated planes for {5, 3, 4}.

We note that in each of the three-dimensional hyperbolic honeycombs, there is icosahedral symmetry of the cells, or the vertex figures, or both. Consequently, there is also (global) fivefold symmetry. We examine how this fivefold symmetry manifests itself in each case. The edges of the "cubes" of {4, 3, 5} are fivefold axes since five cubes fit together about an edge. The fivefold axes of {5, 3, 4} are just the fivefold rotation axes (connecting centers of opposite faces) of the dodecahedral cells. The honeycomb {5, 3, 5} has *both* of these fivefold axes: the edges and the fivefold axes of the dodecahedral cells (two kinds of fivefold axes would be expected since there are two 5's in the symbol {5, 3, 5}). In {3, 5, 3}, the fivefold axes are just the fivefold axes (connecting opposite vertices) of the icosahedral cells.

So, as in the two-dimensional case, three-dimensional repeating hyperbolic patterns can have global fivefold symmetry, whereas this cannot happen in Euclidean 3-space. In fact, three-dimensional repeating hyperbolic patterns can have global icosahedral symmetry that can only exist locally in three-dimensional Euclidean patterns.

Next, we extend the discussion to four dimensions. Among the six three-dimensional spherical honeycombs (or regular polytopes), there are two with fivefold symmetry (in fact icosahedral symmetry) — the 120-cell and the 600-cell. The six spherical honeycombs, when combined as cells and vertex figures, yield 11 regular four-dimensional honeycombs. Of these four-dimensional honeycombs, five are hyperbolic: {3, 3, 3, 5}, {5, 3, 3, 3}, {4, 3, 3, 5}, {5, 3, 3, 4}, and {5, 3, 3, 5}. Each of these hyperbolic honeycombs has the 120- or 600-cell for its fundamental cell or vertex figure. Consequently, the group [5, 3, 3] (the group of symmetries of the 120- or the 600-cell) is a subgroup of the symmetry group of each of these hyperbolic honeycombs, and so each one exhibits fivefold (and icosahedral) symmetry. None of the four-dimensional spherical (or Euclidean) honeycombs have fivefold symmetry, so there are no honeycombs with fivefold symmetry in dimension five or greater. In fact, there are no regular hyperbolic honeycombs at all in dimension five or greater. See Sommerville, Ref. 7, pp. 186–191, for the enumeration of all the regular honeycombs.

6. Summary

The main goal of this contribution is to point out that fivefold symmetry is not special among the kinds of hyperbolic n-fold symmetry, as it is in the Euclidean case. We have given examples of hyperbolic fivefold symmetry in two, three, and four dimensions and have shown the relations between repeating patterns of hyperbolic, Euclidean, and spherical geometry from the point of view of n-fold symmetries.

References

1. H.S.M. Coxeter, *Introduction to Geometry*, 2nd Edn. Wiley, New York (1969).
2. B. Grünbaum and G.C. Shephard, Tilings, patterns, fabrics and related topics, *Jahresb. Deutschen Mathematiker-Vereinigung* **85** (1983) 1–32.
3. D. Dunham, Hyperbolic symmetry, *Computers & Mathematics with Applications (1, 2),* **12B** (1986) 139–153; and *Symmetry: Unifying Human Understanding*, I. Hargittai, ed., Pergamon Press, New York (1986).
4. D. Dunham, Creating hyperbolic Escher patterns, in *M.C. Escher: Art and Science* (pp. 241–248, 398), H.S.M. Coxeter et al., eds., North-Holland, Amsterdam (1986).
5. H.S.M. Coxeter and W.O.J. Moser, *Generators and Relations for Discrete Groups*, 4th Edn. Springer-Verlag, New York (1980).
6. H.S.M. Coxeter, *Twisted Honeycombs*, Regional Conference Series in Mathematics, No. 4. American Mathematical Society, Providence, RI (1970).
7. D.M.Y. Sommerville, *An Introduction to the Geometry of N Dimensions.* Dover, New York (1958).

THE PENTASNOW GASKET AND ITS FRACTAL DIMENSION

Robert Dixon

1. Dissecting the Regular Pentagon

To begin with, we observe something of the geometry of the regular pentagon and its dissection.[1] The length of the diagonal of a regular pentagon is τ times the side length, where τ is the the golden ratio 1.61803. . . .[2] Two diagonals drawn from a common vertex dissect the regular pentagon into three isosceles triangles: one acute golden triangle and two obtuse golden triangles (so named because of the ratios of their sides, as shown in Fig. 1).

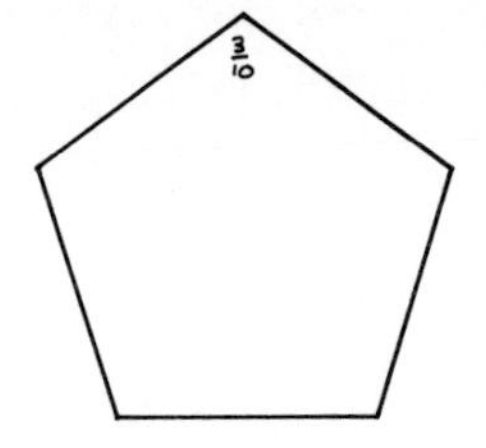

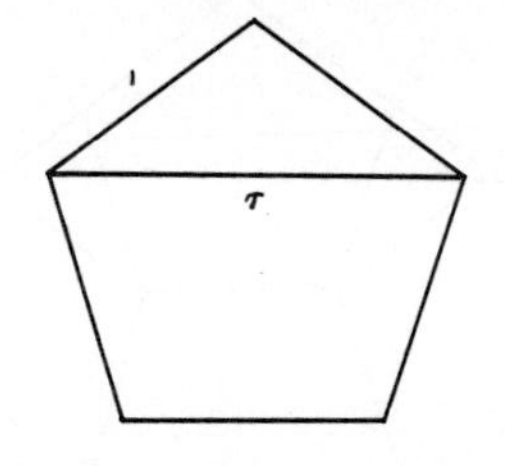

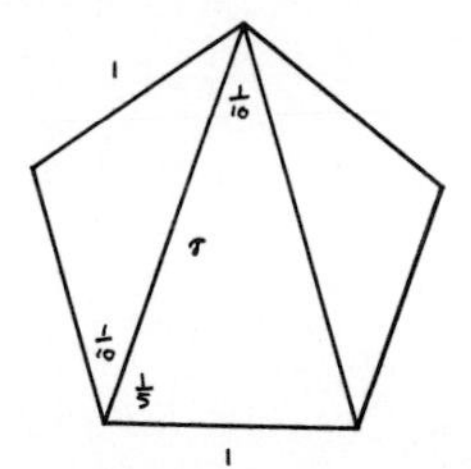

Fig. 1.

The golden ratio has the property that $\tau : 1 = 1 : \tau - 1$, so that either golden triangle may be dissected into two smaller golden triangles, one acute and one obtuse in each case. For this reason golden triangles can be recursively dissected into endlessly smaller golden triangles (Fig. 2).

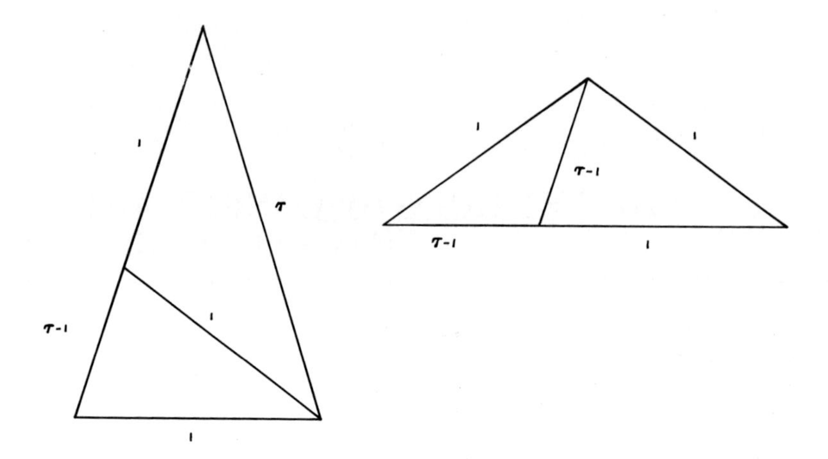

Fig. 2a–b.

A regular pentagon can be dissected into six smaller regular pentagons plus five acute golden triangles, either golden triangle can be dissected into a mixture of one regular pentagon together with two or three acute golden triangles, depending on whether the dissected golden triangle is obtuse or acute, respectively (Fig. 3). For this reason, regular pentagons and golden triangles may be endlessly dissected into smaller and smaller

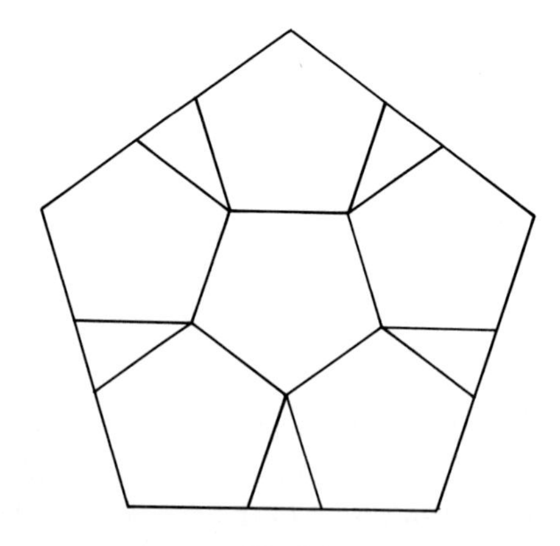

Fig. 3a

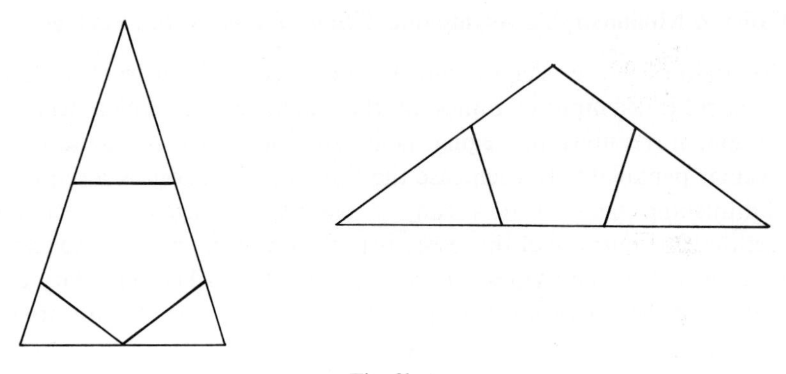

Fig. 3b–c.

regular pentagons and golden triangles. To this list of shapes that can be recursively dissected into similar shapes, we can also add the regular pentagram and the golden rhombi.

2. Definitions

An "asymptotic"[3] tiling of the plane is one in which an infinite sequence of tiles of diminishing sizes are required to fill a finite area. Note that the plane may be filled in this manner with virtually any shape of tile, or mixture of shapes. Such a tiling is "chaotic" however, if no pattern of tiling sequence can be formulated. Tilings that are not chaotic are "ordered." Tilings whose tiles are all of the same shape are "monomorphic."

3. An Ordered Monomorphic Asymptotic Tiling by the Circle

Mandelbrot's *Apollonian set*[4] is a well-known example of an ordered monomorphic asymptotic tiling of the plane. The tile shape in this case is the circle. Starting with three touching circular tiles, the gap between them is partly filled by the circle inside them which touches all three (the problem of Apollonius). This leaves three smaller gaps each to be partly filled in a similar manner, and so on ad infinitum. The process of successive tiling in the Apollonian set leaves an ever-decreasing area of untiled plane whose limiting set of infinitely many connected points has zero area and a fractal dimension estimated to be about 1.3058. This limiting set is called the *Apollonian gasket*.[4]

4. Ordered Monomorphic Asymptotic Tilings Based on the Pentagon

The following tile shapes can all be used to construct ordered monomorphic asymptotic tilings of the plane: either golden triangle, either golden rhombus, the regular pentagon, the regular pentagram, and the regular pentaflake. In each case the tiling construction by a sequence of diminishing sizes of tiles can be readily envisaged by recursive dissection. The purpose of this essay is to illustrate the most complicated case in our list of tile shapes: the regular pentaflake. The boundary of a regular pentaflake (fractal pentagram) is a Koch curve[4] with fractal dimension $D = \log(4)/\log(\tau^2) = 1.44\ldots$ (Fig. 4).

Fig. 4.

5. An Ordered Monomorphic Asymptotic Tiling by Regular Pentaflakes

We shall proceed to explain the tiling construction in, as it were, reverse. That is, we show how to construct the limiting set of untiled points. These points are infinitely many, they are all connected, and the set has zero area. The procedure is as follows:

1. Remove five acute golden triangles from a regular pentagon so that six smaller regular pentagons remain, which are connected and τ^2 times smaller than the initial regular pentagon.
2. Repeat this process on all subsequent pentagons, producing a "paper doily" whose area is reduced at each stage by a factor of $6/\tau^4 = 0.875388\ldots$
3. After three such steps of biting triangles out of the pentagons notice the emergence of a bite whose shape I have called a "flying pig."[3] By filling its belly with a golden rhombus, the bite becomes two touching pentagrams (Figs. 5 and 6).
4. Continue this process ad infinitum to obtain an ordered monomorphic asymptotic tiling of the plane by pentaflakes, called *pentasnow*.[3] In our

Fig. 5.

Fig. 6.

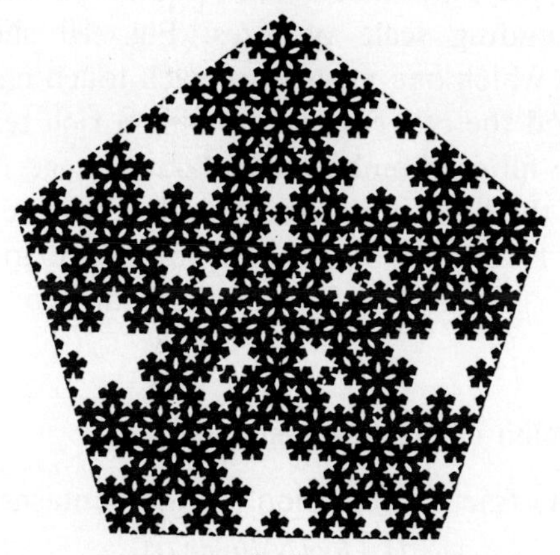

Fig. 7.

diagrams the pentaflake tiles are formed by the white spaces bitten out of the black pentagons. Notice that all pentaflakes occur in isolated touching pairs (Figs. 7 and 8).

Fig. 8.

Figs. 9 and 10 show the process carried to the sixth stage, which mark the limit of graphics resolution available at the time of drawing. The pentagons are still visible to the eye, and it would be nice to meet the artistic challenge of illustrating the tiling process several stages further. The impending blank canvas could be avoided in colour graphics by colouring the pentaflakes of different sizes in a sequence of even steps of colour or tone, or the pentaflake could be coloured alternately black and white in a descending scale of sizes. Fig. 10 shows a sixth-level approximation in which one member of each touching pentaflake pair is coloured white and the other black, yielding a rich texture.

Notice that the filling rhombus described in stage 3 alters the rate at which the area of the doily diminishes in this recursive process, but there is still more than 1/10 lost at each stage, so that the limiting set has zero area. We call this limit set of infinitely many connected points the *pentasnow gasket.*

6. Fractal Dimension of the Pentasnow Gasket

We calculate the fractal dimension[4] of the pentasnow gasket by

$$D = log(N)/log(1/r).$$

Taking $1/r = \tau^2$, we calculate

$$N = 6 + \frac{10}{6^3}\sum_{i=0}^{\infty}\frac{b_i}{6^i},$$

Fig. 9

Fig. 10.

where $b_{i+1} = 2b_i + a_{i+1}$ and $b_0 = 1$ and

$$a_i = \sum_{x=0}^{i} \text{INT}(2^{i-3x}) .$$

The numbers in this calculation have a pleasing pattern:

i		a	b
0	1	1	1
1	2	2	4
2	4	4	12
3	8 + 1	9	33
4	16 + 2	18	84
5	32 + 4	36	204
6	64 + 8 + 1	73	481
7	128 + 16 + 2	146	1108
8	256 + 32 + 4	292	2508
9	512 + 64 + 8 + 1	585	5601
etc.			

The result thus obtained is $D = 1.87968\ldots$.

7. Symmetry

Classical symmetry is defined as invariance of form under one or more congruent transformations (isometry). The concept of symmetry may be generalised in a number of ways, by maintaining the central idea of invariance while extending the scope for permissible transformations. Thus, transformations may be considered which are neither isometries nor one to one. The pattern of pentaflake tiling illustrates such a possibility. Pentasnow is self-similar in a many-to-one manner. The whole is similar to any one of infinitely many different parts by a scaling factor of τ^{2n}, where n is some integer in each case.

The procedure described for constructing the pentaflake tiling and used to calculate the fractal dimension of the pentasnow gasket tells us that *pentasnow is invariant under an infinite sum of dilatations.*

The construction of pentasnow tiling we have described so far involves endless inward extension to obtain a pentasnow tiling of a regular pentagon. But we can just as easily extend the pattern endlessly outward to obtain a tiling of the entire plane. Notice that this procedure yields a pattern with a centre of fivefold mirror symmetry and scaling one-to-one self-similarity. In a pentasnow tiling of the entire plane, however, this centre becomes indeterminate. Only the unobtainable view of the entire plane would distinguish the true centre from any one of infinitely many quasi-centres that occur in an endless hierarchy of scaling sizes.

References

1. A.L. Mackay, De Nive Quinquangula: On the pentagonal snowflake, *Sov. Phys Crystallogr.* **26** (1981) 5; A.L. Mackay, Crystal symmetry, *Physics Bull.* (1976).
2. $$\tau = (1 + \sqrt{5})/2 = 1.61803\ldots,$$

   This is one root of the quadratic equation $1 + \tau = \tau^2$ which arises from the definition of the golden ratio, while the other root is

   $$-1/\tau = -0.61803\ldots = 1 - \tau.$$

   Some writers use ϕ rather than τ to denote the golden ratio. Might we not improve this unhappy state of notation with the useful compromise convention that

   $$\tau = 1.61803389\ldots \text{ and } \phi = 0.61803389\ldots ?$$

   Note, incidentally, that $\tau = 2\cos(1/10) = -2\cos(2/5)$ and that if a "scaling triangle" is defined as one whose side lengths form a geometric progression, than the golden ratio provides the limiting case for scaling triangles.
3. R. Dixon, Pentasnow, *Mathematics Teaching* **110** (1985).
4. B. Mandelbrot, *The Fractal Geometry of Nature*, Freeman (1982).

PENTAGONAL CHAOS

Clifford A. Pickover

1. Introduction

This paper presents an informal survey of several chaotic systems exhibiting pentagonal symmetry. One purpose of this note is to illustrate simple graphics techniques for visualizing a large class of graphically interesting manifestations of chaotic behavior arising from chaotic dynamics. Another goal of this note is to give a flavor of the subject of recurrence relations and chaos.

"A mathematician, like a painter or a poet, is a master of pattern," wrote English mathematician G.H. Hardy. Indeed, well after Hardy's death in 1947, computers with graphics have played a vital role in the visualization and understanding of chaotic phenomena. Today, there are several scientific fields devoted to the study of how complicated behavior can arise in systems from simple rules and how minute changes in the input of a nonlinear system can lead to large differences in the output; such fields include chaos and cellular automata theory.[1]

Chaos theory today usually involves the study of a range of phenomena exhibiting a sensitive dependence on initial conditions.[2] Chaotic behavior in such systems is generally irregular and disorderly — and examples include the weather patterns, some neurological and cardiac activity, and certain electrical networks of computer.[1] Although chaos often seems totally "random" and unpredictable, it actually obeys strict mathematical rules that derive from equations that can be formulated and studied. Some chaotic systems that people investigate are expensive to set up and rather complicated to study.[3] In this paper, simple systems are used from which researchers can easily collect experimental data. Structures produced by the behavior of these systems include shapes of startling intricacy. The graphics experiments presented are good ways to show the complexity of such behavior.[4]

This paper does not attempt to give a detailed historical background to the field of chaos and nonlinear dynamics, even though the field of chaos and the mathematics of iteration have had a rich history. The computer graphical experiments in this section only hint at the important past mathematical work of others from which I have received inspiration. For background, the reader should consult the various books in the reference section, and in particular, the famous books of Gleick,[2] Mandelbrot,[5] and Moon.[16]

2. Background to Iteration Theory

Feedback is a term we often hear today in a variety of settings: for example, amplifier feedback during rock concerts, biofeedback in medicine and psychology, and chemical feedback in the field of biochemistry. Generally, feedback means that the output of a system or machine returns to the input. In electronic amplifiers, feedback can occur if the microphone is placed too close to the speaker. In the world of mathematics, feedback is often produced by an "iteration" or "recursion." By iteration, we mean the repetition of an operation or set of operations. In mathematics, composing a function with itself, such as in $f(f(x))$, can represent an iteration. The computational process of determining x_{i+1} given x_i is called an iteration. Take for example the simple act of squaring a number. Let's presume we have a magic box that squares whatever numbers are fed to it — if we put in 2, out comes 4. Now let's take the output, 4, and feed it to the input. Out comes 16. This self-squaring process is an example of mathematical feedback, and it may be simply described by the following equation: $\zeta \rightarrow \zeta^2$.

This particular squaring feedback seems rather uninteresting. The progress of the iteration is monotonic. Even if we add a constant (which can be designated by the Greek letter μ) to the squaring process, the numbers still follow a rather uninteresting and ever-increasing progression: $\zeta \rightarrow \zeta^2 + \mu$.

Gaston Julia (1893–1978) was one of the first mathematicians to notice that, under certain conditions, this feedback loop produces startling results. These results arise when "complex"[a] z values are used as input (complex numbers are of the form $a + ib$, where $i = \sqrt{-1}$).[b] However, the striking beauty and complexity of "maps" representing such iterative Julia calculations have only recently been explored in detail, due in part to advances in computer graphics. B. Mandelbrot, a mathematician, has extended the theory and graphic presentation of iterated functions as a special class of the new geometry called "fractal" geometry. Fractals represent rough-edged objects or patterns that often appear self-similar; i.e., no matter what scale is used to view the pattern, the magnified portion of the fractal shape looks just like the original pattern.[5]

To familiarize yourself with the appearance of Julia set maps, see Fig. 1. To create this figure, start with an array of complex *values* ζ and have the computer follow the outcome of the process defined by $\zeta \rightarrow \zeta^5 + \mu$ where μ is constant. Once the initial points are selected, each iteration represents a step along a path that hops from one complex number ζ to the next. The collection of all such points along a path constitutes an orbit. The basic goal is to understand the ultimate fate of all orbits for a given system. For example, for certain initial ζ values ζ^5 equation produces larger and larger values; i.e., the function explodes or diverges. For other values, it does not explode (it is bounded). This behavior can be characterized by computer graphics. (See Mandelbrot's book[5] for fascinating computer graphic representations of Julia sets. Also see Brooks and Matelski[21] for some early graphic representation of Julia and Mandelbrot sets.)

[a] If real numbers are taken to represent points on a line, then complex numbers can represent points in a plane. Complex numbers are often represented by the symbols z or ζ.
[b] A note on complex numbers: If you were asked to find an x such that $x^2 + 1 = 0$ you would quickly realize that there was no real solution. This fact led early mathematicians to consider solutions involving the square root of negative numbers. Heron of Alexandria (c. A.D. 100) was probably the first to formally present a square root of a negative number as a solution to a problem (for trivia aficionados, it was: $\sqrt{-63}$). These numbers were considered quite meaningless and hence the term "imaginary" was used. Today, imaginary numbers are indispensable to several branches of mathematics and physics. Carl Friedrich Gauss (1832) coined the word "complex" to describe numbers with a real and an imaginary component.

3. Julia Sets

The previous section gave the reader an introduction to iteration and Julia sets. This section presents similar information in a slightly more mathematical fashion and also introduces another iterative system known as Halley's method.

Julia sets are usually defined as the set of all points that do not converge to a fixed point or finite attracting orbit under repeated applications of a map. Most Julia sets are fractals, displaying an endless cascade of repeated detail. As an alternate definition for Julia sets, consider the repeated applications of a function f that determines a trajectory of successive locations $\zeta, f(\zeta), f(f(\zeta)), f(f(f(\zeta))), \ldots$ visited by a starting point ζ in the complex plane. Depending on the starting point, this results in two types of trajectories, those which go to infinity and those which remain bounded by a fixed radius. The Julia set of the function f is the boundary curve which separates these regions.

3.1 *Maps of Formula/zeta rarrow zeta sup 5 + mu/*

The Julia sets for the iterative process $\zeta \rightarrow \zeta^2 + \mu$, $\mu \in \mathbb{C}$, $\zeta \in \mathbb{C}$ have been widely investigated.[5] Trigonometric relations have also been studied in the past.[6,7] I wish to consider the *iteration* of a function ψ for a complex z plus a complex constant, μ:

$$f(\zeta): \zeta \rightarrow \psi(z) + \mu \,. \tag{1}$$

Insight into the complexity of this nonlinear system may be gained from experimentation on the computer. My goal is to describe the behavior of points under recursion of f. For each selected initial point, ζ_0, the function $f(\zeta)$ is iterated

$$\zeta_n = f(\zeta_{n-1}, \mu);\ n = 1,2,3 \ldots, \infty. \tag{2}$$

For certain values of ζ_0, the sequence ζ_n may diverge (grow increasingly large), and for others the iteration's behavior is bounded. Regions which are stable (do not grow large) are differentiated by regions in the ζ plane which do not explode upon iteration by black and white coloration. In this paper, ψ is simply the function ζ^5 and therefore

$$\zeta_{n+1} = \zeta_n^5 + \mu \tag{3}$$

describes Eq. (1). Lakhtakia and others have observed that, in general if $\zeta_{n+1} = \zeta_n^p + \mu$ is used to generate the Julia set, then the Julia set will have p-fold symmetry.[6,8] A copy of any identifiable structure can be readily obtained through rotation of the Julia set by an angle $2\pi/p$.

Fig. 1 presents a Julia set for Eq. (3). As is true for most Julia sets, the shapes are fractal, and repeated magnification of various parts of the figure reveal increasing details. To produce Fig. 1, the criterion of boundedness allows traditionally divergent points in the ζ plane to be bounded if either the real *or* imaginary component of ζ is small after many iterations:

$$|\zeta_{real}| < \tau \text{ } or \text{ } |\zeta_{imag}| < \tau,$$

where $\tau = 10$. The application of this criterion gives rise to the network of triangular-shaped objects seen in each of the figures. This is useful for artistic regions, but also note that these objects point the reader's eye to regions of stability. The small irregularly-shaped regions correspond to those ζ which explode very slowly or whose trajectories are bounded.

Fig. 1. Julia set representation for $\zeta \rightarrow \zeta^5 + \mu$.

3.2 *Halley Maps for Formula /ς⁵ – 1 = 0/*

Fig. 2 is a Halley map, which can be computed as follows. Let $f(\zeta)$ be a complex-valued function of the complex variable ζ. The *Halley map* is the function

$$H(\zeta): \zeta_{n+1} = \zeta_n - \left[\frac{F(\zeta n)}{F'(\zeta n) - \left(\dfrac{F''(\zeta n) F(\zeta n)}{2F'(\zeta n)} \right)} \right]. \tag{4}$$

Halley's method is used to solve equations of the form $f(\zeta) = 0$. The problem of finding the zeros of a continuous function by iterative methods occurs frequently in science and engineering. These approximation techniques start with a guess and successively improve upon it with a repetition of similar steps. Fig. 2 gives an indication of how well one of these iterative methods, Halley's method, works in order to gain insight

Fig. 2. Halley map for $\zeta = \zeta^5 - 1$.

as to where Halley's method can be relied upon and where it behaves strangely. In particular, Fig. 2 indicates how swiftly the method converges to the five roots of the equation $\zeta^5 - 1 = 0$ by displaying the number of iterations needed for convergence as contour levels. The basins of attractions for the roots of the equation are displayed for various initial values of (ζ_0) in the complex plane (between -2.5 and 2.5 in the real and imaginary directions). The five central white regions contain the roots and correspond to starting points where convergence is achieved rapidly (within three iterations). Initial "guesses" in the tear-shaped basins fanning out from the roots are "safe"; that is, any starting points selected from these regions come close to a root within a small number of iterations. Regions between roots converge much more slowly, and behavior on the radial boundary region is considerably more complicated. These borders consist of elaborate swirls that can pull Halley's method into any one of the five roots. In this vicinity, a tiny shift in starting point can lead to widely divergent results.

The reader may wish to experiment with artistic graphics diagrams for the overrelaxed *Halley map*, which can be defined by the function

$$H(\zeta): \zeta_{n+1} = \zeta_n - \lambda \left[\frac{F(\zeta_n)}{F'(\zeta_n) - \left(\dfrac{F''(\zeta_n) F(\zeta_n)}{2F'(\zeta_n)} \right)} \right]. \tag{5}$$

The coefficient λ in the modified Halley's method is known as a *relaxation coefficient*, and can be greater than 1 in the *overrelaxed* case in order to speed the rate of convergence in some problems.[9] For graphic and mathematical work on a related technique called Newton's method, see Ref. 10.

4. Dynamical Systems with Pentagonal Symmetry

Another deep reservoir for striking images is the *dynamical system*. Dynamical systems are models containing the rules describing the way some quantity undergoes a change through time. For example, the motion of planets about the sun can be modelled as a dynamical system in which the planets move according to Newton's laws. Generally, the pictures presented in this section track the behavior of differential equations. Just as one can track the path of a jet by the smoke path it leaves behind, computer graphics provides a way to follow paths of particles whose motion is determined by simple differential equations. The practical side of dynamical systems is that they can sometimes be used to describe the behavior of real-world phenomena such as planetary motion, fluid flow, the diffusion of drugs, the behavior of inter-industry relationships, and the vibration of airplane wings.

Fig. 3 is a map created by a trajectory governed by three simple functions of x, y and z which I have found particularly useful in demonstrating chaotic dynamical systems[11]:

$$x_{n+1} = \sin \alpha y_n - z_n \cos \beta x_n \tag{6}$$
$$y_{n+1} = z_n \sin \gamma x_n - \cos \delta y_n \tag{7}$$
$$z_{n+1} = \phi \sin x_n \tag{8}$$

($x \in \mathbb{R}$, $Y \in \mathbb{R}$, $z \in \mathbb{R}$). An initial point (0,0,0) is selected and the equations are iterated 5 million times. Fig. 3 is an x-y projection of the resulting

Fig. 3. 3-D strange attractor. This pattern, as well as the patterns for Figs. 4, 5 and 6, represents the trajectory of a single "seed" particle as it moves on a plane. The equations are iterated one million times.

points' trajectory. The darkness of a region of the graph relates to how many times a trajectory crosses a particular pixel in the display. Note that if the dynamical system producing Fig. 3 leads to totally random output, then the plot would be a diffuse random scattering of points in 3-space. If the system were absolutely periodic (like a sine wave), then the figure would consist of a thin curve in 3-space. This curve is delicately poised somewhere between the two extremes and has a potential infinity of values. See Ref. 11 and the Appendix of this paper for precise computational recipes for generating this shape.

In contrast to the asymmetrical chaotic attractor in Fig. 3, in this section, I present symmetrical dynamical systems generated by the Chossat-Golubitsky formula:

$$f(\zeta,\lambda) = (\alpha u + \beta v + \lambda)\zeta + \gamma\bar{\zeta}^{m-1} \tag{9}$$

where

$$u = \zeta\bar{\zeta} \text{ and } v = (\zeta^m + \bar{\zeta}^m)/2. \tag{10}$$

For those readers knowledgeable in group theory, the mapping $f: V \to V$ is equivariant with respect to the group Γ acting on V since $f(\gamma v) = \gamma f(v)$.[12] Figs. 4, 5 and 6 were generated with $m = 5$. The value of λ can be considered a bifurcation parameter, and low values of λ generally correspond to smaller sized attractors having less symmetry than the figures shown here. Figs. 4, 5 and 6 were generated by following the discrete dynamics of

$$f(\zeta,\lambda) = (\alpha u + \beta v + \lambda)\zeta + \gamma\bar{\zeta}^{m-1} + \phi, \tag{11}$$

where ϕ is a symmetry-breaking term which I have introduced and have found to produce interesting dynamics.

5. Conclusions

Among the methods available for the characterization of complicated mathematical and physical phenomena, computers with graphics are emerging as an important tool.[13,14,15] In the present paper, several approaches used to generate pentagonal chaotic structures were described. Julia sets for $\zeta \to \zeta^5 + \mu$, Halley maps for $\zeta^5 - 1 = 0$, and dynamical systems for the Chossat-Golubitsky formula were presented, each with chaotic yet fivefold symmetrical behavior. For the Julia sets of

Fig. 4. Pentagonal chaotic attractor for $f(\zeta,\lambda) = (\alpha u + \beta v + \lambda)\zeta + \gamma\bar{\zeta}^{m-1} + \phi(\gamma = 1, \lambda = -2.6, m = 9, \alpha = 4, \beta = 2, \phi = 0)$. Other parameters yield additional beautiful patterns.

Fig. 5. Same as Fig. 4 except $\phi = 0.1 + 0.1i$. Notice that ϕ breaks the symmetry of the pattern in Fig. 4.

Fig. 6. Same as Fig. 4 except ($\gamma = .5, \lambda = -1.804, m = 3, \alpha = 1, \beta = 0, \phi = 0$).

$\zeta \rightarrow \zeta^5 + \mu$ and the Halley maps of the equation, $\zeta^5 - 1 = 0$, the systems become irregular in well-defined regions. The chaotic portions of these maps, while exhibiting complicated behavior, are composed of various underlying self-similar structures.

The beauty and complexity of these drawings correspond to behavior which no one could fully have appreciated or suspected before the age of the computer. This complexity makes it difficult to objectively characterize structures such as these, and therefore, it is useful to develop graphics systems which allow the maps to be followed in a qualitative and quantitative way. My computer graphics systems with a joystick-driven cursor allow the researcher to interactively magnify chaotic maps.

The richness of resultant forms contrasts with the simplicity of the generating formulas. Apart from their curious mathematical properties and "artistic" appeal, maps with chaotic dynamics and fractal characteristics now have an immense attraction because of the role they may play in understanding the properties of numerical methods and also possibly in shedding insight into heart failure, meteorology, economics, population biology, neural networks, arrays of parallel processors, noisy Josephson junctions, and leukemia (see Ref. 1 for a review). As May[14] points out, the fact that simple, deterministic equations (such as those presented here) can possess trajectories which look like random noise has disturbing practical implications. For example, apparently erratic fluctuations in an animal population may simply derive from a rigidly deterministic population growth relationship.

A report such as this can be viewed as introductory; however, it is hoped that the techniques, equations and system presented will provide a useful tool and stimulate future studies in the graphic characterization of the morphologically rich structures of extreme complexity produced by the iteration of simple transforms. As David Ruelle, chaos specialist, has stressed: "There is a whole world of forms still to be explored and harmonies to be discovered."[1]

6. Directed Reading List

The Julia set fractals in this paper were generated using iterative methods which were first introduced by G. Julia and P. Fatou around 1918.[17,18] However, this field remained somewhat dormant until IBM Fellow Benoit Mandelbrot revealed the striking beauty and intricacy of these shapes in the complex plane (see Mandelbrot's book *The Fractal Geometry of Nature*).[5]

In 1980, Mandelbrot described another self-similar fractal which consists of all values of μ that have connected Julia sets, and John Hubbard subsequently named this fractal the Mandelbrot set. Mathematicians such as J. Hubbard, J. Milnor and H.-O. Peitgen have explored this set's intricacies, developing and proving mathematical conjectures in the course of their computer-aided explorations.[19,20] Several beautiful computer graphics renditions of the Mandelbrot set were printed in *Scientific American* in August 1985.

There are many other references which may be mentioned. For example, Brooks and Matelski presented some early Julia and Mandelbrot set computer graphics in the Proceedings of the 1978 Stony Brook Conference.[21] Robert Devaney of Boston University has studied the dynamical behavior of Julia sets for certain trigonometric functions.[22] Since the time of Brooks and Matelski, many excellent papers and books on the subject have been published, and these are listed in the reference section of the present book. Some of these interesting books are singled out in the following list:

1. H. Peitgen and P. Richter, *The Beauty of Fractals*. Springer, Berlin (1986).
2. M. Barnsley, *Fractals Everywhere*. Academic Press, New York (1988).
3. J. Feder, *Fractals*. Plenum, New York (1988).
4. H. Peitgen and D. Saupe, eds. *The Science of Fractal Images*. Springer, Berlin (1988).

Appendix

To encourage reader involvement, the following pseudocode is given.

ALGORITHM: Complex Iteration Map Generator
The Julia set in Fig. 1 was created using this algorithm.

Variables: *rz*, *iz* = real, imaginary component of complex number
i = iteration counter
u, *z* = complex numbers
u = −.3 + .8 *i*;

```
DO rrz = -4 to 4 by 0.08; /* real axis divided into 800 pixels */
 DO iiz = -4 to 4 by 0.08; /* imag axis divided into 800 pixels */
  z = cplx(rrz, iiz);          /* cplx returns a complex number */
  Innerloop: DO i = 1 to 10;     /* iteration loop          */
   z = z**5 + u;               /* main computation          */
   /* convert to real and imag component */
   rz = real (z); iz = imag (z);
   /* determine if magnitude is above threshold value of 10     */
   if sqrt (rz**2 + iz**2) > 10 then leave innerloop;
  End;              /* InnerLoop              */
  color = i;              /* assign color index based on i */
  /* Plot a colored point using either of two test options       */
  if Convergence Test = a then PrintDotAt (rrz, iiz, color);
  /* Use the test below to create Fig. 1          */
  if Convergence Test = b, then
  if abs (rz) < 10 OR abs (iz) < 10, then PrintDotAt (rrz, iiz, color);
  END;              /* iz loop            */
 END;               /* rz loop            */
```

ALGORITHM: 3-D Strange Attractor Generator
The shape in Fig. 3 can be generated using the approach outlined here.

TYPICAL PARAMETER VALUES:

```
xxmin = - 2; xxmax = 2, yymin = 2, yymax = 2 (*picture boundaries *)
pres   = 1600                  (* picture resolution *)
iter 1 = 1000; iter2 = 5000;
(* iter1*iter2 = total number of iterations *)
```

METHOD: A 5-parameter Dynamical System
OUTPUT: Pixel array containing the output picture intensities.
NOTES: Try experimenting with different values of *e* which can control the degree of randomness of the system.

```
xinc = pres/(xxmax - xxmin);                      (* controls x-pixels position *)
yinc = pres/(yymax - yymin);                      (* controls y-pixel position *)
a = 2.24;b = .43;c = - .65;d = - 2.43;e = 1;(*control parameters              *)
p(*,*) = 0;                                       (* initialize p array         *)
x,y,z = 0;                                        (* starting point             *)
do j = 1 to iter1;
  do i = 1 to iter2;
    xx = sin(a*y) - z*cos (b*x);
    yy = z*sin(c*x) - cos (d*y);
    zz = e*sin(x);
    x = xx; y = yy; z = zz;
    if xx<xxmax & xx>xxmin & yy<yymax & yy>yymin then do;
     xxx = (xx - xxmin)*xinc;          (* scale to range (0, pres) *)
     yyy = (yy - yymin)*yinc;          (* scale to range (0, pres) *)
     p(xxx,yyy) = p(xxx,yyy) + 1;
    end; /* then do */
  end;        (* i          *)
end;          (* j          *)
(*P now contains the intensities for each pixel in the picture*)
```

References

1. C. Pickover, *Computers, Pattern, Chaos and Beauty*. St. Martin's Press, New

York (1990); C. Pickover, *Computers and the Imagination.* St. Martin's Press, New York (1991).
2. J. Gleick, *Chaos: Making a New Science.* Viking, New York (1987).
3. I. Peterson. Toying with a touch of chaos, *Science News.* **129** (1986) 277–278.
4. C. Pickover, The world of chaos, *Computers in Physics* **4(5)** (1990) 460–470.
5. B. Mandelbrot, *The Fractal Geometry of Nature.* Freeman and Company, New York (1983).
6. C. Pickover and E. Khorasani, Computer graphics generated from the iteration of algebraic transformations in the complex plane, *Computers and Graphics* **9** (1985) 147–151.
7. C. Pickover, Chaotic behavior of the transcendental mapping ($z \rightarrow \cosh(z) + \mu$), *The Visual Computer, An International Journal of Computer Graphics* **4** (1988) 243–246.
8. A. Lakhtakia, V. Vasundra, R. Messier and V. Varandan, On the symmetries of the Julia sets for the process $z \rightarrow z^p + c$, *J. Phys. A Math. Gen.* **20** (1987) 3533–3535.
9. C. Pickover, Overrelaxation and chaos, *Physics Letter* **A30(3)** (1988) 125–128; C. Pickover, A note on chaos and Halley's method, *Communications of the ACM* **31(11)** (1988) 1326–1329.
10. H. Benzinger, S. Burns and J. Palmores, Chaotic complex dynamics and Newton's method, *Physics Letters* **A119** (1987) 441–445.
11. C. Pickover, A note on rendering 3-D strange-attractors, *Computers and Graphics* **12(2)** (1988) 263–267.
12. P. Chossat and M. Golubitsky, Symmetry-increasing bifurcations of chaotic attractors, *Physica* **D32** (1988) 423–426.
13. D. Campbell, J. Crutchfield, D. Farmer and E. Jen, Experimental mathematics: the role of computation in nonlinear science, *Commun. ACM* **28** (1985) 374–389.
14. R. May, Simple mathematical models with very complicated dynamics, *Nature* **261** (1976) 459–467.
15. D. Hofstadter, Strange attractors, *Scien. Amer* **245** (1981) 16–29.
16. F. Moon, *Chaotic Vibrations.* John Wiley and Sons, New York (1987).
17. G. Julia, Memoire sur L'iteration des fonctions rationnelles, *J. Math. Pure Appl.* **4** (1918) 47–245.
18. P. Fatou, Sur les equations fonctionelles, *Bull Soc. Math.* **Fr. 47** (1919/1920) 161–271.
19. J. Hubbard, Order in chaos, *Engineering: Cornell Quarterly* **20(3)** (1986) 20–26.
20. A. Douady and J. Hubbard, Iteration des polynomes quadratiques complexes, *Comptes Rendus* (Paris) **2941** (1982) 123–126.
21. R. Brooks and J.P. Matelski, The dynamics of 2-generator subgroups of PSL(2,C), in *Riemann Surfaces and Related Topics: Proceedings of the 1978 Stony Brook Conference.* I. Kyra and B. Maskit, eds. Princeton University Press, New Jersey (1981).
22. R. Devaney and M. Krych, Dynamics of exp(z), *Ergod Th. & Dynam Sys.* **4** (1984) 35–52.

HOW TO INSCRIBE A DODECAHEDRON IN A SPHERE

J. Chris Fisher and Norma Fuller

Euclid's "ruler and compass" construction of a regular dodecahedron inscribed in a sphere (XIII.17) is essentially a two-step procedure:

1. Determine the relative lengths of the sphere's radius and of a face's edge and diagonal.
2. Construct those lengths.

Step 1 is an exercise in trigonometry. One computes (as in Ref. 1, Table 1 on pp. 290 and 293):

the *edge length* = 1

the *diagonal of a face* = τ (where τ is the golden section $\frac{\sqrt{5}+1}{2}$)

the *radius of the circumsphere* = $\frac{\sqrt{3}\tau}{2}$.

Our objective is to demonstrate that step 2 is as easy as inscribing an equilateral triangle in a circle. Indeed, all three lengths appear together in a result of Odom[3] concerning an equilateral triangle and its circumcircle (Fig. 1).

Theorem

$|OC| = 1, |EG| = \tau$, and $|EF| = \frac{\sqrt{3}\tau}{2}$ are, respectively, the edge length, the diagonal of a face, and the circumradius of a regular dodecahedron.

γ is the unit circle with centre O,
ΔABC is an equilateral triangle inscribed in γ,
D is the midpoint of AB,
E is the midpoint of AC,
$F = DE \cap \gamma$,
l is the line perpendicular to EF at F,
$G = l \cap EB$.

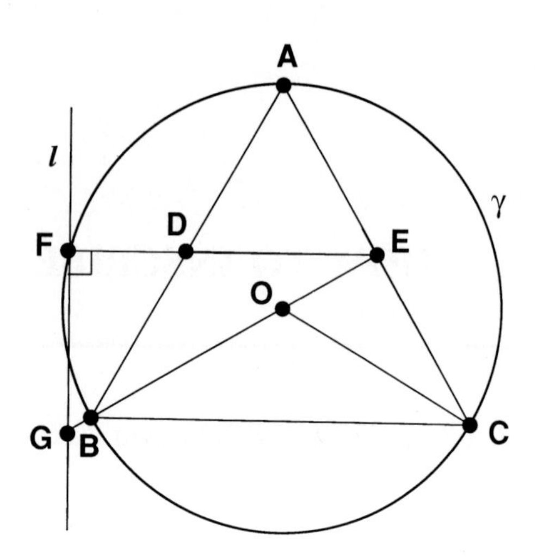

Fig. 1. The construction.

Proof

Problem E3007 in Ref. 3 required showing that $|DE|:|DF| = \tau : 1$. The published solution by Jan van de Craats was a proof without words, with an appropriately labelled figure above the equation $\tau^2 = \tau + 1$. (This equation is the fundamental property of the golden section as explained, for example, by Coxeter, in Ref. 2, p. 161.) We extend his figure to our Fig. 2 using the trigonometry of the equilateral triangle. Again, no words are required.

Remarks

George Odom is an amateur mathematician who sent the basic idea of his lovely discovery to Professor Coxeter in 1982. The latter improved its presentation and encouraged its appearance as a problem in the *American Mathematical Monthly*.[3]

The figures were created using LEGO (LISP-based Euclidean Geometry Operations),[4] an interactive computer graphics system based on Euclidean notions and constructions. LEGO provides the facilities for creating, manipulating and viewing two- and three-dimensional figures. The figures were drawn using ruler and compass techniques, but with greater speed and with considerable accuracy.

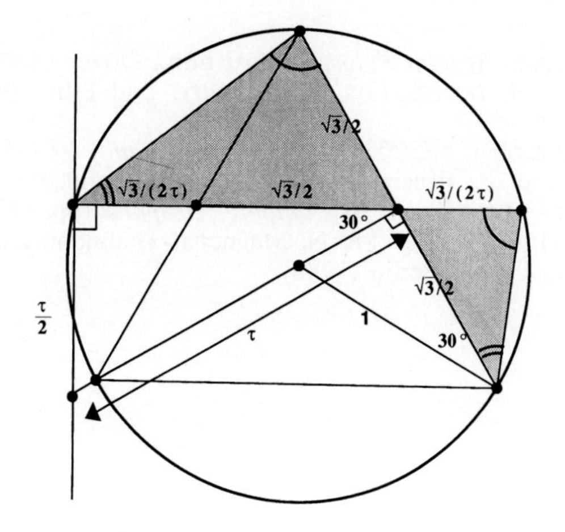

Fig. 2. $\frac{\sqrt{3}\tau}{2} = \frac{\sqrt{3}}{2\tau} + \frac{\sqrt{3}}{2}$.

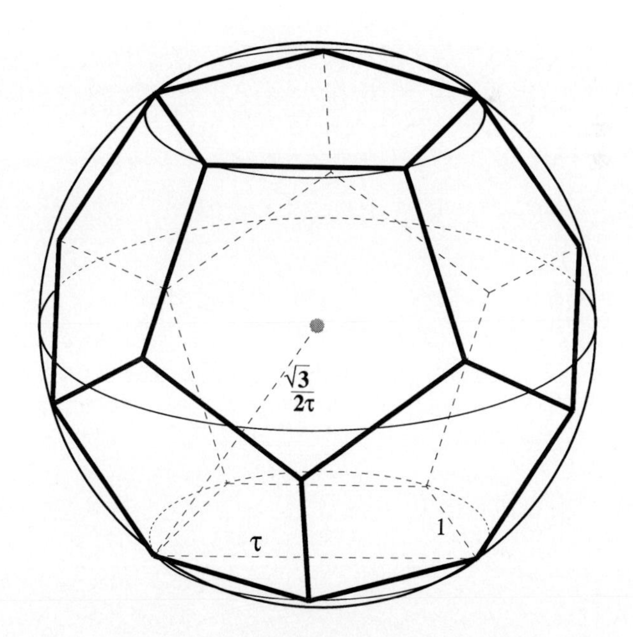

Fig. 3. A dodecahedron of unit edge-length inscribed in a sphere.

References

1. H.S.M. Coxeter, *Regular Polotypes*, 3rd Edn., Dover, New York (1973).
2. H.S.M. Coxeter, *Introduction to Geometry*, 2nd Edn., Wiley, New York (1969).
3. G. Odom, Problem #E3007, *Amer. Math. Monthly* **94** (1986) 572.
4. N. Fuller and P. Prusinkiewicz, Geometric modeling with Euclidean constructions, *New Trends in Computer Graphics* (pp. 379–392). Proceedings of CG International '88, N. Magnenat-Thalmann and D. Thalmann, eds. Springer-Verlag, Berlin (1988)

COMPLETE SYMMETRY OF FIGURES WITH FIVEFOLD SYMMETRY AXES

Ivan S. Zheludev

Complete symmetry,[1] unlike Shubnikov antisymmetry,[2] depicts figures not with the aid of "tetrahedra of general form with symmetry 1" but with spheres with and without a center of symmetry. The former spheres are, in fact, scalars (positive white, +, and negative black, −), whereas the latter are pseudoscalars of two signs of enantiomorphism (positive, left, and negative, right). Such an interpretation of "elementary bricks" in complete symmetry allows it to be successfully used in tensor crystallography. Such spheres may possess not only one quality (i.e., be scalars of pseudoscalars), but can also combine them both (i.e., be simultaneously white and left, black and right, white and right, or black and left.

The essential characteristic of spheres in complete symmetry is their inversion symmetry. Scalar spheres are centrosymmetric (the corresponding operation is denoted as $\bar{1} = C$), *and pseudoscalar spheres are anticentrosymmetric* (*the operation is denoted as* $\underline{\bar{1}} = \underline{C}$). U*nlike antisymmetry, operation* $\underline{\bar{1}}$ *"recolors" a scalar and transforms the right* figure *into the right, the left* figure *into the left one, whereas operation* $\underline{1}$ (*antiidentity*) *changes the sign of both scalar and pseudoscalar quantities* ($\underline{\bar{1}}.\bar{1} = \underline{1}$). I*n complete symmetry, a polar vector is represented by two spheres of "opposite" colors and is anticentrosymmetric; an axial vector is represented by two spheres with opposite signs of enantiomorphism* and is centrosymmetric. The geometric images and complete symmetry of scalars and vectors are illustrated in Fig. 1.

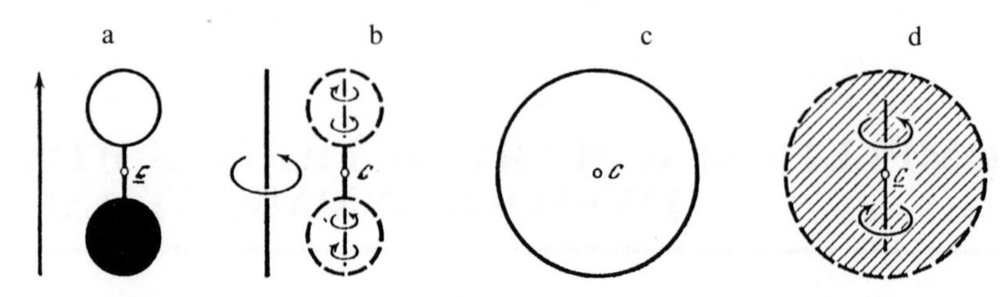

Fig. 1. Geometric images of scalars and vectors and their complete symmetry: (a) a polar vector and two scalar spheres of "opposite" colors, the complete symmetry $\infty/\underline{m}mm$; (b) an axial vector and two pseudoscalar spheres with the opposite signs of enantiomorphism, symmetry $\infty/m\underline{mm}$; (c) a scalar sphere (operation $\bar{1}$ for such a sphere is a symmetry operation; operation $\underline{\bar{1}}$ transforms this sphere into a sphere of the "opposite" color); (d) a pseudoscalar sphere (operation $\underline{\bar{1}}$ for such a sphere is a symmetry operation and operation $\bar{1}$ transforms a left sphere into the right one).

The groups of complete symmetry can be represented graphically with the aid of scalars, pseudoscalars, and polar and axial vectors (Fig. 2, a–f). In some cases such a representation allows the prediction of some physical properties of crystals belonging to some or other symmetry groups. Thus, the crystals described by groups 23 (Figs. 2a,b) may have a piezoeffect, since their space diagonals are polar directions and they have no center of inversion; crystals of classes $\underline{\bar{1}0}\,\underline{m}2$ and $\underline{m}3\underline{m}$ (Figs. 2c, e) may be optically active, etc. Thus, complete symmetry is closely related to physical crystallography.

Altogether, there are 90 groups of complete symmetry for crystals. Of these, 32 coincide with groups of conventional symmetry. They describe scalar one-color crystals; in other words, the geometric image of any of such groups includes a certain scalar component. In a similar sense, of the 58 groups of complete symmetry (90 − 32 = 58), 32 groups are pseudoscalar, i.e., the geometric images include one definite (of one sign) component of a pseudoscalar. Complete symmetry has no "gray" groups in the sense that they exist in antisymmetry, but they have "gray" groups in the sense that either pseudoscalar or scalar components are absent. The latter case is illustrated by Figs. 2d, e.

There are 14 limiting groups of complete symmetry. In addition, there are 21 almost limiting groups, which, in the final analysis, may be

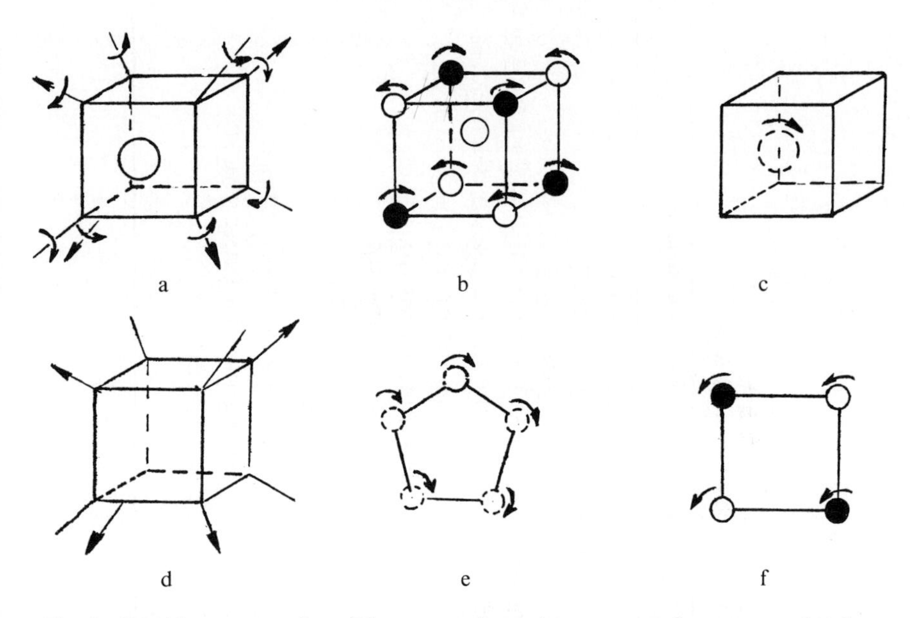

Fig. 2 Graphic representation of the groups of complete symmetry by vectors and scalars: (a) group $\underline{2}3$, (b) group $2\bar{3}$, (c) group $\underline{m}3\underline{m}$, (d) group $\underline{m}3$, (e) group $\underline{10}\,\underline{m}2$, and (f) group $\bar{4}2\underline{m}$. Scalars are depicted by circles with solid contours, pseudoscalars by circles with dashed contours. The sign of a scalar is specified by the color, and that of a pseudoscalar by the direction of circular arrows.

reduced to the limiting groups. Groups describing isometric figures are reduced to 11 limiting groups and those describing other figures are reduced to the remaining 24 groups — 35 groups altogether. Table 1 lists the symmetry groups with fivefold symmetry axes. Of them, 24 are the groups describing figures with one fivefold symmetry axis and three groups describe isometric figures with fivefold axes (eight almost limiting groups describing isometric figures cannot describe figures with fivefold axes). The group notation in Table 1 is analogous to that accepted in the International Tables for X-ray Crystallography. The bar above the group symbol corresponds to an operation of inversion, the bar over symmetry elements denotes mirror reflection. Table 1 also indicates the order of each group. The first seven groups of Table 1 and groups $\bar{1}03m$ and 532 coincide with the conventional symmetry groups.

Among the groups of complete symmetry, special attention is attracted to the groups describing the symmetry of three-dimensional isometric

Table 1 Groups of Complete Symmetry of Figures with Fivefold Symmetry Axes[a]

Numbers	Group Symbol	Group Order	Elements of Complete Symmetry	Numbers	Group Symbol	Group Order	Elements of Complete Symmetry
1	5	5	(1)5	15	$\bar{5}\underline{m}$	20	(1)5, (1)$\bar{10}$, (5)$\underline{2}$, (5)$\underline{m}$, C
2	$\bar{5}$	10	(1)5, (1)$\bar{10}$, C	16	$\underline{\bar{5}}m$	20	(1)5, (1)$\underline{\bar{10}}$, (5)$\underline{2}$, (5)m, $\underline{C}$
3	$\bar{10}$	10	(1)5, (1)$\bar{5}$, m	17	$\underline{\bar{5}}\underline{m}$	20	(1)5, (1)$\underline{\bar{10}}$, (5)2, (5)$\underline{m}$, $\underline{C}$
4	5m	10	(1)5, (5)m	18	$\underline{\bar{10}}/\underline{m}\underline{m}\underline{m}$	40	(1)$\underline{10}$, (1)$\bar{10}$, (1)5, (1)$\underline{\bar{5}}$, (5)2_ , (5)2, (5)m, (5)$\underline{m}$, $\underline{2}$, $\underline{m}$, C
5	52	10	(1)5, (5)2	19	$\underline{10}/mm\underline{m}$	40	(1)$\underline{10}$, (1)$\underline{\bar{10}}$, (1)5, (1)$\bar{5}$, (5)$\underline{2}$, (5)2, (5)m, (5)$\underline{m}$, $\underline{2}$, m, $\underline{C}$
6	$\bar{10}m2$	20	(1)5,(1)$\bar{5}$, (5)m, (5)2, m	20	$\underline{10}/m$	20	(1)$\underline{10}$, (1)$\underline{\bar{10}}$, (1)5, (1)$\bar{5}$, (1)$\underline{2}$, (1)m, $\underline{C}$
7	$\bar{5}m$	20	(1)5, (1)$\bar{10}$, (5)m, (5)2, C	21	$\underline{10}/\underline{m}$	20	(1)$\underline{10}$, (1)$\bar{10}$, (1)5, (1)$\underline{5}$, (1)$\underline{2}$, (1)$\underline{m}$, C
8	$\underline{\bar{5}}$	10	(1)5, (1)$\underline{\bar{10}}$, $\underline{C}$	22	$\underline{10}m\underline{m}$	20	(1)$\underline{10}$, (1)5, (5)m, (5)$\underline{m}$, (1)$\underline{2}$
9	$\underline{\bar{10}}$	10	(1)5, (1)$\underline{\bar{5}}$, $\underline{m}$	23	$\underline{\bar{10}}22$	20	(1)$\underline{10}$, (1)5, (5)$\underline{2}$, (1)2, (1)$\underline{2}$
10	$5\underline{m}$	10	(1)5, (5)$\underline{m}$	24	$\underline{10}$	10	(1)$\underline{10}$, (1)5, (1)2
11	$5\underline{2}$	10	(1)5, (5)$\underline{2}$	25	$\bar{10}3m$	120	(6)5, (6)$\bar{5}$, (10)3, (10)$\bar{3}$, (15)2, (15)m, C
12	$\bar{10}\underline{m}\underline{2}$	20	(1)5,(1)$\bar{5}$, (5)$\underline{2}$, (5)$\underline{m}$, m	26	$\underline{\bar{10}3m}$	120	(6)5, (6)$\underline{\bar{5}}$, (10)3, (10) $\underline{\bar{3}}$, (15)2, (15)m_ , $\underline{C}$
13	$\underline{\bar{10}}m\underline{2}$	20	(1)5,(1)$\underline{\bar{5}}$, (5)$\underline{2}$, (5)m, $\underline{m}$	27	532	60	(6)5, 10(3), 15(2)
14	$\underline{\bar{10}}\underline{m}2$	20	(1)5, (1)$\underline{\bar{5}}$, (5)2, (5)$\underline{m}$, $\underline{m}$				

[a] The number of the corresponding complete symmetry elements is indicated in brackets.

figures. There are finite numbers of such groups. First of all, these are 11 groups of cubic crystals: $m3$, $\underline{m}3$, $\underline{\bar{4}}3m$, $\bar{4}3m$, $\underline{4}3\underline{2}$, $m3\underline{m}$, $\underline{m}3m$, $m3m$ (scalar groups), $\underline{m}3\underline{m}$ (Fig. 2, pseudoscalar group), $\overline{432}$ (the combination of the scalar and pseudoscalar groups), and 23.

Altogether, there are three groups of complete symmetry for three-dimensional isometric figures with fivefold axes. These are groups $\bar{1}03m$, $\underline{\bar{1}0}3\underline{m}$, and 532 (see Table 1 and Fig. 3, a–c). In accordance with the notation of the complete symmetry, the first group may be called a

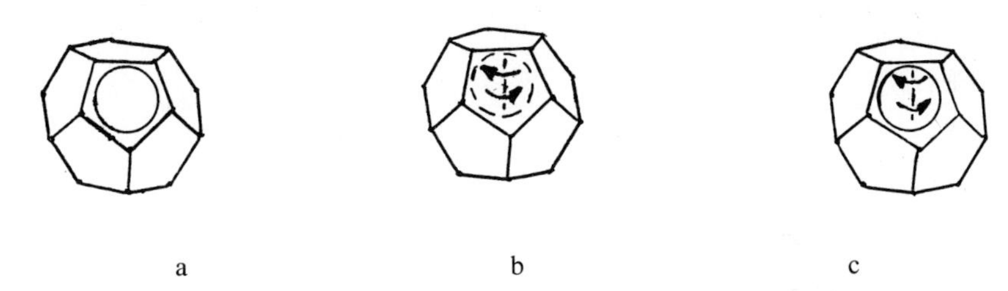

Fig. 3 A representation of the groups of complete symmetry of isometric figures with fivefold axes: (a) a regular dodecahedron combined with a scalar, group $\bar{1}03m$, (b) a regular dodecahedron combined with a pseudoscalar, group $\underline{\bar{1}0}3\underline{m}$, and (c) a regular dodecahedron combined with scalar and a pseudoscalar, group 532.

scalar group (having one definite color) and the second one a pseudoscalar group (having one definite sign of enantiomorphism), whereas the third one is the combination of the scalar and pseudoscalar groups. The geometric representations of the indicated groups include no vector quantities. The figure shown in Fig. 3a can be represented by a positive and negative scalar. The figure shown in Fig. 3b can be represented by two signs of enantiomorphism as "left" and "right" variants, and that shown in Fig. 3c by four pair combinations of scalars and pseudoscalars of two signs. The figure shown in Fig. 3a changes the sign upon the operation $\underline{\bar{1}} = \underline{C}$, the figure shown in Fig. 3b upon the operation $\bar{1} = C$, and the figure in Figs. 3c upon operation **1**.

From the standpoint of physical properties, only figures shown in Figs. 3b and 3c are of interest here, since they describe symmetry of optically active crystals. Note also that in antisymmetry the figure shown in Fig. 3b is depicted as a polyhedron with 60 white and 60 black faces. In such a description, the optical activity of the figure is not obvious.

Polar groups (i.e., the group of complete symmetry of "one-sided" figures) describe the symmetry of "crystal" faces. There are only five groups describing figures with fivefold symmetry axes — these are groups 5, $\underline{10}$, 5m, 5$\underline{m}$, and $\underline{10}$m$\underline{m}$.

References

1. I.S. Zheludev, *Symmetry and Its Applications*. Energoatomizdat, Moscow (1983) (in Russian).
2. A.V. Shubnikov. *Symmetry and Antisymmetry of Finite Figures*. Izdatel'stvo AN SSSR, Moscow (1951) (in Russian).

ICOSAHEDRAL MORPHOLOGY

Gábor Gévay

1. Introduction

The regular icosahedron and the symmetry it represents, banned by classical crystallography, is the starting point for building beautiful morphology. This can be done without reference to lattice (or quasi-lattice) concept provided we consider our morphology in terms of pure geometrical discipline. In this case, the crystallographic limitations, formulated by the Theorem of Barlow (1901)[1] among others, do not prevent us from using the tools of classical crystallography.

In spite of the approach chosen here, our hope is that the topic will be interesting not only for crystallographers but for a much wider audience (including such diverse fields as virology, architecture, etc.)

Independently of their possible material realizations, the polyhedral forms presented here are called quasi-crystal forms, which is justified by the close connection of quasi-crystals and icosahedral symmetry.

2. Simple Quasi-Crystal Forms

A simple (closed) crystal form is nothing but a convex polyhedron whose symmetry group is transitive on its faces; that is, for any two faces there is a group element (a transformation) that maps one onto the other (and the polyhedron as a whole onto itself).

In our case, the group of transformations is the group I (the "proper" icosahedral group containing only rotations), or I_h (the full icosahedral group). Note that they correspond to a "hemihedral" and a "holohedral" "quasi-crystal class," respectively, and form together an icosahedral "quasi-crystal system."[2] The corresponding stereographic projections are shown in Figs. 1 and 2. The rotation axes are depicted in perspective view in Fig. 3. (For details of these groups see for example Refs. 3–7.)

2.1 *Face Positions*

To construct simple forms, we apply the following generating principle. Take the system of symmetry elements corresponding to a point group and place a plane in some definite position with respect to the system (of course, it should avoid the origin). The plane will then be "multiplied" in all possible ways by the symmetry element of the group in question. The desired form will be the smallest convex polyhedron enclosed by the mutually intersecting planes. This process can easily be traced in stereographic projection.

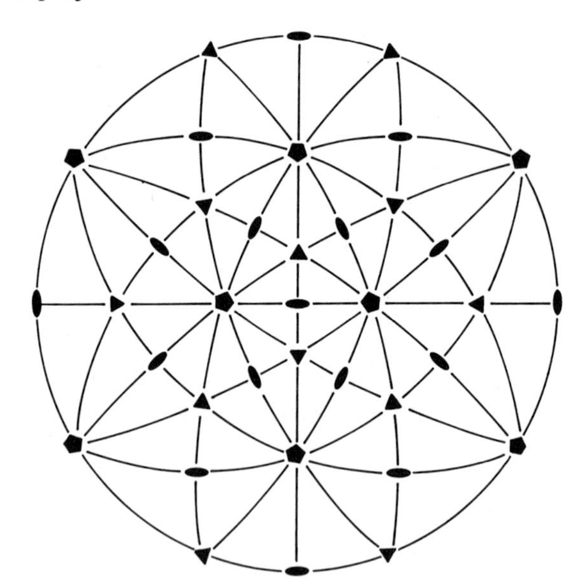

Fig. 1. Rotation axes of the icosahedral symmetry in stereographic projection. The pentagonal, trigonal and "digonal" symbols show the position of the corresponding axes, while the arcs correspond to the planes spanned by the axes.

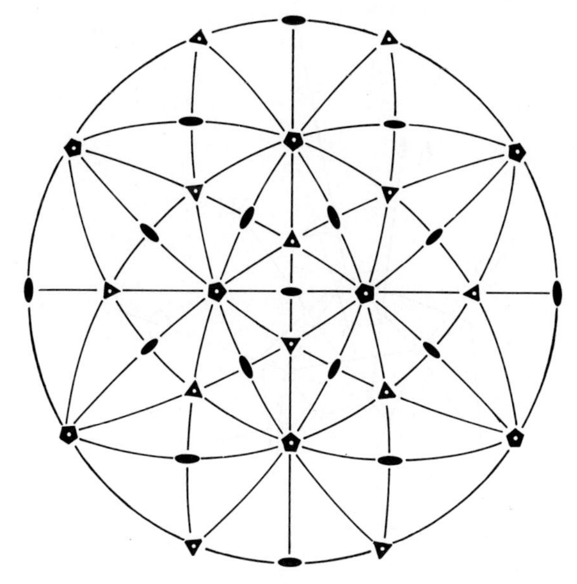

Fig. 2. Stereographic projection of the full icosahedral group. The punctured trigonal and pentagonal symbols refer to rotoinversion axes, the arcs denote mirror planes.

Of course, the result of generating the faces of a polyhedron in this way depends on the chosen initial position. For example, a rotation axis does not multiply at all a plane that is perpendicular to it (the axis represents an element of the "stabilizer subgroup" of the plane within the symmetry group).

Note that, in this context. the position of a face is understood as "modulo parallel translation" (i.e., is identified with the direction of its normal vector). Hence the set of all possible positions is a doubly infinite set corresponding to the points of unit sphere.

This set of face positions (and hence the set of forms) can now be divided into three classes: (1) orthogyral positions, (2) zonogyral positions, and (3) perigyral positions.

The first category means that the face is perpendicular to some rotation axis (to a "gyre"). It is easy to see that in this case every face of the form will be perpendicular to some rotation axis of the same order.

"Zonogyral" means that the normal vector of the face lies between two neighbouring rotation axes in the plane spanned by these axes (note that the set of faces, the normal vectors of which lie within the same plane,

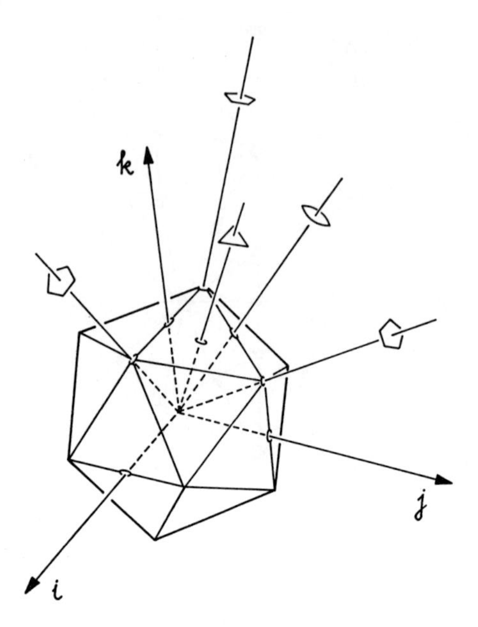

Fig. 3. Rotation axes of the regular icosahedron, with standard positioning in a Cartesian set of basis vectors.

is called zone). If this plane is a mirror plane (as, for example, in the point group I_n), then the face obviously will not be doubled by it.

All of these are special positions. The third category refers to the totally general positions not coinciding with any of the former ones. In the stereogram, it is represented by points inside a minimal spherical triangle determined by three rotation axes. The faces in such position will be multiplied by all symmetry elements of the point group; their number will be equal to the number of transformations (i.e., to the order of the group). Classically, such forms are the general forms, with an indefinite (*hkl*) index, for example, the hexakisoctahedron in the class $m\bar{3}m$ (the order of the point group $m\bar{3}m$ is just 48).

2.2 *Indices*

To give these positions for our icosahedral forms, we use (generalized) Miller indices. They are generalized in the sense that each *h*,*k*,*l* value is not necessarily a rational number. Indeed, at least one of them is necessarily irrational: the old law of rationality, already known to Haüy

(1743–1822), is not fulfilled. The reason is that the icosahedral groups are noncrystallographic (i.e., are not from among the 32 crystallographic groups). This statement is a morphological counterpart of Barlow's Theorem, and it can be shown that the common root of both is the mathematical fact that the equation

$$n \cos \frac{2\pi}{m} = 1$$

has no integral solution m,n other than those given for $m = 2, 3, 4, 6$.

The reference system for the indices is the same as in the cubic crystal system (the Cartesian frame, or orthonormal vector basis). The basis vectors are located along twofold rotation axes so that k bisects the acute angle between two fivefold axes lying in the plane (i, k) (cf. Figs. 2 and 3).

Recall that in such a system, the index of a face can be identified with the coordinate triple of a normal vector of the face. We do not distinguish here in notation between such a normal vector, the "initial face" (or representative plane) of the crystal form and the set of equivalent faces of the form.

2.3 *Listing the Simple Forms*

Let us see first the orthogyral icosahedral forms. We have three choices. Take first the fivefold axes: the corresponding form is the regular pentagon-dodecahedron with index (01τ) (Fig. 4). The faces are located at the 12 axis "ends". Observe that only one fifth of all the possible faces that could be generated by the group I of order 60 occurs in accordance with the "stabilizer" property of the fivefold axes.

Choosing threefold axes, the corresponding form is the famous regular icosahedron (Fig. 5). The index is $(0\tau^2 1)$, but with another choice of representative face it may be (111) as well. A similar "stabilization" as before occurs here also $(60 : 3 = 20)$.

The faces of the third orthogyral form are attached to the 30 twofold axis "ends" $(60 : 2 = 30)$, thus its index is (100). It is the rhombic triacontahedron discovered by Kepler about 1611[8] (Fig. 6).

We do not give separate stereograms for these three forms because they can be studied simultaneously as in Fig. 1, considering that the position of faces and of rotation axes coincide in the projection. (A face position is given by the face pole that is the point where the face normal meets the

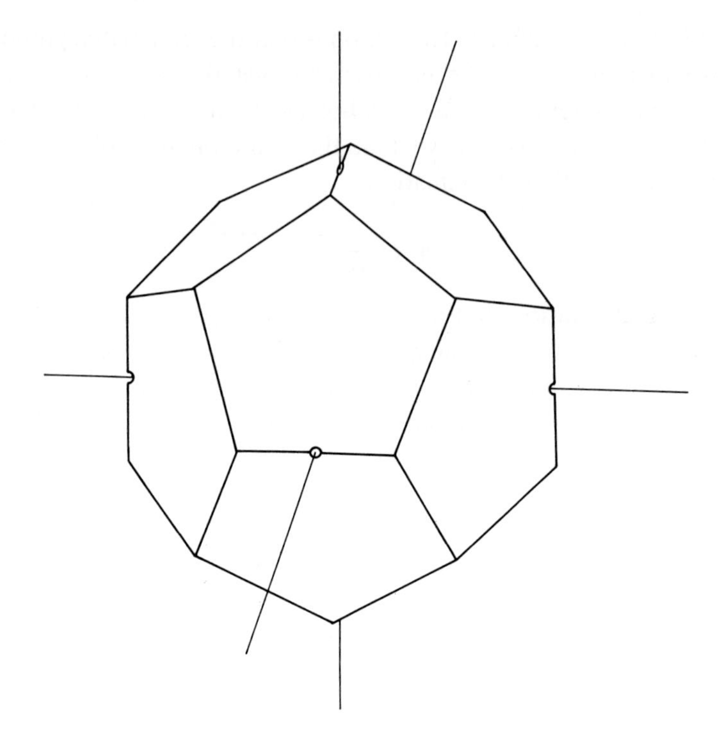

Fig. 4. The regular pentagon-dodecahedron.

surface of the reference sphere, the latter containing the polyhedron located in the centre; in the stereograms the projections of these pole points will occur.)

The situation is slightly different in the zonogyral case when we rely much more on stereographic projection. Consider first an arc of a great circle between a pair of five- and twofold axes. If the face pole is shifted from either of the endpoints of this arc inward (that is, from orthogyral to zonogyral position), then it will be multiplied by five or two, respectively (i.e., it will occur within every equivalent arc, altogether in 60 copies) (Fig. 7a). The form obtained is the pentakisdodecahedron, consisting of 60 isosceles (acute-angled) triangles (Fig. 7b). Its index is $(0kl)$, where $|l|>\tau|k|$. It is analogous, in a sense, to the cubic form called the tetrakishexahedron.

Choosing a position between a pair of three- and twofold axes, the triakisicosahedron (or trisicosahedron) is obtained in an analogous way

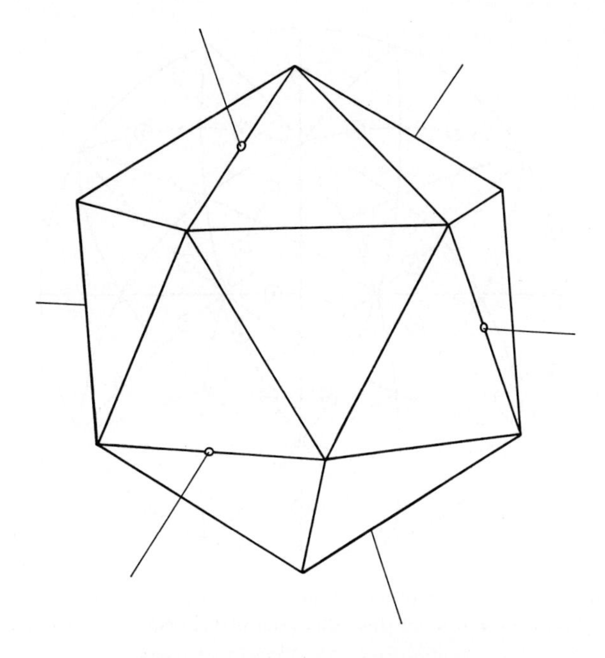

Fig. 5. The regular icosahedron.

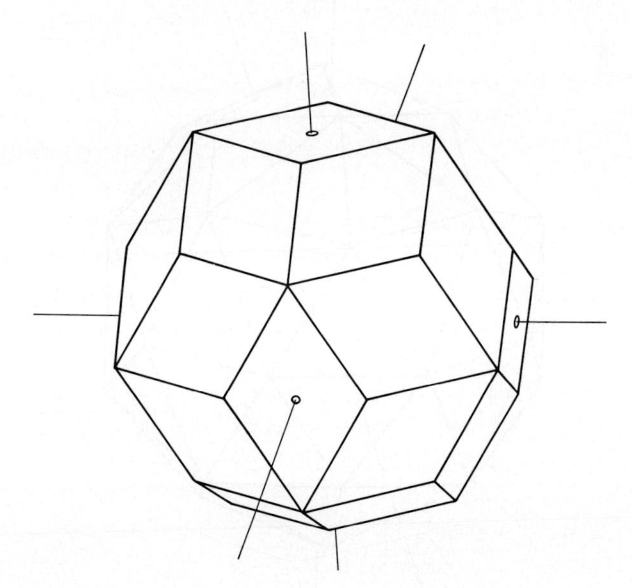

Fig. 6 Rhombic triacontahedron.

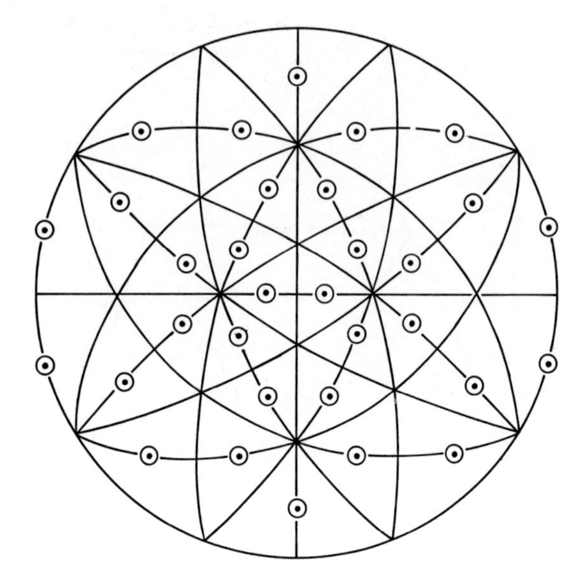

Fig 7a. Pentakisdodecahedron in stereographic representation. Dots and circles denote face poles over and under the equatorial plane, respectively (here coinciding).

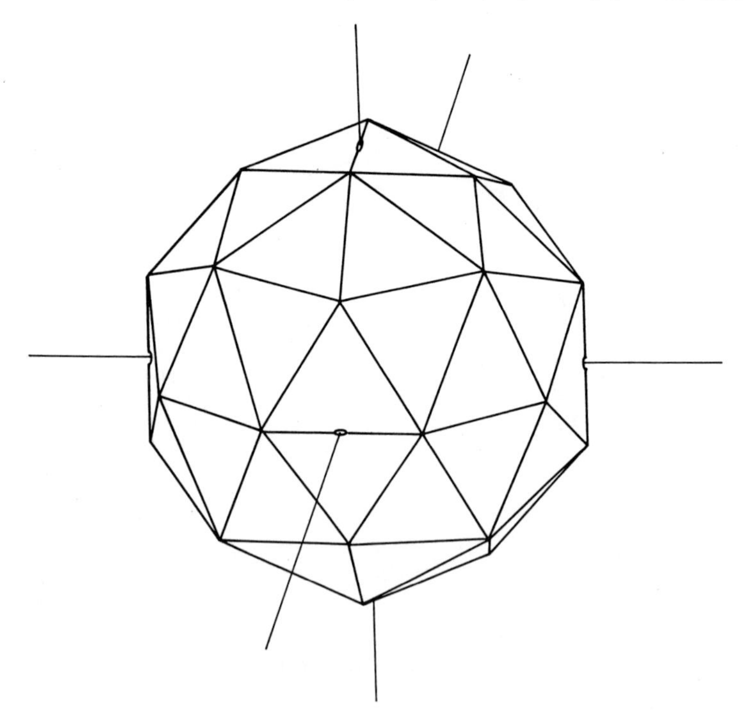

Fig. 7b. A (015) pentakisdodecahedron.

(Fig. 8a). It consists of 60 isosceles obtuse-angled triangles (Fig. 8b), and can be considered as an analogue of the triakisoctahedron. The index is $(0kl)$ with $|k|>\tau^2|l|$.

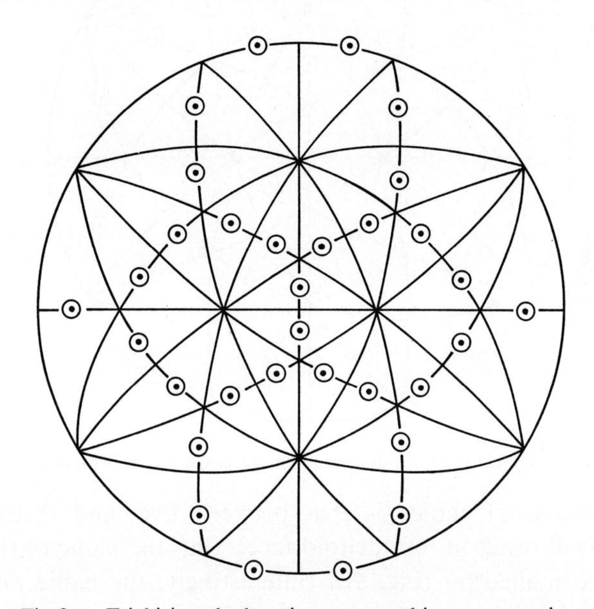

Fig 8a. Triakisicosahedron in stereographic representation.

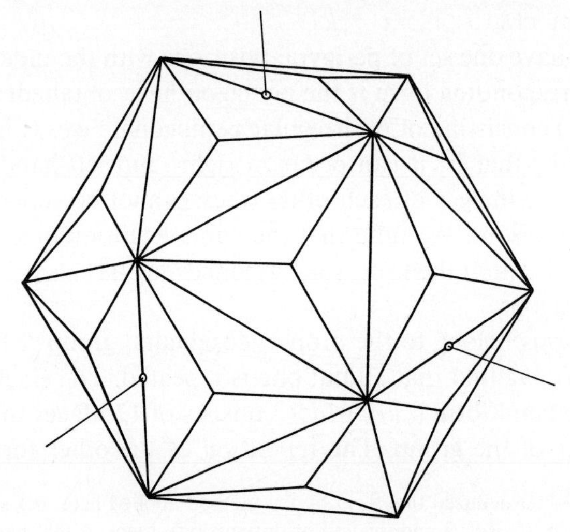

Fig 8b. A (051) triakisicosahedron.

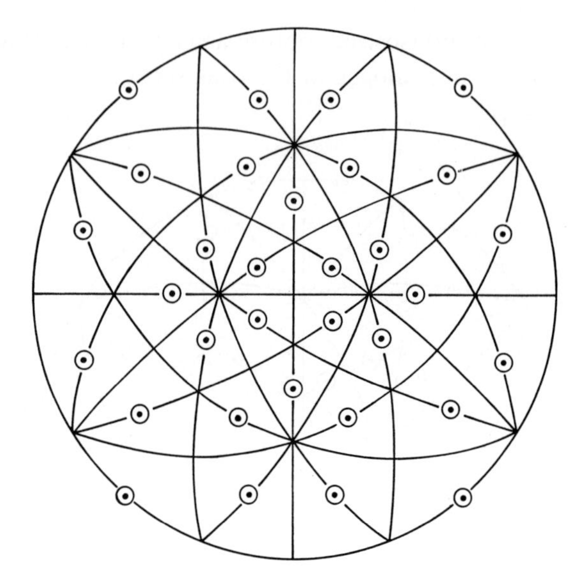

Fig. 9a Deltoid-hexecontahedron in stereographic representation.

The third zonogyral choice is that between five- and threefold axes (Fig. 9a). It will result in (60) deltoid faces, thus the name of the form is deltoid-hexecontahedron (Fig. 9b) (interestingly, the cubic analogue is bounded also by deltoids: it is the deltoid-icositetrahedron). Its index is $(0kl)$ such that $\tau|k|>|1|>\tau^{-2}|k|$.

Finally, we have one set of perigyral positions with the index (hkl) (Fig. 10a). The corresponding form is the pentagon-hexecontahedron (in short, martahedron[a]) consisting of 60 irregular pentagonal faces (Fig. 10b). It is enantiomorphic, that is, it can occur in right- and left-handed varieties which are mirror images of each other (they cannot be superimposed by a continuous motion). We note that the cubic enantiomorphic analogue, the pentagon-icositetrahedron (or, gyrohedron) is also bounded by irregular pentagons.

All these forms belong to the proper icosahedral group I. Passing over to the group I_h, we find that all but one is repeated. The single exception must be the general form (hkl), which consists of 120 faces in accordance with the order of the group. The repetition of the other forms is easily

[a]Actually, when I first deduced this form by the method applied here, not knowing that it was already known (cf. Ref. 9 and the end of this section), I gave it that name in addition to the longer and more exact one.

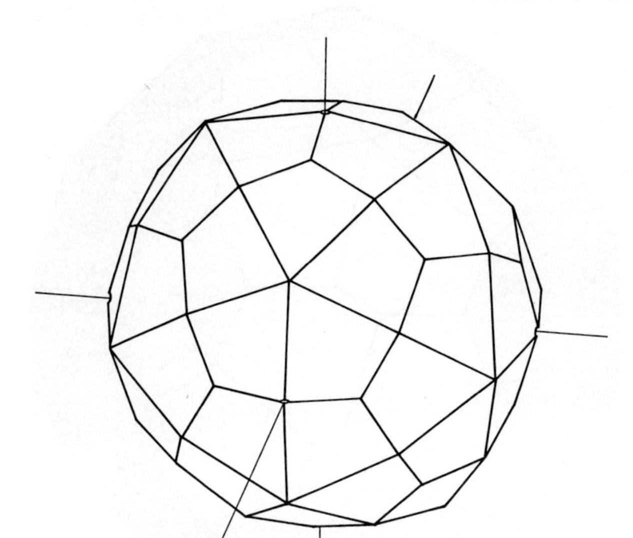

Fig. 9b. A $(0\tau 1)$ deltoid-hexecontahedron with $\tau = (1 + \sqrt{5}) / 2$.

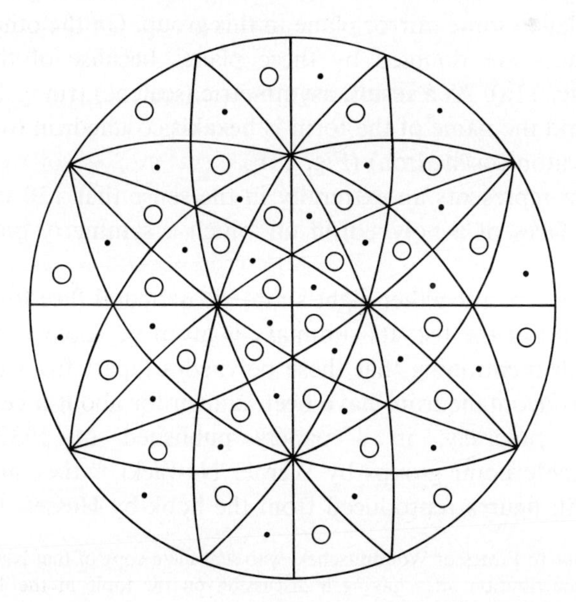

Fig. 10a Pentagon-hexecontahedron in stereographic representation.

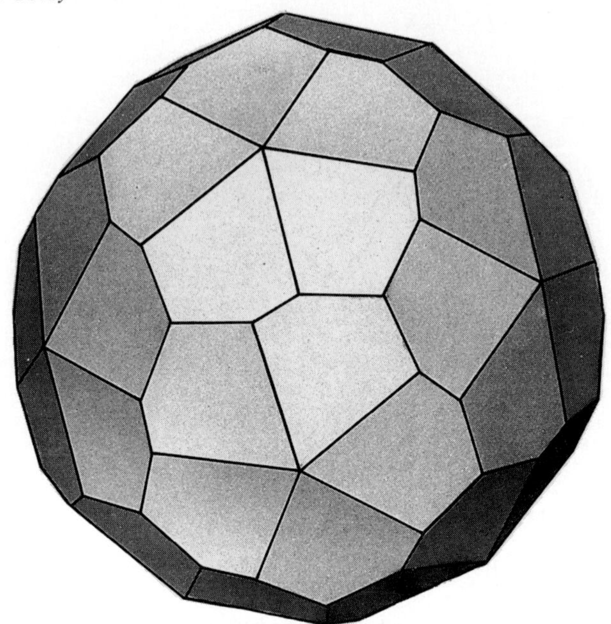

Fig. 10b. A $(1\bar{2}7)$ pentagon-hexecontahedron.

seen, taking into account that all orthogyral and zonogyral faces are perpendicular to some mirror plane in this group. On the other hand, the perigyral faces are doubled by these planes because of their general position (Fig. 11a). As a result, asymmetric (scalene) triangular faces are obtained, and the name of the form is hexakisicosahedron (or hexaicosahedron, hecatonicosahedron) (Fig. 11b).

This form represents an extremity in the sense that 120 is the largest number of faces of a polyhedron on which a symmetry group can act transitively.

Thus, there are altogether eight simple icosahedral face forms, and we are at the end of the list. Its summary is given in Table 1.

It is worth mentioning that these polyhedra (apart from the Platonic solids and triacontahedron) have been known for about a century and a half. More precisely, in a paper[b] published in 1933 on non-crystallographic point groups by Werner Nowacki,[10] they are described together with figures reproduced from the book by Hessel, 1830.[11]

[b]Thanks are due to Professor Wondratschek, who sent me a copy of that paper while I was writing this contribution after having a discussion on the topic at the 12th European Crystallographic Meeting, Moscow.

Table 1 The List of Simple Icosahedral Forms

Class	Name (and Face Position)	Index	Number of Faces	Number of Edges	Number of Vertices
Class $I - 235$	Regular pentagon-dodecahedron	(01τ) with $\tau = \frac{1+\sqrt{5}}{2}$	12	30	20
	Regular icosahedron	$(0\tau^2 1)$ or (111)	20	30	12
	Rhombic triacontahedron	(100)	30	60	32
	Pentakisdodecahedron (between fivefold and twofold axes)	(0kl) with $\lvert l \rvert > \tau \lvert k \rvert$	60	90	32
	Triakisicosahedron (between threefold and twofold axes)	(0kl) with $\lvert k \rvert > \tau^2 \lvert l \rvert$	60	90	32
	Deltoid-hexecontahedron (between fivefold and threefold axes)	(0kl) with $\tau \lvert k \rvert > \lvert l \rvert$, $\lvert l \rvert > \tau^{-2} \lvert k \rvert$	60	120	62
	Pentagon-hexecontahedron	(hkl)	60	150	92
Class $I_h - m\bar{3}\bar{5}$	The first six forms are the same as above				
	Hexakisicosahedron	(hkl)	120	180	62

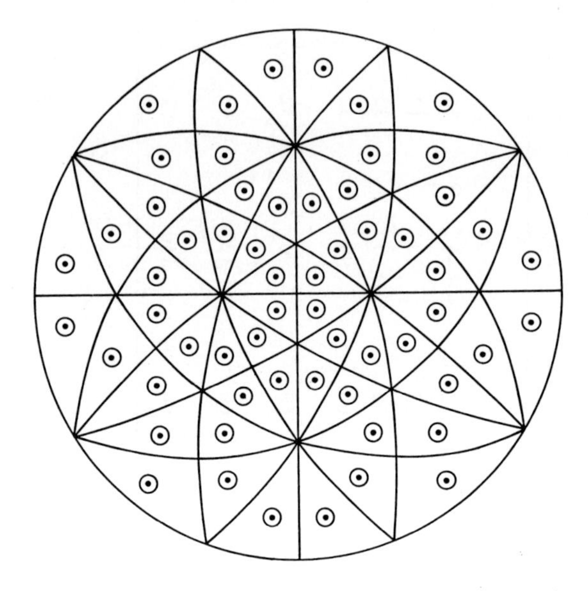

Fig. 11a Hexakisicosahedron in stereographic representation.

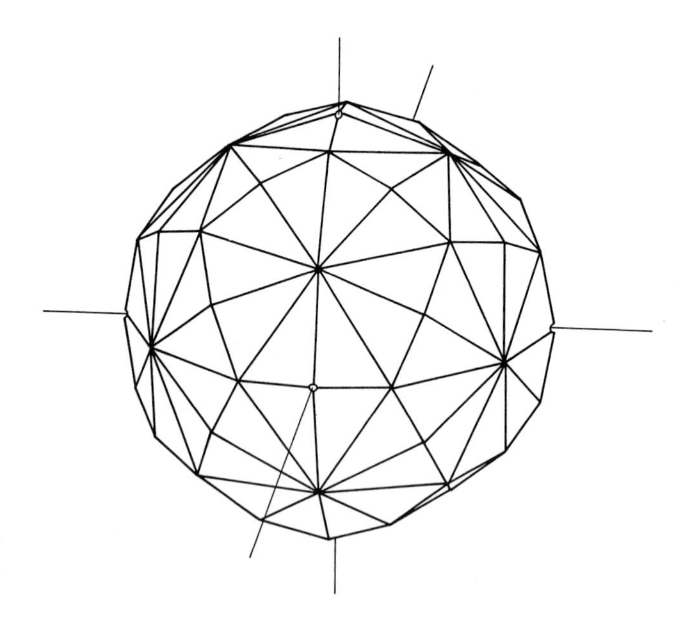

Fig. 11b. A (126) hexakisicosahedron.

2.4 *Connections*

In his book[8] on the icosahedron and dodecahedron (and the three other Platonic solids), Coxeter remarks that "the early history of these polyhedra is lost in the shadows of antiquity. To ask who first constructed them is almost as futile as to ask who first used fire." As a matter of fact, one of our earliest source of knowledge on them is Euclid's Elements.[12]

This knowledge is not restricted to an abstract level, nor is it a European privilege. In her valid contribution to the subject of symmetry. Jean Pedersen presents an interesting example from the Far East: the "Sepak Tackraw ball" from Malaysia, which was mentioned in ancient books concerning leather braiding.[13]

Perhaps the best known material realizations of these solids in Nature are the radiolaria skeletons, which are microfossils of marine origin, e.g., those of Circogonia icosahedra and Circorrhegma dodecahedra.[3,8]

After the light microscope was succeeded by the electron microscope, the icosahedral shape of a whole series of viruses was revealed. One of the earliest reports on this form of viruses, that of the adenovirus, can be found in the famous book *The Thread of Life* by the Nobel Laureate John Kendrew.[14] The idea that certain "spherical" viruses must have this high degree of symmetry appeared in a paper by Watson and Crick[15] (also cited in Ref. 16). The frequent occurence of the icosahedron is now commonplace in virology, see e.g., Tipula iridescens[17] (also mentioned in Ref. 14), the Herpes simplex,[17,18] the satellite tobacco necrosis virus, the southern bean mosaic virus, the alfalfa mosaic virus,[19] etc. On the other hand, the causes of this phenomenon have not yet been sufficiently clarified. It is a challenge not only for biologists, but also for crystallographers (cf. Ref. 16), geometers, . . . in a word, it is a truly interdisciplinary problem! It is a manifestation of the unity of structure and function: perhaps the latest powerful tools of molecular biophysics such as the synchrontron X-ray Laue technique[20] will produce the data set that will be necessary for understanding this unity.

The next level is that of the geometry of the molecule. Here boron chemistry provided a series of examples, starting from the $[B_{12} H_{12}]^{2-}$ dianion with its regular icosahedral skeletal framework as a basic structural bonding unit.

From organic chemistry, a good example is the recently synthesized dodecahedrane $(CH)_{20}$ with a skeleton dual to that of the former example, i.e., the carbon atoms are located at the face centers of an icosahedron, or, what is the same, at vertices of a (regular) dodecahedron.[21]

Perhaps such "Platonic chemistry" will not stop at this point. A very recent attempt to explain some features of biopolymer organization in a spore-pollen wall is based on the supposition that the dodecahedrane molecule may serve as building block for a "pentagonal hierarchy" of structures.[22] Here the hypothetical pentasporane unit plays a crucial role (Fig. 12).

The B_{12} icosahedron is characteristic not only of discrete molecules, but also occurs as structural element of condensed matter. Examples are the allotropes of the crystalline boron itself,[18] the (rhombohedral) $B_{12}C_3$[23] and various other boron containing structures (of infinite extension in principle). In the cubic ($m\bar{3}$) mineral skutterudite, (Co_4As_{12}), the As_{12} icosahedral units are analogous examples.[18]

In supercooled liquids and metallic glasses, recent investigations revealed extended icosahedral bond orientational order.[24,25] (Historically, the idea of local icosahedral order in these substances goes back to Sir Charles Frank, 1952.[26]) In fact, these observations led Levine and Steinhardt to their theoretical model[27] for the "shechtmanite"[5] struc-

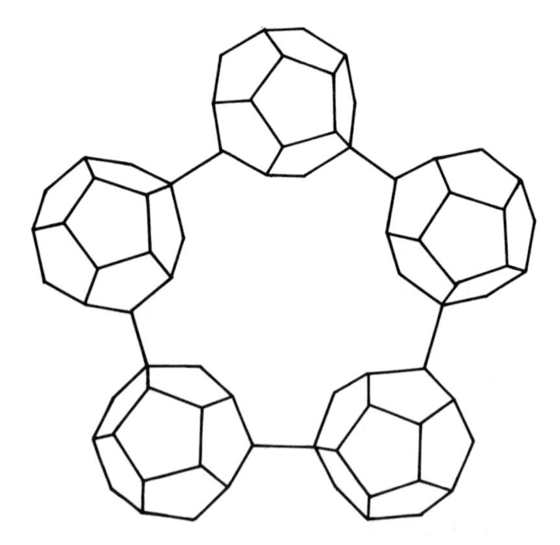

Fig. 12 Pentasporane framework composed of dodecahedrane.

tures. In several models related to icosahedral quasi-crystals, the rhombic triacontahedron appears as a particularly important structural unit due to intriguing geometric properties discovered by Kowalewski, Mackay, Ammann and others (see e.g. the review in Refs. 26 and 28: a brief summary of its geometrical data can be found in Ref. 30). In icosahedrally symmetrical space filling models, the icosahedron and dodecahedron also occur, see e.g. Ref. 31 (some geometrical data for these forms are also given).

Last, but not least — is there any relevance of our morphological considerations for the physical appearance of icosahedral quasi-crystals? It seems that the answer is yes. Here we restrict ourselves to mentioning merely some initial results. The first report on the preparation of quasi-crystals, in which the triacontahedral form could be recognized was that of Dubost et al., 1986.[32] Dodecahedral forms have also been observed.[33] In transmission electron microscope images obtained on quasi-crystal particles, Nishitani et al.[34] recognized dodecahedra with protuberances in the direction of threefold axes (their model drawings show stellated – concave – dodecahedra which we did not deal with here). Theoretical investigations also confirm the possibility of the occurrence of flat face forms, and the first steps towards generalization of the morphological theories of classical crystal physics have been taken as well.[33]

3. Combinations

One may also assign a definite symmetry (group) to the faces of a polyhedron in the case when not all its faces are equivalent. The set of faces then decomposes into disjoint equivalence classes, each equivalence class being a simple form: this is the case of combination of forms (composite crystal forms). Evidently, within an equivalence class the symmetry group acts transitively (though it is intransitive on the whole of the polyhedron).

The simplest combinations of forms are those built of the dodecahedron and icosahedron. The polyhedra in Figs. 13a–c differ from each other in the "relative weight" of the combining forms, which can be expressed practically in terms of the ratio of norms of the face normal vectors. The icosidodecahedron (Fig. 13b) is known as one of the quasiregular polyhedra.[8] The combination in Fig. 13c is well known: it is

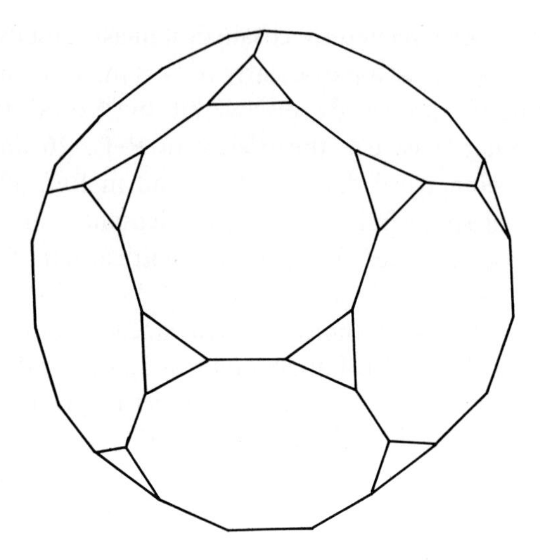

Fig. 13a Dodecahedron-icosahedron combination, dodecahedron dominating.

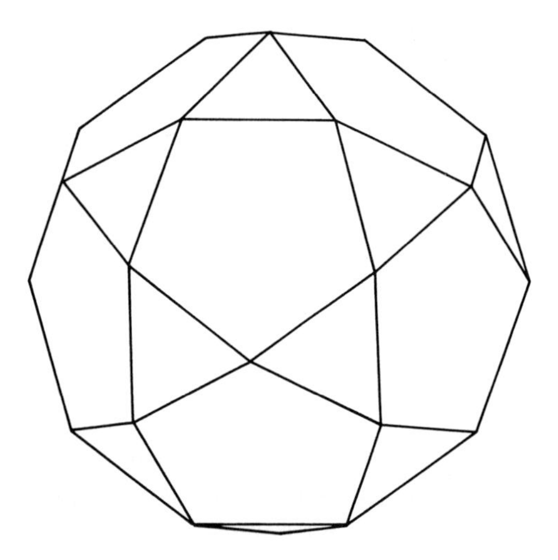

Fig. 13b Icosidodecahedron.

the soccerball shape. A more interesting realization has been found recently: with carbon atoms in its vertices, it is just the famous footballene (or buckminsterfullerene) molecule C_{60} (see the contribution by S. J. Cyvin et al.[39]).

Fig. 13c Dodecahedron-icosahedron combination, icosahedron dominating.

Adding the triacontahedron as a new component, the rhombicosidodecahedron (Fig. 14) is obtained ("triorthogyral form").

The dodecahedron-icosahedron combinations in Fig. 13a–c can be obtained by truncation, that is, by cutting off all the corners of the starting form (Sz. Bérczi has arranged such a truncation series related to the Platonic and Archimedean solids and tessellations[35] into a periodic table. For selected truncation we have the following examples. Let us start

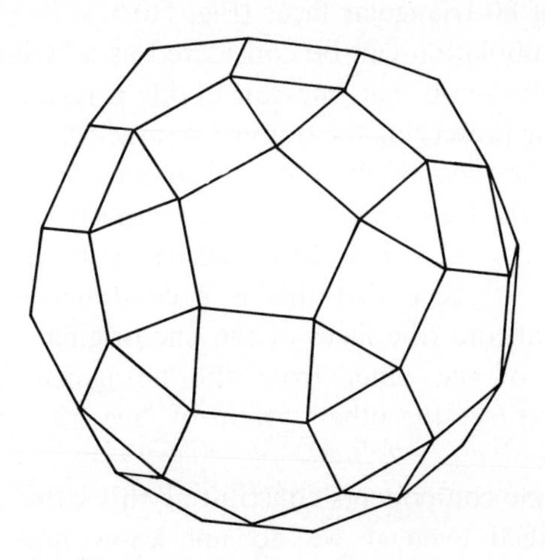

Fig. 14 Rhombicosidodecahedron.

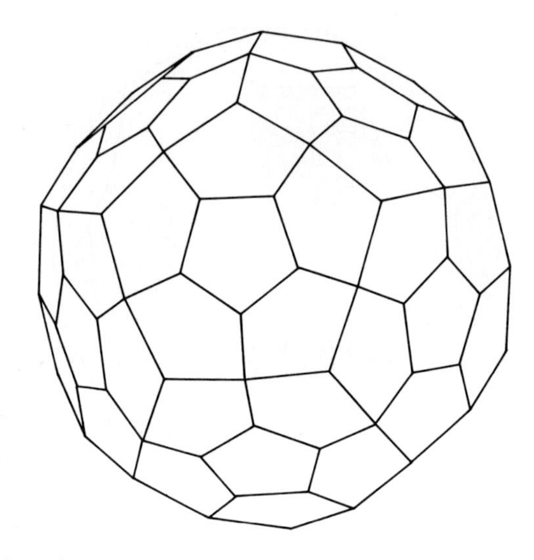

Fig. 15 Combination of dodecahedron and (011) deltoid-hexecontahedron.

from a deltoid-hexecontahedron (cf. Fig. 9b) and cut off the corners with fivefold symmetry. The result is shown in Fig. 15: it is a dodecahedron-deltoidhexecontahedron combination consisting of 72 pentagonal faces (12 regular and 60 symmetric pentagons). Truncating the same starting form at the vertices with trigonal symmetry, at appropriate truncation depth the icosahedron-deltoidhexecontahedron combination is obtained, which consists of 80 triangular faces (Fig. 16).

This latter combination can be considered as a "triangulation" of an icosahedron in the sense that one can divide a regular triangle to four smaller ones while preserving the trigonal symmetry. As another example, let us start with the "soccerball" polyhedron and consider a triangulation fit everywhere to the local symmetry of the respective faces. We obtain a polyhedron with $5 \times 12 + 6 \times 20 = 180$ triangular faces (Fig. 17). It is not too difficult to check that this is a combination of two distinct deltoid-hexecontahedra (the faces of the one originate from the pentagonal faces and of the other from the hexagonal faces) and of a triakisicosahedron (see the other system of face triplets in the place of hexagonal faces).

The three simple components constituting this latter combination can easily be identified even if we do not know how it was derived (triangulation, etc.). The first step is to localize the exit points of the

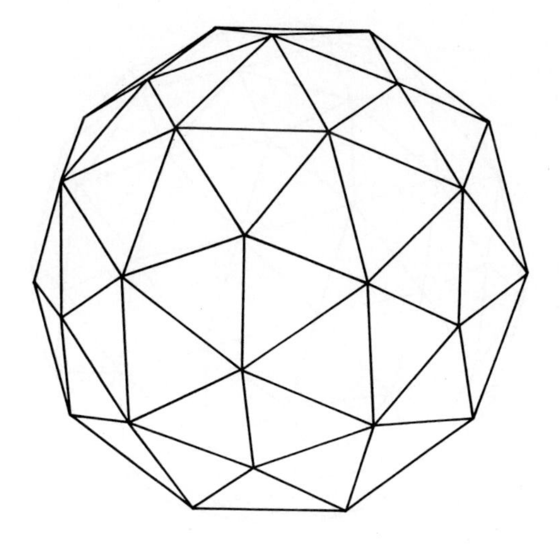

Fig. 16 Combination of icosahedron and (011) deltoid-hexecontahedron.

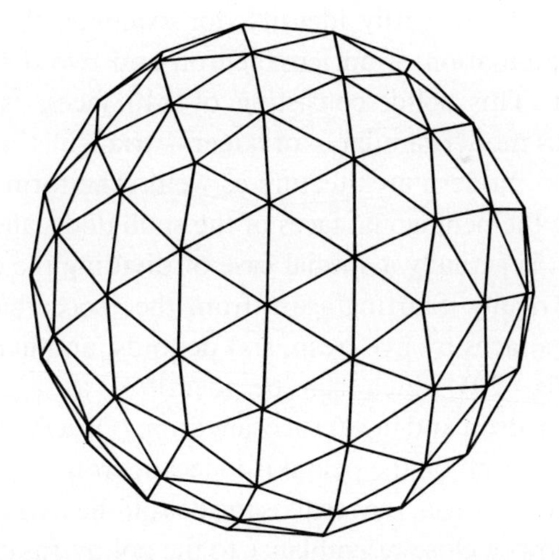

Fig. 17 Combination consisting of a triakisicosahedron and two deltoid-hexecontahedra.

rotation axes: the fivefold axes are directly seen (of course, the icosahedral symmetry is taken for granted), and "half-way" between them, one can find the twofold axes, etc. (cf. the stereograms for icosahedral groups, Figs. 1 and 2). We can then establish the face position with respect to the axes (using the classification scheme introduced in Sec. 2.1).

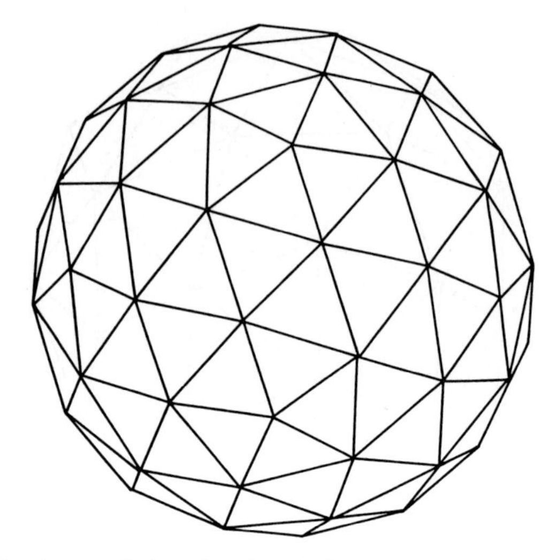

Fig. 18 Combination consisting of an icosahedron and two pentagon-hexecontahedra.

In this way, one can easily identify, for example, the polyhedron in Fig. 18 as a combination of an icosahedron and two distinct pentagon-hexecontahedra. This solid, consisting of 140 faces, is considered in Ref. 36 as a virus model (similar — or larger — triangulations of the sphere are often used in modern architecture as well). The form can be derived by triangulating the pentagonal faces of the snub dodecahedron (Fig. 19).

Triangulation is actually a special case of dividing the faces of a polyhedron symmetrically. Starting again from the "soccerball" polyhedron and replacing its faces by five (resp. six) deltoids, another 180-face form is obtained (Fig. 20) which can be identified as a combination of pentakisdodecahedron and hexakisicosahedron. (Observe that the starting form could just as well be the pentakisdodecahedron; applying analogous process to the icosahedron, the result is the deltoid-hexecontahedron.) This combination shows a close resemblance to the poliovirus capsid structure.

A different method of deriving new forms is "edge truncation" or equivalently, placing planes on the edges of a polyhedron. For example, one truncation series is as follows: dodecahedron (or icosahedron); triacontahedron; deltoid-hexecontahedron (what will be the next member of the series?). By this method, a polyhedron consisting of 180 deltoid faces, different from that in Fig. 20, can be obtained (Fig. 21). The starting form is the hexakisicosahedron and the derived combination is

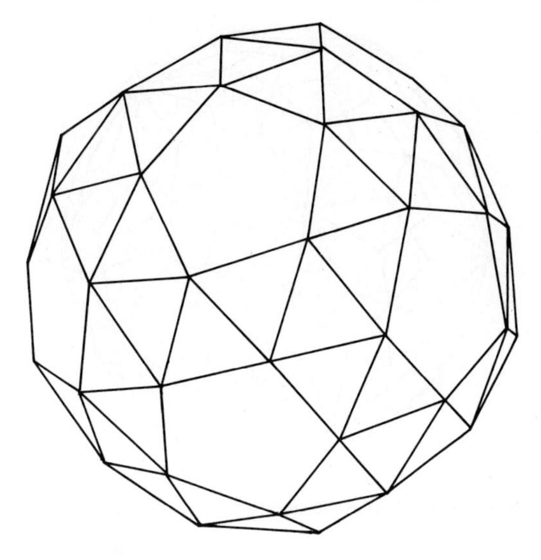

Fig. 19 Snub dodecahedron (combination of dodecahedron, icosahedron and pentagon-hexecontahedron).

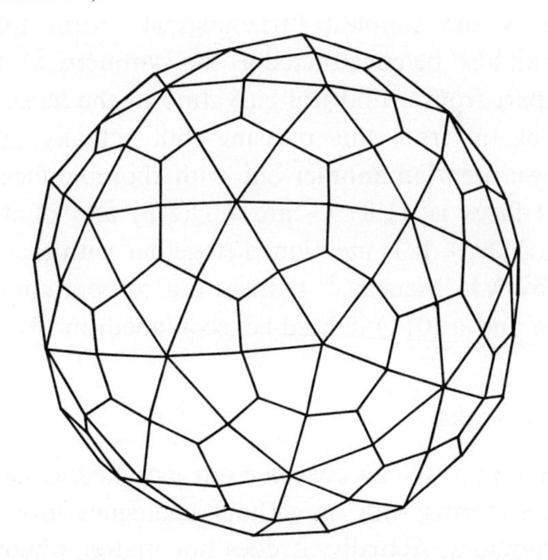

Fig. 20 Combination of pentakisdodecahedron and hexakisicosahedron.

composed of one pentakisdodecahedron, one triakisicosahedron and one deltoid-hexecontahedron (in the figure the corresponding faces can be seen as deltoids of distinct shape — in fact, three systems of them can be

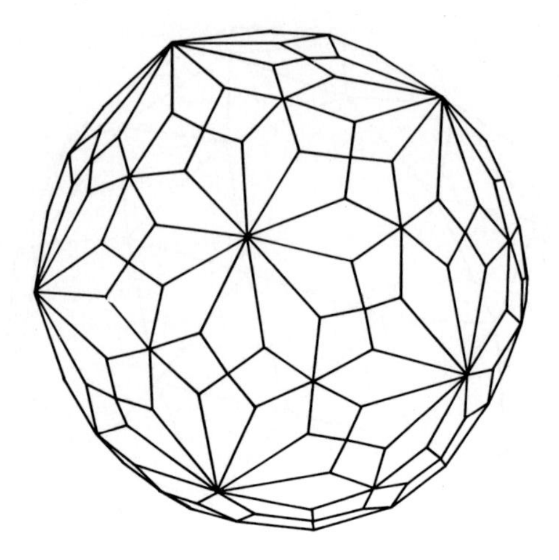

Fig. 21 Combination of pentakisdodecahedron, triakisicosahedron and deltoid-hexecontahedron.

recognized). It is the simplest "trizonogyral" form (of icosahedral symmetry). It can also be constructed as the symmetrical intersection of 15 cylinders (apart from cylindrical curvature of the faces and edges).[37]

Before moving on from this pleasing (but actually infinite) list of combinations, we mention another one with rhombic faces. The vertex angle of its 30 faces is 41.81° = arc sin (2/3) and of its 60 faces is 70.53° = arc cos (1/3). It is mentioned (together with these data) in the famous paper by A.L. Mackay.[38] It turns out to be a combination of a triacontahedron and a (011) deltoid-hexecontahedron (Fig. 22).

4. Concluding Remarks

It is easy to see that we can continue our icosahedral series, inventing newer and newer forms with or without analogues in our natural (or artificial) environment. Actually, it does not matter whether or not we know natural analogues at the moment: science is an interplay between imagination and reality, in which the former may sometimes outstrip the latter. Moreover, fantasy need not be limited by any rule — but in this case we arrive at another territory which is not that of science (Fig. 23) . . .

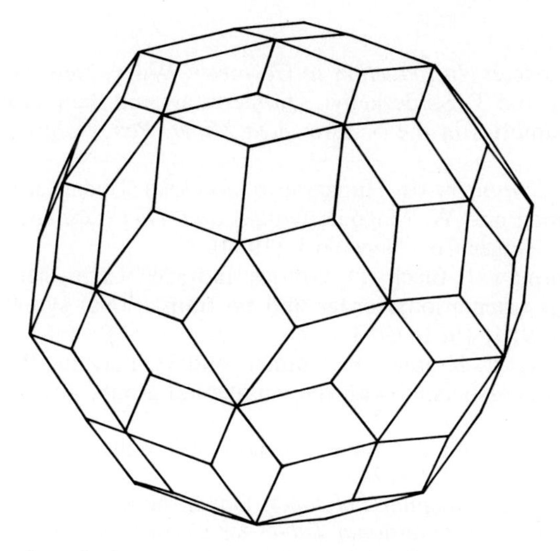

Fig. 22 Combination of triacontahedron and (011) deltoid-hexecontahedron.

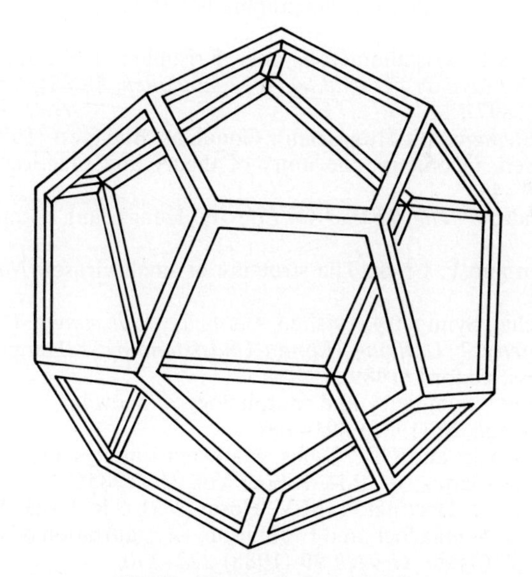

Fig.23 "Dodecahedral tangle".

Acknowledgement

I am grateful to Professor Marjorie Senechal, who read and made corrections to the first version of the manuscript.

References

1. H.S.M. Coxeter, *Introduction to Geometry*. Wiley, New York (1961).
2. G. Gévay and T. Szederkényi, Quasicrystals and their spontaneous formation possibilities in the Nature, *Acta Miner. Petr. Szeged*, **29** (1987–1988) 5–12.
3. H. Weyl, *Symmetry* (in Hungarian). Gondolat, Budapest (1982).
4. I. Grossman and W. Magnus, *Groups and Their Graphs*. Random House, The L.W. Singer Co., New York (1964).
5. D. Shechtman, I. Blech, D. Gratias and J.W. Cahn, Metallic phase with long-range orientational order and no translational symmetry, *Phys. Rev. Lett.* **53** (1984) 1951–1953.
6. R.W. Haase, L. Kramer, P. Kramer, and H. Lalvani, Polyhedra of three quasilattices associated with the icosahedral group, *Acta Cryst.* **A43** (1987) 574–587.
7. C.J. Cummins and J. Patera, Polynomial icosahedral invariants, *J. Math. Phys.* **29** (1988) 1736–1745.
8. H.S.M. Coxeter, *Regular Polytopes*. Dover, New York (1973).
9. T. Hahn, (ed.), *International Tables for Crystallography*, Vol. A. D. Reidel, Dordrecht (1983).
10. W. Nowacki, Die nichtkristallographischen Punktgruppen, *Z. Kristallogr.* **86** (1933) 19–31.
11. J.F.C. Hessel, Kristallometrie, oder Kristallonomie und Kristallographie (*Ostwald's Klassiker der exakten Wissenschaften*, 88/89), Verlag Engelmann Leipzig (1897).
12. Euclid, *Elements* (in Hungarian). Gondolat, Budapest (1983).
13. J. Pedersen, Geometry: the unity of theory and practice, *Math. Intell.* **5** (1983) 37–49.
14. J.C. Kendrew, *The Thread of Life* (in Hungarian). Gondolat, Budapest (1968).
15. J. Watson and F. Crick, The structure of small viruses, *Nature* **177** (1956) 473–475.
16. M. Senechal, Symmetry revisited, *Comput. Math. Applic.* **17** 1–2. Reprinted in *Symmetry 2: Unifying Human Understanding*, I. Hargittai (ed.). Pergamon Press, Oxford (1989).
17. C.J. Schneer, Symmetry and morphology of snowflakes and related forms, *Can. Mineral.* **26** (1988) 391–406.
18. M.B. Boisen Jr. and G.V. Gibbs, Mathematical crystallography. *Reviews in Mineralogy* (Series ed., P.H. Ribbe) Vol. 15 (1985).
19. C. Sehnke, M. Harrington, M.V. Hosur, Y. Li, R. Usha, R.C. Tucker, W. Bomu, C.V. Stauffacher, and E. Johnson, Crystallization of viruses and virus proteins, *J. Crystal Growth* **90** (1988) 222–230.
20. J. Hajdu, Laue diffraction studies on macromolecules using synchrotron X-radiation. Plenary lecture held at the Twelfth European Crystallographic Meeting, Moscow, Aug. 20–29, 1989 (ECM-12 Collected Abstracts, Vol 1, p. 7).
21. A.T. Balaban, Symmetry in chemical structures and reactions, *Comput. Math. Applic.* **12B** 999–1020. Reprinted in *Symmetry: Unifying Human Understanding*, I. Hargittai (ed.). Pergamon Press, Oxford (1986).

22. G. Gévay and M. Kedves, A structural model of the sporopollenin based on dodecahedrane units. *Acta Biol. Szeged* **35** (1989) 53–57.
23. A.L. Mackay and J. Klinowski, Towards a grammar of inorganic structure. *Comput. Math. Applic.* **12B** 803–824. Reprinted in *Symmetry: Unifying Human Understanding*, I. Hargittal (ed.). Pergamon Press, New York (1986).
24. P.J. Steinhardt, D.R. Nelson and M. Ronchetti, Icosahedral bond orientational order in supercooled liquids, *Phys. Rev. Lett.* **47** (1981) 1297–1300.
25. P.J. Steinhardt, D.R. Nelson, and M. Ronchetti, Bond orientational order in liquids and gases, *Phys. Rev.* **B 28** (1983) 784–805.
26. D. Levine and P.J. Steinhardt, Quasicrystals. I. Definition and structure, *Phys. Rev.* **B 34** (1986) 596–616.
27. D. Levine and P.J. Steinhardt, Quasicrystals: A new class of ordered systems, *Phys. Rev. Lett.* **53** (1984) 2477–2480.
28. J.E.S. Socolar and P.J. Steinhardt, Quasicrystals. II. Unit-cell configuration, *Phys. Rev.* **B 34** (1986) 617–647.
29. M. Rochetti, Quasicrystals. An introductory overview, *Phil. Mag.* **B 56** (1987) 237–249.
30. A. Katz and M. Duneau, Quasiperiodic patterns and icosahedral symmetry, *J. Physique* **47** (1986) 181–196.
31. P. Kramer, Non-periodic central space filling with icosahedral symmetry using copies of seven elementary cells, *Acta Cryst.* **A 38** (1982) 257–264.
32. B. Dubost, J.-M. Lang, M. Tanaka, P. Sainfort, and M. Audier, Large AlCuLi single quasicrystals with triacontahedral solidification morphology, *Nature* **324** (1986) 48–50.
33. T. Janssen, A. Janner and P. Bennema, On the morphology of quasicrystals, *Phil. Mag.* **B 59** (1989) 233–242.
34. S.R. Nishitani, H. Kawaura, K.F. Kobayashi and P.H. Shingu, Growth of quasicrystals from the supersaturated solid solution, *J. Crystal Growth* **76** (1986) 209–214.
35. Sz. Bérczi and D. Nagy, Periodicity of extremal geometric arrangements (densest packings, thinnest coverings, tessellations), *Acta Geologica Acad. Sci. Hung.* **23** (1980) 173–200.
36. I. Rayment, T.S. Baker, D.L.D. Casper, and W.T. Murakami, Polyoma virus capsid structure at 22.5Å resolution, *Nature* **295** (1982) 110–115.
37. I.O. Angell, Symmetrical intersections of cylinders, *Acta Cryst.* **A 43** (1987) 244–250.
38. A.L. Mackay, De Nive Quinquangula: On the pentagonal snowflake, *Sov. Phys. Crystallogr.* **26** (1981) 517–522.
39. S.J. Cyvin et al.'s contributions in *Quasicrystals, Networks, and Molecules of Fivefold Symmetry*. I. Hargittai (ed.), VCH, New York (1990).

THE DISCOVERY OF SPACE FRAMES WITH FIVEFOLD SYMMETRY

Stephen C. Baer

1. Fivefold Symmetry

Fivefold symmetry was presumed to be impossible for crystals. Try placing pentagonal tiles on a floor and you will find that you can't cover the floor with them; they either leave large cracks or overlap. Likewise, if you attempt to stack icosahedra or dodecahedra to fill space you find the same problem – they won't fit together without leaving gaps. These are frustrating shapes for the packer.

Why should anyone care about the failure of the pentagon to tile or the dodecahedron to pack? There are many two-dimensional forms that won't tile and many three-dimensional shapes that won't pack. You can appreciate the disappointment over the pentagon if you consider that its neighbouring polygons will tile. The triangle, square, and hexagon tile, but not the pentagon. Our disgust over its failure is diluted when we see that the heptagon and many other regular polygons also will not tile. With three-dimensional regular polyhedra the situation is different. There are only five regular polyhedra: the tetrahedron, cube, octahedron, dodecahedron, and icosahedron. These are the five Platonic solids. You can't fit hexagons together to make a regular polyhedron, nor will any other polygons with more than six sides fit together. Cubes pack together to fill space; octahedra won't and tetrahedra won't but a combination of tetrahedra and octahedra will. The dodecahedron and the icosahedron will not pack — neither one with any others like it, nor any combination of the two. They are stubborn. It is as if God made five animals, three of which could breed, but the two most beautiful and complex could not.

I first ran into this packing failure of dodecahedra when I was a student. I put three truncated dodecahedra together with cardboard and noticed with a fourth that parts were strained (Fig. 1).

The residents at Drop City in Colorado, with my help, built a cluster of three exploded dodecahedra (a triple-fused cluster of rhombo-icosa-dodecahedra). We simply extended panels, ruining the perfect symmetry, to make them fit (Fig. 2).

The 12-sided rhombic dodecahedra will pack to fill space, but they have diamond faces and the corners are of two different types — one bringing three polyhedra together and the other four.

I used these in the first zomes I constructed.[1] I studied zonohedra in Coxeter's [2,3] and A.D. Alexandrow's[4] books. My first inkling of joints with fivefold symmetry came in 1967 when laying out an enneacontahedron made of chopped car and van tops from the junkyard. In calculating some of the face angles, I found I was using the supplement to the dihedral angle of a regular icosahedron; the joints used in constructing an enneacontahedron could be icosahedra. Of course, for this skin structure there were no joints necessary other than the joining of the various vertices of the panels. I also realized that a triacontahedron

Fig. 1 Fused rhombo-icosa-dodecahedra. They don't really fit.

Fig. 2 Cluster at Drop City, Colorado, 1967.

could be made with dodecahedral joints. This realization made it doubly exciting to read in Coxeter's book[3] about Gerhard Kowalewski's subdivision of the triacontahedron into 20 parallelepipeds. I knew that this maze of lines could have tiny dodecahedra at each of their joints, on the surface and within.

When Ed Heinz, Berry Hickman and I started Zomeworks Corporation in 1969, I suggested that triacontahedron with this subdivision, all made of dodecahedral joints is a fascinating structure for a playground climber. We constructed a model using bundles of surgical tubes for the joints, then built a full-scale climber using dodecahedral joints made of washers welded together (Fig. 3). At some point in our work, I realized that these hubs with fivefold symmetry could be used like any other space frame hub. They weren't restricted to filling the inside of a triacontahedron.

For some reason the 6-zone system of the edges of triacontahedron did not seem exciting, but the 10-zone system of lines perpendicular to the faces of an icosahedron was exciting.[5] It was so versatile and lovely, I spent several days in the clouds; Buck Dant and I built a trussed enneacontahedron with the same surgical tube joints, only this time using the lines perpendicular to the faces of an icosahedron rather than dodecahedron.

In both these structures, we were using braces drawn from the 15 zones through the edges of an icosahedron. During that first summer I never

Fig. 3 Triacontahedron with 20 inner cells. Ed Heinz, top; Berry Hickman, right; Holiday Baer, left; José Baer, front; Didier Raven, within.

appreciated the 15-zone system as much more than a brace for the 6-zone or the 10-zone. During the summer of 1969, our progress was much improved by Berry Hickman's introduction of the plastic golf ball with holes drilled in it for a joint. This proved far better than the joints of surgical tubing. Another beautiful joint, which had hexagonal nuts welded together to make a truncated icosahedron, was invented by Ed Heinz to make the 10-zone system.

The entire system began to fit together in March 1970 when I realized that lines always seemed to cut other lines in the divine proportion. Looking back on this, I find that I was resisting having everything fit together in the beautiful simplicity that it does. It was almost as if the beauty and joy of this complete system was too exquisite for a mind to bear (Fig. 4).

During this time, we built models of fivefold symmetry structures of all kinds. I manned the bulky calculator that could do square roots and presided over all authoritative mathematical pronouncements concerning the structures. Berry Hickman and Jim Welty kept urging me to put all of

Fig. 4 Truncated icosahedron.

it together in one grand system, Berry suggesting that "the divine proportion would be everywhere." When I finally realized how to proportion the members that emerge from the vertices, edges, and faces of the icosahedron, it turned out that Berry was right — the divine proportion in power after power along the powers of 2 teemed throughout space.[6]

There is something special about the 31-zone truss, a thrill in the multiplicity of coincidences. Everything fits together in more ways than it seems reasonable to expect. There is a productive and healthy, yet incestuous relationship among the components that acts on the mind of the model builder like a drug. I understand today's scientists who seem so slow to appreciate what this system is. Here, the mind should be guided by the hand because the mind needs the body's fingers to guide it and calm its vertigo upon beholding unexpected coincidences and symmetry.

The initial work on the 31-zone truss yielded nothing except a few buildings and playground climbers. Mathematicians and engineers, with a few exceptions such as David Booth and Reuben Hersh, were uninterested. Then in 1977, Marc Pelletier, a high school student, hitchhiked from New Hampshire to Albuquerque to talk to me about the system. I remember him arriving barefoot, but that's impossible since it was well into fall. For years Pelletier wandered the country extending knowledge of the system and sharing it with all who were interested. Now he and Paul Hildebrandt are working on the plastic molds to produce an advanced modeling system for the 31-zone truss.

2. Nature Builds Crystals; Men Need Not Follow The Same Rules

I was obedient to the mistaken notion that crystals could not have fivefold symmetry. At Zomeworks, we were not concerned whether atoms in crystals could arrange themselves in enneacontahedra, triacontahedra, etc. If one studies zonohedra for use in buildings, as I did at some length,[1] one is inevitably led to questions of packing, then to the discovery that with zonohedra there are no vexing problems in packing. Zonohedra are born to pack (Fig. 5). There are no better tools for designing structures than those developed by geometers and crystallographers such as Federov and A.D. Alexandrow[4] about the notions of zonohedra. The architect using these tools is not restricted to reconstructing salt or garnet crystals. He can tailor-make his zonohedra on any whim — the lines of the zonohedron's star have the symmetries he puts there. Men designing things for their own use need not confine themselves to the shapes of minerals. Perhaps it is sad that minerals cannot follow all the shapes that zonohedra invite, but so what? There is no reason for geometers, crystallographers, architects and engineers to addle themselves to the level of the dumbest mineral. If you look at any texts on space frames, you will notice that there is no use made of the symmetries of the icosahedron.

The 31-zone space frame system provides greater construction versatility with a smaller inventory of parts than any other system.[7] The

Fig. 5 Eleven-zone zonohedra cluster.

cube-based space frame system has 13 zones, so its edges can point in 13 directions while the 31-zone points in 31. Using these different edges one can form 13 × 12 / 2 × 1 = 78 different pairs of lines in the 13-zone system, and 31 × 30 / 2 × 1 = 465 different pairs in the 31. Not every pair of lines defines a new plane. The 31-zone system has 25 different planes while the 31-zone has 121.

There are six different triangles formed in the 13-zone system and 20 in the 31-zone (this counts right- and left-handed versions of the same triangle as one).

You can see that while the 31-zone system has only 31 / 13 = 2.4 times as many zones, it forms 20/6 = 3.3 times as many triangles and 121/125 = 4.8 as many planes. We are getting more than our money's worth in versatility from the added lines. In three dimensions the reward expands. Taking the three-dimensional version of the triangle, I have counted nine different quets (Martin D. Richman's terms for tetrahedra) in the 13-zone system and 65 quets in the 31-zone system. Again, this count combines right and left versions as one: 65/9 = 7.2. This versatility should appeal to engineers, architects, and builders (Fig. 6).

3. Are Looks Misleading?

All the pictures I have seen of quasicrystals look like the zome structures we built 15 and 20 years ago. The facets are the faces of triacontahedra with angles 63.435° and the faces are angle-regular, but not regular otherwise. The edge lengths (zone lengths) vary to change the shapes of the figures just as we did in manufacturing, to produce a great variety of shapes without departing from a small number of angles and components.

The forms are not separate polyhedra tangent to each other; rather, they are polyhedra fusing with one another — just the technique we used to get the best forms for people to live in (Fig. 7).

When I read an article about quasicrystals (Fig. 8), I don't read that the shapes resemble our structures, although many workers in the field use our structural system (U.S. Patent No. 3,722,153) to make their models and are very well aware of the prior work. Instead I read about brilliant theoreticians who have devised clever aperiodic tiling schemes. I can't see their patterns in nature's production; I just see forms that look like our buildings. I hope someone explains why the logic behind our

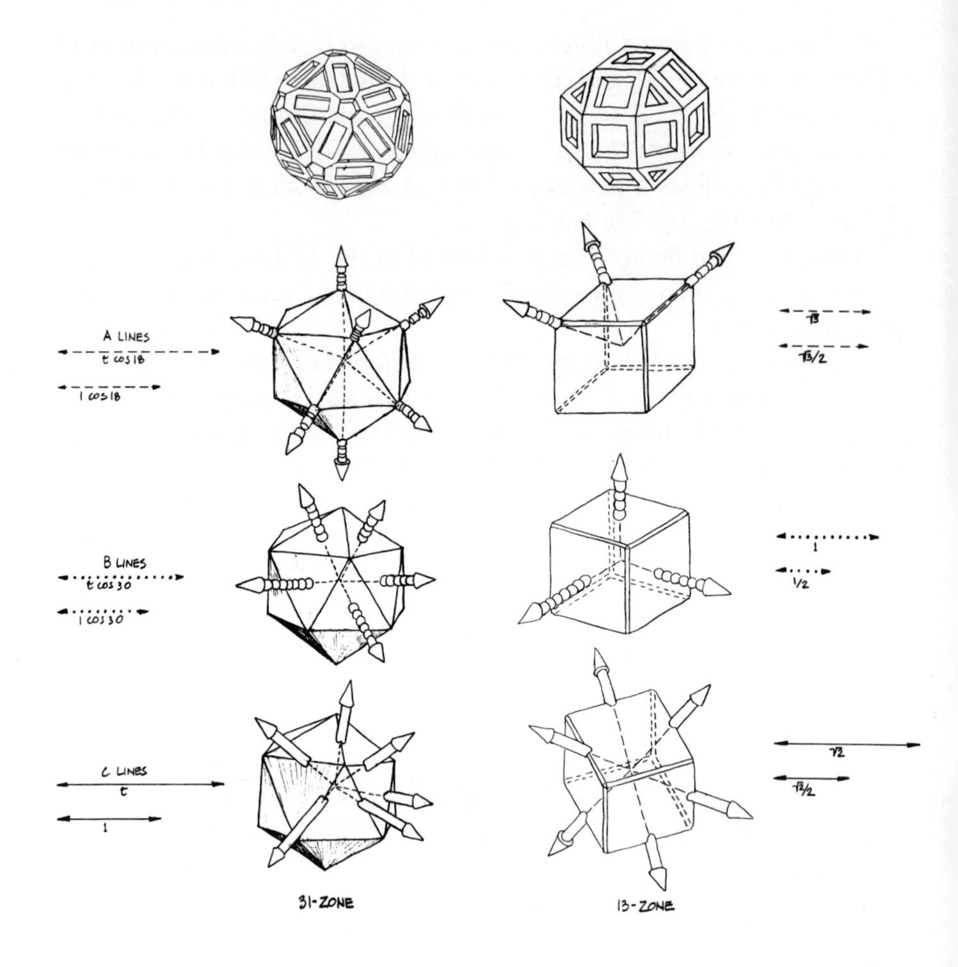

Fig. 6 The systems.

zome houses explains nothing, while the Penrose darts and kites will probably turn out to explain everything.

Ignored, boiling in rage over the slow reception of the 31-zone system we introduced directly with buildings, modeling kits, and playground climbers, I have come to see our powerful industrial society as a collaboration of perverts. To impress those in positions of power, don't show them beautiful objects and simple parts that fit together in mysterious ways, show them bibliographies, footnotes, graphs, computer

Fig. 7 Kittle Zome, Santa Barbara County, California, 1975.

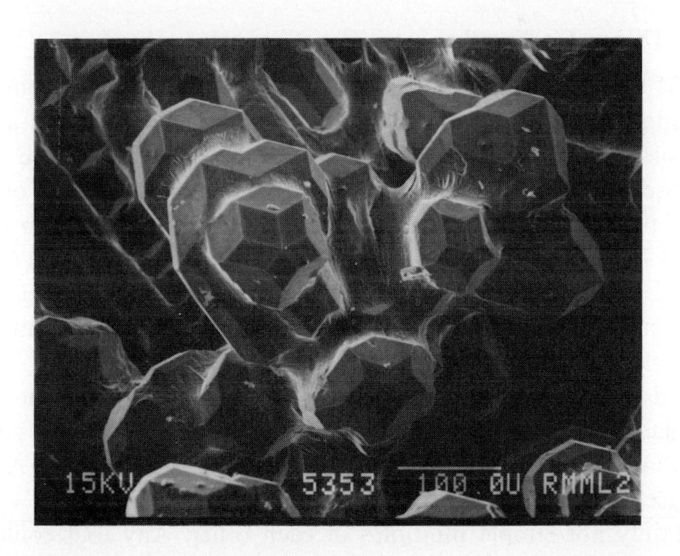

Fig. 8 Quasicrystals, Ref. 8. (Photograph courtesy of Frank W. Gayle, National Institute of Standards and Technology.)

images, matrices, and lists of coordinates. Be sure to stay in two dimensions. Realize that this audience feels more comfortable this way — like men who prefer photographs of women to real women who breathe and have opinions.

4. An Overlooked Treasure

A discussion of space frames with fivefold symmetry quickly turns to a disturbing subject — ignorance and dull-wittedness on the part of generations of scientists and mathematicians. How could this simple topic, which yields rich rewards and requires little education to understand, have been overlooked? Perhaps it was a hierarchy of ignorance.

The crystallographers guessed at the construction rules that molecules follow. The mathematicians seized on these rules and elaborated and refined all their possible outcomes. The crystallographers then looked only for the forms predicted by the mathematicians, who had by then devised an elegant prison in which to cower. Architects and engineers working with tubing and welders obediently confined themselves to forms devised by mathematicians. Early mathematical omissions left enormous unoccupied territories in our world of forms.

5. How Things Fit Together

The 31 lines in space emerge from the vertices, face midpoints, and edge midpoints of an icosahedron. Every pair of lines defines a plane. Sometimes more than two lines exist in a plane. Because of the symmetries found in the icosahedron, the pairs of lines lie in only five kinds of planes. Three of these are the planes perpendicular to the $6A$, $10B$, and $15C$ planes. We call a plane perpendicular to a C line a C plane.

The C planes teem with triangles. Fig. 9 shows a sample of an icosahedral joint (31-zone) with the lines of a C plane. The pattern has the triangles in the C plane.

It is disappointing, after our talk of simplicity, to see what a mess nature has left us with. Look at some of the angles in these patterns: 31.71747441° . . .20.90515744° . . . . Hardly an even degree. And the three base lengths $A = 0.95105652, \ldots B = 0.86602540, \ldots C = 1.000$ are certainly not simple multiples of each other. Any fool could have advised mother nature to use multiples of 30° (the sensible manufacturers of drafting equipment commonly place ruts every 15° in protractors

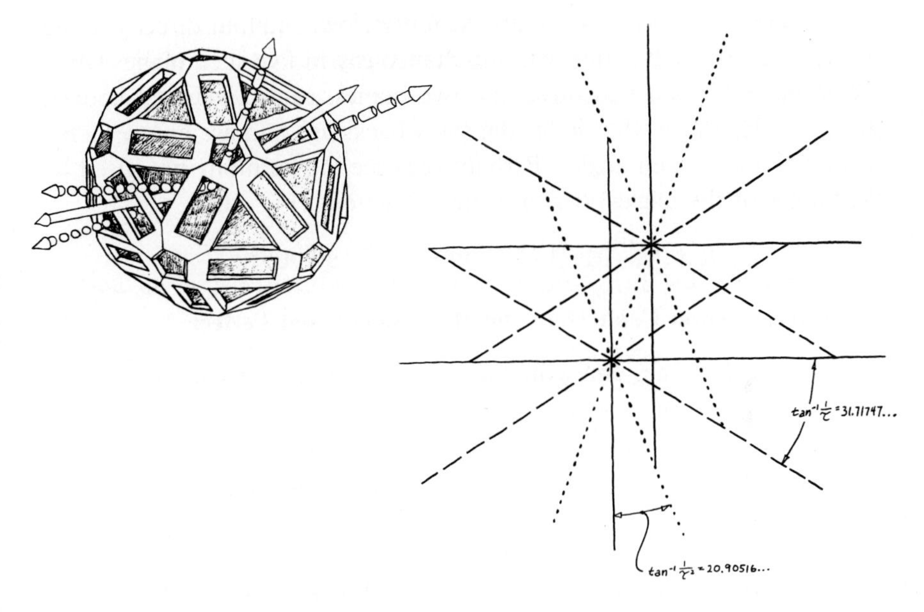

Fig. 9 C plane pattern.

to help draftsmen on legitimate projects — angles that differ slightly become impossible to draw, since the tool soon finds the happier nearby position). We should remember that pulling this plane out of the womb of the icosahedron and nagging at the peculiar angles it has is not fair. Looking at the *C* plane from above, as we see in Fig. 9, is like looking behind stage at a magic show.

Look at a *C* plane from along any of the four *A* lines that rise out of the plane when the rest of the 31 zones are added. The angles appear in a different light. They are all multiples of 36° (all but one *C* line that breaks our rule), and the lengths from these views are all powers of $\pi = (\sqrt{5} + 1) / 2$. Every triangle (that does not include the one *C* line) is a golden triangle. Fig. 10a illustrates the view from along the two *A* lines that make an angle of 58.285° with the *C* plane, while Fig. 10b show that of the other two *A* lines whose angles are 31.717. . .°.

It is comforting to return to these vantage points and simply forgive nature the strange lengths and angles she employs behind stage to make her actors behave regularly for the 12 views from the vertices of an icosahedron. It is quite a feat, considering that each plane puts on a perfect show for 12 views (four above, four below, and four directly in the plane) and that each *A* line acts simultaneously in five such planes while the *B* and *C* lines act in three and two planes, respectively. Of course, there are also those who prefer the show backstage and are happy facing irrational lengths and angles. Perhaps they see the harmony promised by the values of the angles' trigonometric functions (Fig. 10):

$$\tan^{-1}\pi = 31.717°\ldots \quad \tan^{-1}\pi^{-2} = 20.905°\ldots$$

6. An Icosahedron Can't Help But Be Complex and Perfect

How is it that lines emerging out of the corners, faces, and edges of a regular icosahedron could be up to so many tricks at once? The icosahedron simply could not help itself. Having arranged its 20 triangular faces in the most uniform and regular fashion, it was helpless to do otherwise.

The midpoints, corners and edges of the 20 triangles find that they are unable to avoid perfection in their relationship with each other.

Space is extremely docile and can meet many different demands simultaneously, as long as one conforms to the simplicity of the regular figures.

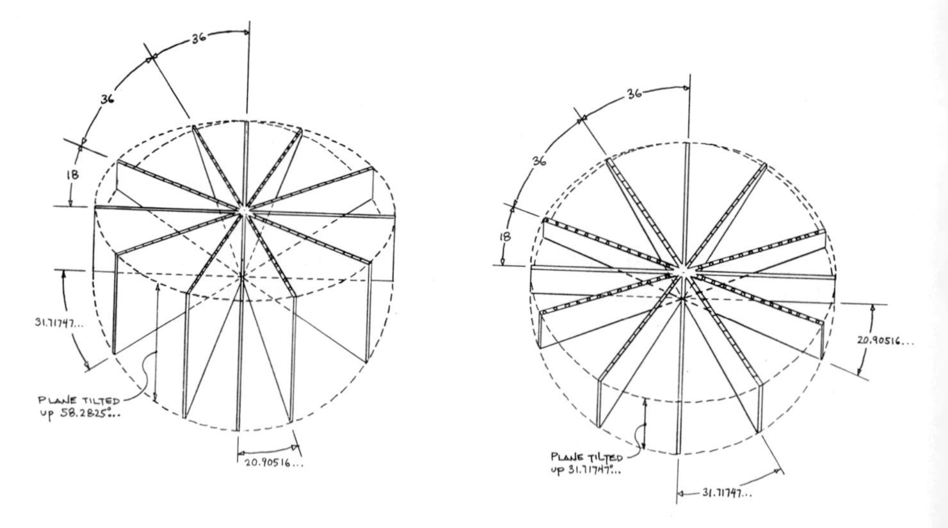

Fig. 10 C plane projections (*A* and *B*).

The exhilaration of using the 31-zone construction system after a life of using a cube-based system is like what I imagine the poet would feel on discovering five or six letters in the alphabet, which could be sprinkled on words to make words with new meanings and new rhymes.

7. Fibonacci Space

The 31-zone truss is very closely related to the Fibonacci numbers and the golden triangles.

Are there other series and triangles that fulfill themselves in spaces we cannot imagine?

The Fibonacci series is intimately related to the triangles shown in Fig. 11. The numbers label how many tenths of a circle the angles are. Triangles ABC and BCD are similar. We want to know the ratio R between the short and long sides of such triangles. Our key is the line CB, which is at once the short side of the large triangle and a long side of the small triangle. Thus:

$$\frac{1}{R} = \frac{R}{1+R} \qquad R^2 = R + 1 \qquad S_n = S_{n-1} + S_{n-2}.$$

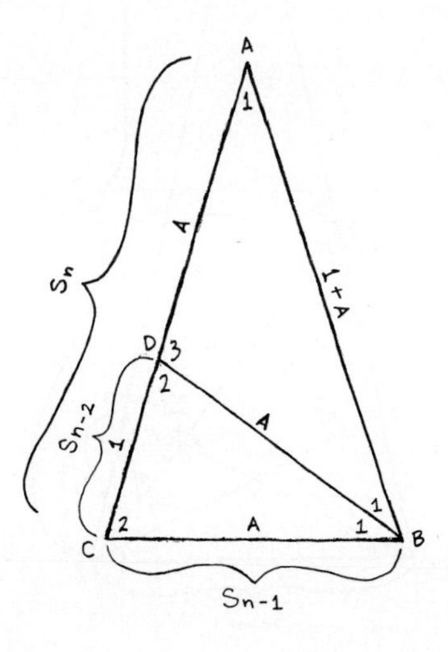

Fig. 11 Fibonacci five-line series.

This exercise is familiar to almost everyone acquainted with mathematics. Now, let's look at triangles made with angles that are multiples of $\pi/7$. In Fig. 12, there are four similar triangles descending in size. Each triangle uses the small side of the next larger triangle as one of its long sides: *ABC* largest; *CDB*, middle size; *BGD*, small; *GDJ*, smallest.

The long sides of the descending triangles are labeled $S_n, S_{n-1}, S_{n-2}, S_{n-3}$. How do you set S_n, equal to a sum of the smaller components?

It can be seen that $S_{n-1} = CF$ while $CA = S_n$. Also by observation, $FA = FB$. How do we set FB equal to other terms? We note that $FH = S_{n-1}$. Also $HB = S_{n-2} - S_{n-3}$, so $S_n = 2S_{n-1} + S_{n-2} - S_{n-3}$ and the ratio R between the long and short sides of such a triangle is the solution to the equation $R^3 = 2R^2 + R - 1$.

If we start with any three numbers S_1, S_2, and S_3 and form the succeeding terms, the ratio between consecutive terms quickly ap-

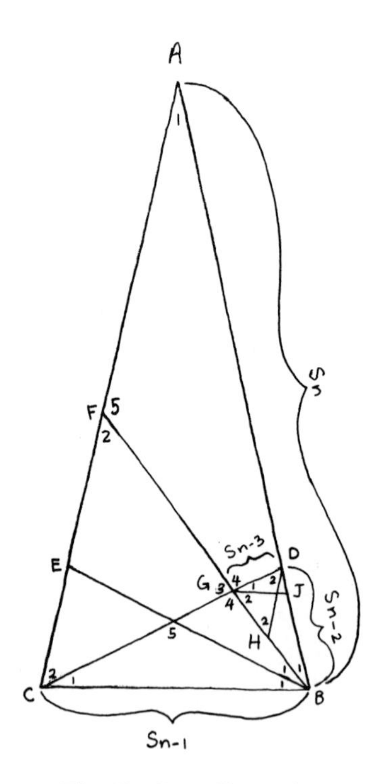

Fig. 12 Seven-line series.

proaches 2.247 . . . – the solution to the equation. Any polynomial equation can be solved in a similar manner.

Although this method of solving equations has been known for a long time among students of algebraic number theory, few math teachers are aware of it and a beautiful connection between patterns and algebra is missed.

I have tried in vain to find series for 9, 10, and 11 lines like those for 5 and 7 lines.

References

1. S. Baer, *Dome Cookbook.* Lama Foundation, Corrales, NM (1968).
2. H.S.M. Coxeter, *Regular Polytopes* (pp. 27–32), 2nd Edn. Macmillan, New York (1963).
3. W.W. Rouse Ball, revised by H.S.M. Coxeter, *Mathematical Recreations and Essays* (pp. 140–143). Macmillan, New York (1962).
4. A.D. Alexandrow, *Konvexe Polyeder* (pp. 308–316). AkademieVerlag, Berlin (1958).
5. Supplement to the *Whole Earth Catalog,* Extra (pp. 23–29). Portola Institute, Menlo Park, CA, July 1969.
6. S. Baer, *Zome Primer.* Zomeworks Corporation, Albuquerque, NM (1970).
7. S. Baer, The 31-zone structural system, in H. Nooshin (ed.), *Third International Conference on Space Structures* (pp. 872–875). Elsevier Applied Science Publishers, London and New York (1984).
8. F.W. Gayle, *J. Mater. Res.* **2** (1987) 1–4

THE NEW ZOME PRIMER

David Booth

1. Introduction

"For the last thirty or forty years, most Americans have somehow lost interest in geometry," wrote H.S.M. Coxeter in the preface to *Introduction to Geometry*[1] in 1961. In the Waldorf School movement geometry is important in maintaining the interest of students in mathematics throughout their twelve school years, grades one through twelve. The geometry of the pentagon is characteristic of this subject as a whole. The following passage from an article of H. von Baravalle[2] describes the pedagogical situation.

> "The geometry of the pentagon has become almost a foster-child beside other chapters of geometry, as for instance the geometry of the triangles or of the quadrilaterals. Considering terminologies, we find the whole field of trigonometry deriving its name from the geometry of triangles and the 'quadrature of areas' from the regular representative of quadrilaterals, all units for measuring areas being also squares.
>
> The characteristic elements of the geometry of the pentagon are neither related to the trigonometric reproduction of forms nor to measuring areas . . . It is the particular appeal of the pentagon to the sense of beauty and the unique variety of mathematical relationships connected with it which are the characteristics of the geometry of the pentagon. This geometry is therefore particularly fit to stimulate the mathematical interest and investigations."

In 1985, one of the Waldorf School's projects involved obtaining a large Zometoy kit and using it to acquaint students with the icosahedral system in space; the ordinary plane geometry of the pentagon appears on certain of its plane sections. The Zometoy was manufactured for architectural studies by Zomeworks Corporation and was briefly sold on the market from 1970–1973. Its developer, Steve Baer, described the kit in *Zome Primer.*[3] The few kits sold were gradually lost, scattered, and broken so that by 1980, one could acquire parts only with considerable difficulty. Biocrystal Corporation reproduced the Zometoy in order to help develop a version which would be free of clumsy features, and when Biocrystal had finished its studies, a large kit became available for classroom use.

To described the kit, let τ and $\bar{\tau}$, respectively, be the positive and negative roots of $x^2 - x - 1$. The kit contains three kinds of strut: call them "red" (r), "yellow" (y), and "blue" (b), although the colors are not necessarily intended to suggest an analogy with color theory. The red struts are held rigidly along a fivefold axis of symmetry of an icosahedron by a connector. The yellow struts are held on a threefold symmetry axis. Finally, the blue struts lie along a twofold symmetry axis.

Let r, y, and b be the lengths of the shortest red, yellow, and blue struts, respectively. These lengths are chosen so that $r:y:b$ is $\sqrt{\tau+2}:\sqrt{3}:2$ or equivalently cos(18): cos(30): cos(0). The other struts to be kept in inventory have lengths $\tau^n r$, $\tau^n y$, $\tau^n b$ for some n. Because $\tau^{n+2} = \tau^{n+1} + \tau^n$, it is theoretically necessary to keep only two lengths of each color in inventory, but for practical reasons the Zometoy supplied four different lengths of each color.

High school students have produced drawings, performed calculations, assisted pupils in other schools, and carried out these tasks under a wide variety of classroom circumstances, sometimes formal and sometimes informal. The original Zometoy was not, however, primarily intended for educational purposes. The use of the kit and indeed the introduction of pentagonal symmetry itself cannot be rigidly prescribed in an educational syllabus. Nevertheless, the lines of thought described below have all been tested in the classroom.

2. Zonohedra

A quarter century ago, Buckminster Fuller's lectures stimulated a tide of popular enthusiasm for geometry. Many of Fuller's young admirers set out to build geodesic domes for themselves; but among these eager young craftsmen, the rhombic triacontahedron, which played a role in certain of Fuller's designs, had a mysterious reputation. The *Zome Primer*[3] clarified the triacontahedron by presenting it as a member of the more general family of zonohedra, a line of thought which had appeared in H.S.M. Coxeter's *Regular Polytopes.*[4] "Zome" was coined as a portmanteau of "zone" and "dome." There is still no more reliable way to introduce the icosahedral system to novices than by beginning with the rhombic triacontahedron. The following description of a triacontahedron can easily be turned into building instructions: it is a convex body whose edges are all red struts of equal length and which meet either three or five to a vertex; no vertex of valency three can be adjacent to one of valency five.

Once a triacontahedron has been built, it can be stretched and shrunken along its set of parallel edges. Mature students can be given the general idea of zonohedron, of which the triacontahedron is an instance. The fact that stretching and shrinking is possible without tearing the basic shape is based on the affine character of these polyhedra families.

The algebraic description of zonohedra begins with a set of vectors $\{a_1, \ldots, a_n\}$ in space (to make the rhombic triacontahedron $n = 6$ and the vectors are of equal magnitude and point along the fivefold axes of an icosahedron centered at the origin). Next, form each of the 2^n sums obtained by adding vectors whose indices are taken from subsets of $\{1, \ldots, n\}$.

Let s be one of the sums of the $\{a_i\}$ and let b, c be members of $\{a_i\}$ which have not contributed to s. Then $\{s, s + b, s + c, s + b + c\}$ form a parallelogram. The triacontahedron is the boundary of the resulting lattice, or, to be more general, we could form the sums of the vectors $\{a_i\}$ and then regard the actual zonohedron generated by $\{a_i\}$ as the convex hull of the resulting lattice of vector sums.

Zonohedra thus fall into affine families whose defining vectors share the same directions and whose members share common angles, like natural crystals, in spite of their variations in actual shape. These families can be classified by the "first projective diagram" of Coxeter,[5] which arises when the plane at infinity intersects the lines of action of the defining vectors $\{a_1, \ldots, a_n\}$.

Pairs of points in the plane at infinity characterize a family of parallelograms having common angles. Triangles in the diagram at infinity give families of parallelopipeds.

Zonohedra of four directions produce a quadrilateral on the plane at infinity. Since all non-degenerate quadrilaterals are projectively equivalent, there are only three distinct families of zonohedra with four directions. When these four directions are independent, polyhedra of twelve sides result. The most regular of these is the rhombic dodecahedron, whose faces are rhombi having the ratio $1 : \sqrt{3}$ along the diagonals and whose edges are related as the four threefold axes of symmetry of a cube.

This relatively familiar zonohedron, the rhombic dodecahedron, might be considered a natural first construction project. It is, however, so simple that it does not present as clear a picture of zonohedra to the mind as does the triacontahedron. In addition, it possesses the static aspect found in the cubic system and is thus less stimulating to the artistic sensibilities. Finally, it is inconvenient for beginners with the Zometoy to construct the rhombic dodecahedron because its defining vectors lie along but four of the ten yellow three-way axes.

At Green Meadow Waldorf School, we keep a reasonably large collection of cubic parts manufactured by Synestrutics Corporation and described by Peter and Susan Pierce in their book[6] on the applications of the cubic system in space. The Synestrutics Kit has three kinds of strut, having lengths in the ratio $1 : \sqrt{2} : \sqrt{3}$ and lying along the fourfold, twofold and threefold axes of the cube, respectively. Using this kit, it is very easy to produce a rhombic dodecahedron, because the four vectors lie along all four of the cube's three-way axes.

Returning now to the general theory, the rhombic dodecahedron is the most regular representative of the family of zonohedra whose projective diagram is a quadrilateral.

A curious feature of the rhombic dodecahedron renders it atypical among the zonohedra whose diagrams are quadrilaterals. It frequently happens that teenagers develop a fascination with the idea of higher dimensional spaces. One should give them some encouragement in the hope that it will provide a motive for further study. Now, the zonohedra generated by four independent vectors $\{a, b, c, d\}$ could be seen as projections into three space of the four-dimensional zonotype (a higher dimensional analog of a zonohedron) whose projective diagram is a non-degenerate tetrahedron in the hyperplane at infinity of projective four-space. If one builds a zonohedra of four vectors $\{a, b, c, d\}$ and includes all the 16 points obtained by the sums $\{0, a, b, c, d, a + b \ldots, a + b + c + d\}$, then one sees something readily recognized as the projection of a four-dimensional object. With the rhombic dodecahedron, however, the four vectors $\{a, b, c, d\}$ have the property that $a + b = c + d$. Thus there are only 15 rather than 16 points in the lattice (two coincide) and the figure requires effort to be seen as a projection from a higher dimensional space.

One can easily check that this phenomenon takes place a dimension lower using just pencil and paper. Take three arbitrary vectors $\{a, b, c,\}$ in the plane. Form the sums $\{0, a, b, c, a+b, b+c, a+b+c\}$. Connecting the edges parallel to the three generating vectors, one has what appears to be the projection of a box. Now repeat the experiment but with a, b, c of equal length and standing 60 degrees apart; let us say that the angle from a to c is 120 degrees. Since $a + c = b$, the eight vectors will reduce to seven. The resulting zonogon is a regular hexagon divided into equilateral triangles and now lies flat to the eye.

We eventually arrive at six generating vectors, as well as the triacontahedron which they generate. It thus requires some time for the mind to arrive at the first construction project, already familiar to the hands.

An interesting degenerate case of a six-vector zonohedron can be built from the Synestrutics Kit using the six lines of the twofold symmetry axes of the cube as vector directions. This produces a truncated octahedron whose faces are eight regular hexagons and six squares. Divide the hexagons into triples of $1 : \sqrt{3}$ rhombi, to produce the thirty cells of a distorted triacontahedron. Viewed in this way, the truncated octahedron stands as a sad, tragic looking projection of a formerly active, lively form. This shows the beauty and dynamism of the pentagonal system, the ultimate reason for its great educational potential.

Fig. 1 shows a cluster of congruent rhombic triacontahedra, pairs of which can be joined along askew hexagons. These can also be joined to similar figures whose struts are one golden section multiple different in size. The details need not be given here; but is should be obvious that the triacontahedron alone offers very rich opportunities for construction projects.

Fig. 1

The second zonohedra family to enter the classroom is the enneacontahedron family. Fig. 2 shows a cluster of enneacontahedra of two different sizes, the larger being a golden section multiple of the smaller. These were built by different groups of students who quickly saw how to join them.

The rhombic enneacontahedron is not degenerate: it is generated by 10 vectors lying on the threefold axes of symmetry of an icosahedron, no three of which fall into a plane. There are, however, two distinct angles among these lines, neglecting supplements. Consider an icosahedron with a fivefold axis vertical. It has a pentagonal cap, top and bottom, and a drum about the center in the form of a pentagonal antiprism.

Fig. 2

Now fix any threefold axis through the center of a triangle on the top cap. The first angle is formed between this ray and another through an adjacent triangle on the drum: the second is formed with a threefold axis through a drum triangle that meets the original cap only at a vertex rather than along an edge. One can check the remaining threefold axes to see that they produce angles either congruent or supplementary to these. Because there are two different kinds of angles, the ten vector zonohedron is bounded by two different kinds of rhombi. One of them is the rhobus to be found bounding the rhombic dodecahedron.

3. The Blue Lines

Perpendicular to a red, fivefold axis of symmetry of an icosahedron lies a red plane. The icosahedron bears a pentagonal cap on top and an inverted cap on the bottom when it is placed with a vertical fivefold axis, so the red plane through its center slices the drum in half, which is an anti-prism of ten triangles that separates the two caps. This plane cuts each of the ten edges of the drum in half, so it must contain five blue, twofold directions. There are no red or yellow lines on these planes; thus a red plane contains only blue lines. These blue lines form the characteristic pentagonal forms that we would expect from viewing an icosahedron along a fivefold axis: pentagons, pentagrams, 36-72-72 isosceles triangles, 108-36-36 isosceles triangles and the associated rhombi.

The geometry and trigonometry of these pentagonal forms have been omitted from the modern textbooks so that most school teachers must study the matter for themselves. These figures are the ultimate test of the pupils' ability with straightedge and compass: younger pupils merely follow the instructions given them, but older students can analyze the constructions from the arithmetic of the golden section ratio.

Not all forms constructible with straightedge and compass are Zometoy constructible. Pen and ink can strike out in any direction, but the connectors allow only angles which are integral multiples of 36 degrees between blue lines. For example, a decagon can be constructed with the Zometoy, but a pentagon obtained by skipping the decagon's alternate vertices cannot, since this would create an angle of 18 degrees. Other pentagons are, of course, constructible, but not ones concentric with a constructible decagon.

One soon becomes used to these limitations; school pupils scarcely seem to notice them when they find the rich structures constructible on the six different directions along which red planes lie. Regular dodecahedra, icosahedra, stellations of both, and golden section rectangular solids are just a few of the things that can be found among the blue line forms.

Fig. 3 shows one of the rich variety of three dimensional stars constructible with blue lines along the red fivefold planes of the icosahedral system.

Besides the ten blue lines in each red plane, there are also two mutually perpendicular blue lines in each blue plane. That is to say, there are triples of mutually perpendicular twofold symmetry axes in an icosahedron. Select one such blue line. The plane orthogonal to it through the icosahedron's center is a blue plane. It contains two perpendicular blue lines.

Thus the structures constructible with blue lines fall into two general kinds: the pentagonal system in planes perpendicular to a fivefold axis and the cubic system in planes perpendicular to a twofold axis.

Let $F_0, F_1, F_2, \ldots$ be the Fibonacci numbers with $F_0 = 0, F_1 = 1$. This is the first, most natural, example of a recursive sequence, and can

Fig. 3

be used with good effect in junior high schools. There is a very rich variety of numerical patterns among these numbers. They provide numerous elementary examples of arguments through mathematical induction, and they provide a means by which students who have become discouraged in mathematics can renew their friendship with the subject through experimental verification of numerical laws while more advanced students pursue the same topics. The equation $\tau^{n+1} = F_{n+1}\tau + F_n$ will be discovered by students themselves under the proper circumstances. It is wonderful to discover that this can be extended to negative values of n by the natural extension of the Fibonacci sequence into negative indices.

It is a simple matter to check that

$$\frac{\tau}{2} = \frac{2\tau + 1}{2\tau + 2}$$

using the identiy $\tau^2 = \tau + 1$. The left side arises from calculating cos (36) using the law of cosines on a 36, 72, 72 triangle while the right side arises from the law of cosines on a 36, 36, 108 triangle.

One can review other trigonometric identities by applying them to pentagonal angles. It may happen that τ disappears entirely from such calculations in favor of $\sqrt{5}$ and can be introduced again later.

Baravalle's paper[1] provides many other suggestions. In his lectures on the subject, Baravalle showed how a teacher could appear like a magician to his class. To learn the secret of the magician's tricks, however, which are based on the properties of Fibonacci sequences and the golden section ratio, one must study mathematics.

4. Kit Constructible Trusses

A kit is a collection of components of three kinds: point components or nodes, line components, and planar components or panels. The Zometoy and the Synestrutics kits are both kits of this general kind; others have been produced for scientific studies, technical research, education, or simply for amusement. The theory of kits is fragmentary at present and has spread from two opposite foci. Flexible kits allow the use of concepts from projective geometry, Grassman algebra, and invariant theory: rigid kits, whose angles and strut lengths are fixed, permit the use of metric concepts.

The study of flexible kits has been driven by the need to discover how rigid structures can be formed from flexible components or, if rigidity is not attained, what motions are possible for a complex structure which is locally flexible.

The study of rigid kit centers on the problems of constructibility: what can actually be built? One would like a kit to have a simple inventory of few different kinds of parts, yet it must not be so simple that it cannot be used for important purposes. The construction of trusses to brace polyhedra is just such an important purpose. School pupils recognize this immediately. They become more and more eager when they are able to construct double layered, trussed polyhedra. Their work seems architectural, the models are of striking beauty, and it is inspiring to see complex forms arising from clear, simple ideas.

In the older engineering literature, trusses consisted of planar frameworks which were placed in parallel planes so that the copies were translated with respect to each other in the direction normal to the plane of each. More recently, trusses are copies of identical frameworks placed in parallel planes, as before, but the copies may be shifted so that the translation from one to the other need not be perpendicular to the planes of each.

When trusses are constructed, the frameworks in adjacent planes are linked by struts. In the older conception of a truss, the linking struts were all parallel to each other and perpendicular to the planes of the frames. Thus quadrilaterals were formed which had to somehow be braced by members outside of the framework in order to make the whole structure rigid.

In the more recent conception, the frameworks have been shifted so that connections between the planes can be more easily inserted.

A trussed zonohedron will have double layers on each of its faces. The faces are divided into a grid of small rhombi similar to the original. The two layers of each face have their smaller rhombi offset so that centers of small rhombi in one layer correspond to vertices in the other.

Fig. 4 shows a student assembling a trussed triacontahedron. The panels were constructed in a class period prior to actual assembly.

To construct such a trussed triacontahedron, one uses $(n - 1)^2$ rhombi on the inner plane and n^2 on the outer one. The inner face must be somewhat smaller than the outer one. If $n = 1$, there is no inner layer, and we have a simple triacontahedron. As n increases, the thickness of the shell decreases relative to the diameter of the whole.

It remains to be shown that one can position these two grids so that vertices on one can be joined to the four neigboring vertices of the other. These joints form the bracing of the truss.

This bracing must, of course, involve only the angles available to the kit. The two layers cannot be set at an arbitrary distance apart and still use only the available angles. It is a fact that the required braces are constructible with the kit.

The only previously published description of these zonohedral trusses seems to be the brief account of Baer in Ref. 7.

5. Projection from Hyperspace

It has been discovered by Marc Pelletier of Biocrystal that the most commonly pictured projections of the four dimensional, the regular polytopes {5, 3, 3} and {3, 3, 5}, are kit constructible with the Zometoy.

Fig. 4

There are photographs in Coxeter[4] of these polytope projections. The actual models were built by Paul Donchian for an exhibit more than fifty years ago, but have since fallen into disrepair.

A full account of these projections will not be given here, though they are of practical importance. Students who build freely with the kit may easily stumble upon peculiar forms. Fig. 5 shows a group of students who have gone exploring on their own. One often finds that they have foreshortened versions of icosahedra, dodecahedra, regular stars, and so on, embedded in their creations. A master of the icosahedral system needs to be well acquainted with these kit constructible projections.

6. The New Primer

Marc Pelletier and Paul Hildebrandt of Biocrystal have developed a new version of the Zometoy which is simple to use and eliminates the need for both written instructions and experienced teachers. As this new version of the kit becomes available, many more people will become

Fig. 5

familiar with the icosahedral system. There could be a stimulation of spatial imagination should the icosahedral system spread widely. "An instruction manual in every ball" is one of their slogans. That means that the kit itself would be its own primer.

References

1. H.S.M. Coxeter, *Introduction to Geometry*, Wiley, New York (1961).
2. H.Võn Baravalle, The geometry of the pentagon and the Golden Section, *The Math. Teacher* **XLI** No 1 (1948) 22–31.
3. Steve Baer, *Zome Primer*, Zomeworks, Albuquerque (1969).
4. H.S.M. Coxeter, *Regular Polytopes*, 2nd Edn. Macmillan, New York (1963).
5. H.S.M. Coxeter, The classification of zonohedra by means of projective diagrams, *Twelve Geometric Essays*, So. Illinois Univ., Carbondale (1968).
6. Peter Pierce and Susan Pierce, *Experiments in Form*, VanNostrand, New York (1980).
7. Steve Baer, Ten zone truss, *Whole Earth Catalog*, July 1969, 24–29.

A UNIQUE FIVEFOLD SYMMETRICAL BUILDING: A CALVINIST CHURCH IN SZEGED, HUNGARY

Szaniszló Bérczi and László Papp

1. Biblical Traditions

"A star appears from the tribe of Jacob,
Mace shoots from Israel."
Prophecy of Baalam. Moses: The Book of Numbers, 24:17

This prophecy predicts the arrival of Jesus Christ metaphorically in the symbol of a star. "I see him but not now, I see him but not closely. A star appears from the tribe of Jacob, mace shoots from Israel," says prophet Baalam. Central spaces may express this symbol in their forms and in the structure of their ceilings appearing in many synagogues and churces all over the world.[1]

One of the star symbols, the fivefold star symbol, has a deep root in Greek mathematics and philosophy. The pentagram was the secret sign of the Pythagoreans, who overemphasized the role of numbers in construction and deciphering the structures of creatures. Recent construction of a building with fivefold symmetry needs more living traditions in a community.

2. Medieval Arts and Architectural Traditions in Hungary

In its last splendid decades of independence, the medieval Hungarian Kingdom was characterized by peace and the flourishing of the Renaissance under the rule of king Mathias Hunyadi. His royal court was most active in promoting the arts and sciences in Europe in the 15th century. The spirit of the royal court radiated throughout the country well into the next century. Among other structural and symbolic (design) elements, ceiling cassettes in country castles and in new and rebuilt churches have been preserved. Among surviving design traditions, the fivefold symmetry of flower rosettes of ceiling cases have remained in the greatest number all over the country. They were preserved in spite of the frequent destructions of the Turkish wars in the 16th and 17th centuries, mainly in more obscure parts of the country. The church in which an ensemble of the most remarkable examples of Renaissance cases can be seen is in Gelence, Székelyland, Transylvania, Romania. (The medieval Hungarian Kingdom was divided into many parts in the Trianon Peace Treaty after World War I.) The cases of the church of Gelence were painted in the early decades of the 17th century (Fig. 1).

The other root of architectural tradition in the Szeged Calvinist Church can be found in fortress constructions, which were built against Turkish attacks. A chain of fortress defence was first built by Sigismund I (King of Hungary and later Emperor of the German-Roman Empire) in the 15th century. Along the changing front line between the Hungarians and Turks more and more, fortifications were necessary to strengthen the defence. In the 16th century, our Habsburg kings ruled in one third of the country, Transylvanian princes in another third and Turks in the remaining third. Italian architects were invited to build these fortifications for the Habsburg-Hungarian Kingdom and Transylvania. Many fortresses were strengthened by earthworks. (Turks had the strongest artillery in Europe at that time.) Mounds surrounding the old castles were built in a regular star-shaped form, frequently with fivefold symmetry. The fortresses with fivefold structure are shown in Fig. 2; a representative example is the ground-plan of Lenti (see Fig. 3).[2]

3. Border Conditions for Fivefold Symmetrical Central Building in Szeged

Besides biblical, design and fortress architectural traditions, a very practical condition played a dominant role in planning the Szeged Calvinist Church with fivefold symmetry. This had to do with the church grounds: the form of the plot which was given to the Calvinist Church by the town council. The plot was a triangular district, where streets meet

Fig. 1 Renaissance design preferred the pentagonal forms of flowers. Castles and churches were frequently decorated by painted ceiling cases in the 16th to 18th centuries. One of the oldest case ceilings with many pentagonal flowers can be found in Gelence, Székelyland, Transylvania (now Romania). (It was built in 1628. Donators: Zsigmond Csoma and Mihály Nagyercsei Tholdalagi.)

in an acute angle near 36 degrees. This is the angle between two non-neighbouring sides of a regular pentagon (Fig. 4).

The formation of this plot goes back to the general replanning of Szeged after the Great Flood of 1879. On March 12 of that year, the icy

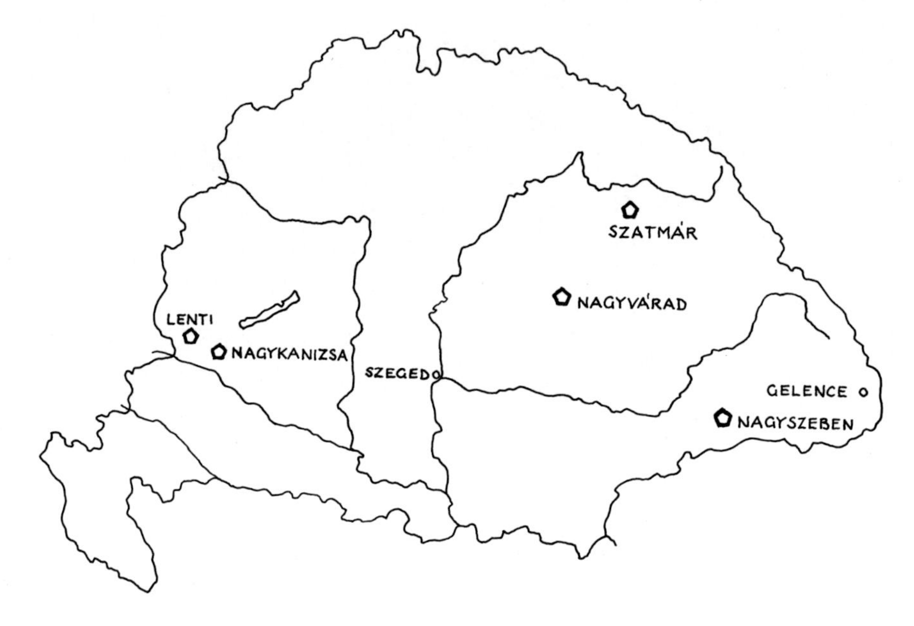

Fig. 2 Fortresses with fivefold symmetry could be found all over Medieval Hungary. They were fortified by earthworks and walls in the Turkish wars which lasted more than one and a half centuries (during the 16th to 17th centuries). Szeged, where the Calvinist church with fivefold symmetry was built in the early forties of our century, and the place of Gelence, where the cases of Fig. 1 can be found are also indicated.

flood from the river Tisza inundated the town, which had a population of about 60,000 at that time. The water remained in the town for months and destroyed about 6,000 houses. Only the old churches, the old fortress and some communal buildings — built from brick — such as schools, the city hall, the railway station and some factories survived. Partly because of the tabula rasa created by the flood in the district, and partly because of the flourishing economy of the Austro-Hungarian Monarchy, it became possible and necessary to plan and build the town in modern urbanic structure. A cobweb system of avenues and two rings were planned for Szeged similar to the earlier city-plan of Budapest (Fig. 4).

In this cobweb system, where the small ring (Tisza Lajos körut) met the basically square-cross street system, triangular areas took shape. Three of these triangular areas were given to different Reform churches. The Honvéd Square Calvinist Church conducted a competition for the design

Fig. 3 Plan of fortification by earthworks of the fortress of Lenti, Western Hungary, from the 17th century (from Ref. 2).

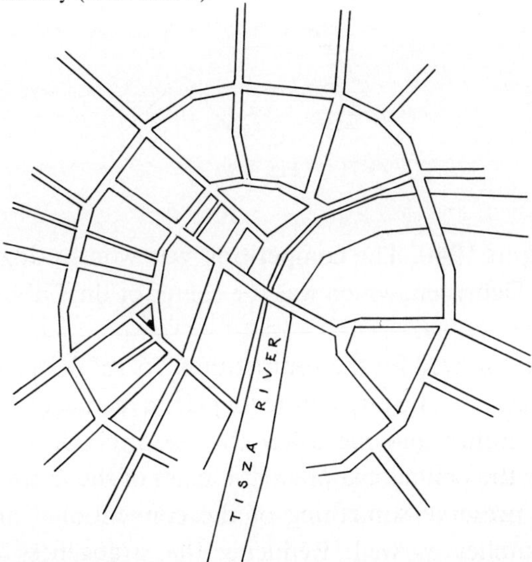

Figs. 4a and b. The network of the main streets (avenues and rings) in the replanned Szeged (after the destructive icy flood of the Tisza river in 1879). a: (see above) The black corner shows the site of the Honvéd Square Calvinist Church. b: (see next page) The enlarged plot shows the main mass of the building at an acute angle meeting Tisza Lajos Körut and Földvári utca.

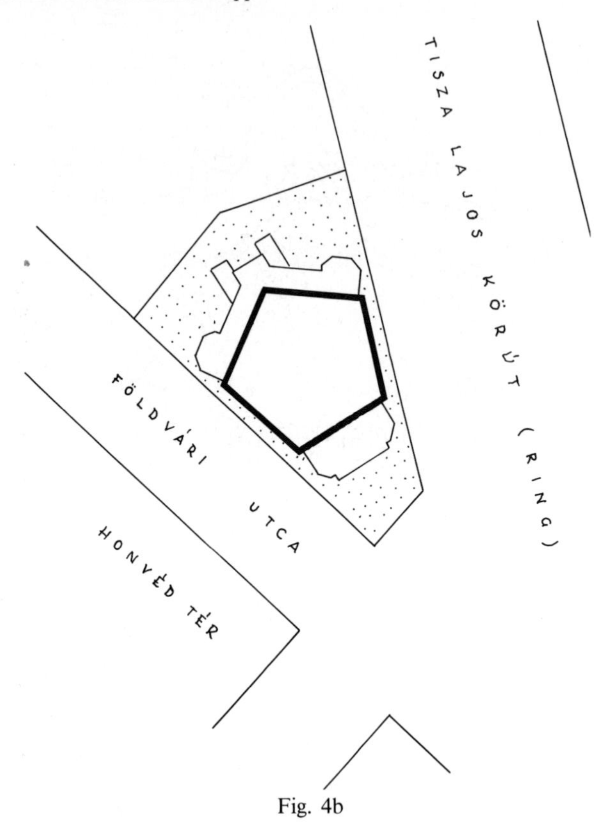

Fig. 4b

of the building in 1940. The competition was won by József Borsos, an architect from Debrecen, which was the center of the Calvinist Church in Hungary.

In his work submitted for the competition, József Borsos described that the angle of the area inspired him to plan a pentagonal building. He planned the entrance and the tower (in the direction of the 36-degree acute angle) by the pentagonal prismatic mass of the main building. This way he could preserve something of the conventional arrangement of Hungarian churches as well. Reducing the strangeness of pentagonal space by this conventional arrangement was so successful that — ironically enough — most people in the town do not recognize the extraordinary shape of this building, even if they walk by it every day (Figs. 5 and 6).

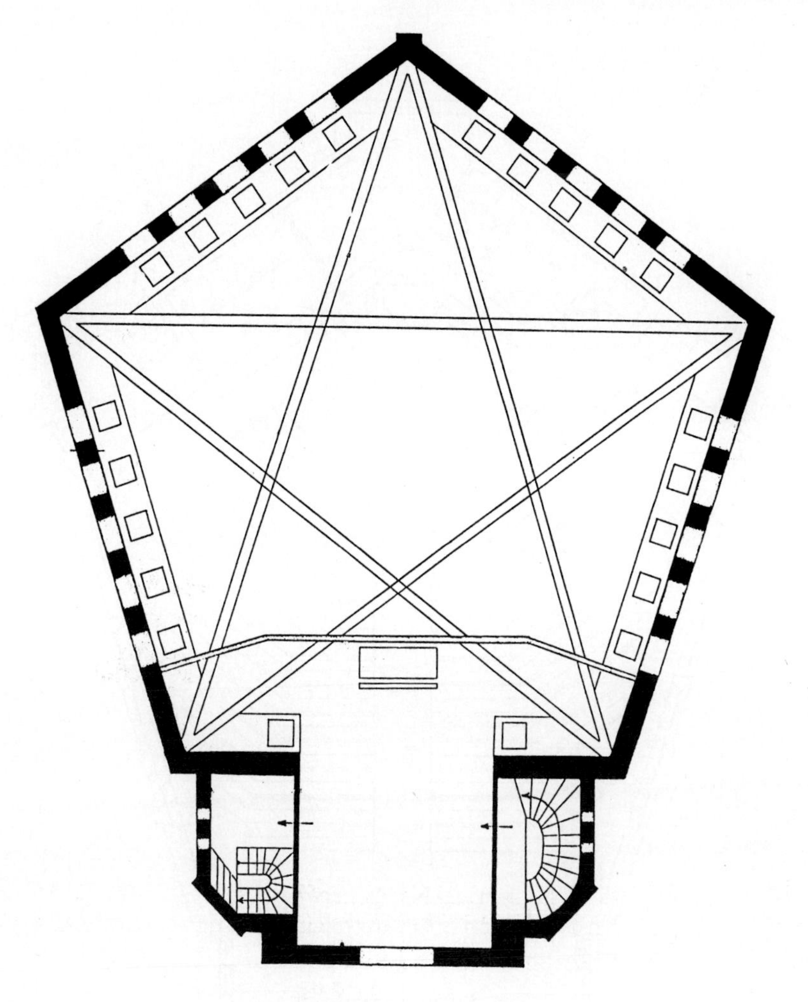

Fig 5a. Overview plan of the Honvéd Square Calvinst Church made by József Borsos, architect of the church (from Ref. 3). The pentagram shows the five main arched rib-console which hold the ceiling.

In summary, we can conclude that although different old traditions such as the Biblical star symbol of Christ, the pentagonal architectural and Renaissance design might have motivated the architect, József Borsos designed the Honvéd Square Calvinist Church in pentagonal form primarily because of the shape of the area given to the church by the Szeged town council after the Great Flood of 1879.

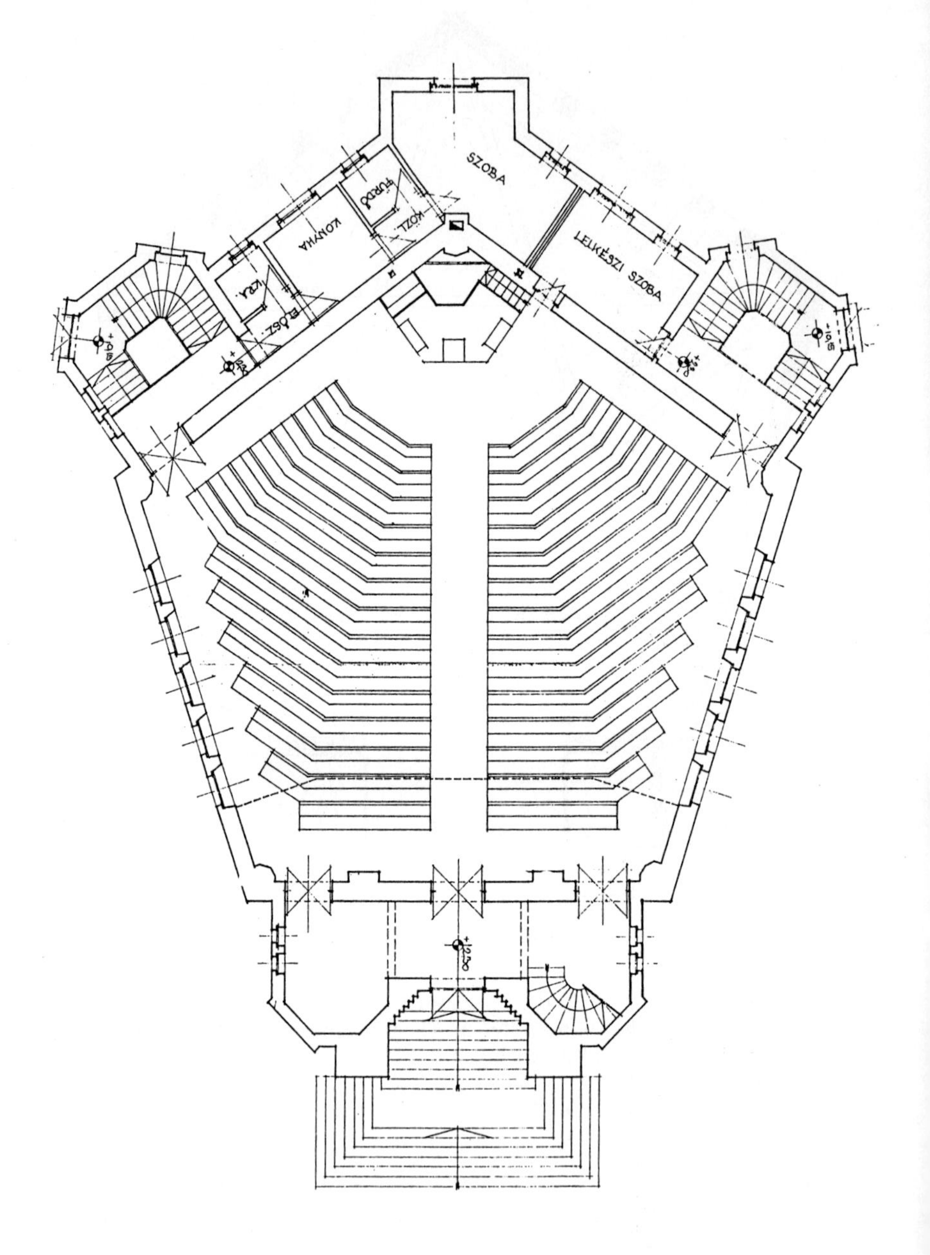

Fig. 5b. Detailed plan of the church santuary.

Fig. 6 The side view of the Honvéd Square Calvinist Church looking from SW to NE (1942).

References

1. S. Ritz, A Templom: Mult, jelen és jövö felülmulhatatlan alkotása; Az Örök-Templom Santo Stefano Rotondo, az Apokalipszis uj Jeruzsálemének szent városa, a magyarok nemzeti temploma Rómában. [The Church: unsurpassable creation of past, present and future; The Eternal Church Santo Stefano Rotondo, the Saint Town of the New Jerusalem of the Apocalypse, the National Church of Hungarians in Rome.] Edition of the author, Rome, pp. 70 (W.D.)
2. L. Gerö, Magyarországi várépitészet [Fortress building in Hungary]. Budapest (1955).
3. J. Borsos, A Szeged Honvéd téri Református Templom pályaterve [Tender plan of the Szeged Honvéd Square Calvanist Church]. Szeged (1940).

FIVEFOLD SYMMETRY AND (BASKET) WEAVING IN VARIOUS CULTURES

Paulus Gerdes

1. Introduction: Why Did Fivefold Symmetry Arise in Human Culture?

Symmetry is such an overwhelming phenomenon both in nature and in culture that it is easy to forget to question: Why? Why do fans all over the world have a line symmetry? Why do most cooking pots display a rotational symmetry? Why do many baskets show four-, six-, or fivefold symmetry patterns?

At first sight one might think that (fivefold) symmetry or other basic geometrical ideas arose in human culture as a blind copy of symmetry and physical forms in nature. For example, Eves[1] writes:

> The first geometrical considerations of man ... seem to have had their origin in simple observations stemming from human ability to recognize physical form and to compare shapes and sizes... Many observations in the daily life of early man must have led to the conception of curves, surfaces and solids. Instances of circles were numerous — for example, the periphery of the sun or the moon, the rainbow, the seed heads of many flowers, and the cross section of a log. A thrown stone describes a parabola; an unstretched cord hangs in a catenary curve;... spider webs illustrate regular polygons... (pp. 165–166).

In this way, might the idea of a regular pentagon have been suggested to early man by a seastar (Fig. 1)?

Recent studies show that symmetry and other regular forms of early human artifacts like hand axes[2] and baskets[3] were not the result of a conscious use of geometrical ideas, abstracted (as Eves supposes) from

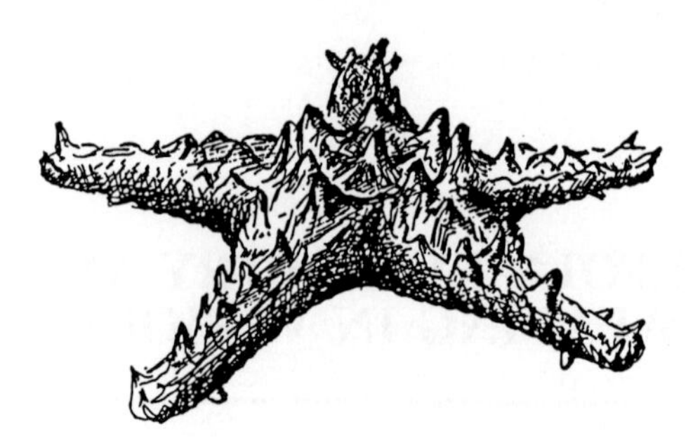

Fig. 1 Pentagonal seastar.

nature. On the contrary: regularity and symmetry of man-made objects are the result of creative human labor, as we will illustrate in this paper for the case of fivefold symmetry.

As there are so many different forms in nature, it has to be explained why man gradually became capable of observing certain forms in nature. There are no forms in nature that are a priori destined for human observation. The capacity of man to recognize geometrical forms in nature and in his own products has been developed through his labor activity ("*Tätigkeit*").[3,4] The real, practical advantages of an invented regular and symmetrical form for an artifact lead to a growing consciousness of this regularity and symmetry. The same advantages stimulate man to compare this artifact with other labor products and with natural phenomena. The regularity and symmetry of a product generally simplifies its reproduction, and in this way both consciousness of its form and interest in it are reinforced. With growing consciousness and interest, there develops at the same time a positive valuing of the invented form: the form is also used where it is not necessary for material, objective reasons; it becomes experienced as beautiful.

Fivefold symmetry patterns in woven baskets, fishtraps, hats, brooms, cages, thimbles, etc. may appear at first sight to be the result of instincts or of an innate feeling for these forms, or, mechanically, to represent imitations of natural phenomena. In reality, however, these forms have been created by man to satisfy his daily needs. Working with the materials at his disposal, man learned to understand which were the forms necessary to produce something useful.

In this paper we show examples of how fivefold symmetry emerged quite "naturally" when artisans were solving some problems that arose in (basket) weaving.

2. Pentagonal–Hexagonal Baskets

Mozambican peasants weave their light transportation baskets (*litenga*) and fishermen their traps (*lema*) with a pattern of regular hexagonal holes. One way to discover this pattern is the following.

How can one fasten a border to the walls of a basket, when both border and wall are made out of the same material? How does one wrap a wallstrip around the borderstrip? What is the best initial angle between the borderstrip and the wallstrip? (see Fig. 2).

When both strips have the same width, one finds that the optimal angle measures 60°. By joining more wallstrips in the same way and then introducing more horizontal strips, one gets the *litenga* pattern of regular hexagonal holes (Fig. 3) or, when one omits alternate horizontal strips, one gets a pattern of semiregular pentagonal holes (Fig. 4) as did the Karaibe Indians.[5] Both patterns are plane patterns.

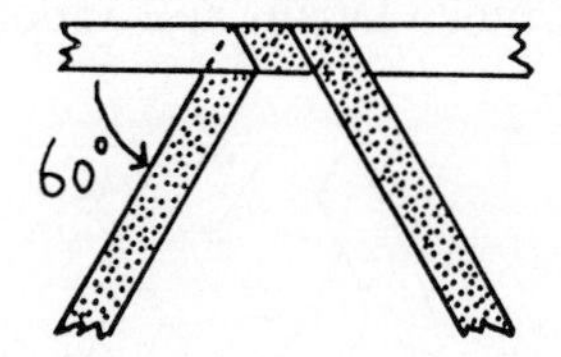

Fig. 2 Best initial angle between borderstrip and wallstrip.

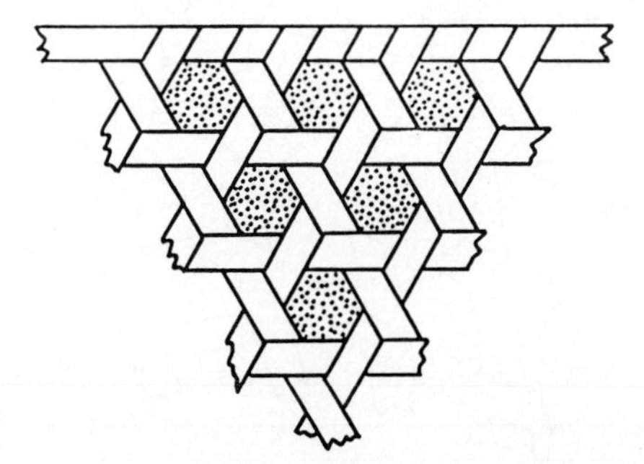

Fig 3. *Litenga* pattern of regular hexagonal holes.

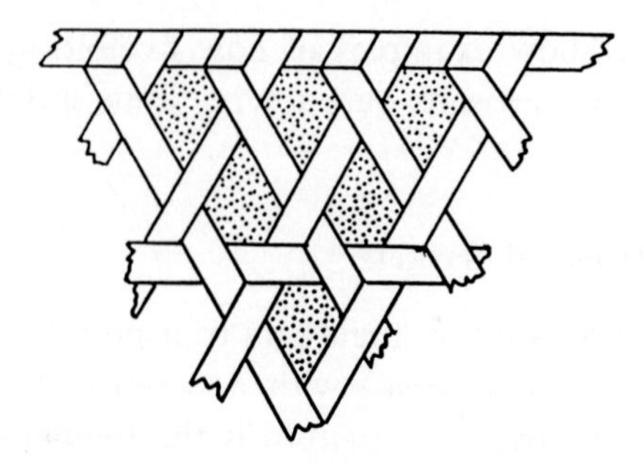

Fig. 4 Pattern of semiregular pentagonal holes.

Only the faces and edges of the *litenga* basket display the regular-hexagonal-hole pattern. The artisans discovered that to be able to get a basket that "curves" the faces at its "vertices," it is necessary to reduce the number of strands. At these "vertices" of the basket, the six strips that "circumscribe" a hexagonal must be reduced to five. That is why one encounters at these "vertices" little, slightly curved regular-pentagonal holes (Fig. 5). These pentagonal–hexagonal baskets are, for instance, also woven by the Ticuna and Omagua Indians (northeastern Brazil), by the Huarani Indians,[6,7] by the Kha-ko in Laos,[8] and by the Menda in India.[9] One sees them also in China, Japan, and Indonesia.[10–12] These woven

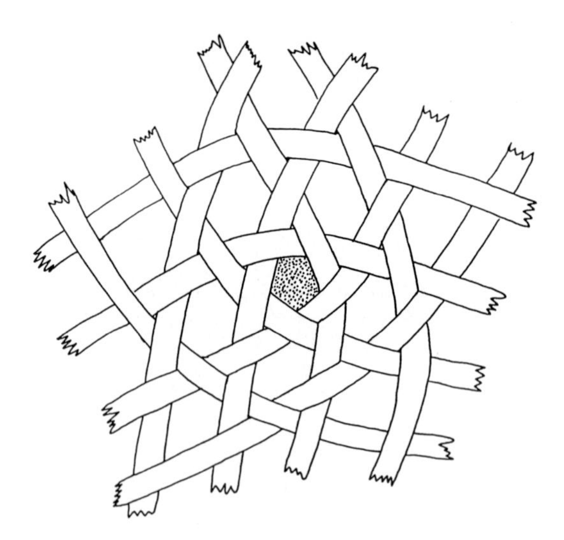

Fig. 5 "Vertex" with regular-pentagonal hole.

pentagonal basket "vertices" may have served as a suggestion for the well-known symbol of a woven pentagram (see Fig. 6), for example, in Japan.[13]

Fig. 6 Woven pentagram.

What types of pentagonal–hexagonal basket can be woven? Is it possible to weave closed baskets? We find that the smallest possible "basket" (Fig. 7), made out of six strips, is similar to the well-known modern soccerball made out of pentagon and hexagonal pieces of leather. The "*sepak tackraw*" ball from Malaysia and of ancient origin is woven in the same way.[14]

Fig. 7 Woven ball with pentagonal holes.

There exists also a "compact" way to weave in three directions. This time no holes appear (Fig. 8), as encountered, for example, in the Palau Islands (east of the Philippines) and in South Africa. A face of a basket woven in this way displays regular hexagonal stars (Fig. 9). At the

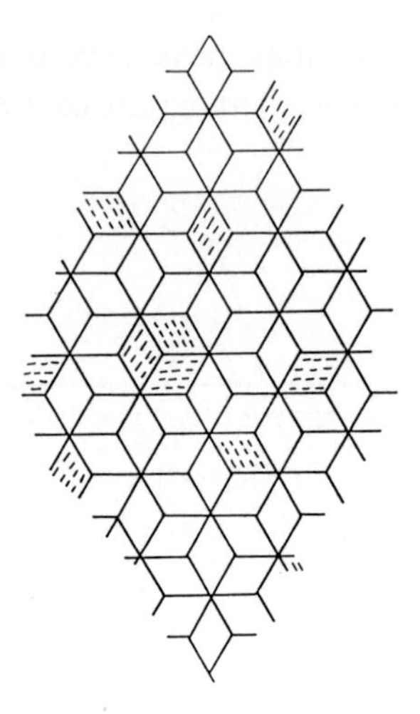

Fig. 8 Compact three-directional weave.

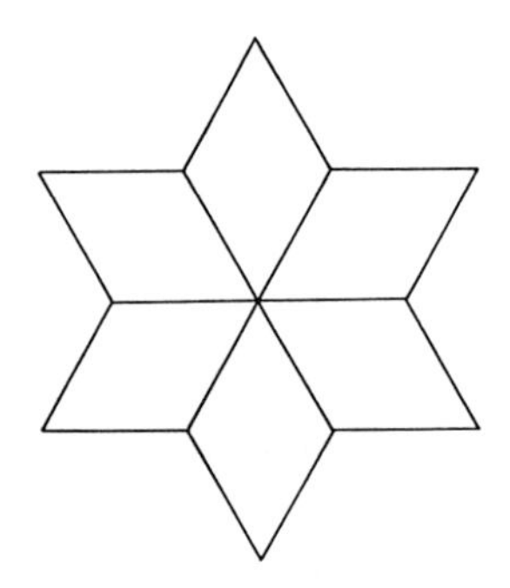

Fig. 9 Regular hexagonal star.

"vertices" it becomes once more necessary to reduce the number of strands. That is why one encounters at these "vertices" *regular pentagonal stars* (Fig. 10).

3. A Pentagonal Thimble (cf. Ref. 4)

A well-known way to fold a regular pentagon is to tie a knot in a strip of paper and press it flat (Fig. 11). If one folds over one end of the strip and holds the knot up to a strong light, one sees a pentagram (Fig. 12). Fig. 13 shows two traditional Japanese crests where the pentagonal knot is applied.[13]

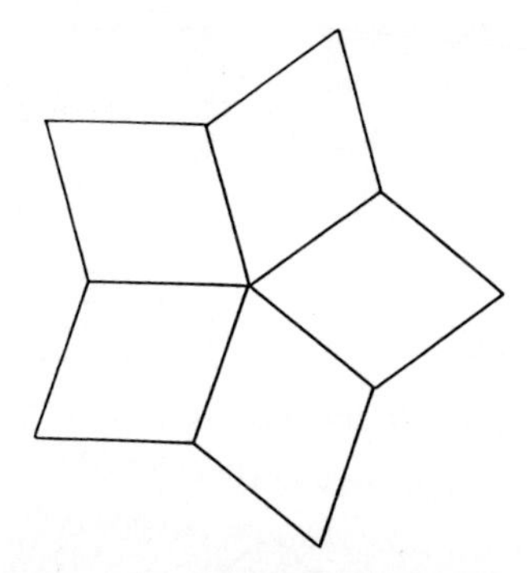

Fig. 10 Regular pentagonal star.

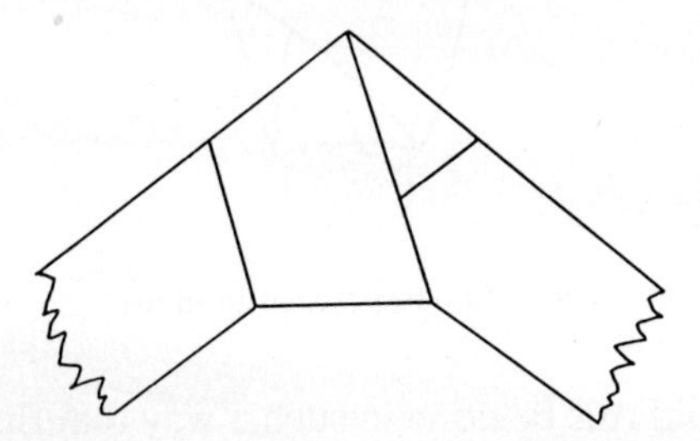

Fig. 11 Folding of a regular pentagon.

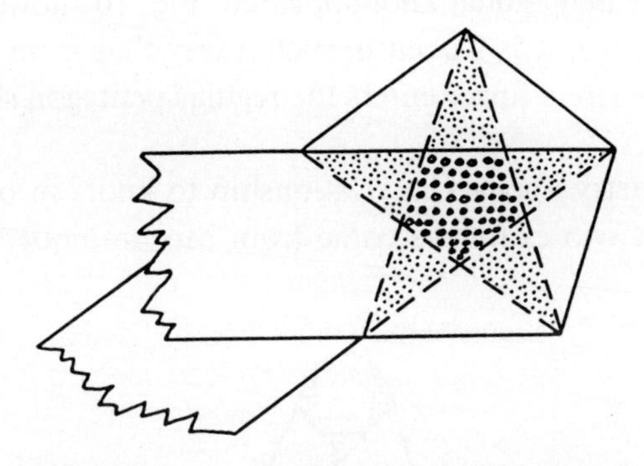

Fig. 12 Woven pentagon with regular pentagram.

A drawing of a thimble (Fig. 14) used by peasants of the Indonesian island of Roti[15] suggests a possible context for the discovery of the above-described way to fold a regular pentagon. As the islanders intended to avoid hurting their fingers when they were stripping off grains of corn, they tried to protect the thumb and forefinger with a strand of leaf

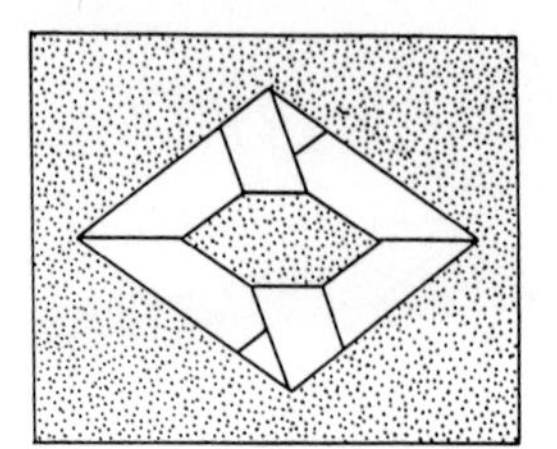

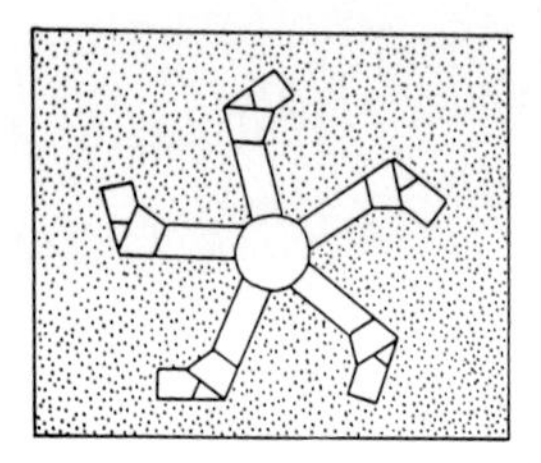

Fig. 13 Traditional Japanese crests.

Fig. 14 Pentagonal thimble.

(Fig. 15). How could this be done in such a way that the "thimble" would not fall off? It would be necessary to tie at least one knot in the strand. Automatically the pentagonal knot appeared. Fig. 16 shows further steps in creating a thimble; it is woven in such a way that it remains open on (only) one side, where a finger enters the regular-pentagon sheath that has been created.

Fivefold symmetry appears in relationship to knots in other contexts, as well. The next two examples come from Mozambique.

Fig. 15 How to protect a finger.

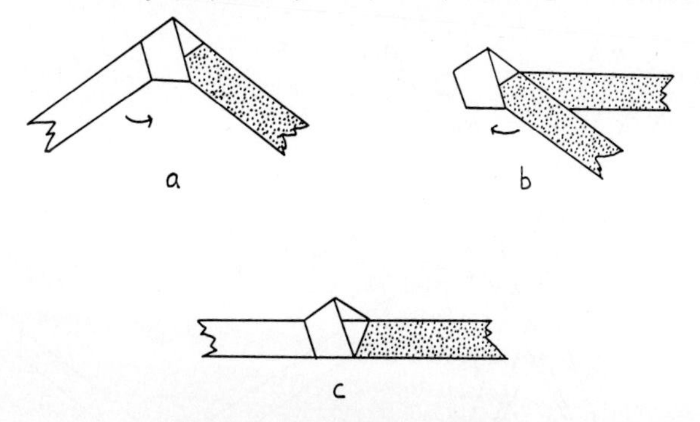

Fig. 16 Weaving of a thimble.

4. A Semiregular Pentagonal Knot (Ref. 4)

Artisans in the south of Mozambique start the weaving of *huama* handbags (Fig. 17) and *funeco* shoulderbags as follows. In perhaps the simplest possible way, two palm strands – or, more likely, two pairs of two strands each – are tied together in a knot as in Fig. 18. Then the wall of a bag is woven, as in Fig. 19.

The knots are pentagonal. The angles that appear in the flattened knot measure 108°, 90°, 90°, 126°, and 126° (Fig. 20a). When such a knot is produced from two thin strands (or strips of paper) and held up to a strong light, a regular pentagon and an almost complete pentagram are seen (cf. Fig. 20b).

5. Fivefold Symmetry and a Broom

Brooms of the type shown in Fig. 21 are very common in Mozambique. The (vertical) sheaves of strips are held together by horizontal

Fig. 17 *Huama* handbag from Mozambique.

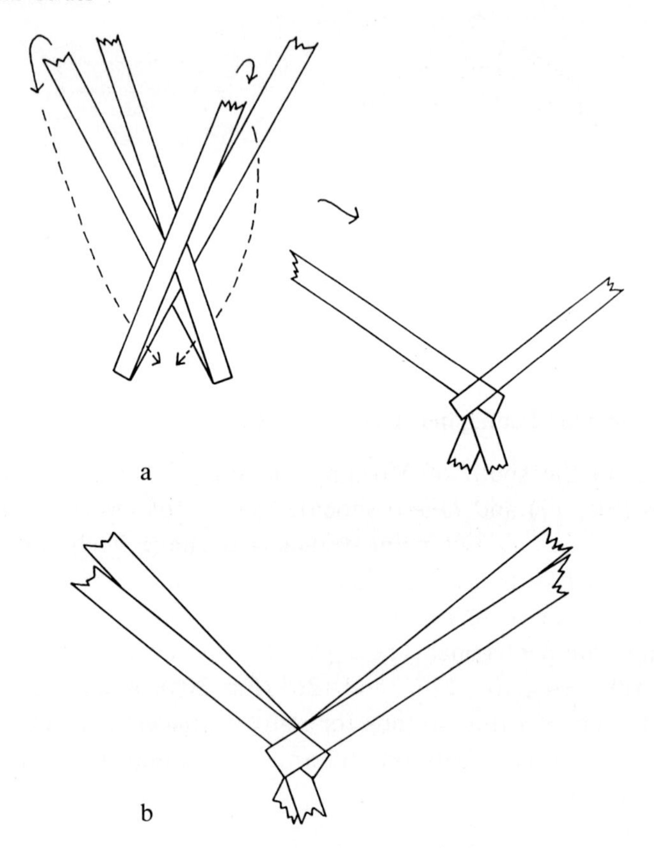

Fig. 18 Weaving of knots for *huama* handbags: (a) with two strands, (b) with two pairs of strands.

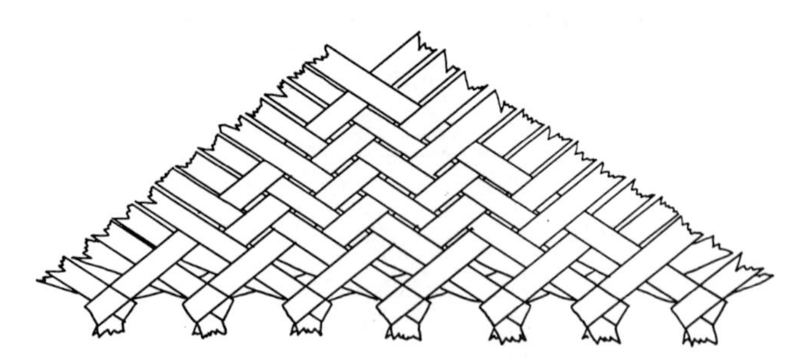

Fig. 19 Part of the wall of a *huama* handbag.

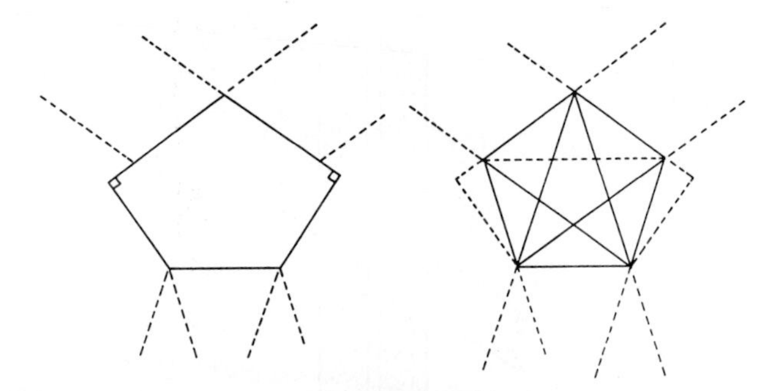

Fig. 20 Flattened knot.

Fig. 21 Traditional Mozambican broom.

layers of strands about twice as wide as the sheaves. The strands are interlaced with two sheaves at a time as in Fig. 22. A chain of flattened, (almost) regular pentagonal loops is the result (Fig. 23).

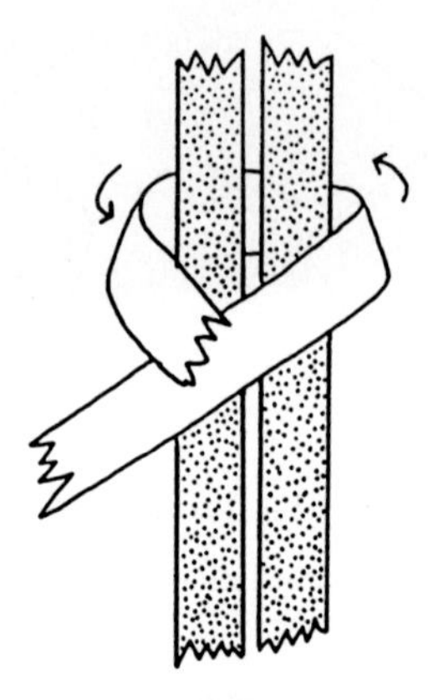

Fig. 22 Starting the interlacing of two sheaves.

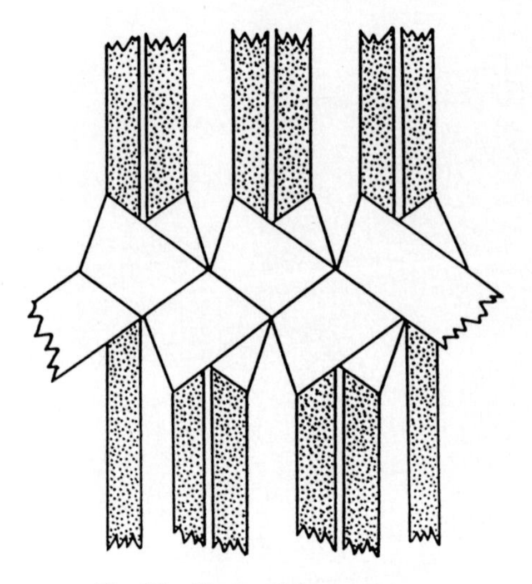

Fig. 23 Chain of flattened loops.

In Angola a similar method, although leading to less regular pentagonal loops, is used when binding together the bars of a cage (Fig. 24; cf. Ref. 16, Plate 7).

6. Woven Hats With Fivefold Symmetry

The Belu of Central Timor weave hats with three- and fivefold symmetries (Fig. 25; cf. Ref. 17, Plate XVIII). Their basic technique is

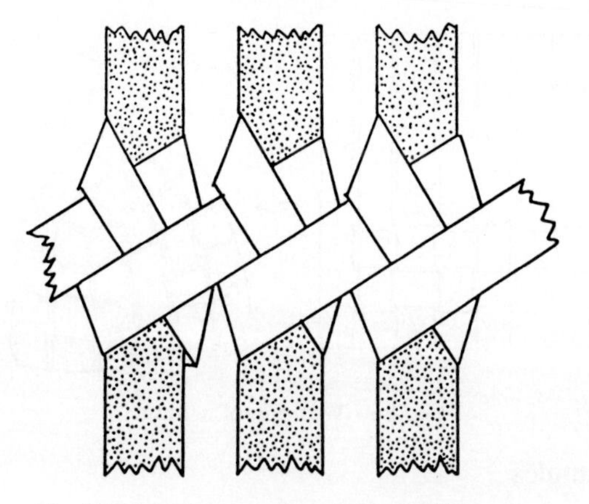

Fig. 24 Binding together the bars of a cage.

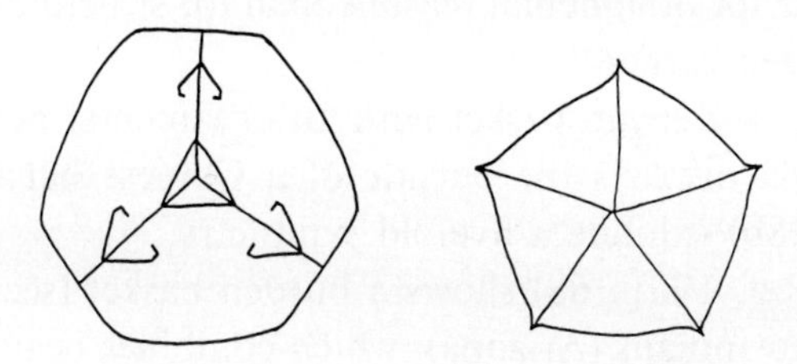

Fig. 25 Woven hats seen from above.

apparently the same as the one employed by artisans in the north of Mozambique and in the south of Tanzania (cf. Ref. 18, Plate 19), and by Kuba artisans (Congo) when they weave their funnels. In the case of funnels and the hats with threefold symmetry, one starts a square mat (weaving in two perpendicular directions) but does not finish it: with the strands in one direction (horizontal in Fig. 25), the artisan advances only until the middle. Then, instead of introducing more horizontal strips, he interweaves the vertical strands on the right with those on the left. In this way, the mat does not remain flat, but is transformed into a "basket." The center goes downward (or upward) and becomes the vertex of the funnel or the top of the hat. Fig. 26 illustrates this production process. When one weaves in three directions (with angles of 60° between them) instead of two orthogonal directions, one arrives in a similar way automatically at the Belu hats with fivefold symmetry. At the top of such a hat, there appears once again a regular pentagonal star (Fig. 10).

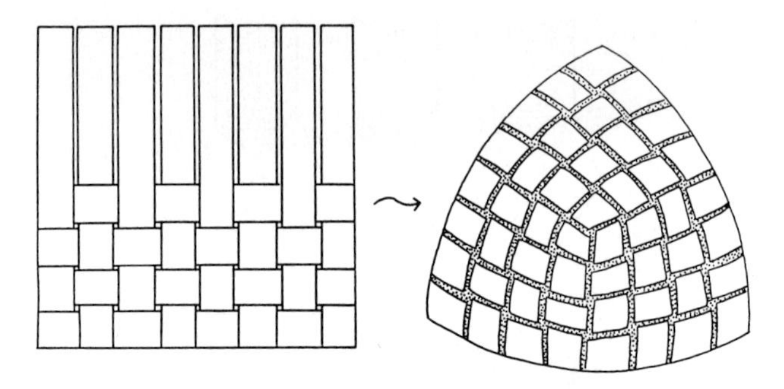

Fig. 26 Weaving of a funnel.

7. Other Examples

There are other instances of fivefold symmetry appearing in basketry, sometimes more for ornamental reasons than for structural reasons, as in the cases analyzed earlier.

Fig. 27 shows a Kenyan basket with an ornamental pentagonal spiral pattern, Fig. 28a displays the outside of a Chinese hat whose interior weaving (Fig. 28b) exhibits a fivefold symmetry.

Fig. 29 (cf. Ref. 19, p. 68) shows a burden basket (seen from above) from the Papago Indians (Arizona), which combines beautifully a global sevenfold symmetry with local fivefold symmetry. Fig. 30 displays the center of the bottom of a Japanese basket, which combines global ninefold symmetry with local fivefold symmetry.

Fig. 27 Kenyan basket.

Fig. 28 Chinese hat: (a) exterior and (b) interior. (From the collection of the author.)

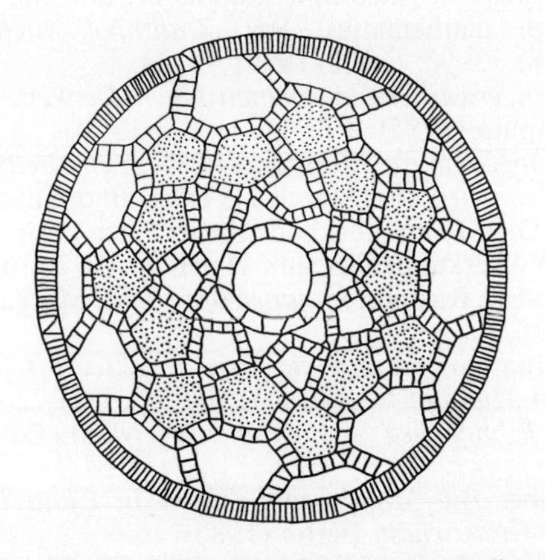

Fig. 29 Burden basket of the Papago Indians.

Fig. 30 Center of the bottom of a Japanese basket. (From the collection of the author.)

References

1. H. Eves, *The History of Geometry. Historical Topics for the Mathematics Classroom.* NCTM Washington, DC (1969), pp. 165–191.
2. B. Frolov, Numbers in paleolithic graphic art and the initial stages in the development of mathematics, *Sov. Anthropol. Archeaol.* (1977–1978) 142–166; (1978) 73–93; (1978–1979) 41–74.
3. P. Gerdes, *Zum erwachenden geometrischen Denken.* Eduardo-Mondlane University, Maputo (1985).
4. P. Gerdes, *Ethnogeometrie: Kulturanthropologische Beiträge zur Genese und Didaktik der Geometrie.* Franzbecker Verlag, Bad Salzdetfurth (1990).
5. K. Kästner, Die Pflanzstockbodenbauer der tropischen Waldgebiete Südamerikas, Völkerkundemuseum, Herrnhut (1978) pp. 94–107.
6. P. Neumann and K. Kästner, *Indianer Brasiliens*, Museum für Völkerkunde, Dresden, (1983).
7. I. Wustman, Huarani Indianer im Oriente Ekuadors, *Kleine Beiträge aus dem staatlichen Museum für Völkerkunde* **5** (1982) 22–29.
8. V. Grotanelli, *Ethnologica – l'uomo e la civilita.* Edizioni Labor, Milan (1965).
9. L. Icke-Schwalbe, *Die Munda und Oraon in Chita Nagpur: Geschichte, Wirtschaft und Gesellschaft.* Berlin (1983).
10. T. Bodrogi, Bestattungsbräuche der Sadang-Toradja, *Kleine Beiträge aus*

sem staatlichen Museum fur Völkerkunde, **2** (1978) 17–21.

11. H. Roth, *The Natives of Sarawak and British North Borneo*. University of Malaya Press, Singapore (1896, 1968).
12. J. Faublée, *L'ethnographie de Madagascar*. Bibliothèque d'Outre Mer, Paris (1946).
13. F. Adachie, *Japanese Design Motifs*. Dover, New York (1972).
14. J. Pedersen, Geometry: The unity of theory and practice, *Math. Intell.* **5:4** (1983) 37–49.
15. W. Hirschberg and A. Janata, *Technologie und Ergologie in der Völkerkunde*. Biographisches Institut, Mannheim (1966).
16. M. Areia, M. Martins, and M. Miranda, *Cestaria tradicional em Africa*. Coimbra University, Coimbra, Portugal (1988).
17. B.A.G. Vroklage, *Ethnographie der Balu in Zentral-Timor*. Brill, Leiden (1953).
18. K. Weule, *Wissenschaftliche Ergebnisse meiner Ethnographishen Forschungsreise in den Südosten Deutsch-Ostafri-kas*. Berlin (1908).
19. C.L. Tanner, Papago burden baskets in the Arizona State Museum, *Kiva* **30** (1965) 57–76.

PENTAGON AND DECAGON DESIGNS IN ISLAMIC ART

Gilbert M. Fleurent

1. Patterns of the Plane Group cmm with Stars or Rosettes

"We conjecture that the Euclidean plane admits no tiling in which each tile has fivefold symmetry."[1] However, there are Islamic patterns which nearly cover the plane by means of decagons; the same holds for 14-gons. Mostly these patterns belong to the symmetry group cmm, and nearly always the same method is used to construct the polygons. In the present article, we limit our attention to patterns with decagons.

The plane symmetry group cmm is one of the 17 two-dimensional crystallographic groups.[2,3,4] The lattice is rhombic, as seen in Fig.1. The diagonals of the rhombus are axes of reflection. Centers of twofold rotational symmetry lie upon the point of intersection of the diagonals, upon the four corners of the rhombus and on the mid-points of the sides of the rhombus as well. The lines, connecting the mid-points of the sides, are axes of glide reflection, which are parallel to the diagonals. The rhombus is divided by the diagonals into four congruent, right-angled triangles, each of which is to be regarded as a fundamental or generating region.

How are the stars or rosettes generally constructed in a rhombus of a pattern of type cmm? Let the right-angled triangle *ABC* be one fourth of the rhombus of a lattice (see Fig. 2a). This triangle is a fundamental region. *D* is a point of twofold rotational symmetry on the hypotenuse. This is a

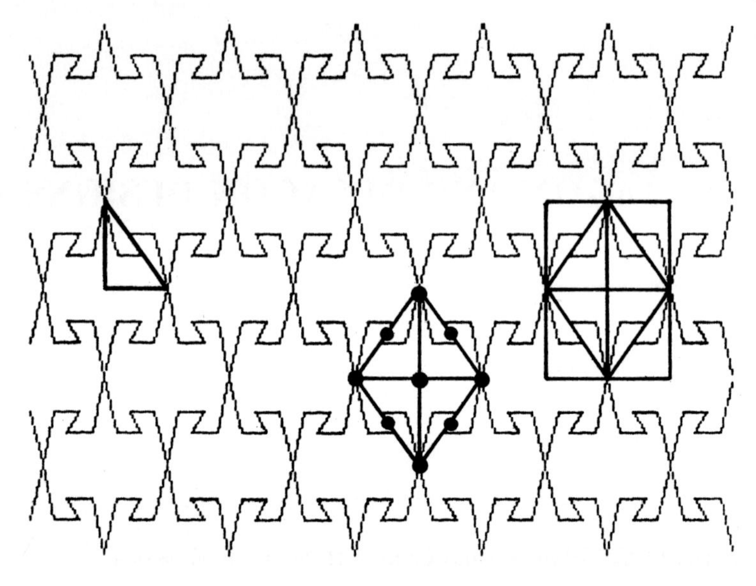

Fig. 1 An arbitrary non-Islamic pattern illustrates the type cmm.

significant point for the symmetry type cmm. In order to stress this, we take *A'*, so that *A'* and *A* are symmetrical with *D*. The rectangle *ABA'C* is a suitable figure to recognize patterns of type cmm; in all patterns such a rectangle will be drawn. The stars which we are studying are regular decagons, thus their number of corners is even, and therefore we describe these polygons as $2n$-gons (if $n = 7$ or $n = 9$, we get regular 14-gons or regular 18-gons. These motifs have n axes of reflections, therefore an axis of reflection cuts every corner into two equal parts. Thus we get $4n$ parts. *B* and *C* are the centers of the $2n$-gons and contain $4n$ parts. Our basic angular unit is one part. Therefore our unit of measure is $\pi/2n$. The measure of the right angles *ABA'* and *ACA'* is n. The measure of the angle *ACB* is k and of the angle *ABC* is l, so that $k>l$. All the angles we need are measured by the letters n, l and k. We will use the configuration "n, l-k" as a characteristic for a given lattice and thus for a pattern with stars or rosettes of the type cmm. The dimensions of the rhombus are determined by $AB/AC = \tan(k.\pi/2n)$.

Because of the construction of the pattern, the fundamental region can mostly be subdivided by equal partitions. This is worked out in Fig. 2b

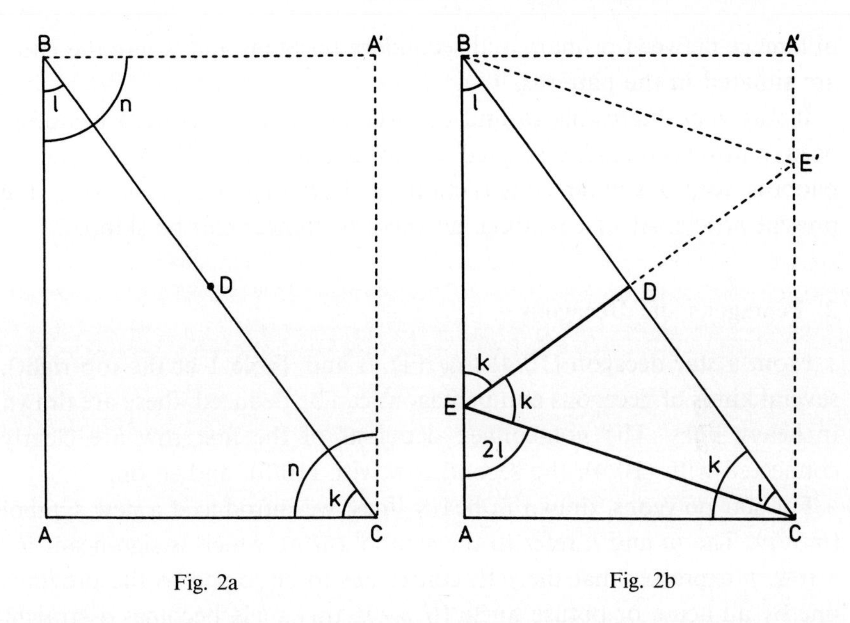

Fig. 2a

Fig. 2b

for a lattice 5, 2-3. We construct EC so that the angles ECB and EBC are equal and their measure is l. In E angles are formed the measures of which are k and $2l$.

Stars or rosettes constructed in B and C are called primary polygons; stars or rosettes constructed elsewhere are called secondary polygons. The latter are chiefly constructed in E, except in the case where n and k are both odd.

The just-described construction of stars or rosettes in a lattice of type cmm has mainly been applied by the Islamic artist on patterns with regular decagons ($n = 5$) and regular 14-gons ($n = 7$). Exceptionally he applied it also on patterns with regular polygons for which n is even. We have to notice that Christie[5] had already observed that much Islamic patterns can be drawn in a grid of diamonds (see Ref. 5, p. 258). The construction can also be extended to lattices of type cmm, which are not used by Islamic artists, e.g., patterns with regular polygons for which $n = 9$.

In this article, in view of the purpose of the present book, we restrict ourselves to the description of patterns in which pentagons and decagons occur. In Sec. 2, we enumerate all possible polygons with star shape that are derived from a regular star decagon. In Secs. 3 and 4, we describe the Islamic patterns by means of that enumeration. We hereby emphasize the difference between primary and secondary polygons and where the latter are situated in the patterns.

It was very interesting to find out whether or not a pattern has often been reproduced and discussed by former authors. We also refer to patterns with a similar construction as these that are depicted in the present article. At first reading, all these references can be skipped.

2. Pentagons and Decagons

From a star decagon (10/4) (see Fig. 3 and Table 1 at the top right), several kinds of decagons and pentagons can be deduced; these are drawn in heavy lines. The non-convex decagons of the first row are clearly connected with (10/4), the second row with (10/3), and so on.

For star polygons, drawn in heavy lines, we introduced a new symbol ($m/n/p$). The m and n refer to the symbol (m/n), which is significant for a row; p expresses that the p-th corner has to be joined to the previous one by an acute or obtuse angle (if $p = n$, this angle becomes a straight one). If m and n have a common divisor, the non-convex polygon is a compound of two or more components. From each star decagon of the second or fourth column of Fig. 3, two pentagons or two star pentagons, which are aligned on the lowest row, can be deduced. One of the two is drawn in heavy lines. They have the reference symbol ($m/n/p$)*c*, which refers clearly to the original non-convex decagon ($m/n/p$); *c* means "component." In the table, all the non-convex decagons and the deduced pentagons are tabulated according to their place in Fig. 3. The symbol ($m/n/p$) is followed by the shorter, classical symbol, if possible (see Ref. 3, pp. 82–83 and 88 and Ref. 6, p. 93).

The following polygons occur frequently in Islamic designs: (10/4/2),

Table 1. Symbols of Pentagons and Decagons

(10/4/1) [10/4] (10/3/1) [10/3] (10/2/1) [10/2] (10/1/1) (10) decagon	(10/4/2) (10/3/2) (10/2/2) (10/2)	(10/4/3) (10/3/3) (10/3)	(10/4/4) (10/4)
(10/2/2)c (5) pentagon	(10/3/2)c	(10/4/2)c [5/2] pentacle	(10/4/4)c (5/2) pentagram

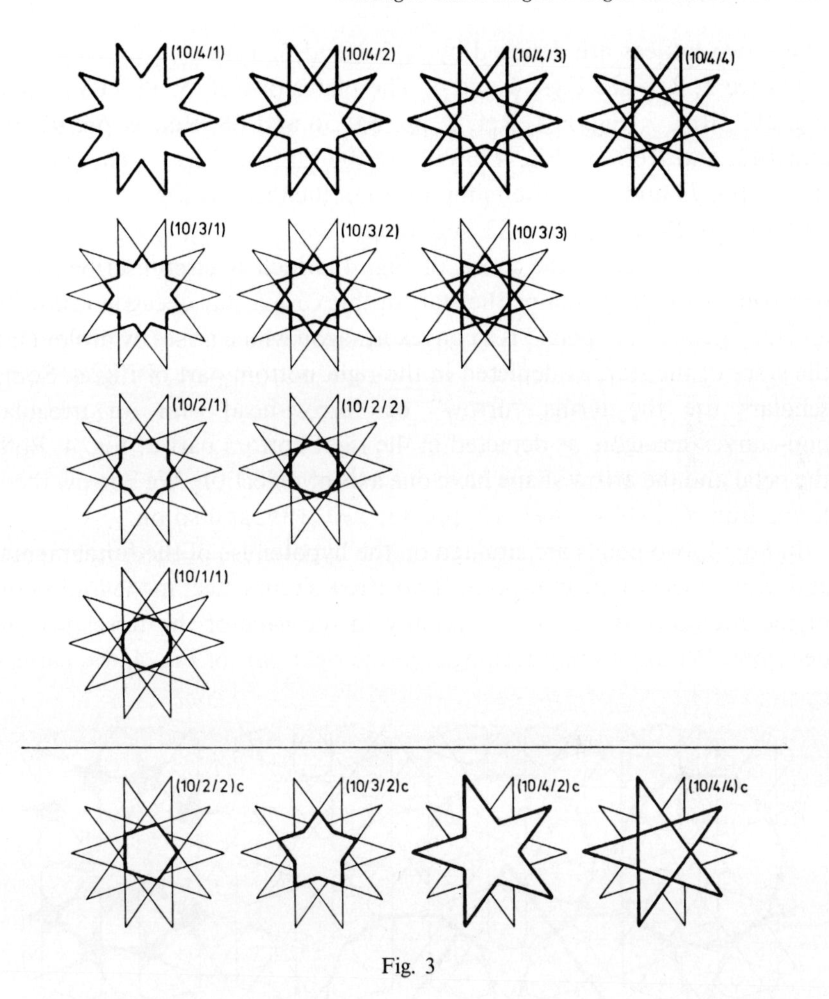

Fig. 3

(10/3/2), (10) and the first three pentagons on the lowest row. These three are also Penrose prototiles (see Ref. 3, pp. 539–545 and 558–566).

3. Patterns With Star Decagons (10/4/2)

In the case of $n=5$, there are two lattices (see Fig. 2a):

$$l=2,\ k=3, \text{ angle } ABC = 36°, \text{ angle } ACB = 54°;$$
$$l=1, k=4, \text{ angle } ABC = 18°, \text{ angle } ACB = 72°.$$

These two lattices are denoted by 5, 2-3 and 5, 1-4.

Lattice 5, 2-3: see Figs. 4 and 5. The plates of Ref. 5, Pl.XLVIII and Fig. 292, Ref. 7, Fig. 114, Ref. 8, pp. 54, 56 and 88, Ref. 9, pp. 92, 93 and 142, and Ref. 10, Fig. 45, left, are the same as Fig. 4. The plate of Ref. 8, pp. 74 and 76 is after simplification the same as Fig. 5. Analogous patterns are Ref. 11, Pls. 182 and 186a.

We call a "rosette" the motif of Fig. 4, which is circumscribed by a decagon. A rosette contains the star; in this case a star decagon (10/4/2), and the "petals." A "petal" is a convex hexagon which arises by prolonging the sides of the star, as depicted in the right bottom part of Fig. 4. Some scholars use the terms "arrow" or "arrow-head" for an irregular non-convex hexagon, as depicted in the right bottom part of Fig. 4. Both the petal and the arrow shape have one axis of reflection. We borrow these terms from Critchlow, Ref. 12, pp. 83, 130, 139, and so on.

In Fig. 4, two petals are situated on the hypotenuse of the fundamental region and meet in the mid-point. Two arrow shapes meet in the mid-point of the rhombus and they lie partially in the gaps of the tesselation of decagons. Within a small rectangle, in the right part of Fig. 4, the pattern

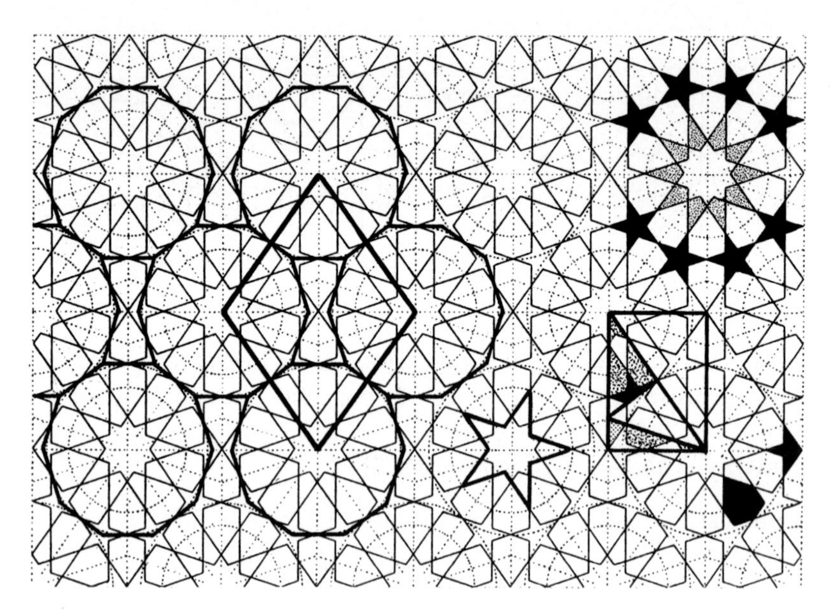

Fig. 4 B. 171[11]

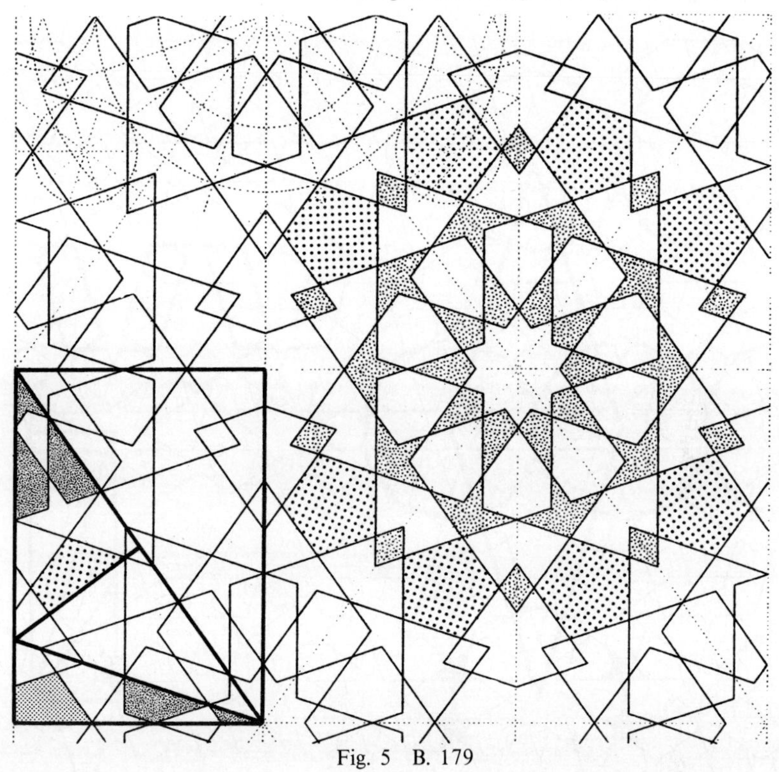

Fig. 5 B. 179

is analysed according to Fig. 2b in order to elucidate that it is of the type cmm. Each rosette is surrounded by eight secondary pentacles.

The rosette of Fig. 5 is smaller than the one of Fig. 4, but the petals of both are situated in a same way with respect to the fundamental region. The rosette of Fig. 5 is surrounded by ten arrow shapes, eight of them are used to produce secondary pentagons. After omitting the rosettes in Fig. 5 and after simplifying the remaining pattern, we get the plate of Ref. 11, P1. 175.

Lattice 5, 1–4: see Fig. 6. Compare with Ref. 11, Pl. 189. The decagons, circumscribed about the rosettes, touch (in contrast with Fig. 4) one another in the right angle of the fundamental region but they do not meet in the mid-point of the hypotenuse. These decagons do not belong to a tesselation with gaps as in Fig. 4, but they are aligned in parallel strips. The secondary polygons are pentagons and pentacles. The

Fig. 6 B. 174

pentagon is constructed in point E of the fundamental region (k is even); see Fig. 2b.

The rosettes of Fig. 7 are very small, and they are constructed in an anomalous lattice. The pattern is peculiar because of the secondary star pentagon (10/3/2)*c*; see also Fig. 12 at the right, where the same star pentagon is drawn in heavy lines.

Rosettes with star pentagon (10/4/2) may be constructed in a pattern of type pmm; see Ref. 8, Pl. 89; for this design we suppose that it is prolonged to the right and to the left by a translation and that the ribbons are replaced by lines. Compare with Ref. 8, Pl. 141.

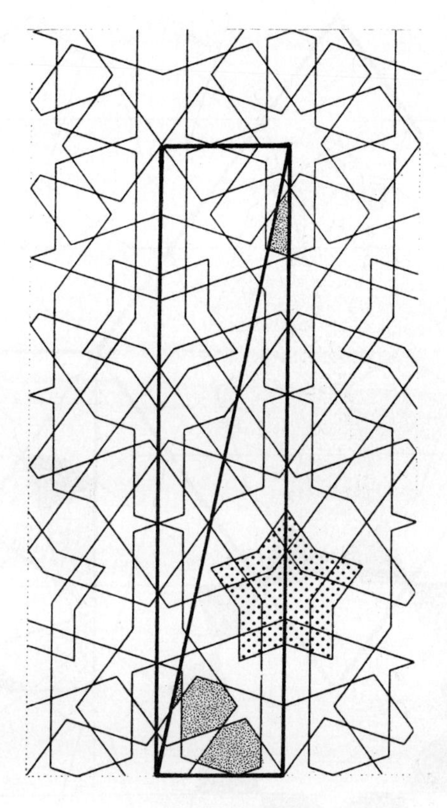

Fig. 7 B. 185

4. Patterns With Star Decagons (10/3/2) and (10/3/1)

Lattice 5, 1-4 with star decagon (10/3/2): see Fig. 8. This pattern is used in different ways on the frontispiece of Qur'ans; see Ref. 7, Figs. 25, 32, 41 and 146. The plate of Ref. 10, Fig. 48 is a drawing of the original plate of Fig. 146 of Ref. 7. By drawing rosettes into the star decagons of Fig. 8, we obtain the pattern of Ref. 11, P1. 182 after some modifications; for a ribbon version of the last plate, see Ref. 12, pp. 96–99.

The rectangle in dotted lines of Fig. 8 delineates the design, as it appears on the frontispiece of the Qur'ans.[7] The directions of all the lines correspond to the directions of the five sides of a regular pentagon: 0°, 36°, 72°, 108° and 144°. E. Wilson also observed this; see the figure legend of Ref. 10, Fig. 45. This holds also for Figs. 4–8 and 11–16;

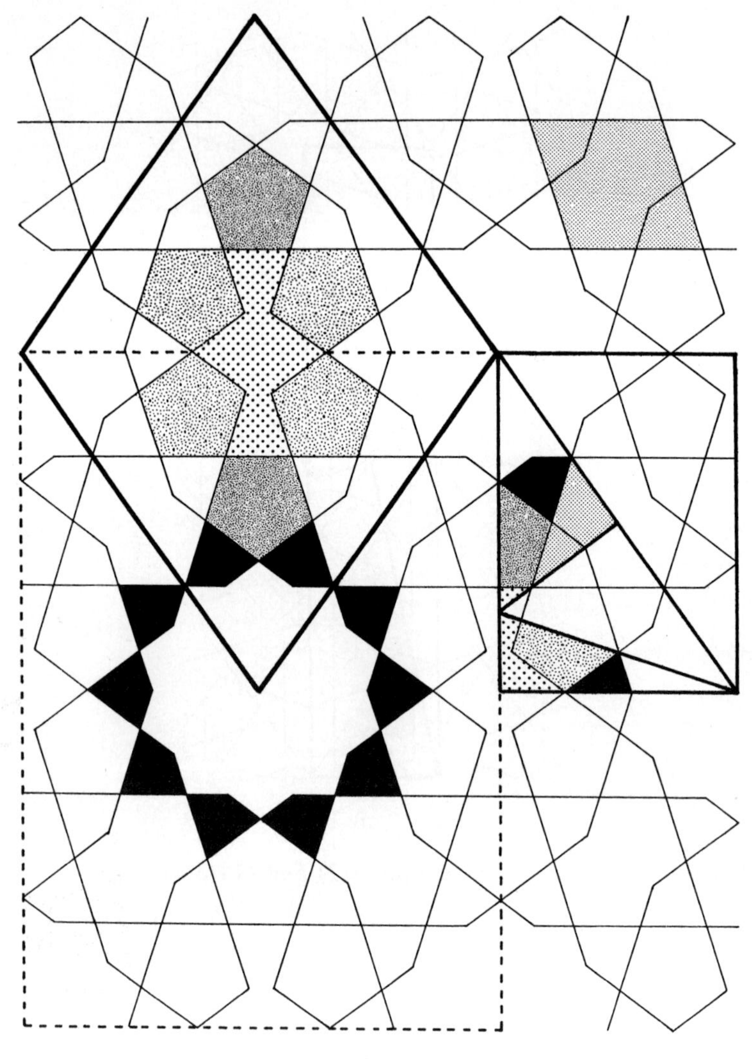

Fig. 8

sometimes the directions are rotated through 18° and are 18°, 54°, 90°, 126° and 162°. A designer who is creating a large pattern will achieve this accurately when he draws a grid according to these five directions, adapted to each pattern.

Perhaps the origin of the pattern of Fig. 8 can be explained by Figs. 9 and 10; compare with the star device of Ref. 5, Fig. 289. The

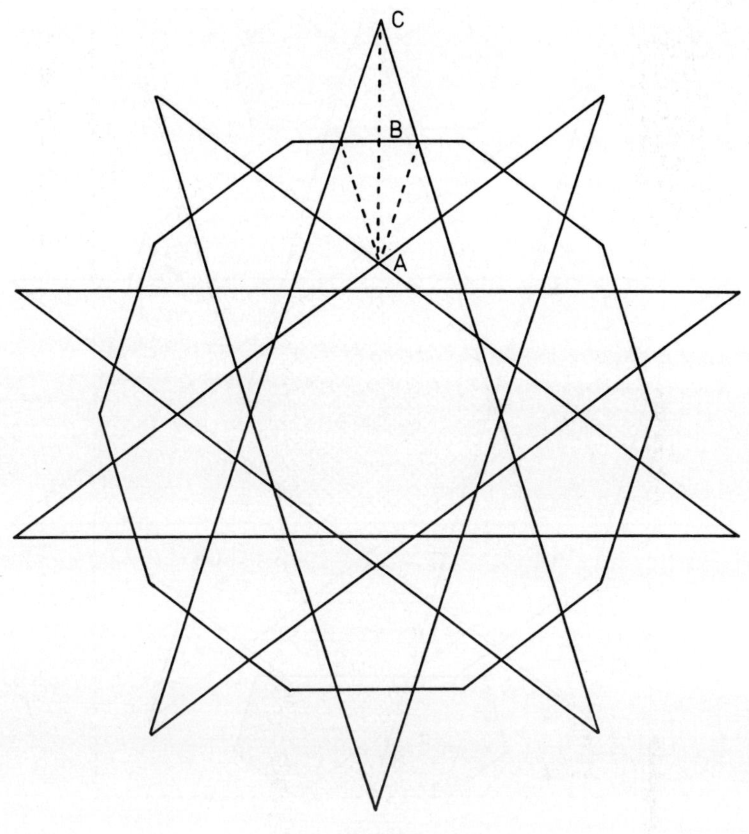

Fig. 9

starting-point is a star decagon (10/4). We construct an interior decagon so that $AB = BC$. By this, a pentagon appears in each of the ten points of (10/4). At the bottom and at the top of Fig. 10, we draw outside the two points four pentagons of the same size. In dotted lines, we delineate the rectangle and now every shaded part can be situated in Fig. 8. In the latter design, there are ten secondary pentagons connected with one star decagon (10/3/2).

We establish that Fig. 25 of D. James[7] is the most precise drawing of his four original plates because it corresponds closely to Fig. 8, looked at as its geometrical ideal. Is it possible that Islamic designs were copied by craftsmen who were not completely aware of the geometrical construction? Or were the drawings adapted (e.g., elongated) to the available text area?

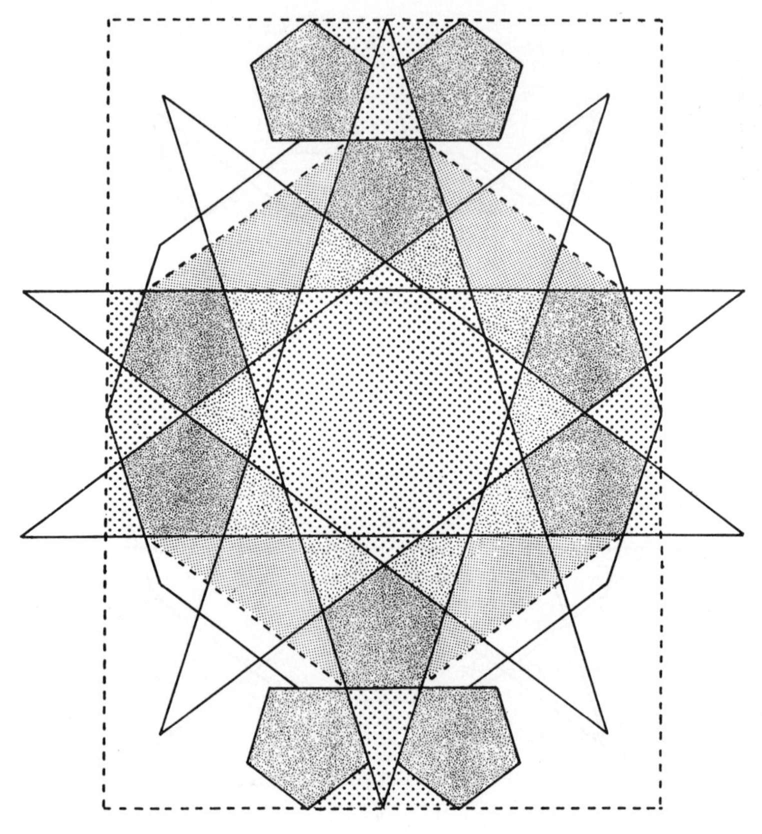

Fig. 10

Lattice 5, 2-3 with star decagon (10/3/1): see Fig. 11. This pattern appears on the double frontispiece of a Qur'an; see Ref. 7, Fig. 135 and on the jacket and Ref. 10, Fig. 49.

The original patterns of D. James[7] and the drawing of E. Wilson[10] do not correspond completely, neither are they drawn in all details according to the five directions of a regular pentagon. E. Wilson indicates the connection of this pattern with the drawing of our Fig. 9. In order to give in Fig. 11 an impression of this beautiful pattern, I followed my own method. This design is completely set up according to the five fundamental directions. Therefore, two regular pentagons are drawn about every point of the star decagon of Fig. 10, as they are drawn at the top

Fig. 11

and at the bottom of the plate. This yields two overlapping pentagons between two consecutive points. By means of this grid, it is easy to accomplish the pattern.

Within each star decagon, there is a decorative motif of a pentagon with five rhombi (in Fig. 11, I have drawn them according to the method of the plate of E. Wilson; see there at the top left); this motif has not a tenfold but a fivefold rotational symmetry. If this central motif would be cancelled, the pattern would be of the type cmm. The central motif can be rotated about 36°. This occurs once in the pattern of D. James and once in Fig. 11: have a look at the right top angle of the small rectangle.

For another pattern of latice 5, 2-3 with star decagon (10/3/1), see Ref. 11, Pl. 180.

Several patterns with star decagons (10/3/2), namely Figs. 12–15, have the same tiles as the pattern of Fig. 8. It is possible to recognize in several of them an underlying strip of regular pentagons of the same decomposition. Critchlow[12] called attention to this. This strip of overlapping pentagons is drawn in heavy lines at the top of Fig. 12 (lattice 5, 1-4; see also Ref. 5, Figs. 290 and 291). The pentagons of the same decomposition form a zigzag strip in Fig. 13 (lattice 5, 2-3). A strip of the same pentagons, arranged edge-to-edge, is drawn in heavy lines at the bottom of Fig. 14 (lattice 5, 1-4). Observe that such a pentagon can rotate. In Fig. 14, all those pentagons can rotate through different angles. In Figs. 12 and 13, only those pentagons not overlapping each other can rotate simultaneously. By such a rotation, a pattern is no longer of type cmm. In Fig. 13, the design within the rectangle, drawn in dotted lines, is a frontispiece of Qur'ans; see Ref. 7, Figs. 34, 39 and 57. It is surprising that the fundamental region cannot be divided in the usual manner as in Fig. 2b.

Figs. 8, 12, 13 and 15 are composed of the same tiles. This persuaded me to make a jigsaw puzzle in cardboard. By means of this I found

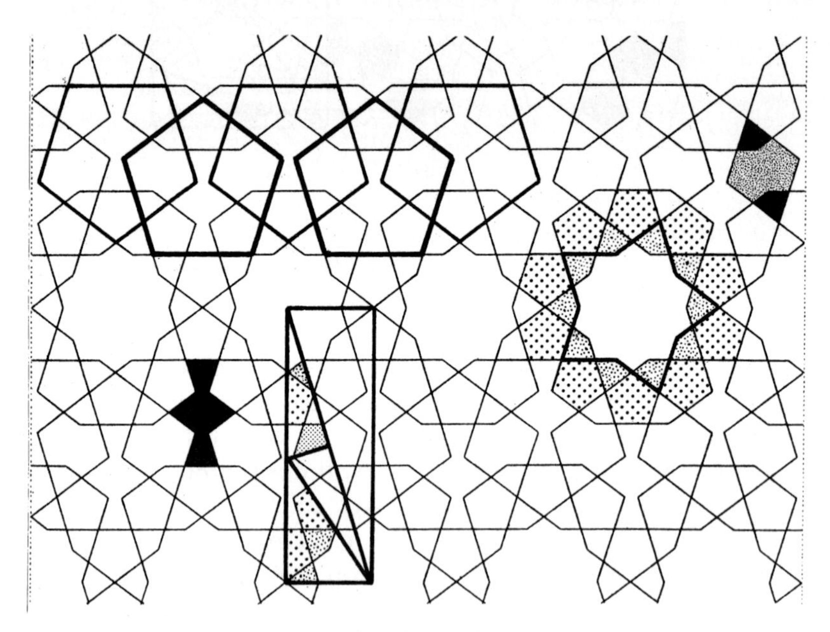

Fig. 12 B. 173

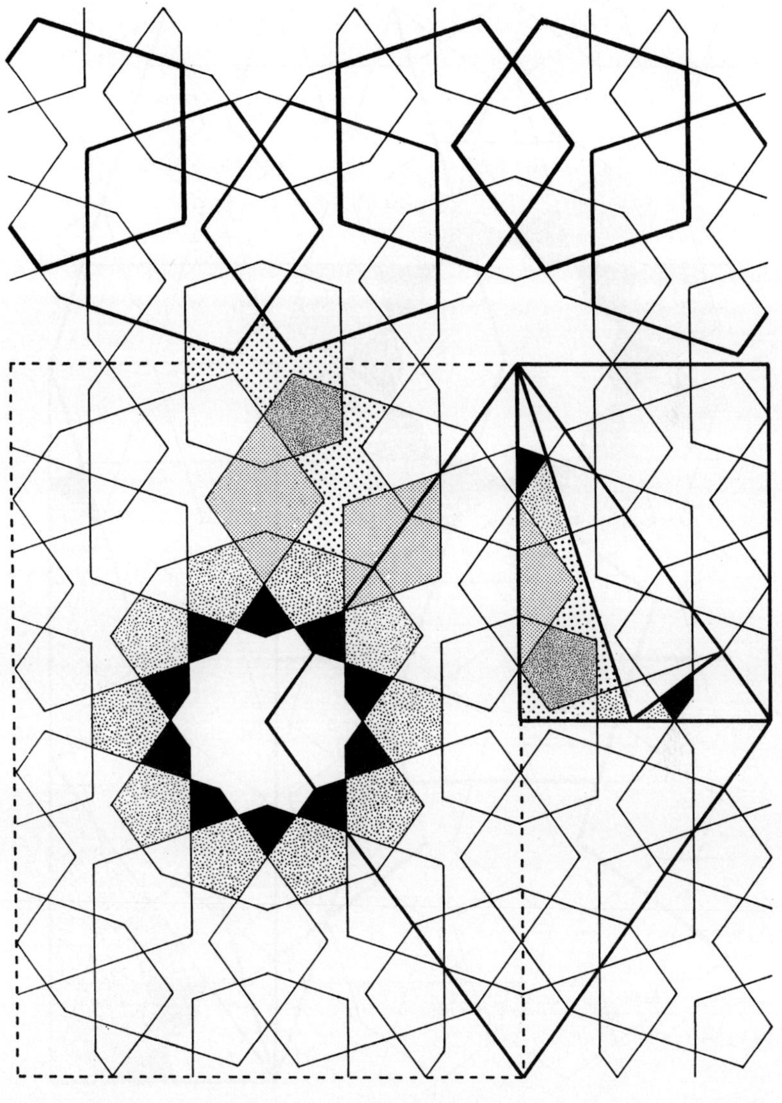

Fig. 13

several new patterns, and among them the one of Fig. 14. The tiles of Figs. 8 and 14 are arranged in the same way on the small diagonals of the rhombus, but elsewhere the patterns are different, because they belong to two distinct lattices. It is possible to construct many patterns with the set

Fig. 14

of these tiles. When omitting one tile, you can construct the well-known patterns of Ref. 8, pp. 55, 57; Ref. 9, pp. 88–89; Ref. 10, Fig. 47; and Ref. 11, Pl. 175 (lattice 5, 2–3).

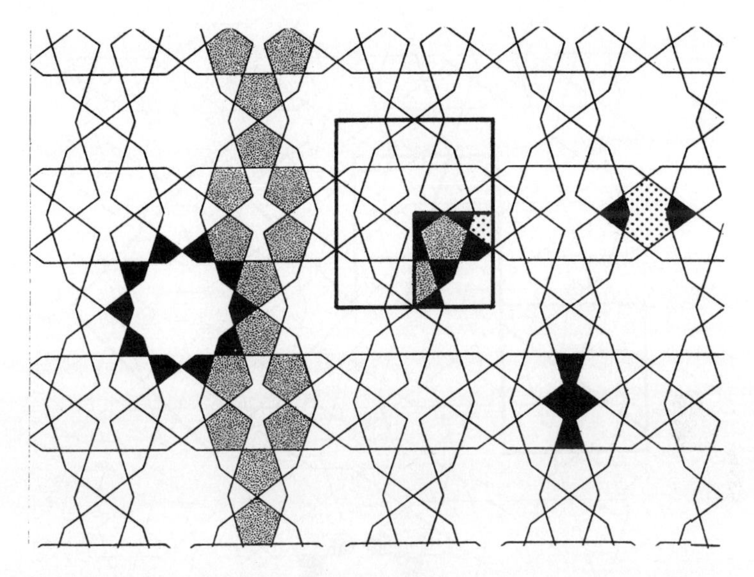

Fig. 15 B. 176

To accomplish the set of patterns with star decagons (10/3/2), note two peculiar designs. Fig. 15 is a pattern of type pmm; see Ref. 9, pp. 85–87 and Ref. 10, Fig. 46. Fig. 16 is a pattern of type cmm but it is quite different from all previous patterns. At a first glance, you get the impression that the rhombus of the lattice is a square. Bourgoin in his original book calls it a *"Plan carré."*[13] However, calculation proves that the rhombus is not a square. An analogous pattern is Ref. 11, Pl. 188a.

Secondary pentagons occur in many Islamic designs of which the primary polygons do not have a fivefold rotational symmetry, e.g., Fig. 17. Pentagons surround the 16 pointed rosette and the 12 pointed rosette as well. A (7/2) heptagon is recognizable. By using such different kinds of polygons in one design of type p4m, the Islamic mathematician or artist proves his playfulness, ability and mastership.

Finally, we have to mention that one of Kepler's tilings is of type cmm with lattice 5, 1–4; see Ref. 3, Fig. 2.5.10(c), left. Supplementary information on pentagons and decagons in Islamic designs is given in Ref. 9, Ch.II.4, pp. 82–97 and Ref. 12, Ch. 5, pp. 74–103.

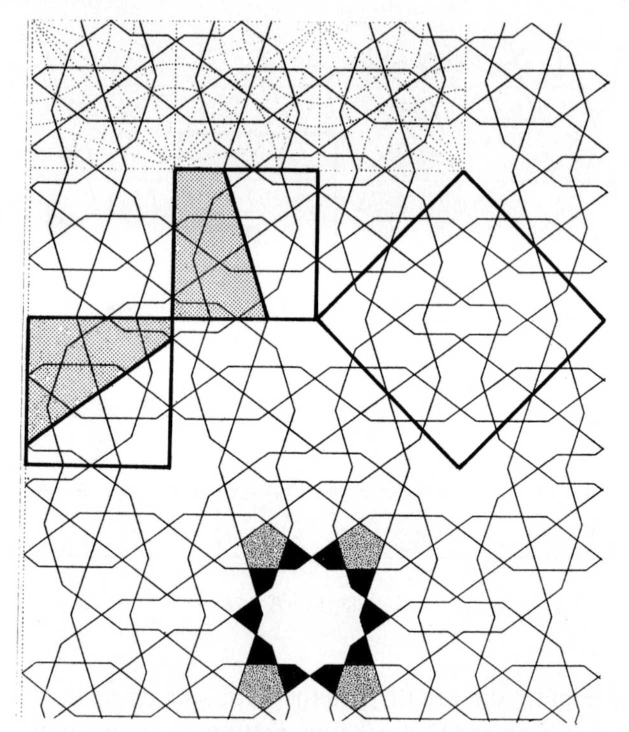

Fig. 16 B. 177

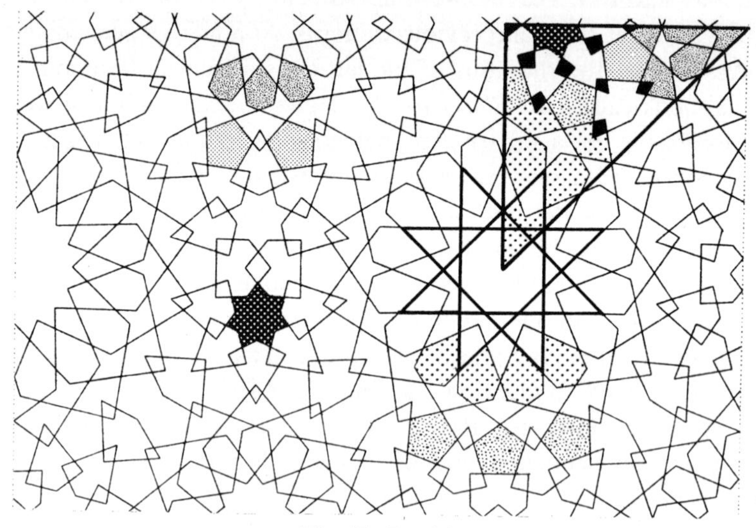

Fig. 17 B. 134

Acknowledgement

The patterns of eight illustrations, marked by "B.", are reproduced from J. Bourgoin, Arabic Geometrical Pattern and Design, Dover Pictorial Archive Series, Dover Publications, New York (1973).

I express gratitude to J. Wouters and Dr. J. Cuypers for critical reading of the manuscript.

References

1. L. Danzer, B. Grünbaum and G.C. Shephard, Can all tiles of a tiling have fivefold symmetry? *Amer. Math. Monthly* **89** (1982) 568–570 and 583–585.
2. B. Grünbaum, Z. Grünbaum, and G.C. Shephard, Symmetry in Moorish and other ornaments, *Comp. Maths. Applic.* Reprinted in *Symmetry: Unifying Human Understanding,* I. Hargittai (ed.). Pergammon Press, New York (1986). **12B** (1986) 641–653.
3. B. Grünbaum and G.C. Shephard, *Tilings and Patterns.* Freeman, New York (1987).
4. D. Schattschneider, The plane symmetry groups, their recognition and notation, *Amer. Math. Monthly.* **85** (1978) 439–450.
5. A.H. Christie, *Pattern Design.* Dover, New York (1969).
6. H.S.M. Coxeter, *Regular Polytopes*, 2nd Edn. Macmillan, New York (1963).
7. D. James, *Qur'ans of the Mamluks.* Alexandria Press — Thames and Hudson, London (1988).
8. D. Wade, *Pattern in Islamic Art.* Overlook Press, Woodstock, NY (1976).
9. I. El-Said and A. Parman, *Geometric Concepts in Islamic Art*, 2nd Edn. Scorpion Ltd. and World of Islam Festival Trust (1988).
10. E. Wilson, *Islamic Designs.* British Museum Pattern Books, London (1988).
11. J. Bourgoin, *Arabic Geometrical Pattern and Design.* Dover, New York (1973) (paperback reprint of plates).
12. K. Critchlow, *Islamic Patterns, An Analytical and Cosmological Approach.* Thames and Hudson, London (1976).
13. J. Bourgoin, *Les Eléments de l'art arabe: le trait des entrelacs.* Firmin-Didot, Paris (1879).

Acknowledgment

The patterns of eight [illegible] by [illegible] are reproduced [illegible] J. Bourgoin, Arabic Geometrical Pattern and Design (Dover, [illegible]) [illegible]

References

1. [illegible] and other [illegible] [illegible] (1985) [illegible]
2. B. Grünbaum and G. C. Shephard, Tilings and Patterns (Freeman, New York, 1987).
3. [illegible] perception, [illegible] recognition and [illegible]
4. [illegible] (Dover, New York, [illegible])
5. [illegible] New York, 1969.
6. [illegible]
7. [illegible]
8. [illegible] (1985).
9. [illegible]
10. [illegible] (Dover, New York, [illegible]) [illegible] reprinted [illegible]

AN ISLAMIC PENTAGONAL SEAL (FROM SCIENTIFIC MANUSCRIPTS OF THE GEOMETRY OF DESIGN)

Wasma'a K. Chorbachi and Arthur L. Loeb

1. The Manuscripts

In 1971, Wasma'a Chorbachi[1] discovered the existence of several manuscripts on geometric design in the Khudabakhsh Library in Patna, India, a number of which were copied in the city of Mosul in northern Mesopotamia early in the thirteenth century A.D. Follow-up research on people working in the sciences and teaching them in the early thirteenth century in Mosul led to Kamal ad-Din Musa Ibn Yunus Ibn Man'a (555–639 A.H./1156–1242 A.D.), who had been teaching and writing on mathematics and geometry there. One of his surviving works on geometry is now in Iran (*Mashhad: Kitabkhana Astan-i Quds-i*).

This manuscript turned out to be a commentary on a much earlier work, a geometry text by Abu'l-Wafa' al Buzjani, a well-known mathematician living in Baghdad between 945 and 987 A.D., who wrote one of the earliest surviving manuals of practical geometry for artisans and architects. In examining a Persian translation of Abu'l-Wafa'al Buzjani's treatise in the Biblitheque Nationale in Paris, Chorbachi also came upon a manuscript apparently without title or author, mentioned in the catalog only as a manuscript of geometry problems with geometric figures.

These manuscripts are bound in one volume now deposited in Paris Bibl. Nat. ancien fond Persan Ms # 169. Starting on folio 141 verso is the Persian translation of Abu'l Wafa'al Buzjani, *Kitab fima Yahtaju Ilayhi al-Sani'u min A'mal al Handasa* named *A'mali Handasiya wa Pirkar wa Kunya* (*The Book on What Artisans Need from Geometry Problems*). This translation is followed by another document contained between folio 180 recto and folio 199 verso. In the middle of the right-hand side margin of folio 180 recto is the title *Fi Tadakhul al-Ashkal al-Mutashabiha aw al-Mutawafiqa (On Interlocking Similar or Congruent Figures)* (Fig. 1), dating from some time after the original of Abu'l-Wafa'al Buzjani.

Fi Tadkhul al-Ashkal al-Mutashabiha aw al-Mutawafiqa can best be decribed as a series of problems in the design of geometric patterns and their geometric constructions. We shall discuss here two of these problems, found in folios 180a and 184b. Both of these problems deal with issues which have recently resurfaced as a result of publications by Roger Penrose and Martin Gardner.[2] The first problem (Fig. 1) deals with interlocking convex decagons and pentagonal stars (concave decagons); Chorbachi[3] has pointed out the similarity between the early Islamic scientific illustration in the Paris manuscript and geometric drawings of the *Scientific American* of January 1977, illustrating what has recently become known as Penrose Tiling (Figs. 2a and b).

Although the approach to the generation of this pattern in the Paris manuscript is quite different from that taken by Penrose, it is notable that these "quasi-periodic" patterns were already of interest at least in the thirteenth century A.D. The manuscript stresses the uniqueness of the fivefold center of rotational symmetry in the pentagonal seal, thus implying the lack of translational symmetry in the pattern, but does not explicitly deal with the matter of non-periodicity.

Before considering the problem of folio 180a, however, we shall consider folio 180b, in which there is a problem dealing with the construction of the so-called *Kunya-5*, a right triangle having an angle of 36°. Obviously, such a triangle will provide a convenient tool in the artisans' workshops for the construction of regular pentagons and decagons (cf Sec. 3 and Fig. 3d below).

2. The *Kunya* 5

Fig. 3a shows the construction of the Kunya in the Paris manuscript, together with a transcription into modern notation (Fig. 3b). The reconstructed procedure is as follows: (Figs. 3c ff.) On the circumference of a circle having its center at C and diameter AB, two points, D and E

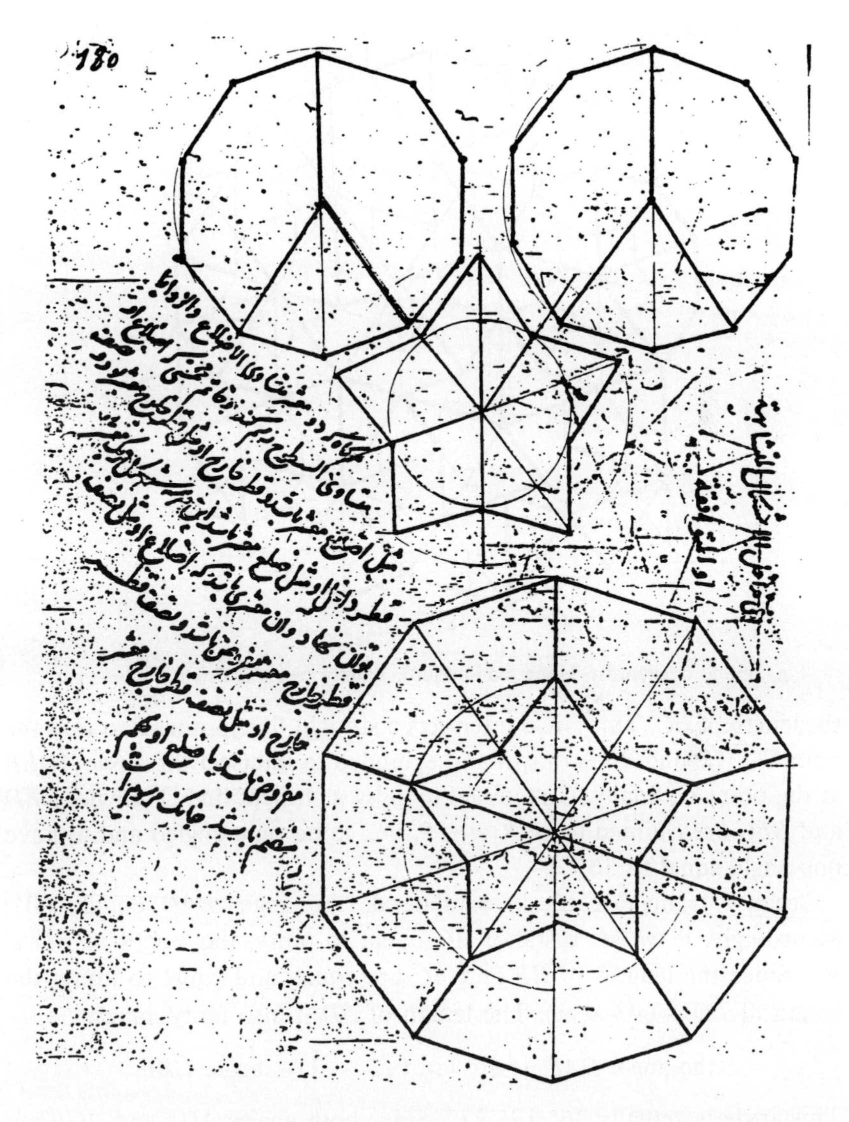

Fig. 1 An Islamic Pentagonal Seal: geometric problem from folio 180a in the manuscript *On Interlocking Similar or Congruent Figures*, Paris, Bibliotheque National ancien fond Persan Ms# 169.

are located such that $AD = AC = BC = BE$. A is joined to D by a straight line. A circular arc is drawn having its center at B, and radius equal to

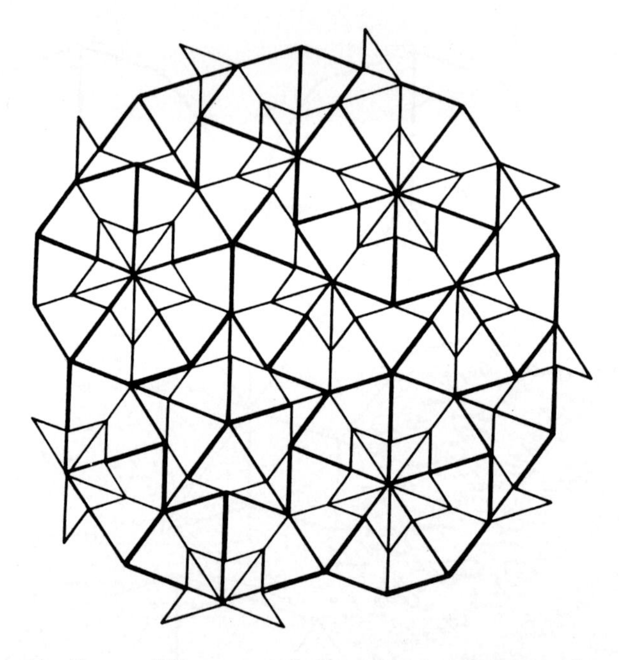

Fig. 2a and b Penrose Tiling: non-periodic pattern (c 1977, *Scientific American*).

the length of *BC*. This circle intersects a straight line joining *B* to *D* at the point *F*. The line *AF* is drawn and extended to intersect the circle *ADEB* at the point *G*; finally, *G* is joined to *B* by a straight line. Triangles *AFD* and *BFG* are claimed to be *Kunya* 5, i.e., to be right-angled and to have one angle equal to 36°.

Because triangle *CAD* is equilateral, the angle *BAD* equals 60°; accordingly, if we set distance *AC* equal to unity, the length of *BD* is $\sqrt{3}$. Since the lengths of *BF* and *BC* are equal, and equal to unity, the length of *FD* is $(\sqrt{3} - 1)$. The length of *AD* is also unity, hence:

$$\text{the angle } DAF = \text{arc tan } (\sqrt{3} - 1) = \text{angle } GBF.$$

This angle is actually 36° 12′ 22″. Since both angles *ADF* and *AGB* are right angles, both triangles *DAF* and *GBF*, shaded in Fig. 3b are good approximations to the *Kunya* 5.

The construction was therefore remarkably accurate, though not correct. Kamal ad-Din Musa Ibn Yunus Ibn Man'a in his thirteenth-century commentary on Abu'l-Wafa' al Buzjani's book on the geometry

Fig. 2b

of construction, with whom this construction may well have originated, actually was quite explicit in cautioning that some of his constructions, in particular of the heptagon, were practical, but not mathematically exact. They can be used in small-scale designs without noticeable discrepancies, which however become manifest on a larger scale.

3. The Pentagonal Seal

Our problem in folio 180a (Fig. 1, transcribed in Fig. 4a) deals with interlocking decagons and fivepointed stars (cf. also Fig. 2). The aim of this problem is to develop a tile inlay pattern of a large regular decagon with an interstitial pentagonal star such that the length of the side of the decagon equals the edgelength of the interstitial star; the stars have internal angles which are alternately larger than and smaller than 180°.

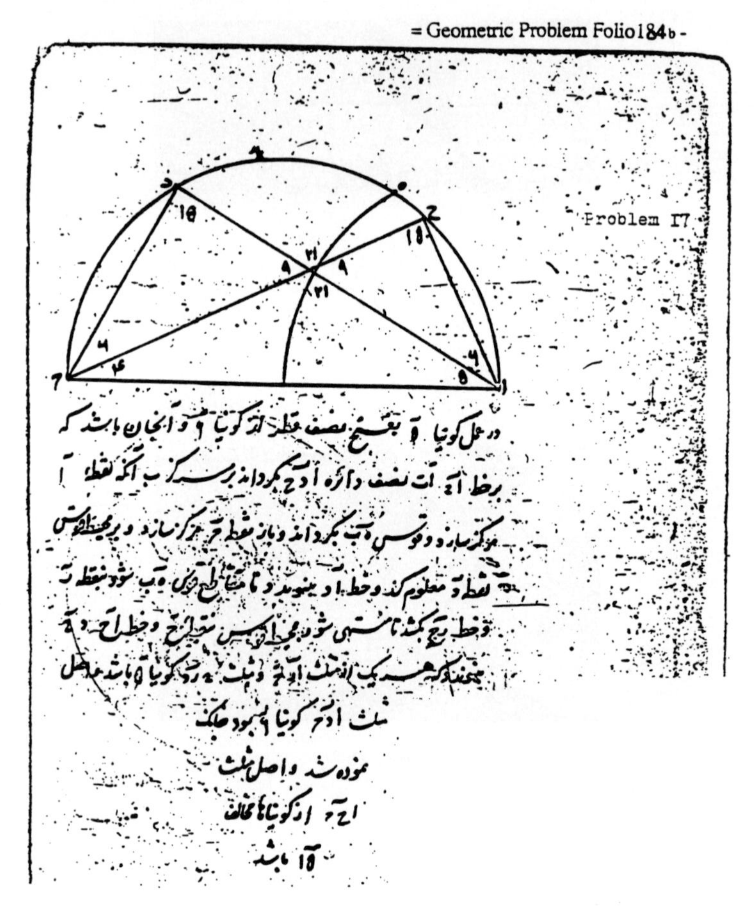

Fig. 3a The *Kunya* 5 problem from folio 184b in the manuscript: *On Interlocking Similar or Congruent Figures*, Paris, Bibliotheque National Ancien fond Persan Ms# 169.

It is clear that the term "interlocking" means what we call "tessellating," i.e., covering the plane without overlap or interstice.

The text of the problem is translated as follows:

"Whenever two decagons are drawn with equal sides, angles and areas, and one Khatam-Mukhammas pentagonal inlay star (pentagonal seal) is drawn whose:

(i) sides are equal to the sides of the decagon,
(ii) outer diameter equals the outer diameter of the decagon,
(iii) half the inner diameter is the same as the side of the decagon, then all these three figures can be put together in one [larger] decagon (Figs. 1 and 4b), [and] that decagon would be a decagon wherein:

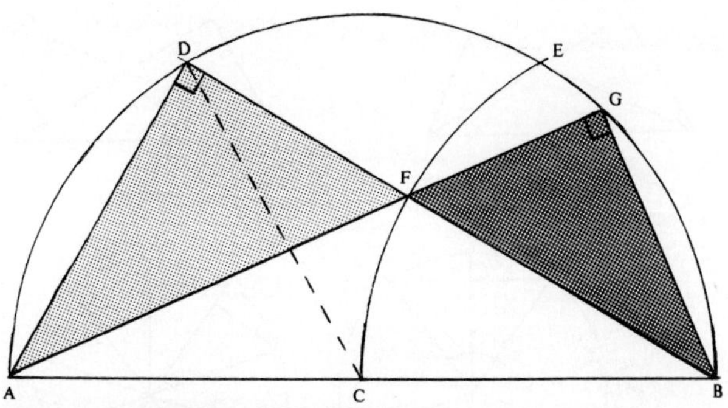

Fig. 3b Modern transcription of the *Kunya* 5 problem.

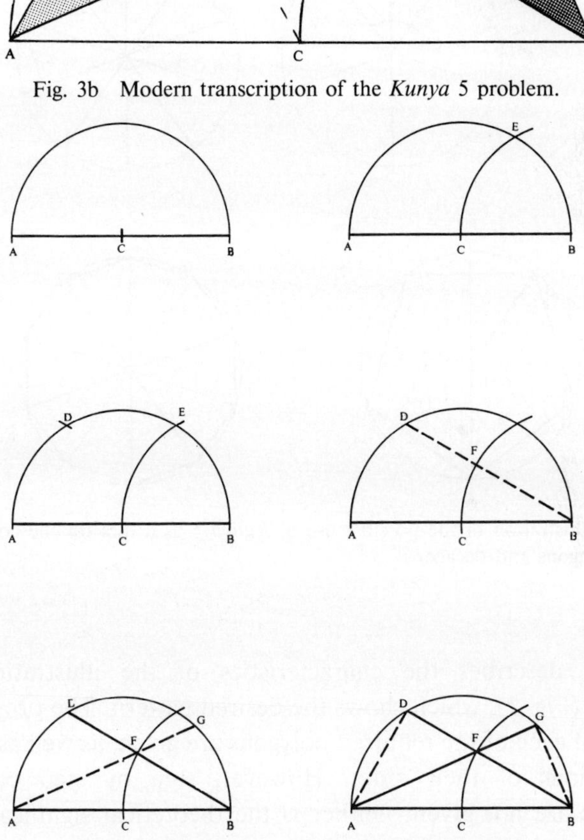

Fig. 3c Step-by-step reconstruction of *Kunya* 5.

(iv) each of its sides would be equal to half the outer diameter of the original [smaller] decagon,

(v) half of its outer diameter would be equal to half the outer diameter of the original decagon plus the length of one side.

(vi) In this way it is drawn in the illustrated figure."

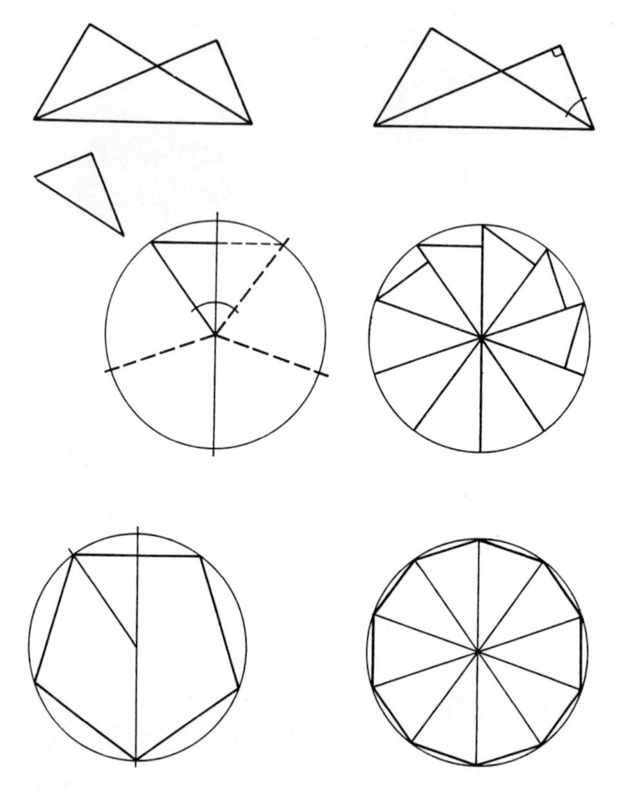

Fig. 3d An illustration of the possible use of *Kunya* 5 as a tool for the construction of pentagons and decagons.

The text describes the characteristics of the illustration in the manuscript (Fig. 1), which shows the desired pattern. The proportions of the radius of each of the required polygons are given, as well as the length or proportions of their sides. However, step by step construction instructions are not given; neither is the theoretical significance of the given proportions.

4. Golden Triangle and Golden Gnomon

Before outlining the re-construction of the pentagonal seal, it will be helpful to recall the modern definitions of the *golden triangle* and the

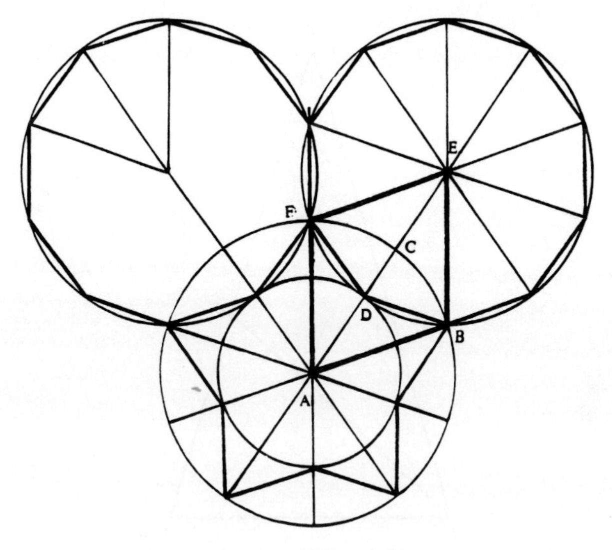

Fig. 4a The Islamic Pentagonal Seal: interlocking decagons and fivepointed star.

golden *gnomon*;[4] these concepts were used by the Islamic mathematicians/designers, but not explicitly so designated. The former (Fig. 5) is an isosceles triangle having angles equal to 36°, 72° and 72°; its base-length

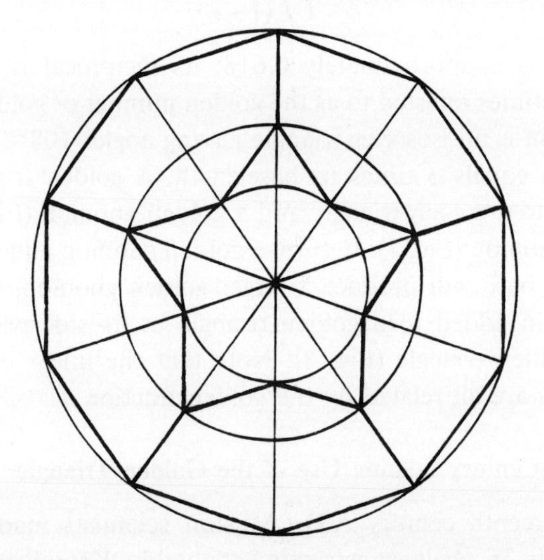

Fig. 4b The two decagons and one fivepointed star seal combining to form one larger decagon.

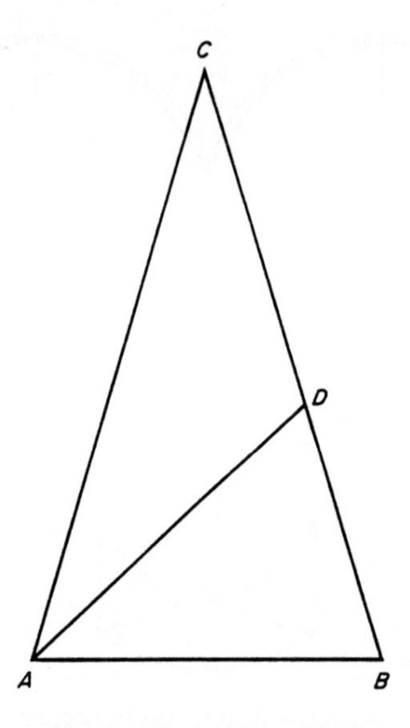

Fig. 5 The golden triangle.

equals ϕ times its side-length, where ϕ is the golden fraction defined by the equation

$$\phi = 1 / (1 + \phi).$$

The value of ϕ is approximately 0.618; its reciprocal is about 1.618, which is sometimes referred to as the golden number or golden ratio. The golden gnomon is an isosceles triangle having angles 108°, 36° and 36°; its side-length equals ϕ times its baselength. A golden triangle may be subdivided into a golden triangle and a golden gnomon (Fig. 6), as may be a golden gnomon (Fig. 7). In turn, a golden gnomon added to a golden triangle at its base will produce a larger golden gnomon, but a (larger) golden gnomon added to a golden triangle at its side will once more produce a golden triangle (Fig. 8). Note that the linear dimensions of these triangles are all related by the golden fraction ϕ.

5. Thirteenth-Century Islamic Use of the Golden Triangle

By the thirteenth century A.D., Muslim scientists made use of the golden triangle in their constructions, notably Kamal ad-Din in his commentary on the construction of the pentagon by Abu'l Wafa'

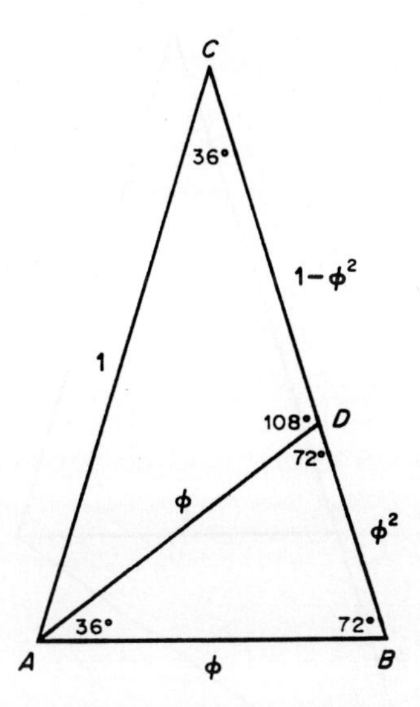

Fig. 6 The subdivision of a golden triangle into a golden triangle and golden gnomon.

al-Buzjani (Fig. 9). Moreover, in the second half of the thirteenth century (ca. 1259) in the town of Marāgha, which became a center of scientific activities and contained the famous observatory, another illustrious mathematician, Nasir ad-Din at-Tusi, wrote commentaries on Euclid, in which he made obvious use of the golden triangle (Fig. 10), labels transcribed in Fig. 11. Although he is renowned for his work on algebra and astronomy, he was also interested in geometry, and his commentaries on Euclid included a short treatise dealing with the inscription and circumscription of polygons within the circle: *Sittat Maqalat min Kitab Tahrir Uqlidis: Six Books/Articles from Euclid's Book of Elements.*

In the fourth article in the section entitled "How to construct a pentagon inside a circle," he clearly uses the golden triangle, saying: (cf. Figs. 10–11)

> "For example in circle *ABG*, we construct a triangle of the pentagon labeled *KER*, and within the circle *ABG*, we construct a triangle whose angles are equal to the angles of triangle *KER*; this is triangle *ABG*. We bisect the angles *ABG* and *AGB* with the two lines *BH* and *GT*, and we connect *AH*, *HG*, *AT*, and *TB*. The figure *ATBGH* is a pentagon, and this is because the five angles *BAG*, *GTH*, *HBA*, *AGT* and *THB* are equal and their arc segments are equal. Therefore the sides of the pentagon are equal and each of the angles is subtended by one of the five equal arcs, so that the angles are equal also; this is what we desired to construct."[5]

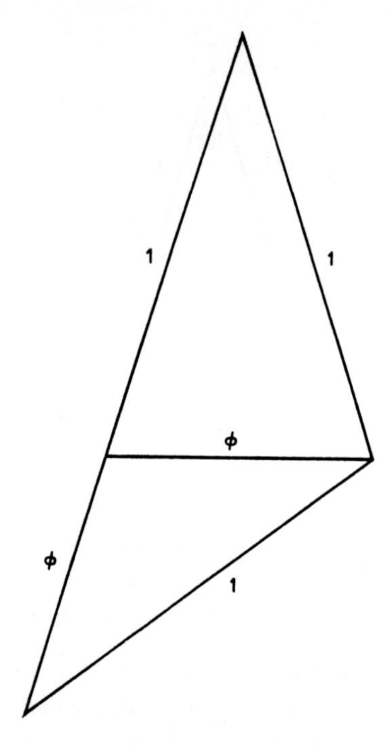

Fig. 7 The subdivision of a golden gnomon.

6. Reconstruction

Using these properties of the golden triangle and golden gnomon, we have transcribed the instructions for constructing the pentagonal seal given in Sec. 3 as follows:

Step 1: Draw a circle with any compass opening; we will see that this radius is $1 + \phi$ (Fig. 12).

Step 2: Draw five lines through the center of the circle, making angles of $36°$ with each other. (The construction of the $36°$ angle may have used the method of folio 184b, discussed above. However, there is no specific reference to folio 184b in folio 180a; any other method, for instance, the drawing of specific polygon angles given and used in other problems in the same manuscript might equally well have been used.) Draw a decagon joining the points where the radial lines intersect the circumference of the circle. The center of the circle is then the apex of ten *golden* triangles, whose base-length will be 1 (Fig. 13).

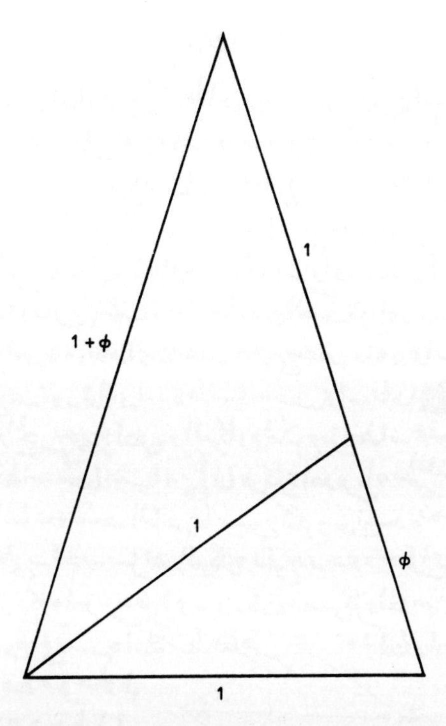

Fig. 8 The addition of a golden triangle to the golden gnomon to produce a golden triangle again.

Step 3: Draw a second circle, whose radius again equals $1 + \phi$ (Fig. 4a). The distance between the centers of the first and second circle (AE in Fig. 4a) is $(2 + \phi)$. Draw a third circle, concentric with the second one, having a radius equal to the edgelength of the decagon, i.e., 1. Again draw five spokes at 36° angles through the common center of the second and third circles. These spokes intersect each of the concentric circles twice, marking ten mutually equally spaced points on each circle (Fig. 14).

Step 4: The pentagonal star is then constructed by joining each of alternate points on the outer circle to a pair of nearest points on the inner circle (Fig. 15). One of the spokes constructed in Step 3 connected the centers of the first two circles. There will be four additional circles, each centered on one of the additional spokes. All of these circles have their centers located at distance $(2 + \phi)$ from A, the center of circles 2 and 3 (Figs. 16, 17, and 18). Accordingly, the center of the concentric circles is a base vertex of each of ten golden-gnomon triangles which are pairwise

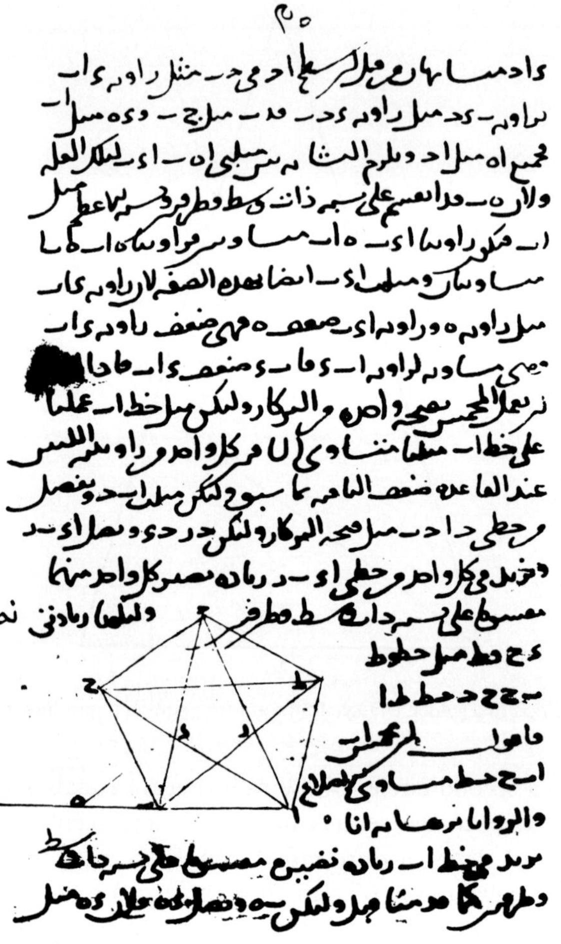

Fig. 9 The late twelfth-century use of a golden triangle by Kamal ad-Din Musa Ibn Yunus Ibn Man'a.

joined at their base, whose length equals the distance AE. Since these gnomon triangles are isosceles, the edgelength of the star equals the edgelength of the regular decagon and also half of the star's inner diameter, which is set equal to unity. Its outer diameter equals the diameter of the regular decagon, namely $(1 + \phi)$, as claimed in the original instructions (Figs. 16, 17, and 18).

(١١٦)

زاوية ح ک ا مشترکا فزاوية ب ک ا اعني زاوية ب

مثل زاويتي ح ک ا ح ا ک اعني زاوية ب ح ک الخارجة

فب ک اعني ا ح مساو لح ک وبالجملة فزاوية ا مساوية

لزاوية ح ک ا وکانت مساوية لزاوية ح ک ب فکل واحدة

من زاويتي ا ب ک ا ک ب مثلا زاوية ا وذلك ما اردناه

وهذا المثلث يعرف بمثلث المخمس .

يا

نريد ان نعمل في دائرة مخمسا ونعني بالمخمس.

والمسدس وامثالهما متساوي الاضلاع والزوايا

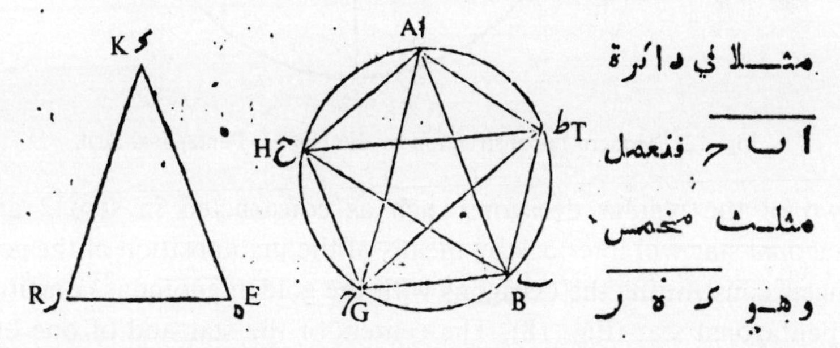

وفي دائرة ا ب ح مثلثا يساوي زواياه زوايا مثلث ک ه ر

وهو مثلث ا ب ح ونصف زاويتي ا ب ح ا ح ب بخطي

ب ح ح ط ونصل ا ح ح ح ا ط ب ط فسطح

Fig. 10 Mid-thirteenth century use of the golden triangle by Nasir ad-Din at-Tusi.

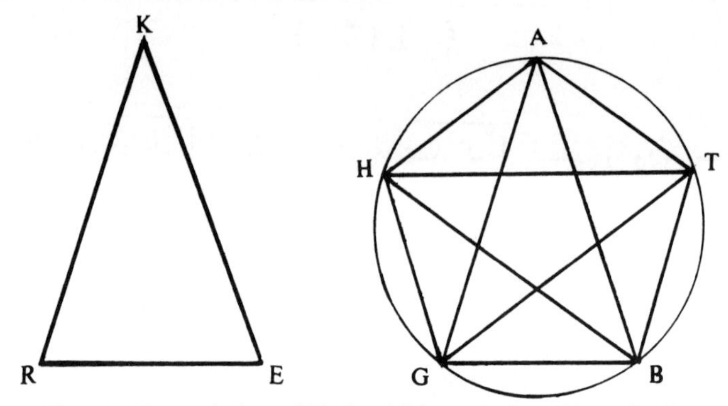

Fig. 11 Transcription of Nasir ad-Din at-Tusi's golden triangle.

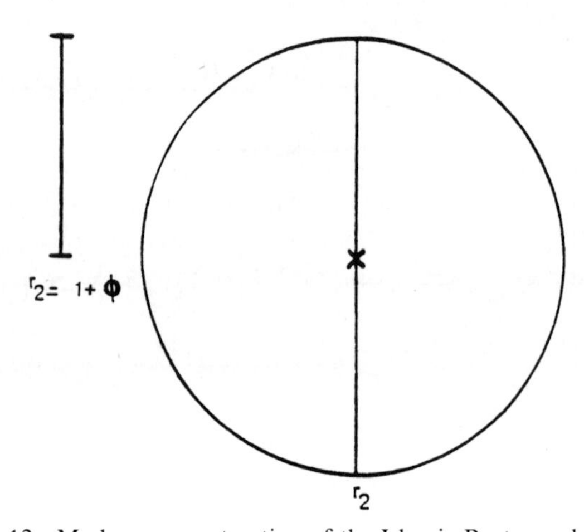

Fig. 12 Modern re-construction of the Islamic Pentagonal Seal.

Two of the regular decagons such as constructed in step 2 and a pentagonal star will interlock by means of the juxtaposition of the golden triangles constituting the decagons with the golden gnomons constituting the pentagonal star (Fig. 18). The centers of the star and of one of the decagons form the base of a golden gnomon whose apex is formed by a vertex common to the star and the decagon.

It has been shown[6] that not only the linear dimensions, but also the areas of a golden triangle and a golden gnomon which together constitute a larger golden gnomon are related by the golden fraction, ϕ. Accordingly, the area of the pentagonal star equals ϕ times that of each of the smaller decagons.

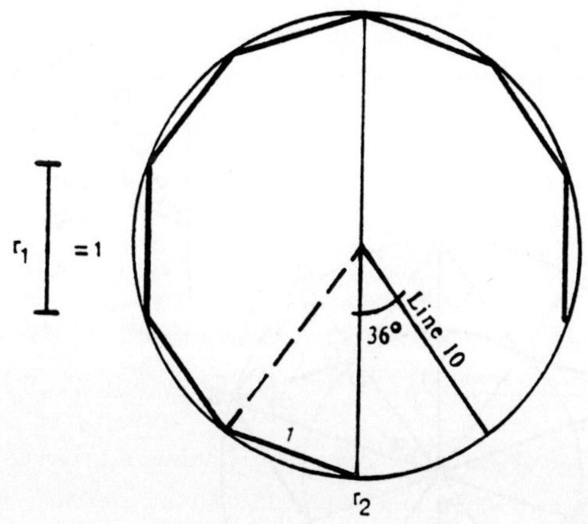

Fig. 13 Modern re-construction of the Islamic Pentagonal Seal.

7. Conclusions

The construction of the interlocking decagon and fivepointed star in the Paris manuscript is significant for several reasons. The fivepointed star drawn in the problem of folio 180a of the Paris manuscript is a very particular one, constituted of ten golden gnomons which exactly match the ten golden triangles which constitute the decagon. Thus the complementarity of the particular pentagonal star and decagon used in folio 180a, whose areas are related by the golden ratio, is particularly significant, and in some respect analogous to the *Yin-yang* symbolism in Chinese design in its complementary and interlocking features.

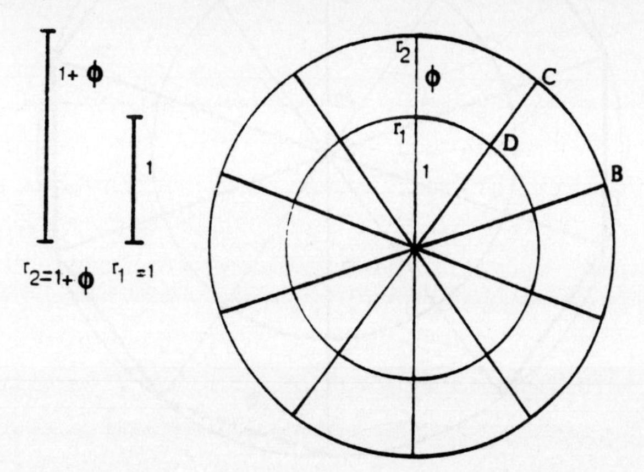

Fig. 14 Modern re-construction of the Islamic Pentagonal Seal.

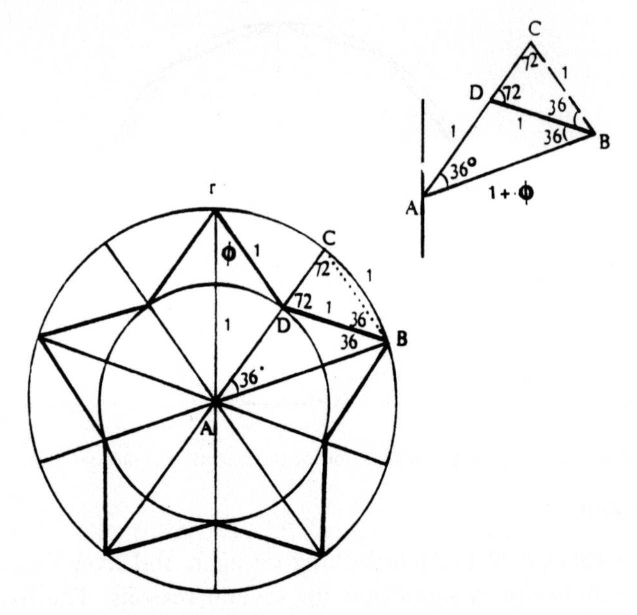

Fig. 15 Modern re-construction of the Islamic Pentagonal Seal.

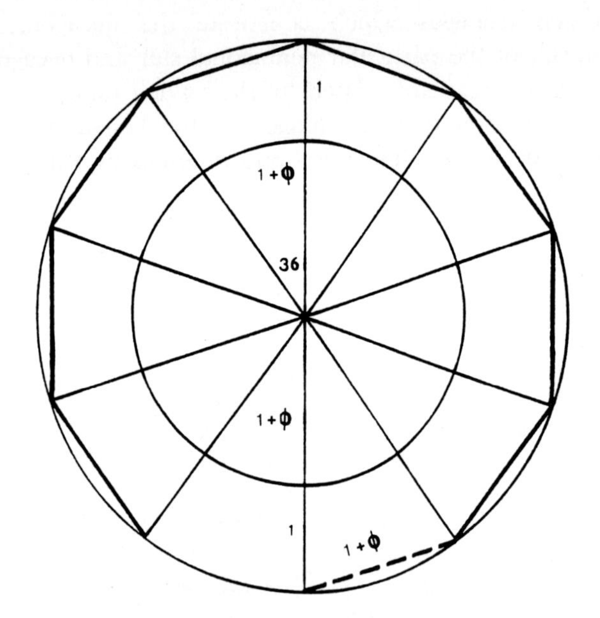

Fig. 16 Modern re-construction of the Islamic Pentagonal Seal.

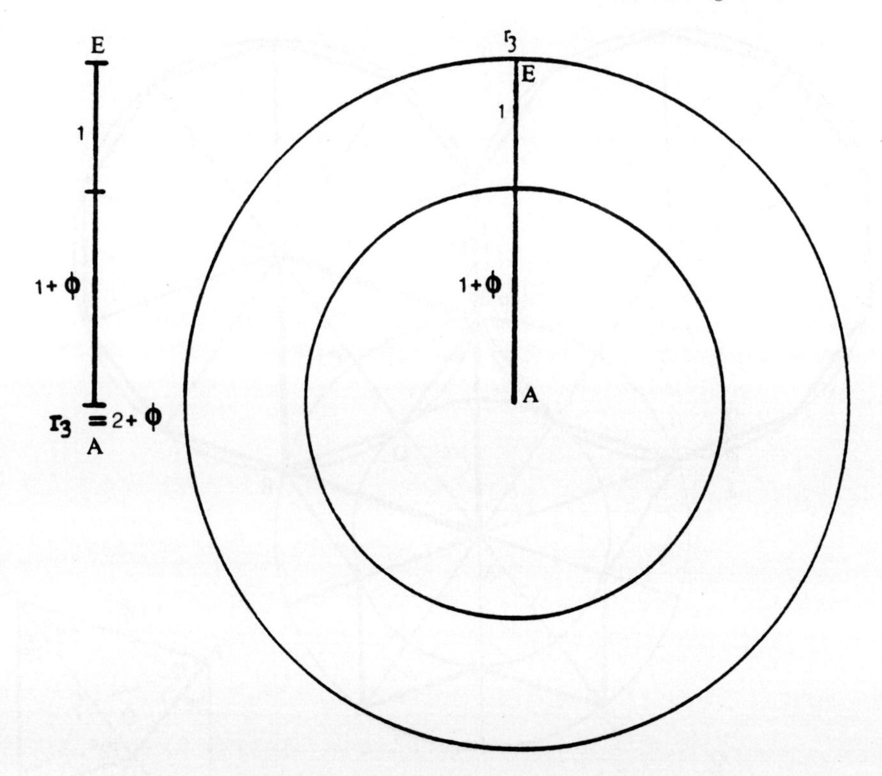

Fig. 17 Modern re-construction of the Islamic Pentagonal Seal.

It is historically significant that as early as the thirteenth century A.D., it was known that what we presently call the golden triangle and golden gnomon are together capable of tessellating the Euclidean plane, and that during the Middle Ages, Islamic design continued in the tradition of the Alexandrian and other eastern mediterranean schools of mathematics. The use of this fivepointed star appears to have stimulated mathematicians to work on these practical problems in design. The importance of this problem to the Muslim scientists may be inferred by the fact that they tried over the course of several centuries to find the perfect solution.

The Islamic Pentagonal Seal was not only intended for scientific illustration in geometric instruction, but also to be part of the Islamic design repertoire, to be used in ornaments and decoration. An example is the Pentagonal Star Seal delineated in brick decoration in the Abbasid

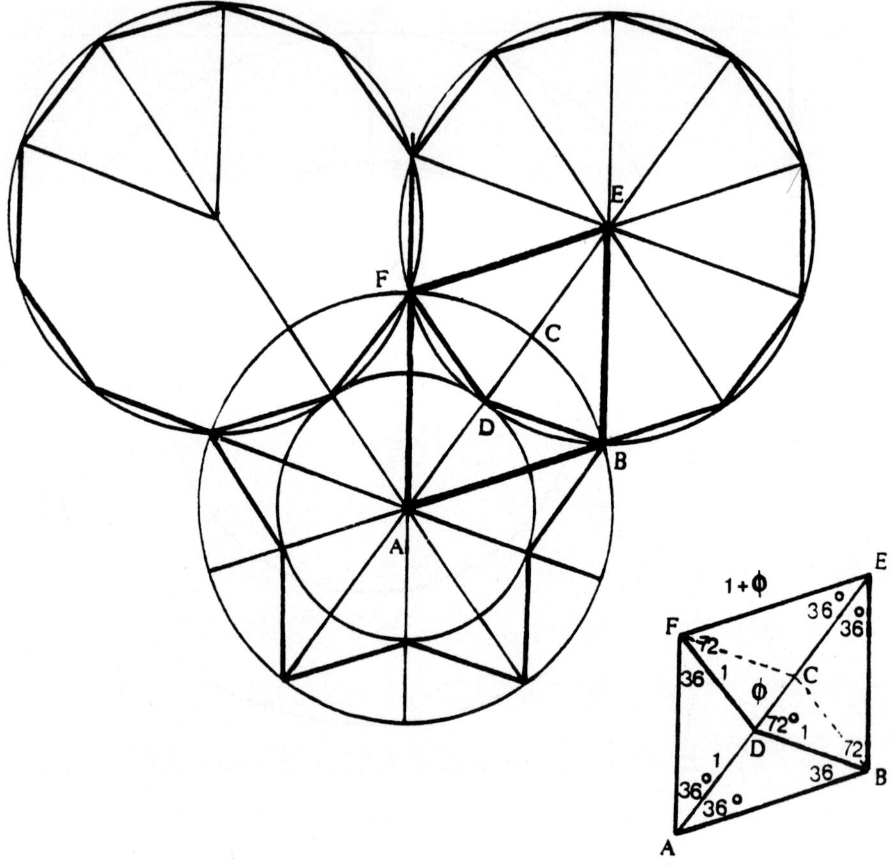

Fig. 18 Modern re-construction of the Islamic Pentagonal Seal.

palace in Baghdad, built between 1180 and 1230 A.D. (Figs. 19 and 20). In this instance, the Pentagonal Star Seal also contains a smaller five-pointed star at its center. Although the details of the decoration were heavily restored in the twentieth century, and cannot be evaluated completely accurately in every detail, enough traces remain to permit the assumption that the details are semi-authentic. Since this architectural decoration dates from the period 1180–1230 A.D., we may conclude that knowledge about the Pentagonal Seal had by that time filtered down to artisans, so that the material presented in folio 180a may be dated to before 1200 A.D., circa 1180–1230 A.D.

Fig. 19 Brick-work decoration in the ceiling of an alcove from the Abbasid Place, Baghdad.

Fig. 20 Detail of the central medallion decoration, in the ceiling of an alcove from the Abbasid Place in Baghdad circa 1180–1230 A.D., with a delineation of the Islamic Pentagon Seal.

References

1. W.K. Chorbachi, In the Tower of Babel: Beyond symmetry in Islamic design, *J. Computers Math. Applic.* **17** (1989) 751–789 and in *Symmetry 2:*

Unifying Human Understanding, I. Hargittai, ed. Pergamon Press, Oxford (1989).

2. M. Gardner, Theory of tiles: Extraordinary non-periodic tiling that enriches the theory of tiles, *The Scientific American* (January 1977) 110–121.
3. W.K. Chorbachi, op. cit. p. 772.
4. A.L. Loeb, The magic of the pentangle: Dynamic symmetry from Merlin to Penrose, *Computers Math. Applic.* **17** (1989) 33–48 and *Symmetry 2: Unifying Human Understanding*, I. Hargittai, ed. Pergamon Press, Oxford (1989).
5. Nasir al-Din at-Tusi, "*al-Magala al-Rabi's*," *Sitat Magalat min Kitab Tahrir 'Uqlidis. al-Matba'a al-Hindiya*, 116–117, Calcutta (1924).
6. A.L. Loeb, op. cit.

NOTES ON SOME PENTAGONAL "MYSTERIES" IN EGYPTIAN AND CHRISTIAN ICONOGRAPHY

Lima de Freitas

1. Number as Symbol

It is the quarter full of characters; they collide like protons and electrons, always in a five-dimensional world whose fundament is chaos.

Henry Miller (*Plexus*)

It is not my intention to dwell on the usual reiterations about the remarkable mathematical and geometric properties of the pentagon, highlighted by historians of thought as well as by art history scholars with special prominence when studied in association with the "golden ratio," of which the figure proves to be an inexhaustible purveyor. "According to Plato and to ancient cosmology in general the world is a certain proportional entity subjected to the law of the harmonic division, that is the golden section," writes the historian of philosophy and thought A. F. Losey.[1]

It is a well-established fact that a great number of what we call art creations of the past, in so far as they were monuments, or works associated to monuments, of religious purpose, and were devised by highly complex cultures rich in old traditions, show a profound concern with measure, proportion, number and geometry. However it must be said that many of the most reliable studies tend (yielding to a restrictive fashion) to elaborate the subject from an utilitarian viewpoint, i.e., concentrating their attention on the rules of symmetry or the technicalities of systems of measurement and practical ways of achieving proportion in architectural planning, in the distribution of "aesthetic" motives decorating vases, walls, etc., thus often leaving untouched or vaguely mentioned the philosophical, religious and symbolical grounds, and their numerical expressions, on which such forms of art were founded. Numbers and proportions or *ratios* were of a sacred nature, but this did not induce obscurity and imprecision, as is clearly shown in the greater creations of Christian sacred art and the art of ancient Egypt, classical Greece and other civilizations. But this area of investigation is vast.

Here I shall restrict myself to a few remarks on certain aspects of pentagonal symmetries relevant to the traditional arts, viewed not from a strict mathematical vantage point, for which I have no competence whatsoever, but in the light of an hermeneutic attempt to grasp the ideas embodied by certain symbols and their meaningful concatenations, at once spiritual, metaphysical, religious as well as what we term "artistic," as they appear in our culture.

The aspects I want to decode are related to the symbology of the pentagon and depend on the "magic" or "mystic" value ascribed to some of its geometrical properties. They deal with the inner coherence of some symbolic images and representations of life, death and resurrection, which naturally occupy a central place in the permanent concerns of mankind. To the best of my knowledge, the relation between geometry, number and myth and religion that I will try to determinate in the present paper have never been pointed out before, although obviously belonging to the traditional common treasure of ancient mystagogues and masters of arts.

Before speaking of those symbolic aspects and their numbers, let us recall some basic notions concerning the pentagon and the number five, not as a mere geometrical and mathematical object in itself, but as a symbolic and metaphysical one, at the same time trying to retrieve what sort of thought-chains and assumptions may be presumed to lie behind the wisdom that shaped so many marvellous creations of religious art.

From the outset we shall keep in mind that geometrical figures were contemplated as "beings" in themselves, intimately related to numbers (figure and number taken as expressions of the same reality, much as the written and spoken word are taken as describing the same object). Those "beings" were thought to possess their own specific forms of energy and intrinsic "qualities," both physical and meta-physical, each kind of shape structure being conceived as derived from a divine archetype to which individual irregular shapes should conform, thus attaining (or having the possibility to attain) the perfection pertaining to their true "essence." Such perfection corresponded to the regular polygons and to integer numbers (the only numbers created by God, according to Kronecker, all the others being the work of man). We consider, of course, bidimensional representations. The inscription of regular polygons in the circle or the division of the circle by integers was, according to Ghyka, a seemingly initiatory requirement in the mysteries of Delphos,[2] and also one of the fundamental bases from which, according to other authors, evolved the notions of proportional numbers, of inner structure and "secret" value of the integers and the numerical connotation of sacred letters and alphabets.[3]

The circle, as it were, stood, at least since the Egyptians and the Greeks, for the idea of Totality, Wholeness, as enfolding all possibilities of manifestation (space, time, matter, consciousness) past, present or future, conceivable and inconceivable. It stood also for Emptiness (the "zero" that the notation evokes by its form) assumed as the invariable vacant region that contains all latent or potential events, as well as for Fullness of what there is, and of what there is not, or not yet, or never will be, the eternal Being of all transient beings. According to such a way of thinking, any particular figure contained in a circle manifests a particular possibility of being, which ontologic intensity, so to speak, increases as it proves itself geometrically ordered (maybe one should say "ordained"), and even more so if it shows the quality of regularity; inscribed regular polygons constitute the divine perfection of each family of polygonal structures. In this context, for example, the equilateral triangle reflects in its geometrical perfection the divine idea of Three, the perfect unity of threefold being, such as the Holy Trinity of Christianism; the square exhibits the perfection of Four, in which the divine unity is concealed as ten hence the pythagorean cult of the *Tetraktys* or "sacred monad," and the religious relevance of the *Tetramorphos* in Christian iconography.[4]

2. The "Somber" and "Luminous" Pentagons

The truth in two halves:. . . half self, half light.

Norman MacCaig

Now, as we come to consider the pentagon or the five, we face a more puzzling and paradoxical entity. This is not the place to elaborate on number as a psychological entity; let us admit, now, without further examination, that psychologically five seems to express the unity of four, a difficult one, for the passage of three to four is full of tension and conflict, the triad representing a dynamic relative unity of spirit or instinct, and the tetrad the reconquered stable unity of the previously conflicting forces; the five (4 + 1) stands psychologically for the "unity of four."[5]

This presumably brings into the nature of five some traces of a "dangerous" duality, as can be detected in the Egyptian *ta dwat*, meaning "pentad," where *dwa*, two, still echoes; and again in the fact that *Dwat* was the name of the Egyptian Hades, the realm of the dead, which had to be crossed by the sun every night with terrible risks and where Osiris, as god of the dead, sitting in a cavern of the underworld, judges the deceased. Similarly, the fifth *Sephira* of the Jewish Tradition carries with it the notions of justice, severity, trial and also fear. Five appears thus as a number which is imbued with very "human" contradictions as well as rich in portentous possibilities. Death implies resurrection; trial and judgement imply anguish, fear, punishment, but also the triumph of righteousness, joy and compensation. Again, five shows a magic power of growth, becoming and individualization (implied in the sanskrit *prapantschi*, expressing the expansion of individual dynamism, where *pantscha*, like the Greek *pente*, means "five"), either in "good" or in "evil." Coincidentally, the Arabic alphabet, consisting of 28 letters (twice the number of the phalanges of the hand) divides into two *hurüf* (2 × 14), one "luminous," corresponding to the right hand, and one "dark," corresponding to the left hand. As Schiller wrote, five is sacred because it is like the human soul; five is the first number formed by both even and odd numbers, the human soul contains in itself both the good and the evil. Modern science seems to meet this kind of paradoxical combination of opposing polarities — forces of stability together with forces of instability — in many fields, from fractals to the study of life and nature (how to combine the laws of thermodynamics with purpose and awareness?; "how a purposeless flow of energy can wash life and consciousness into the world?"[6]); nature appears more complicated in the face of the recent notions of "chaos" and "complexity." In defying previous assumptions of ecology, for instance, "chaos" seems to William M. Schaffer "both exhilarating and a bit threatening. . ."[7]

An analogous feeling of uneasiness has been growing lately in the area of the study of fivefold symmetries, where some puzzling contradictions and a combination of "good" — or order — and "evil" — or chaos — evolve from various attempts to try pentagonal tilings of the plan. Professor Penrose's ultimate shapes, named "darts" and "kites" by John Horton Conway, are derived from a rhombus with angles of 72 and 108 degrees (we will meet these angles again in this paper) by dividing the long diagonal in the golden ratio of $(1 + \sqrt{5}) / 2 = 1.61803398\ldots$, then joining the point to the obtuse corners (Fig. 1). The surprising phenomenon discovered by Penrose, called "inflation" and "deflation," occurs when trying to prove that the number of "darts" and "kites" tilings is unaccountable. As Martin Gardner writes:

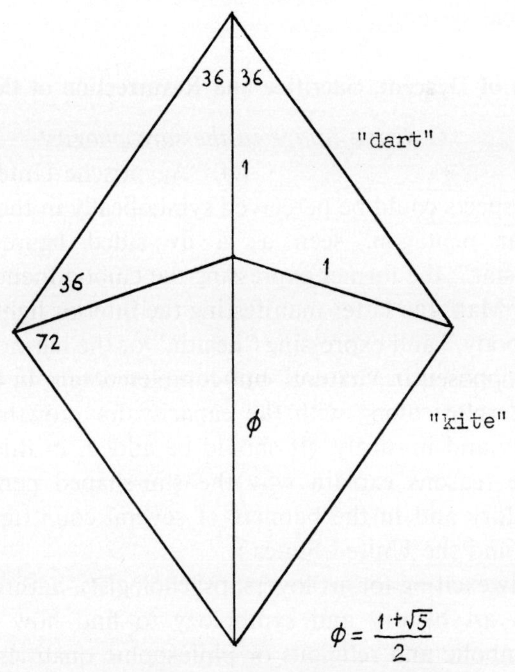

Fig. 1 Professor Penrose's "dart" and "kites."

> "Although it is possible to construct Penrose patterns with a high degree of symmetry (an infinity of patterns have bilateral symmetry), most patterns, like the universe, are a mystifying mixture of order and unexpected deviations from order... There is something even more surprising about the Penrose universes. In a curious finite sense, given by the 'local isomorphism theorem', all Penrose patterns are alike. Penrose was able to show that every finite region in any pattern is contained somewhere inside every other pattern. Moreover, it appears infinitely many times in every pattern."[8]

To make this "crazy situation" worse — not forgetting that Penrose universes are *not* periodic — the truly unexpected Conway theorem shows that any region of the tiling is never away of its exact replica by more than the diameter of the chosen region!

To Pythagoreans five was, indeed, the number of Man. The pentagon, when represented in the star-shaped version, was the symbol of the *Anthropos*, the archetype of human perfection, partaking of both the "perfection of flesh" and the immortal essence of his divine origin. To put it in another way, both the laws of necessity, which all things obey, and the freedom of unpredictable creation (or, paraphrasing the title of a scientific paper, "the unpredictable behaviour of deterministic non-linear dynamical systems."[9]

3. The Angles of Descent, Sacrifice and Resurrection of the Godhead

O divine image in the sarcophagus!

(cf. "Ägyptische Unterweltbücher")

These two aspects could be perceived symbolically in the double shape of the regular pentagon, seen as a five-sided figure and as the "pythagorean star"; the former expressing the embodiment of the "idea" or "model" of Man, the latter manifesting the interior light of the divine or "diamond body," and expressing "health," or the harmonious balance of tendencies opposed in "nature" but complementary in their vital and everlasting interplay, along with the capacity for growth without end, both outwardly and inwardly. (It should be added, in this respect, that such symbolic reasons explain why the star-shaped pentagon can be found in heraldry and in the banners of several countries, such as the Soviet Union and the United States.)

It is naturally exciting for art lovers, psychologists, anthropologists and researchers in art history and symbology to find how strikingly the traditional symbolic and religious or philosophic qualities attributed to pentagonal numbers and symmetries resemble the unexpected properties of the recently found "new state of matter" displaying an *impossible* fivefold symmetry. In fact, as Manfred Schroeder puts it, "until recently, few, if any, people suspected that there could be another state of matter sharing important aspects with both crystalline *and* amorphous substances. Yet, this is precisely what D. Shechtman et al. discovered when they recorded electron diffraction patterns of a special aluminium-manganese alloy (Al_6 Mn). The diffraction pattern, i.e., the two-dimensional Fourier transform, showed sharp peaks, just like those for a periodic crystal. But the pattern also showed a fivefold symmetry that periodic crystals simply *cannot* have . . . it turns out that an aperiodic self-similar sequence derived from the Fibonacci numbers is a good model of this new state of matter . . ."[10] Thus, five again exhibits its double and paradoxical nature and also its "transgressive" character — and science seems to undergo, as A. P. Stakhov[11] puts it, a process of "fibonaccisation"!

Let us proceed a little further. In a regular pentagon ABCDE, we trace the line EB (Fig. 2), one of the five diagonals that form the star-shaped

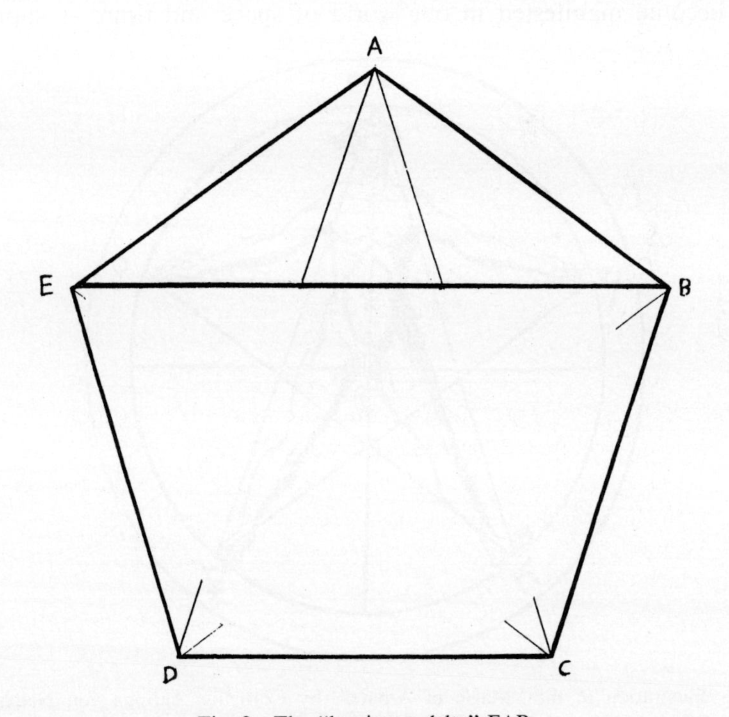

Fig. 2 The "luminous delta" EAB.

polygon within the former pentagon. If we keep in mind the anthropomorfic symbolism of our geometrical figure, illustrated by the famous drawing of Leonardo da Vinci, in which the extremities of the limbs of a man touch a circle, or even more clearly (if less artistically) in one of the drawings inserted by Cornelius Agrippa von Nettesheim (1486–1535) in his book on the "Magic of Arbatel" (Fig. 3), chosen among a large number of such kind of representations, it becomes natural to describe the triangle ABE as the "head" of the pentagon. In fact, it was known to members of medieval confraternities of master builders and craft-guilds, and has been adopted in modern masonic symbolism, under the name of "luminous delta." Light is obviously an attribute of the intellect related to clarity, vision, awareness, consciousness. One may be permitted, following the same path of reasoning, to ascribe to the "luminous delta" the symbolic equivalence of the Trinity of God as it manifests itself to the spirit of man, not as He can be conceived in His eternal and pure perfection (to which the equilateral triangle corresponds) but such as He may become manifested in our world of space and time, as supreme

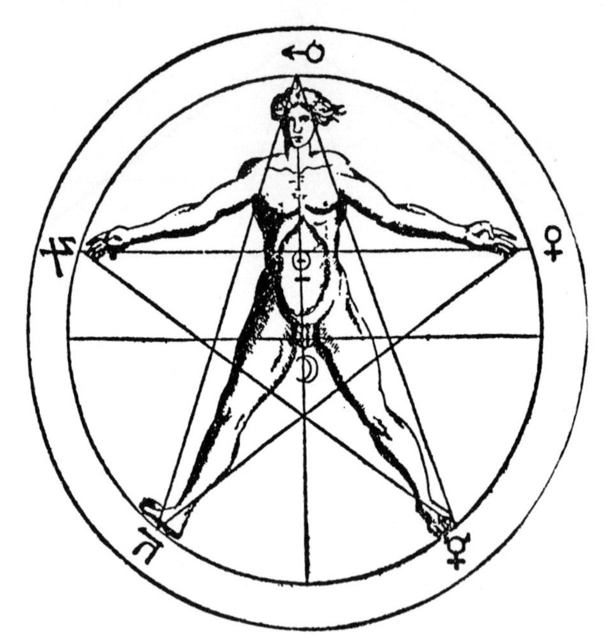

Fig. 3 Illustration of the "Magic of Arbatel" by Cornelius Agrippa von Nettesheim (1486–1535).

intellect, spiritual light, seed of all life, principle or archetype of human perfection (the triangle ABE). The three angles are not equal any more; still, the perfect trinity of the divine principle can be perceived by the mind's eye in the subtle allusion proposed by the angle values, for they prove to partake of the "trinitarian essence": the 108° angle in A divides by 3 in 36, precisely the value of the two other angles in B and E, as well as the value of the "golden triangle" ACD that fascinated the Pythagoreans. A further step allows us to recognize in the 108° angle (A) the "head" of the "luminous delta" (ABE), the divine unity before it is polarized into the existential level in the two angles of 36° (B,E). In other words, let us say that the divinity, simultaneously (and intrinsically) One and Three, "sacrifices" — or polarizes — itself in order to beget the world, of which Man is the epitome (as "image of God") and this "death" of the divine being takes place, according to the geometric symbolism sketched above, through the "descent" of the unitarian divine principle, expressed in the number 108, into the duality, on the material level, given by 2×36. 72 thus becomes the number that symbolically expresses the life-giving fragmentation of the supreme Being, his sacrifice or "passion," and also, as a consequence, the symbolic number of mankind's "nations," languages or "interpreters" of the Word, as well as the number of those who "trapped" the godly principle. So the divinity "dies." But as the seed of plants buried in the earth give birth, in due time, to new life-giving crops, so the divine "body" will resurrect in the perennial cycle of manifestation; or again, following our pentagonal code, the archetype of preternatural Unity encoded in the number of degrees (108) of the upper angle of the "light triangle," ABE, shall be restored through the mysterious operation of the regenerative life-force dispensed by the "dead" hypostasis of the Divine, i.e., through the reconstitution of that number from the parts into which it was divided.[12]

This may seem a far-fetched, improbable, too esoteric way of reasoning to a modern mind nurtured by rationalistic hard-edged habitudes of thought, but in fact exactly corresponds to what certain ancient creators of religious works of art most probably thought, as can be inferred from their creations, which give us proof that the use of such symbols and numbers could not be the whim of some eccentric artist or architect, but necessarily corresponded to the religious and metaphysical philosophy of the initiates, ordained priests and learned spirits of the time.

How can we ascertain and prove the truth of the "pentagonal" symbolic meanings adumbrated above? By examining the existing monuments and works of art that relate to the subject, namely the use of fivefold symmetric devices, figures and numbers, in theological dicta pertaining to the creation of Man, the triumph over death and darkness, the sacrifice and dismemberment of the God-made-man and his Resurrection as promise of divine perfection and eternal life. We shall turn our attention to the iconography of Christ, and especially the representations of the Resurrection, and at the same time search for significant analogies in other cultural areas. We have, in fact, a precious reference in Egyptian art and religion, not only central to the shaping of Greek civilization and hence our own, but also documenting, among its most important spiritual conceptions, the belief in a god's life-giving sacrifice and subsequent resurrection.

My first example will concern the symbology of 36 and 72 linked, as we have seen, to the pentagonal polarization of the 108 archetype of divine unity. We find in the New Testament a versicle stating that "the Lord appointed seventy-two other [disciples] and sent them before Him, two by two, unto all towns where He should have to go" (. . . *designavit Dominus et alios sptuaginta duos et misi illos binos ante faciem suam, in omnem civitatem et locum, quo erat ipse venturus, Lucas, X,1*). This passage received the following commentary from the Jesuit theologian Antonio Vieira (one of the most eminent writers and exegetes of the XVII century): "Christ made dispositions for view of the preaching of the Gospel and the conversion of the world, and after the twelve Apostles had been appointed, corresponding to the twelve sons of Jacob and the twelve tribes of Israel, He chose seventy two envoys." Why seventy two? Antonio Vieira proceeds: "If we search in the sacred writers the mystery and proportion of this number, we find the answer in St. Hieronymus, and with him the current sentence of the interpreters, that those forerunners and ambassadors of Christ were seventy two for as many (as we said before) are the nations of the World, that the Lord, by means of preaching and doctrine, wished to bring (as He did) to the knowledge of the Faith."[13] A little further the same author sees a wondrous meaning in the fact that the translators of the Bible into Greek — who worked in Alexandria in 282 or 283 B.C. — numbered seventy two and translated what read *juxta numerum filiorum Israel* into *juxta numerum Angelorum*

Dei, in this way "calling the children of Israel *angels* or *ambassadors* of God, for that was the end and the goal for which they were commited unto all nations and taken and parted according to their number."[14]

The incomprehensible wording in "before him, two by two" (*misi illos binos ante faciem suam*) of Luke 10:1, becomes clear in the light of the polarization mentioned earlier.

4. The Osiris Legend as a Pentagonal "Mystery"

To the ancient Egyptians as to the first alchemists the "tomb of Osiris" was the mysterious place where the corpse of the dead may become the source of a new life.

Marie-Louise von Franz (*Traum und Tod*)

If we turn to the Egyptian legend of Osiris — a deity which presents important analogies with Christ — we find the same number with a similar connotation. In his *De Iside et Osiride*, Plutarch writes:

> "In the very distant past, after having completely organized and civilized the land of Egypt, Osiris confided the care and governing of the country to Isis, his sister, and left for the south in order to teach agriculture, the laws of harmony, and the ways of worshipping the divine powers, to the still savage peoples of those regions. When, after a long absence, Osiris returned to Egypt, Seth and his *seventy-two* accomplices trapped him in an ambush. They enclosed him in a coffin made to fit his dimensions and threw it into one of the arms of the Nile. It was carried out to sea and floated northward until the waves washed it ashore at Byblos in Lebanon. A magnificent tree grew up around the coffin, and the king of that country, when he heard of this marvel, had the tree cut down to make a column in his palace. Isis, meanwhile, having heard from the whispers of the winds what had become of Osiris, set out in search for him. Arriving one day at Byblos, transformed into a swallow, she flew round and around the column. Finally she succeeded in bringing the sarcophagus back to Egypt, where she hid it in a remote part of the Delta. But one night when Seth was hunting by the full moon he discovered it, seized the body of Osiris, cut it into fourteen pieces and scattered them all over the country. So Isis, aided by her sister Nepthys, set out in quest of the parts of Osiris' body. As she found each part, she buried it in the place where it had come to rest, as a sacred relic."[15]

I think it useful to add a few remarks calling attention to some aspects of this myth that may otherwise pass unnoticed but which seem to confirm, with beautiful precision, my "pentagonal" interpretation. One can detect, from the onset, a telling allusion to the golden *ratio* in the fact that Osiris is described, in Plutarch's words, as the "teacher of the laws of harmony," hinting beyond the notion of measure to that of proportion or *ratio.* This does not preclude other interpretations of the same legend, as it pertains to the essence of symbol and myth; but in such cases plurality is not exclusive of singular forms or levels of comprehension.[16] The "geometric" encoded meaning is obviously a "secret" one. These are, as briefly as possible, my comments:

(i) The legend says that Osiris "civilized Egypt" and left its government to his sister Isis. Again one feels that the northern region of the Nile Delta (Osiris is a deity of the North, Seth is a deity of the South) constitutes a sort of "luminous triangle," where the light of the solar divinity is reflected by the feminine or "lunar" hypostasis, as application of the laws. Then Osiris proceeds to the South in order to bring his light to the "uncivilized" spaces of the cosmos, i.e., to extend the pentagonal structure to the remaining open area of the circle (Fig. 4).

(ii) On his return, Osiris is trapped by seventy two accomplices of Seth and put in "a coffin made to fit his dimensions" (Fig. 5). Seth, of course, cannot be counted together with the "accomplices," because he is *Osiris' counterpart* (a complex system including both stability and "chaos"). The seventy two accomplices act under his orders, in

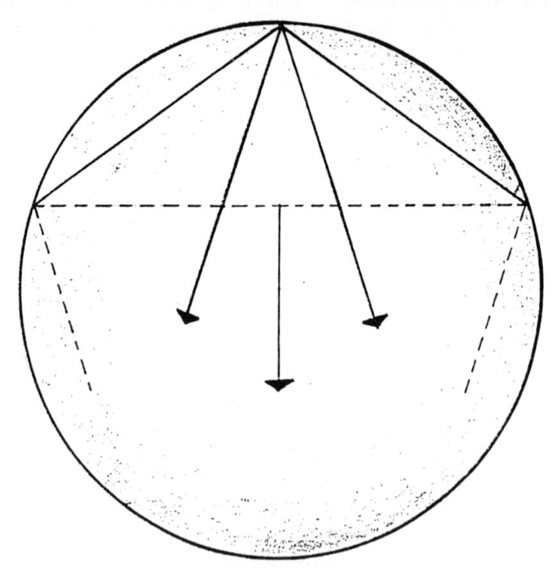

Fig. 4 The voyage of Osiris to the South.

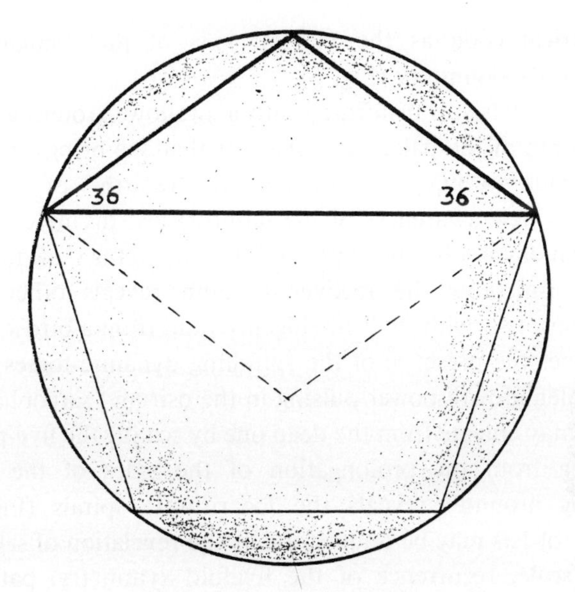

Fig. 5 The coffin of Osiris.

other words, they are both the agents of the god's confinement and its numerical effect, fixing the "dimensions" of the regular pentagon. In the interplay of polarities, Seth behaves like Osiris' "shadow," as a "chaos-order" introducing instability in the very processing of "order" – a creative behaviour, in fact, for its end product will be the generation of higher degrees of complexity.

(iii) The Sethian power to upset, contradict, and increase chaos, acts upon the Osirian power to organize, develop, and harmoniously impart structure, balance, proportion, rhythm and beauty. Now Osiris is "trapped" in the "dimensioned" pentagon, contracted to a "dead" body whose angles are known but whose divine light, precisely because of this knowledge, suffers the effects of information-entropy: the "coffin" is abandoned to the "sea." But somehow the portentous achievement of defining the regular pentagon and its angles cannot disappear forever in the unconscious; the coffin reaches land again (not unlike Jonas in the Bible) and takes root, so to speak, in a country to the North – the direction where Osiris comes from originally. The resulting Tree or Column of the King's Palace, symbol of the *Axis Mundi*, can be translated to the geometrical code as the vertical axis of the regular pentagon, stressing its symmetry (Fig. 6).

(iv) The flight of Isis, transformed into a swallow, "round and around" the column of the palace and the fact that she brings the "sarcophagus" back to Egypt, postulates the "rebirth" of Osiris as the intellect's new awareness of the (almost) lost divine essence of the pentagon (symbolically expressed by the return of the coffin to Egypt). But now the recovered coffin reveals other wonderful properties. Isis, with swift intelligence (and from a bird's viewpoint!) recognizes the presence of the spiralling dynamic forces that reveal the hidden growth-power pulsing in the osirian "sarcophagus." New bodies may emerge from the dead one by scaling the five-pointed star resulting from the prolongation of the sides of the coffin; the "turning around" reveals the logarithmic spirals (Fig. 7). The "flight" of Isis may be understood as the revelation of self-similarity across scale, recurrence of the fivefold symmetry, pattern inside pattern, and the "magic" engendering of golden number cascades. A remarkable confirmation of this approach is the fact that the incarnation of Light, the deity Iwf, when reascending from the

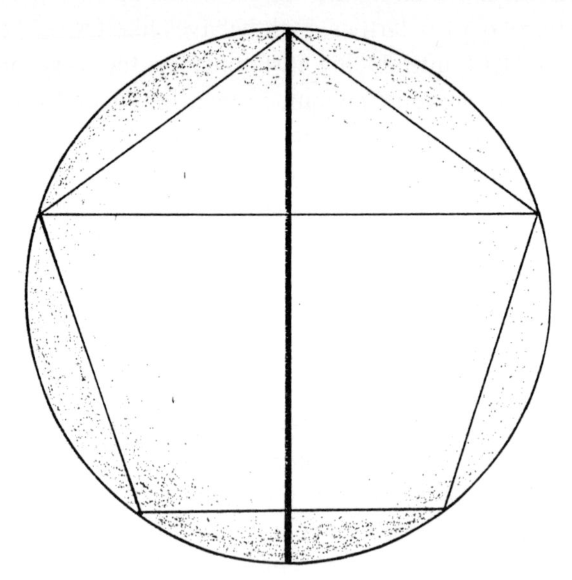

Fig. 6 The Tree, or the Column of the King's Palace.

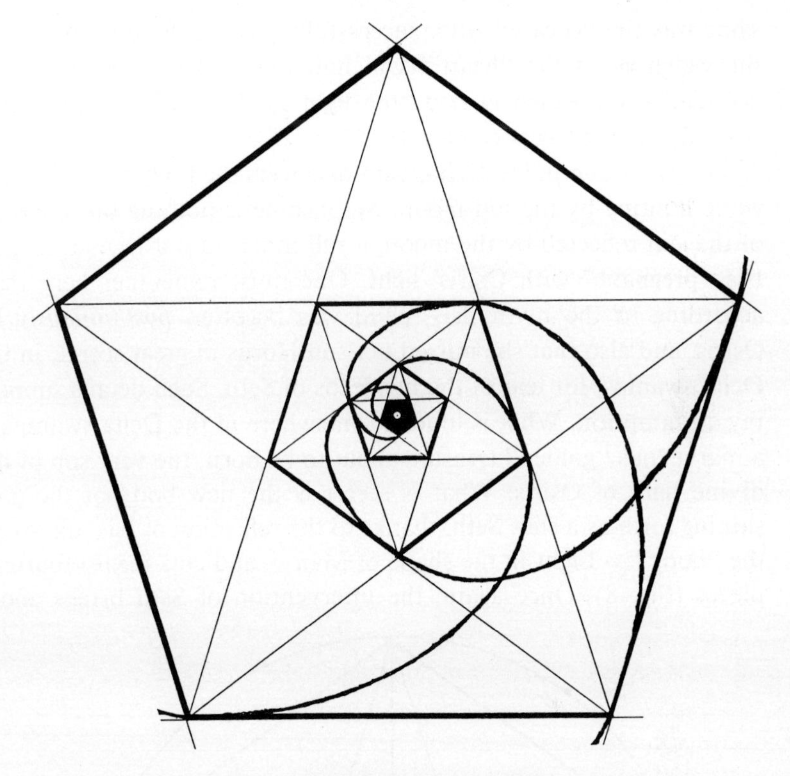

Fig. 7 The "flight of the swallow."

darkness of the Dwat during his nocturnal voyage, stands in the coils of the serpent Mehen, whose name means "spiralling."[17] It is noteworthy that Ra, the Sun god, while crossing the underworld, speaks of the "divine image in the sarcophagus" and calls the Osiris corpse an "image" or "reproduction of an image."[18] Origenes, who lived and taught in Alexandria, certainly inspired in the Eygptian tradition, wrote that the spiritual body springs forth from the dead body by the virtue of an energy (*virtus, dynamis*), a seed, or the invisible principle of a seed, that he also called *spintherismos*, meaning a springing of sparks.[19] It is plausible to understand such "sparks" as "stars," knowing that the eternal and divine soul *ba* was represented by a bird or a star.

The ascending light of Iwf, as we have seen, flows on the coils of the serpent from the underworld to the glory of resurrection. In the same way the deceased, after successfully passing the *psychostasis* or the weighing of the "heart," will find themselves in Sekhem, the celestial town, which is also "the light of the world coming forth from the sarcophagus of Osiris."[20]

(v) Although hidden in the Delta, Seth discovers the body of his brother while hunting by the full moon. Moonshine is nothing but the light of the sun reflected by the moon, a full moon in our legend, i.e., an Isis "pregnant" with Osiris' light. One must remember here that, according to the myth, Isis' child was begotten *posthumously* by Osiris, and also that she nursed her son Horus in great secret, in the Delta swamps, for fear of the misdeeds of Seth. Such details support my asssumption. What is hidden somewhere in the Delta swamps is a marvellous "golden" treasure about to be born, the very son of the divine light of Osiris. What is secret is the new body of the god, shining forth as a star. Seth, thanks to the reflection of Isis, discovers the "body" — Light in the shape of Man — and cuts it into fourteen pieces (Fig. 8). Once again, the intervention of Seth brings about

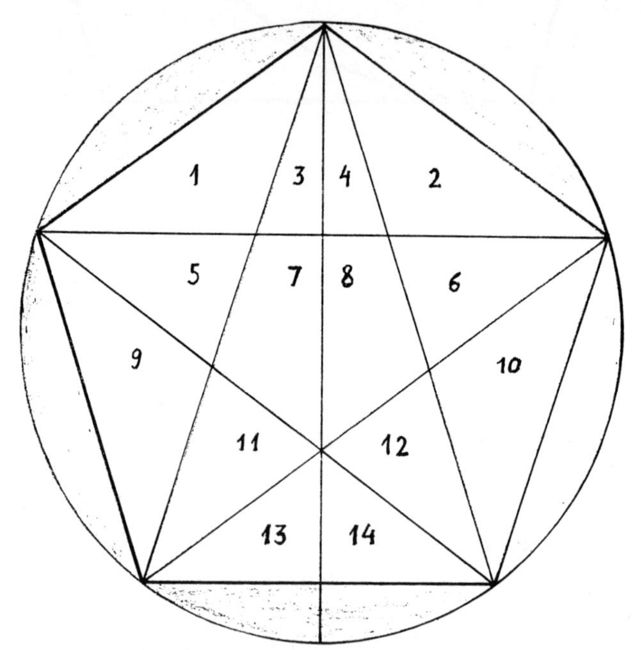

Fig. 8 The Body of Osiris cut in Fourteen Pieces.

numbering. This time the number does not correspond to the angles but to the "pieces of the body" (to which the regions of Eygpt correspond symbolically); fourteen is related — as would be expected — to the lunar calendar[21] and also, as we have seen, to the phalanges of the hand; the number needs to be doubled. In fact, for the Eygptians everything in the world had its double. Along with the actual constricted body divided in separate and numbered pieces — the material reality of the left "hand" of Seth — there is a luminous double of Osiris, simultaneously present in our existential level as the sun's light, and remote from us in his supra-mundane and divine realm. This last luminous body cannot be cut to pieces or divided, in so far as it represents the unity and the indivisibility of the god.

5. The Number of Resurrection

And the soul will cry to the body of light: Wake up, get out of the Hades, stand out of the tomb, wake up and leave darkness. For you are enshrouded in spirit and divinity . . . the house is sealed and the statue (andrias) of light and divinity stands up. The fire has begotten the unity, transforming it, it's due to fire that the One got out of the mother's womb.

Comarios

in *Berthelot, "Collection des anciens alchimistes grecs"*

Let us now consider once more the values of the angles of the "luminous delta." If we assume 72 as representing the first aspect of the death of Osiris, in other words his "reflection" on the horizontal surface of the manifested world, the correlative resurrection must correspond to the restoration of the vertical or standing position to the god, as opposed to the "dead" horizontal position. We have seen verticality introduced in the myth by the Tree or the Column of the King's Palace. We discover in different Eygptian documents an analogous "erection of the column," the Djed Column to be more precise, taking place precisely on the last day of the month of Khoiak, celebrating publicly the resurrection of Osiris. There is a beautiful representation of the raising of the Djed Column in the Temple of Seti I, at Abydos.

The king makes the offering of Maât to Osiris and pronounces an incantation, known from the Berlin Papyrus (Maât is the supreme light of cosmic consciousness, the ultimate goal of creation and of every creature).

Here are some extracts from it: "I have come to you, I am Thoth, *my two hands united to carry Maât* . . . Maât is in every place that is yours . . . You rise with Maât, *you join your limbs to Maât*, you make Maât rest *on your head* in order that she may take her seat on your forehead . . ."[22]

The vertical axis of the pentagon joins the "head" angle in A (Fig. 2), symbolizing the supreme consciousness promised to man by the Light-Deity, or the absolute "North," or again the "seat of Maât"; as we know this angle has 108 degrees. It is not absurd, consequently, to conceive that Osiris' resurrection may be translated symbolically through this same angle. And, in fact, this assumption can be verified in several representations dealing with the resurrection of Osiris or with the resurrective power he embodies. The erection of the Djed Column symbolizes the coming to life of the god, a process that starts with the column lying on the ground and ends when the column stands vertically; the intermediate stage between "death" and full life is the moment of resurrection proper. In the bas-reliefs of the Temple of Seti I, at Abydos (c.1300 B.C.), that moment is represented by an image in which the inclination of the Djed Column is at an angle to the ground-line of approximately 72° on the side of "death" and consequently an angle of 108° on the side of "resurrection" (Fig. 9). Here I call the "side of death" the part of the bas-relief left of the Djed, i.e., the side where the Djed is being raised from the ground, and "side of resurrection" the side where the king stands and the Djed is to reach its vertical position. The same "Angle of Resurrection" can be found in other Eygptian documents, namely in certain ithyphallic images of Min-Amun, of Sekh-met "the powerful," of Osiris himself, where the erect phallus represents the energy of life, active in the process of resurrection. Also in the Tomb of Amenophis II, at Thebes, there is a representation of the incarnation of Light, Iwf, standing for the first time at the Eleventh Hour, just before sunrise, at the prow of the boat drawn by the serpent Mehen; Iwf holds the two wings of a winged serpent, under the eyes of the Sun and the Moon and among two-headed figures symbolizing the union of opposites — a clear representation of the resurrecting Divinity; the angle formed by the wings measures 54°, exactly half of 108° (Fig. 10). The number 108, which geometrically defines the regular pentagon, becomes in such representations the symbol — only visible for those who can measure! — of the perennial renewal of the life-resurrecting principle, intimately conjoined with the light princi-

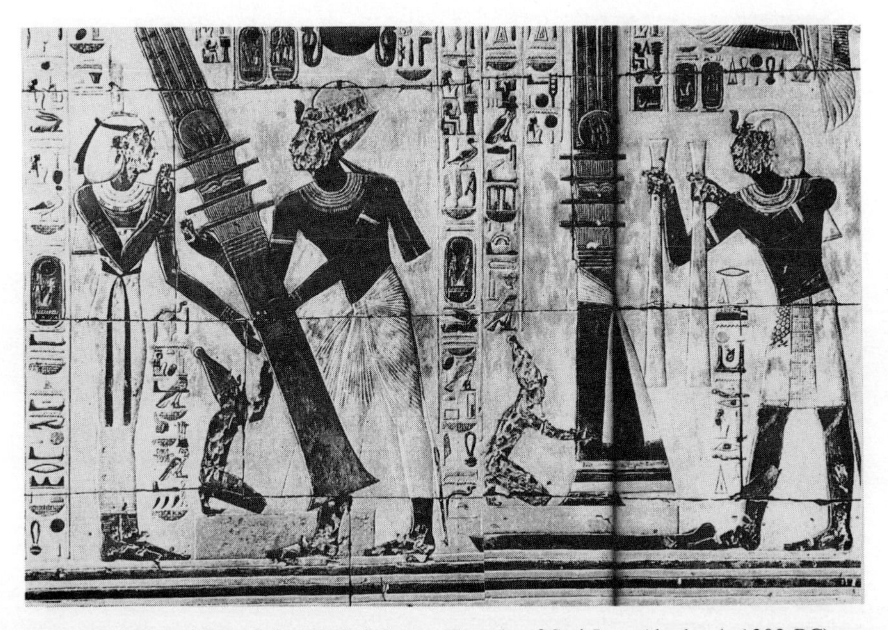

Fig. 9 The raising of the Djed Column. Temple of Seti I, at Abydos (c.1300 BC).

ple, both emerging as a divine Unity from darkness and death. In the *Book of What is in the Dwat*, we read that the solar god, under the name of Ra, while journeying through the regions of the twelve hours of the night, is acclaimed by multitudes of voices each time he passes another gate; among many expressions of joy and praise they sing:

"Thou livest, now flesh, in the earth!" and "Come then, Ra, . . . illumine the primordial darkness that Flesh may live and renew itself!"[23]

I will end these brief glimpses of the universe of pentagonal symmetries in sacred art by stressing a rather unexpected fact — that the Christian iconography proves conclusively that the above-mentioned symbology of the pentagonal angles was known to the true representatives of the Church and to a number of masters of the arts and the crafts of our Middle Ages, at least up to the 16th century. In the course of the last few years, I have collected (without any systematic spirit nor the required persistence) enough examples of the use of the 72 and 108 angles in paintings, stained-glass windows, engravings, etc., to persuade myself that I was not being confronted with meaningless coincidences but, on the contrary, I was discovering a conscious and frequent use of those

Fig. 10 *The Book of What is in the Dwat*, the Eleventh Hour. Tomb of Amenophis II, Thebes (c.1450 BC).

angle-values on the part of ancient artists. The Egyptian conceptions — I'm confining myself to the pentagonal figures and numbers — have survived mysteriously, so it appears, across a vast gulf of centuries, to inspire many works of sacred art in Christian Europe, especially in connection with Jesus Christ's resurrection and related motives. Among the most striking examples that came at random to my knowledge, I shall mention here nine:

(i) The stained-glass window of the Museum of Donaueschingen (c. 1320) representing the "Apparition of Christ to the Virgin" after the Resurrection (Fig. 11);

(ii) The stained-glass window by Pierre Hemmel (1481) in the Museum L'Oeuvre de Notre-Dame, at Strasbourg, representing the descent of Christ to Limbo;

(iii) An engraving by Martin Schongauer (1470–91) representing St. Michael, in the Museum Unterlinden, at Colmar;

(iv) The "Apparition of Christ to the Virgin" in a stained-glass window of the Monastery of Batalha, Portugal (c. 1514) (Fig. 12);

Fig. 11 Detail of a stained-glass window in the Museum of Donaueschingen (c.1320) representing the Apparition of Christ to the Virgin.

(v) A wood-cut from *La Légende Dorée* of Jacques de Voragine (Lyon, 1483) (Fig. 13);

(vi) The painting by Garcia Fernandes representing the same subject of the "Apparition to the Virgin" (1531), in the Museum of Coimbra;

(vii) A wood-cut representing the Resurrection in the *Rosarium Philosophorum* (Frankfurt, 1550) (Fig. 14);

(viii) The "Apparition to the Virgin" by Gregório Lopes (middle of the 16th century) at the Church of St. Quintin (Sobral de Monte Agraço, Portugal).

(ix) The engraving *Melencolia* by Albrecht Dürer (1514), a famous masterpiece on a rather mysterious subject.

There are other examples, sometimes intriguing, sometimes less convincing; many more certainly exist, scattered through the immense *corpus* of European Christian art, awaiting a systematic research and classification far beyond my capacity to perform.

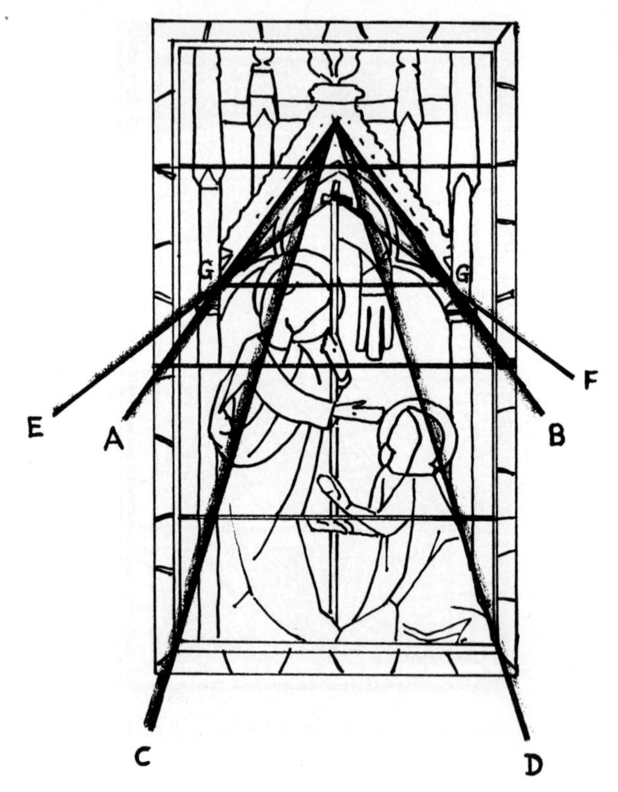

Fig. 11 The angle AB of the vertex of the temple's roof has 72 degrees, the number of "descent" and "dispersion" of the Word. The angle CD has 36 degrees, corresponding to the nucleus of the whole symbolic pentagonal representation of the apparition of the resurrected Christ to the Virgin. The number of resurrection is hidden in the angle EF having its vertex in the center of the *labarum's* cross and connecting it with points GG. In this image resurrection through the cross is linked to the basis of the temple's triangle.

I include here only four of the abovementioned works, the stained-glass window of the Museum of Donaueschingen, the wood-cuts of 1483 and of "Rosarium Philosophorum," and the stained-glass at the Batalha Monastery. In three of these examples, the *labarum* in the hand of the resurrected Christ — similar, in its symbolism of the victory over death, to the Eygptian Djed Column — shows an inclination-angle of 108° (and 72°) to the horizontal line. Thus the Resurrection angle resurrects in European art many centuries after its disappearance from Eygpt! In the *Evangel of Barnabe* one reads: "Fear not, I am Jesus. Shed no tears, I am alive and not dead."

Fig. 12 The Apparition of Christ to the Virgin, stained-glass window of the Monastery of Batalha, Portugal (c.1514). The *labarum* of Resurrection shows the canonical angles.

¶ De la resurrectiō ō nr̄e seigneur ih̄ūcrist

Fig. 13 Wood-cut from *La Légende Dorée* of Jacques de Voragine, Lyon, 1483 (108° angle stressed).

Fig. 14 Resurrection of Jesus Christ. "Rosarium Philosophorum," Frankfurt, 1550. (The 108° angle is stressed by bold line.)

References and Notes

1. A.F. Lesev, *Istoriya Filosofiyi* (1981). Quoted A.P. Stakhov, The Golden Section in the measurement theory, in *Symmetry 2: Unifying Human Understanding*. I. Hargittai, ed. Pergamon Press, Oxford (1989).
2. M.C. Ghyka, *Le Nombre d'Or*. Gallimard, Paris (1931).
3. R. Abellio, *La Bible Document Chiffré*. Gallimard Paris (1950); R. Abellio and C. Hirsch, *Introduction à une Theorie des Nombres Bibliques*. Gallimard, Paris (1984).
4. There is no polygon "manifesting" the essence of the One and Two. The circle itself can be conceived as the geometrical "hieroglyph" of One in its "empty" aspect or as the unity in "absence" and the Two can be imagined as the diameter of the circle, "visible" manifestation of the division of the Whole but not yet a polygonal figure. Only in the Three do we reach representable shapes, namely the Triangle. As Abellio and Lupasco stressed, all manifestation requires the convergence of three principles or forces.
5. See C.G. Jung, *Mysterium Conjunctionis*, French translation by Etienne Perrot, Albin Michel, Paris (1980); also M.-L. von Franz, *Zahl und Zeit*, Ernst Klett Verlag, Stuttgart (1970).
6. P.W. Atkins, *The Second Law* (1984). Quoted by James Gleick, *Chaos* Penguin Books (1988).
7. William M. Schaffer, *Chaos in Ecological Systems* . . . (1986). Quoted by J. Gleick, *op. cit.*.
8. Martin Gardner, Mathematical games, *Scientific American*, January 1977.
9. The paraphrased title is that of a paper by Roderick V. Jensen quoted by J. Gleick, *Chaos. op. cit.*
10. Manfred Schroeder, *Number Theory in Science and Communication*. Springer, 2nd Edn., p. 317–319.
11. A.P. Stakhov, The Golden Section in the Measurement Theory, in *Symmetry 2: Unifying Human Understanding*. I. Hargittai, ed. Pergamon Press, Oxford (1989).
12. Lima de Freitas, 515 — A symmetric number in Dante, in *Symmetry 2: Unifying Human Understanding*. I. Hargittai, ed. Pergamon Press, Oxford (1989).
13. António Vieira, *História do Futuro*. Published for the first time in Lisbon in 1718. Sá da Costa, Vol. 2, Lisbon (1953).
14. *Ibidem.*
15. J.G. Griffiths, *Plutarch's De Iside et Osiride*, with an introduction, translation and commentary. Cardiff (1970).
16. On the multiple levels in the interpretation and hermeneutics of the symbol there is a vast bibliography, including many important authors as Freud, Pierce, Cassirer, Ogden, Jung, Eliade, Ricoeur, Lévy-Strauss, Alleau, Durand, etc. See *The Encyclopedia of Religion*, MacMillan, New York (1987).
17. *The Book of What is in the Dwat*, Eleventh and Twelfth Hours, in the tombs of Amenophis II and Thutmosis III, both at Thebes.
18. *Ägyptische Unterweltsbücher*, Artemis, Zurich/Munich (1972). Quoted by Marie-Louise von Franz, *Traum und Tod*, Kösel-Verlag, Munich (1984).
19. Marie-Louise von Franz, *op. cit.*

20. Albert Champdor, *Le Livre des Morts — Papyrus d'Ani, de Hunefer, d'Anhaï, du British Museum*. Albin Michel, Paris (1963).
21. It is relevant to point out that the Egyptians divided the year in 36 weeks of 10 days each (the *decans*), thus leaving aside five days that were not considered as truly belonging in the "real" year. The same "epagomenic" five days are obtained taking in account the 72 accomplices of Seth (or Typhon) in his fight against Osiris, or again the eating away of 1/72 of the Moon by Hermes-Thot every day, corresponding to 20 minutes of a 1440-minute day. The Moon is totally eaten in 72×20 minutes = 1440 minutes, which corresponds to a day of 24 hours of 60 minutes. The difference between a year of 365 days and a year of 360 "real" days is the same as $525{,}600 - 518{,}400 = 5 \times 1440$. These five days were devoted to the feast of the Osirian circle of gods (5 was the number of Horus, son of Osiris and Isis, as well as the number of Seth). It is interesting to compare this with the year of the Mayas, composed of 360 days and named *tun*. Each *tun* had 18 months of 20 days each (their numerical system being based on 20) and the *katun* were periods of 20 years, comprising 7200 days. The five days "out-of-the-year" were devoted by the Mayas to the feasts of the demoniac gods. We shall not forget that the circle, like the "real" year's days, has 360 degrees.
22. *Berlin Papyrus*. Quoted by Lucie Lamy, *Egyptian Mysteries*, Thames and Hudson, London (1981).
23. *Book of What is in the Dwat*, burial chambers of Amenophis II and Thutmosis III, Thebes. Quoted by Lucie Lamy, *op. cit.*

THE ICOSAHEDRAL DESIGN OF THE GREAT PYRAMID

Hugo F. Verheyen

1. Introduction

The reader may find it quite strange to come across an article on the Great Pyramid at Giza, apparently having fourfold symmetry, in a book on fivefold symmetry. From the outside, indeed, the Great Pyramid has fourfold symmetry about its vertical axis, but the complicated system of inner corridors and chambers has no fourfold symmetry at all. On the contrary, the measures and angles within the Pyramid seem to indicate a conceptual design in terms of fivefold symmetry, suggested by the clear and frequent appearance of the *golden section*.[a]

For over two centuries now, examiners have been surveying and measuring the Great Pyramid and have made conclusions from these figures. Among these, fantastic speculations occur, from the appearance of π all over, to the holy inch and the distance of the earth to the sun. However, the Great Pyramid has been severely damaged in the past centuries, and exact figures for the facial slope and height are virtually impossible. In Chapter 4: the Group of Giza, I. Edwards[2] mentions in his book how the Egyptian Government charged J.H. Cole[3] with the investigation of the measures and orientation of the Great Pyramid by modern instruments. The figures, seemingly accepted now, are:

[a] or *Sectio Divina, Golden Mean* of proportion or *Sectio Aurea*: the most harmonious way to divide a line segment.[1] This proportion is calculated to be $\tau = \frac{1}{2}(\sqrt{5} + 1) \approx 1.618033989 \ldots$ while $\tau^{-1} = \frac{1}{2}(\sqrt{5} - 1) \approx 0.618033989 \ldots = \tau - 1$. Some authors use "φ" for τ^{-1}.

edges of base: 230.4 m (N)
230.6 m (S)
230.54 m (E)
230.5 m (W)
angles of square: 90°03'02" (NE)
89°59'05" (NW)
89°56'27" (SE)
90° 0'33" (SW)
height: 146.8 m, of which the upper 9.45 has vanished.
slope of the faces: ~51°52'

In this paper it will be explained how the Great Pyramid may have been conceived out of the simple shape of the *icosahedron*, and how this concept of geometrical thinking also fits into the esoterical tradition.

2. A Brief History of the Pyramid

The Great Pyramid is believed to have been built in c.2644 B.C., during the reign of Pharaoh Khufu (Cheops) of the IVth dynasty, for whom it would have been erected as a mausoleum. In one of the five cavities above the King's Chamber, a seal of Cheops was discovered, believed to have been roughly drawn by the craftsmen during the construction of the Pyramid. Besides that, no other identification or decoration has been found, and the open sarcophagus of the King's Chamber was empty.

However, certain groups of people are convinced the Pyramid is much older than assumed, and has been used for other purposes than a mere shrine. Rosicrucians[b] speak in terms of "at least 12,000 years of age" and suspect "the Pyramid to have been built by descendants of the sunken Atlantis. Much later, after movements of people, the ancient Egyptians have found this lonesome reminder and imitated its shape in later pyramids." But the fact is that the Great Pyramid, besides being the largest of all pyramids in Egypt, is also unique in the way of its construction: it is the only known pyramid (yet) to have an ascending passage (Fig. 1) and two inner chambers, at a height quite above surface level.

[b] Mystical Society, split up into many fractions, and the origin of Free-masonry in Europe.

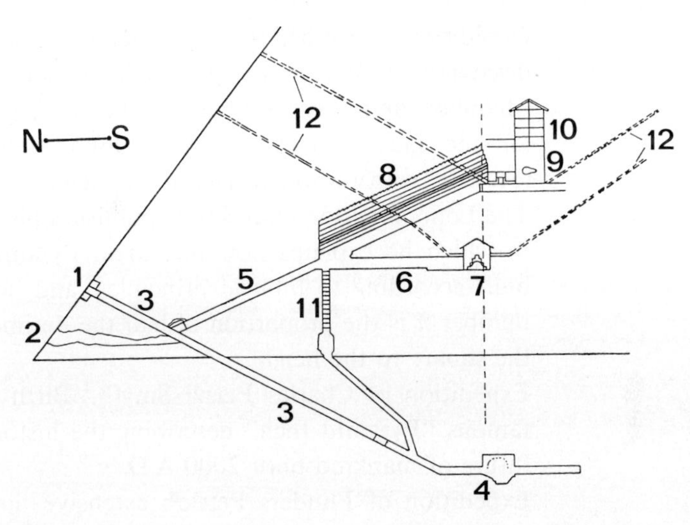

Fig. 1 Passages and chambers inside the Great Pyramid (cross section N-S), based on a drawing by Naber.[12]

1. entrance
2. entrance forced by Al-Mamun
3. descending passage
4. chamber below surface level
5. ascending passage
6. horizontal passage
7. Queen's chamber
8. Grand Gallery
9. King's chamber, containing open sarcophagus
10. roof construction containing 5 cavities (low chambers)
11. rough vertical passage
12. ventilation shafts of King's and Queen's chambers

Some historical facts:

~450 B.C.	Visit of Herodotus to the complex of Pyramids at Giza, and description of the Pyramids in book II. It is assumed that the inner configuration of the Great Pyramid is not revealed yet.
~850 A.D.	The Arab Khalif Al-Mamun forces an entrance into the northern face of the Great Pyramid, and finds his way into the corridors and chambers, which are all completely empty, lacking any kind of decoration.
Late 18th century	Expedition of Napoleon Bonaparte. The geographer E.F. Jomard measures the facial slope to be 51°20'.

1837	Expedition of Howard Vyse,[c] assisted by J.S. Perring: discovery of four more cavities above the King's Chamber[d] and a seal of Pharaoh Cheops. Correction for the facial slope: $\sim 51°50'$,[4] John Herschell's reaction[e] with respect to Herodotus' passage.
1859	The London editor John Taylor publishes his book,[6] in which he explains how the Great Pyramid was built according to biblical principles, and how the number π is the proportion of half the perimeter of the square to the height.
1864	Expedition of Charles Piazzi Smyth.[f] Birth of the famous "Pyramid Inch" describing the history and future of mankind until 2000 A.D.[7,8,9]
1880	Expedition of Flinders Petrie[g]: extensive measurements, discussion of π appearing in the measures and the King's Chamber. Scientific approval of his published results.[10]
1907	The German Hermann Neikes discovers the golden section in the Pyramid, unaware of Herschell's result 80 years previously.[11]
1915	Publication of H.A. Naber's lectures[h] on the golden

[c] British Colonel who financed his own expedition which took seven months. He worked together with the Italian Caviglia, and afterwards with the British engineer J.S. Perring.

[d] The first report of a low room above the ceiling of the King's chamber came from Davison in 1765.[2] Vyse discovered three more similar low rooms above Davison's chamber (Wellington's, Nelson's, and Lady Arbutnoth's chambers) and on top of these a fifth room right under a pointed ceiling (Campbell's chamber), in which the seal of Cheops was found.[5]

[e] See Sec. 3.

[f] by then already 18 year "Astronomer Royal of Scotland" and professor at the University of Edinburgh; a personally financed expedition.

[g] son of an Egyptologist; received a grant for his expendition.

[h] Dr. H.A. Naber, professor of Physics at the Rijkshoogere Burgerschool at Hoorn, the Netherlands, found new appearances of the golden section in the Great Pyramid, and his lectures were published in a book during W.W.I by the Theosophical Society of Amsterdam. Back in the sixties, when I grew up in a suburb of Antwerp and had great interest in Geometry, the book was given to me by a neighbour who possessed an extensive "occult" library. I started to read this book thoroughly after having finished university studies, and, although written in a confusing way, it contains an incredible amount of geometrical information on the golden section, the Great Pyramid, and polyhedra. This work is the very start of my interest in these topics. Written in Dutch, which very few foreign people realize

	section in the Pyramid.[12]
1925	Establishment of the Pyramid's measures.[i]
1978	The author visits the King's Chamber in the Great Pyramid and lies a few minutes alone in the royal sarcophagus for meditation on this paper . . .

3. Interpretation of an Ancient Greek Remark

In about the middle of the 5th century B.C., the Greek historian Herodotus visited Egypt, among other countries, and wrote his travel memoirs in nine books. The chapter on the Pharaohs who built the Pyramids at Giza appears in book II, named *Euterpe*, pp. 124–134.

About this, Martin Gardner[13] mentions:

> "Herodotus explains that the pyramid has been built so that the face of every side is equal to the square whose edge is the height of the pyramid. . . . and the proportion of the height to twice the base is automatically a surprisingly correct value for π (see *Popular Astronomy*, April 1943, p. 185)."

However, Richard J. Gillings[14] mentions this passage of Herodotus quite differently in "Appendix 3: Great Pyramid Mysticism":

> "But Turnbull[j] continues: 'As Professor D'Arcy Thompson has suggested, the very shape of the Great Pyramid indicates a considerable familiarity with that of the *regular pentagon.* A certain obscure passage in Herodotus can, by the slightest emendation, be made to yield excellent sense. It would imply that the area of each triangular face of the Pyramid is equal to the square of the vertical height. If this is so, the ratios of height, slope, and base, can be expressed in terms of the *golden section*, . . .' I am unable to understand exactly what Turnbull means by this last sentence. But whatever it means, with further slight emendations, the dimensions of the Eiffel Tower or Boulder Dam could be made to produce equally and pretentious expressions of a mathematical connotation."

is the same language as Flemish, it may as well be one of the few remaining copies still in existence.

[i] see Sec. 1

[j] Gillings refers to pp. 2f[15]; Turnbull's mentioning of D'Arcy Thompson — as Gillings explains in a footnote — cannot be located in Ref. 16, unless it refers to a "brief reference on p. 931 of Vol. 2": "The *sectio aurea* or "golden mean" of unity ... is a number beloved of the circle squarer, and of all those who seek to find, and then to penetrate, the secret of the Great Pyramid." It needs to be said, however, that Gillings had perhaps a reading problem, for he goes on: "... Nor can I locate the "certain obscure passage in Herodotus", to find out what this slight "literal emendation" might be."

If Gillings was not able to understand what Turnbull meant by his last sentence, this simple calculation may solve the problem: In Fig. 2, TO = h, OA = c, TA = b and the edge of the square a. Then,

$$\frac{ab}{2} = h^2 \quad \text{(``according to Herodotus'')}$$

and since $c = \frac{a}{2}$,

$$cb = h^2 :$$

application of Pythagoras' theorem, $h^2 = b^2 - c^2$:

$$cb = b^2 - c^2$$

$$\Rightarrow \quad cb + c^2 = b^2$$

$$\Rightarrow \quad c(b + c) = b^2$$

$$\Rightarrow \quad \frac{c}{b} = \frac{b}{b + c}$$

which means $c/b = \varphi$, the golden section, and the angle of slope is α where $\cos \alpha = \varphi$. Hence,

$$\alpha = \cos^{-1}\phi \qquad = 51°49'38.3''.$$

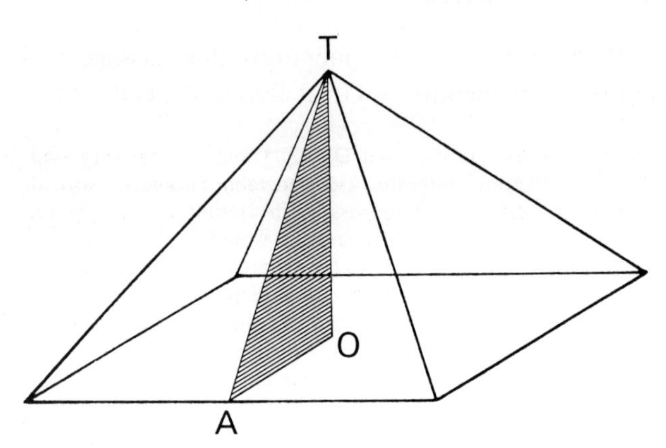

Fig. 2

But let us now look at what Herodotus really has written in *Euterpe* 124, 21–25, the "obscure passage" that Gillings could not locate:

> Τῆ δὲ πυραμίδι αυτῆ χρόνον γενέσθαι είκοσι έτεα ποιευμένη, τῆς ἐστι πανταχῆ μέτωπον έκαστου οκτὼ πλέθρα ἐούσης τετραγώνου κΓὶ ύψος ίσον, λίθου δὲ ξεστου τε καὶ αρμοσμένου τὰ μάλιστα. οὐδεὶς τῶν λὶθων τριήκοντα ποδῶν ἐλάσσων.

A classical translation, such as found with Ph.-E. Legrand,[17] sounds: "Pour la construction de la pyramide même, le temps employé aurait été de vingt ans: *elle est carrée; elle a de tous les côtés un front de huit pléthres, et une égale hauteur;* elle est de pierre polie, exactement jointe; aucun bloc n'y a moins de trente pieds.[k]

It is obvious that the classical translators have difficulties in understanding here, which they are certainly willing to admit.[l] Clearly this translation is partly incorrect, but I agree with D'Arcy Thompson that the text is obscure. In an effort to find out what Herodotus means by his equality (ἴσον), one notices the use of the measure πλέθρον, which apparently throughout the text is used for measuring areas of pyramidal faces. And yet in the translation, the face is put equal to the height in plethrons! The position of the words ἐούσης τετραγώνου may rather indicate a relationship with the word ὕψος (height) in which case then the face is put equal to the square on the height.[m]

Hence, the translation of the passage could be as good as:

"The time spent on the construction of the pyramid itself would have been twenty years; *all the sides have a front of eight plethrons, equal to a square on the height*; it consists of . . ."

This passage describes the pyramid of Mycerinus, the third and smallest one at Giza. Note also the position of the similar words ἐούσης τετραγώνου. The translator[17] writes in footnote 1, p. 159:

"Mot à mot difficile: *manquant de vingt pieds sur chaque côté de trois pléthres, c'est-à-dire sur chaque côté qui, sans cela, mesurerait trois pléthres* (?) ..."

Translation: "He (Mycerinus) also left a pyramid, much smaller than the one of his father; *it fails twenty feet on each face of three plethrons; it is square*, of Ethiopean stones until mid-height."

[k] translation:

"The time spent on the construction of the very pyramid would have been twenty years; *it is square, and of all sides has a front of eight plethrons, and an equal height*; it consists of polished stones, exactly joined and no block is smaller than thirty feet."

[l] compare to Herodotus, Euterpe 134, 1–4:

Πυραμίδα δέ καὶ οὗτος κατελίπετο πολλὸν ελάσσω του πατρός, *είκοσι ποδῶν καταδ έουσαν κῶλον έκαστον τριῶν πλέθρων, εούσης τετραγώνου*, λίθου δὲ ες τὸ ήμισυ αιθιοπικοῦ.

[m] A third description of a pyramid in Herodotus II is in 126, 8–11, where a smaller one facing the Great Pyramid would have been dedicated to the daughter of Cheops. Also here, the measure "plethron" is used for a face:

...εκ τούτων δὲ τῶν λίθων έασαν τὴν τὴν πυραμίδα οικοδομηθῆναι τὴν εν μέσω τῶν τριῶν εστηκυῖαν, έμπροσθε τῆς μεγάλης πυραμίδος, *τῆς εστι τὸ κῶλον έκαστον όλου καὶ ημίσεος πλέθρου.*

Translation: "... and, with these stones, said (the priests), the pyramid was constructed which is in the middle of the group of three, facing the Great Pyramid, *and of which each face measures one and a half plethron*" (towards to Legrand[7]).

From a geometrical point of view, this seems to have been noticed for the first time by John Herschell, an astronomer whose advice was requested for the figures of Howard Vyse, after his return to England in 1837.[n] Vyse had found the facial slope to be $\sim 51°50'$, being approximately $30'$ more than the figure found by Jomard during Napoleon's expedition. And Herschell concluded: "The slope of the Pyramid's faces is $\pm$ 50°50'? Then it is quite clear what Herodotus meant; there has to be read: *surface face* = *square on height*. This results into 51°50'; hence, it is based on a 'well established architectural principle.'"

According to H. A. Naber, "without any doubt Herschell meant the *Sectio Divina*, although modern architecture[o] ignores, or hardly recognizes, this principle."

4. Geometrical Implications for the Great Pyramid

Naber mentions next that the Herschell-Herodotus interpretation results into the equation $\sin\alpha = \cot\alpha$, without further explanation. However, a small calculation results in a more general equation:

since $\tau = 1 + \frac{1}{\tau}$ (see note a on p. 330), $1 = \frac{1}{\tau} + \left(\frac{1}{\tau}\right)^2$, and $\tau^2 = \tau + 1$, implying when $\alpha = \measuredangle OAT$ (Fig. 2):

$$\sin^2\alpha = 1 - \cos^2\alpha = 1 - \left(\frac{1}{\tau}\right)^2 = \frac{1}{\tau} = \varphi$$

$$\tan^2\alpha = \sec^2\alpha - 1 = \tau^2 - 1 = \tau.$$

Hence:

$$\boxed{\cos\alpha = \sin^2\alpha = \cot^2\alpha = \varphi}$$

And let us now see what this may mean for the shape of the Great Pyramid:

[n] like mentioned by Naber[12] on p. 11. Resources are unknown, but the only ref. to Herschell is made in footnote 3 on pp. 15–16, in which he explains how the biography of Herschell *would occur* in the "Tijdschrift der Pyramidalisten" (Magazine of Pyramidalists) Nr. 3 (1884).

[o] i.e. in the time when Naber gave the lecture to the "Dispuutgezelschap 'De Eenhoorn'" at Hoorn, on January 15th, 1914.

(a) $\cos\alpha = \varphi = \frac{c}{b}$. Multiplication by $\frac{a}{2}$

$$\frac{ca}{2} / \frac{ba}{2} = \varphi$$

or: the projected area of a face is in the golden proportion with that face (Fig. 3a).

Hence, *the basic square, which is the projection of the pyramid, is the golden section of the visible surface* (Fig. 3b).

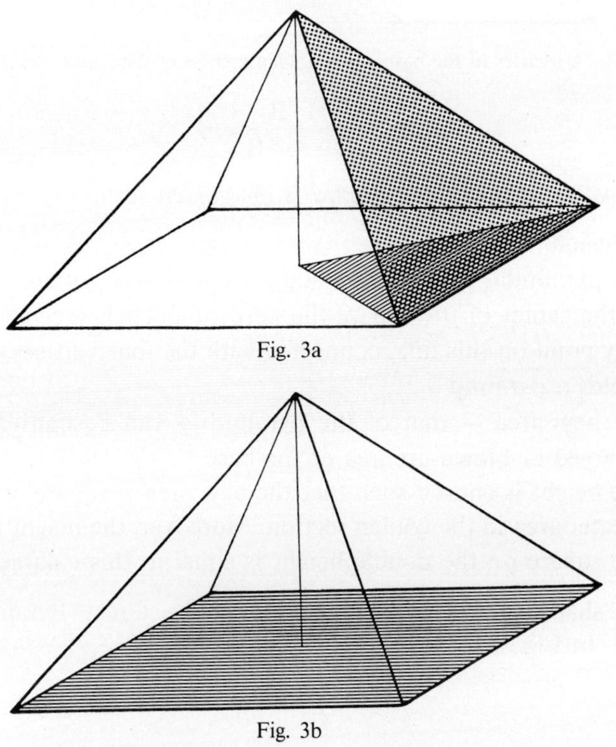

Fig. 3a

Fig. 3b

Fig. 3 a. The projected area of a face is the golden section of the face.
b. The base is the golden section of the pyramid.

(b) $\cot^2\alpha = \varphi$. Or: the square on the half base (a fourth of the basic area) is the golden section of the square on the height (Fig. 4). Hence, *the basic square is the golden section of the square on the double height.*

(c) If the basic area is A and the sum of the facial areas (the visible part of the Pyramid) is B:

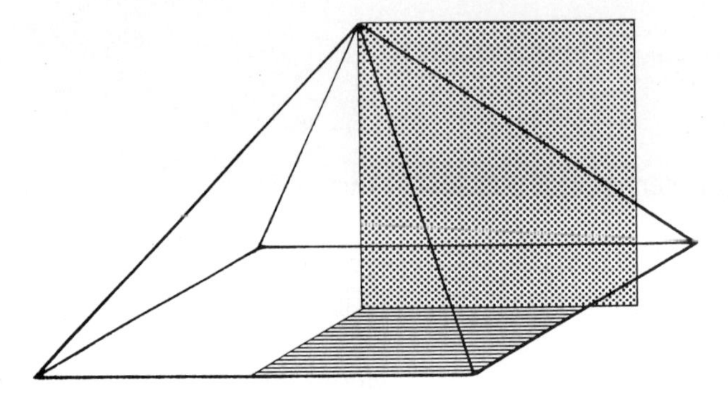

Fig. 4 A quarter of the base is the golden section of the square on the height.

$$\frac{A}{B} = \frac{B}{A + B} = \varphi$$

or: *the visible area of the Pyramid is the golden section of the total area.*

In conclusion:

(i) the pyramidal construction starts from a basic square
(ii) in the center of the square the perpendicular is erected
(iii) any point on this line, connected with the four vertices of the base, yields a pyramid
(iv) the new area — that of the pyramid — can be considered as an enlarged or blown-up area of the base
(v) the height is chosen such that the new area is square × τ, the first greater area in the golden section. Moreover, the height is such that the square on the double height is equal to this enlarged area.

(d) The shape of the triangular face of the Great Pyramid is also revealed. In Fig. 2:

$$c = \frac{a}{2} \text{ and } \frac{c}{b} = \varphi$$

hence

$$\frac{a}{2} = \varphi$$

which means the triangle is a half *golden rhombus*[1] (Fig. 5a). The golden rhombus, whose diagonals are in the golden section, has been split along the shorter diagonal. The top angle of the triangle is:

$$2 \tan^{-1} \varphi = 63°26'06''.$$

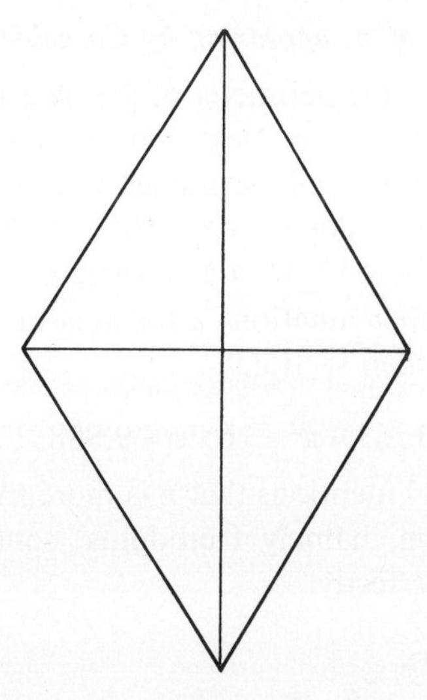

Fig. 5a

Fig. 5b

Fig. 5 a. the golden rhombus whose diagonals are in the proportion of τ.
b. model of the rhombic triacontahedron, the polyhedron composed of 30 golden rhombi about the 15 twofold axes of the icosahedral group of isometries and the dual of the icosidodecahedron.

4.1 *An Approximation of π, appearing by Coincidence*

Taylor[6] observed that the perimeter of the base is approximately equal to the height, multiplied by 2π. Next, Petrie[10] had the idea to choose 7 for the height and 22 for the half perimeter, thus using the approximation 22/7 for π, resulting in a slope $\alpha = 51°50'34''$. But Naber[12] calculated that the slope would be $51°51'14.3''$ when π would be calculated as correct as possible. Naber mentions a formula of Vieta for the approximation of π by the golden section:

$$9.870\ldots = \pi^2 = 16 \times \varphi = 9.888\ldots$$

Hence, when Gardner[13] mentions that π is more likely a coincidence due to another construction, namely Herodotus' concept,[p] the formula of Vieta illustrates this perfectly:

In Fig. 2:

$$\frac{2a}{h} = \frac{4c}{h} = 4 \cot \alpha = 4\sqrt{\varphi}$$

and applying the approximation of Vieta for π

$$\sqrt{\varphi} \approx \pi / 4:$$
$$\frac{2a}{h} \approx \pi.$$

5. The Egyptian Triangle and the Golden Triangle

Some more very clear clues to the golden section were found within the Great Pyramid, and, to a geometer, they are really clear. Hence, in order to understand these, it is necessary to start this section with some elementary but nevertheless remarkable geometrical preliminaries, before returning to the Great Pyramid.

5.1 *Constructing the Golden Section of a Line Segment*

Since $\varphi = \frac{1}{2}(\sqrt{5} - 1)$, a very useful triangle to construct the golden section of a line segment is the rectangular triangle whose rectangular edges are $1 : 2$. Then, its hypotenuse is $\sqrt{5}$. By simply subtracting the smaller edge from the hypotenuse, the line segment of length $\sqrt{5} - 1$

[p] Gardner seems not to have been aware of the interpretation of Herodotus' words, and has taken the translation for granted.

remains, which obviously is the golden section of the larger edge (Fig. 6). This triangle will be further on referred to as the *golden triangle*. The angles of the golden triangle are:

$\beta = \tan^{-1} 0.5 = 26°33'54''$ (in A)
$\delta = \tan^{-1} 2 = 63°26'06''$ (in D).

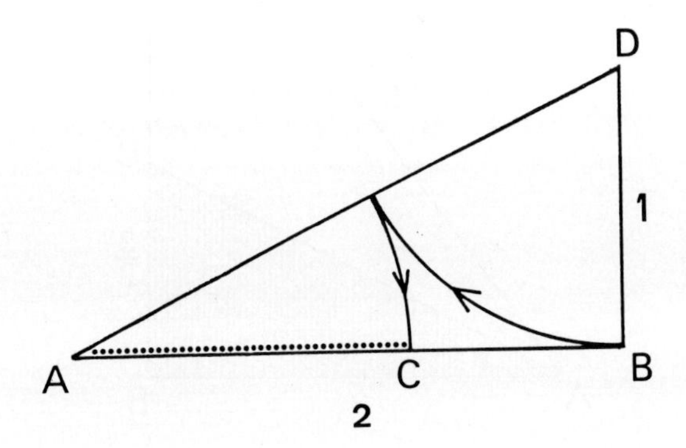

Fig. 6 Construction of the golden section of of AB:

$|AB| = 2,\ |DB| = 1,\ |AD| = \sqrt{5}$

5.2 *The Egyptian Triangle*

Application of $\gamma = 2\beta$ to the goniometrical formula for the double angle:

$$\text{tg}\, 2\beta = \frac{2\,\text{tg}\,\beta}{1 - \text{tg}^2\beta}$$

gives the result:

$$\text{tg}\, 53°07'48'' = \frac{4}{3}$$

which refers to the triangle whose rectangular edges are 3:4 (Fig. 7) and hence, whose hypotenuse is 5. In cap. 56, Plutarch[18] has described this triangle as the symbol of Egyptian Trinity:

$3 \rightleftharpoons$ Osiris
$4 \rightleftharpoons$ Isis
$5 \rightleftharpoons$ Horus

This triangle, which is also mentioned by Vitruvius, Architect of Emperor August (p. 49, Ref. 12) is called the *Egyptian triangle*, and may

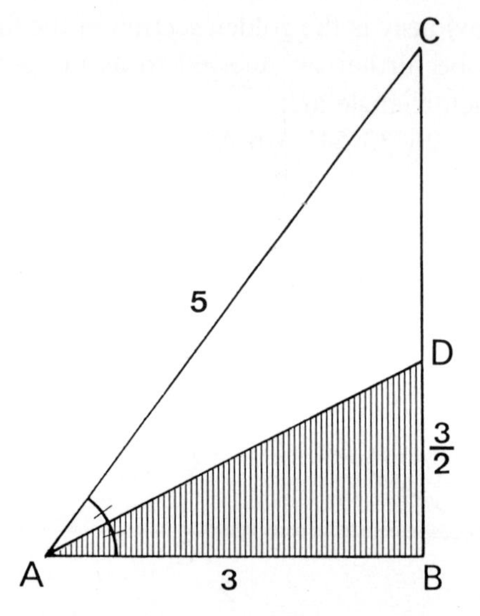

Fig. 7 The hypotenuse of the golden triangle ABD is the bisector of ∡ BAC (γ) of the Egyptian triangle ABC.

have been known to ancient Egyptian geometers after all, despite Gillings' outburst: "It is, however, nowhere attested that the ancient Egyptians knew even the very simplest case of Pythagoras' theorem!"[14]

On p. 34 of his book, Naber[12] found another remarkable aspect of the golden section in the Egyptian triangle, which is quite obvious to prove, but arriving at the idea is more difficult. He constructs cones whose axes are, respectively, the rectangular edges, and folds them open flat. The so-obtained angles are those of the pentagon and the pentagram (Fig. 8).

This is easily understood, for the proportion of the circumferences of each cone's basic circle to the larger circle of unfolding is, respectively, 4/5 and 3/5, resulting into an arc of, respectively, 288° and 216°. The line segments of these arcs are the edges of the pentagon and the pentagram, and are in the golden proportion.

5.3 *Both Triangles in the Great Pyramid*

The golden triangle is found at once in the measures of the floor of the King's Chamber:[2] 5.23 m (N–S) and 10.47 m (E–W). The diagonal of this rectangle measures 11.7 m, and divides it into two golden triangles.

In Piazzi Smyth's second book,[8] the names of a certain Simpson and

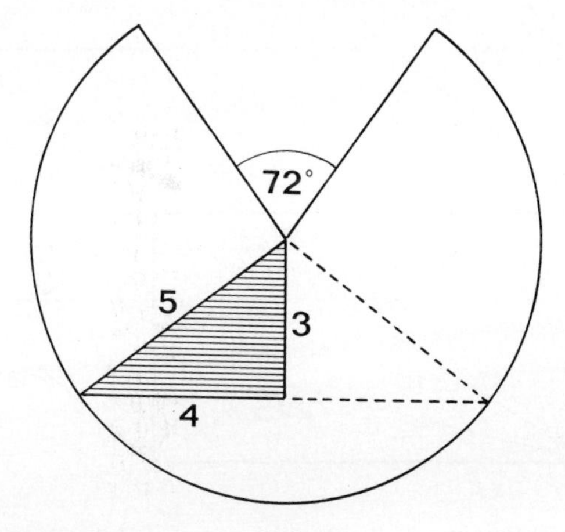

Fig. 8a

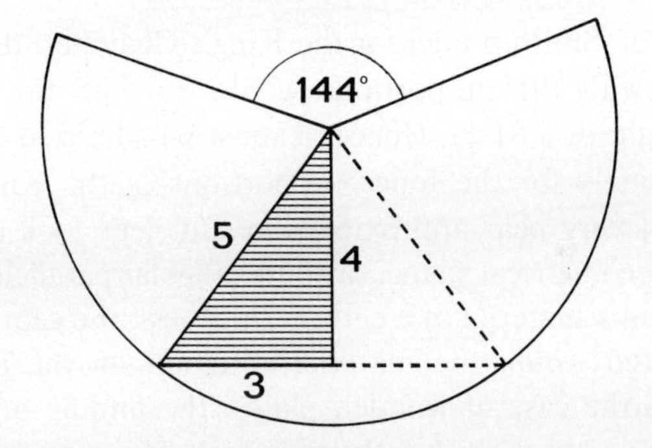

Fig. 8b

Fig. 8 Unfolding cones constructed with the Egyptian triangle:

a. cone about axis "3": the unfolding is a sector of 288°. The remaining sector of 72° corresponds to the central angle of a pentagon;

b. cone about axis "4": the unfolding is a sector of 216°. The remaining sector of 144° corresponds to the central angle of a pentagram.

Hamilton L. Smith occur with respect to an interpretation of the measures of the King's Chamber. When the edges of the base were taken as 2 and 4, the height turned out to be a very close approximation of $\sqrt{5}$, resulting in a geometrical model of the rectangular parallellepiped like is illustrated in Fig. 9:[q]

[q] No source given by Piazzi Smyth, but in footnote 3, p. 19, Naber[12] explains: "I cannot determine whether Simpson and Hamilton L. Smith also have put the attention to the fact the vertical diagonal planes reveal the proportion 1: 2."

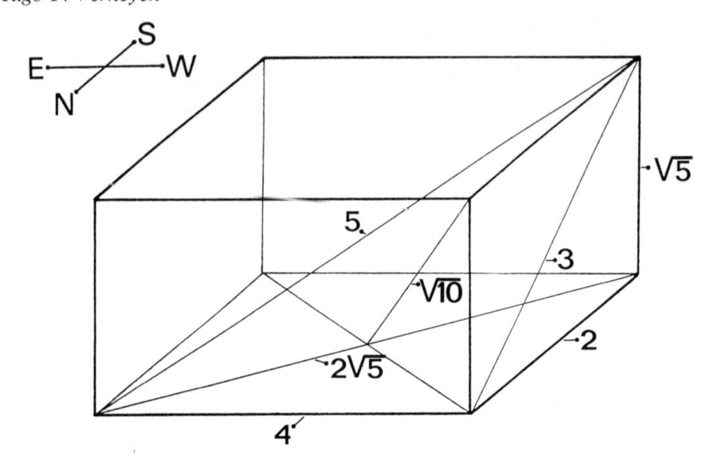

Fig. 9 Simpson's and Hamilton L. Smith's interpretation of the dimensions of the King's Chamber, according to Piazzi Smyth.

In the Simpson-Smith model for the King's Chamber, the exact height corresponding with the proportions would be 5.85 m, but Edwards[2] mentions a height of 5.81 m. Hence, it must be admitted that the model does not fit exactly for the inner proportions of the King's Chamber, although it is a very near approximation. But let's look at it this way: when a certain geometrical shape, say a rectangular parallellepiped, has to be constructed in a material of a certain thickness, the geometrical model has to be located *within* the material, and not on the inner or outer surface; e.g., in the case of wooden plates, the middle of the thickness would be ideal. Could it be that the architects of the pyramid wanted to materialize a *certain* parallellepiped, of well-chosen geometrical proportions? Two discoveries are in favour of this theory. First, Sporry and Tadema[5] mention that the walls of the King's Chamber are constructed completely separately from the inner material of the Pyramid, creating an open space between each wall and a second wall behind the first one. Together with Davison's Chamber above the ceiling, these open spaces turn the King's Chamber *into a room within a room*, or even better: a *house* within a room. Hence, if a geometrical model of well-defined proportions was materialized for certain reasons due to the proportions, the model has to be found *within* the material of the walls and ceiling, and not on the inner or outer measures of the Chamber. Second, there is a good reason to believe that the proportions of the King's Chamber, as well as its location within the Pyramid, were well defined. In 1986, a documentary was made by the German TV station ZDF, called "The Curse of the Pharaoh", in which is explained how the "mysterious" causes of death of grave explorers were due to a deadly fungus. With respect to the King's Chamber in the Great Pyramid, two experiments were described. Sponsored by the Mankind Research Foundation in Washington D.C., Dr. Carl Schleicher proved that meat deteriorized considerably slower in the King's Chamber, due to a reflecting of energy in some form that is not known to us at this point, by which the bacteria was destroyed and a protective fungus was produced. Moreover, in Munich's University of Technology, Prof. Eichmeyer experimented with razor blades in an oriented scale model of the Pyramid of Cheops. One blade was placed in the King's Chamber and one on another spot in the Pyramid. After 8 days, the electron microscope revealed a difference in molecular structure and in width of the sharp edge.

Returning to the Simpson-Smith model for the King's Chamber, the vertical diagonal planes reveal the rectangle 1:2 like the floor, and hence, the golden triangle (Fig. 10a). The Egyptian triangle 3:4:5 appears in side diagonal planes (Fig. 10b).

But Naber's greatest discovery is the appearance of the golden triangle in the passages of the Great Pyramid. By then the slope of the ascending passage and Grand Gallery was measured as $\sim 26°30'$,[r] a remarkably close figure to the angle β of the golden triangle, of which $\tan \beta = 1/2$, and according to Naber[12] (p. 19–20): "an idea which even Flinders Petrie did not entirely reject."

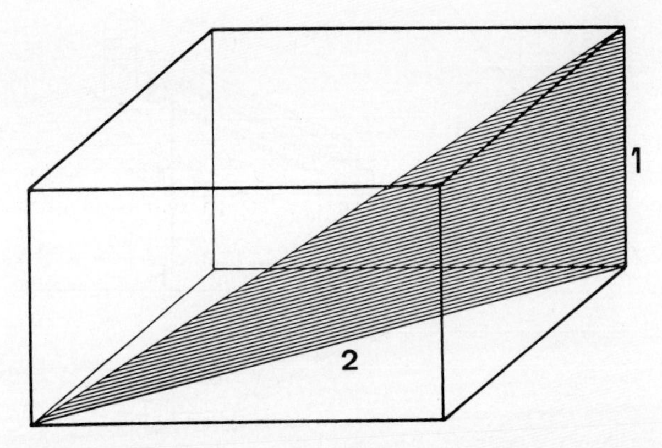

Fig. 10a. Appearance of the golden triangle in the King's Chamber.

[r] Naber mentions 26° to 27°.

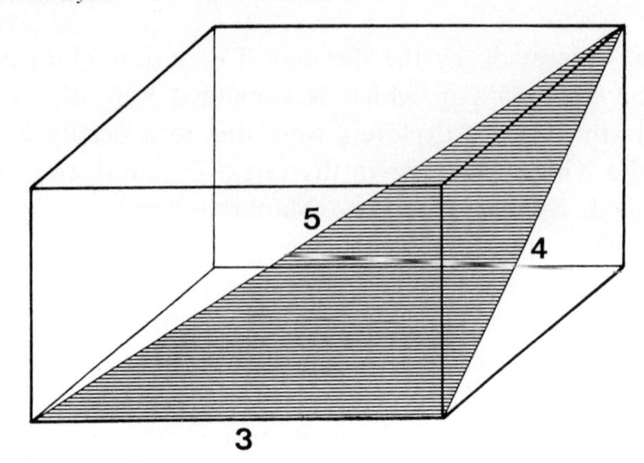

Fig. 10b. Appearance of the Egyptian triangle in the King's Chamber.

Apparently, Naber was not aware of the equality of the top angle δ of the Great Pyramid and the sharp angle of the golden rhombus. Otherwise, he would have noticed the appearance of this angle in G (Fig. 11), for δ is a sharp angle of the golden triangle (4.1). Hence, the triangle EFG has the following sharp angles:

(i) the top angle of the Pyramid, δ, formed by the vertical axis and the Grand Gallery;

(ii) the slope of the ascending passage (and Grand Gallery), illustrating

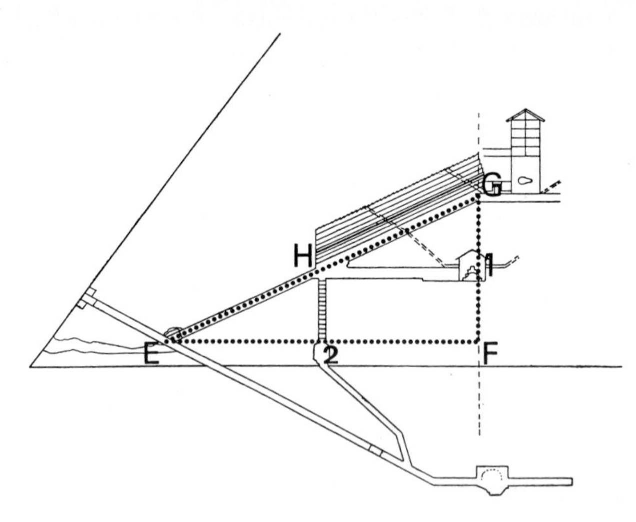

Fig. 11 Naber's explanation for the slope of the ascending passage: the triangle EFG is a golden triangle.

the fact that these are complementary angles. However, Naber's revelation is the construction of the golden section, suggested by the Grand Gallery in his golden triangle EFG within the Great Pyramid. He observed that the passages indicated the construction of the golden section of the base EF, which resulted in *the Grand Gallery*.[s] The slight difference with the construction as explained in 4.1 is the fact that the edge of proportion 1 (FG) is subtracted from the diagonal (EG), *but starting from point E instead of G*, such that the remaining part — the golden section of EF — is the Grand Gallery! (Fig. 12).

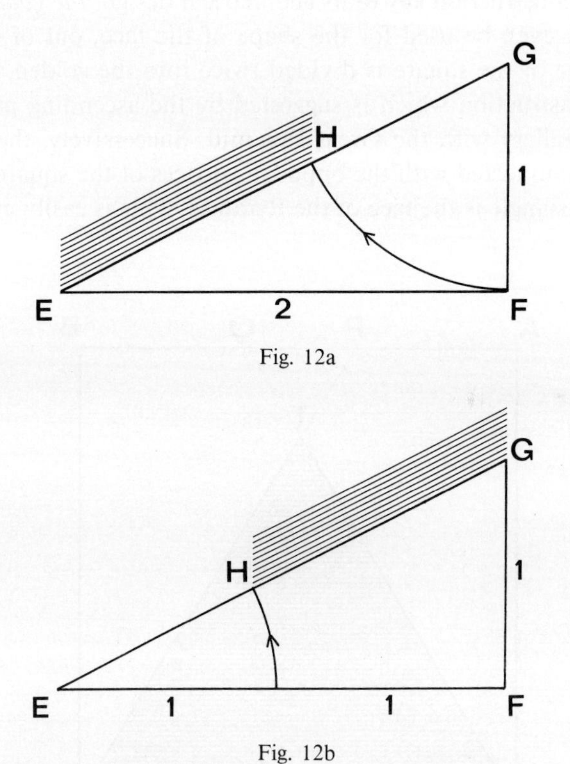

Fig. 12a

Fig. 12b

Fig. 12 a. Construction of the golden section of EF, starting in G.
b. Same construction, but starting in E: the Grand Gallery.

[s] Figures given by Naber: Gallery: 47 m, horizontal distance AB: 76.5 m or more precisely: 1882 and 3062 inches. This would offer 5/8, being a quite crude approximation for the golden section, as this represents one of the first proportions of the series of Fibonacci, while the golden section is the limit of these. But a true remark is necessary here: where does one start measuring within these passages? Hence, the approximation may be accepted after all.

Naber ends his discovery with the words: "... or does everything seem like a coincidence?" And finally, I found a second appearance of the Egyptian triangle by projecting G onto the descending passage. Since the angle formed by ascending and descending passages is $\gamma = 2\beta$,[t] an Egyptian triangle is formed.

6. Base and Face of Great Pyramid

By now, all the measures and angles of the Great Pyramid clearly indicate the construction key to its geometrical design: *the golden section.* This *key* can even be used for the shape of the face, out of the square base: An edge of the square is divided twice into the golden section, by using the construction which is suggested by the ascending passage and the Grand Gallery with the Great Pyramid. Successively, the points of division are connected with the opposite vertices of the square (Fig. 13). The shaded triangle is the face of the Pyramid. This is easily understood,

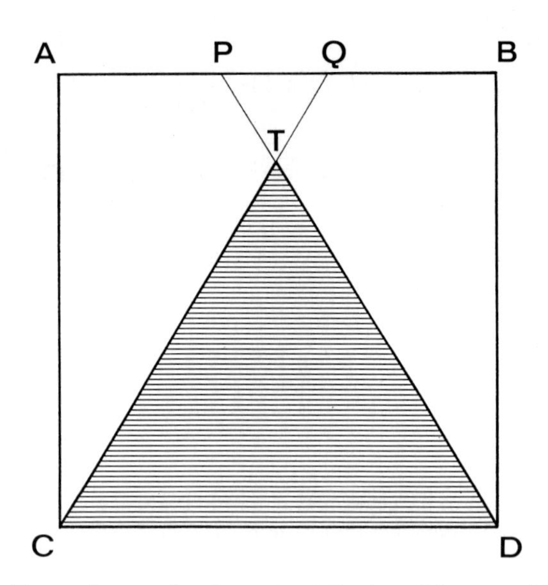

Fig. 13 The golden section as a key to construct the face of the pyramid out of the base.

[t] The slope of the descending passage is nearly equal. Edwards[2] mentions 26°2′30″ for the ascending passage, "while the slope of the descending passage is equal up to the fraction of a degree."

since $\tau = PP'/P'D = TT'/T'D$ (Fig. 14), which means that the triangle is a half golden rhombus.

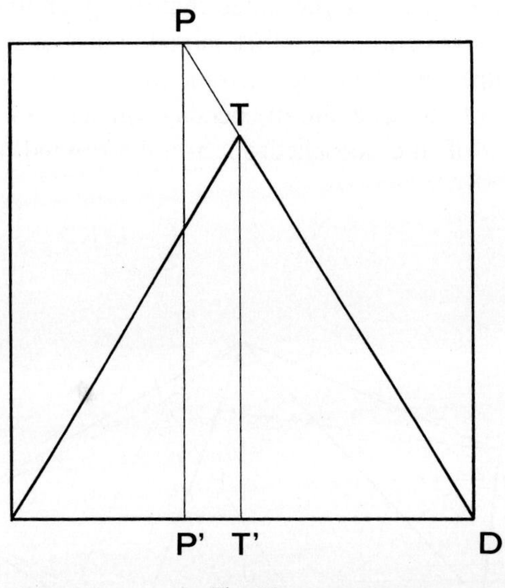

Fig. 14

7. The Icosahedral Design of the Great Pyramid

Being now acquainted with the discoveries of the golden section in the Great Pyramid, and assuming that these have been deliberately applied by the architects, the next step in the reconstruction of the design is the search for the existence of a fundamental idea on which this specific pyramidal shape may have been based. And since there appears to be an interference between ancient geometry and mysticism, I shall try to reconstruct the parallel aspects of both.

7.1 *The Geometrical Concept of the Great Pyramid*

The angle $\delta = 63°26'06''$ of the face of the Great Pyramid, being a half golden rhombus, is the angle between any two of the six fivefold axes of the icosahedral group of rotations. Hence, this angle is also found in the central triangles of the icosahedron. Such a triangle is obtained by connecting an edge of the icosahedron with its center, since the vertices of the icosahedron (12) lie in opposite pairs on the fivefold axes of rotation (Fig. 15).

Without going into the details of a geometrical proof of this property of the golden rhombus, it can be stated that this phenomenon also occurs with the other convex rhombic solids, namely the rhombic dodecahedron (octahedral group), and the cube[u] (tetrahedral group), respectively in the central triangles of the cube and the octahedron. The 30 central triangles, or half-rhombi, of the icosahedron are distributed as side-faces of

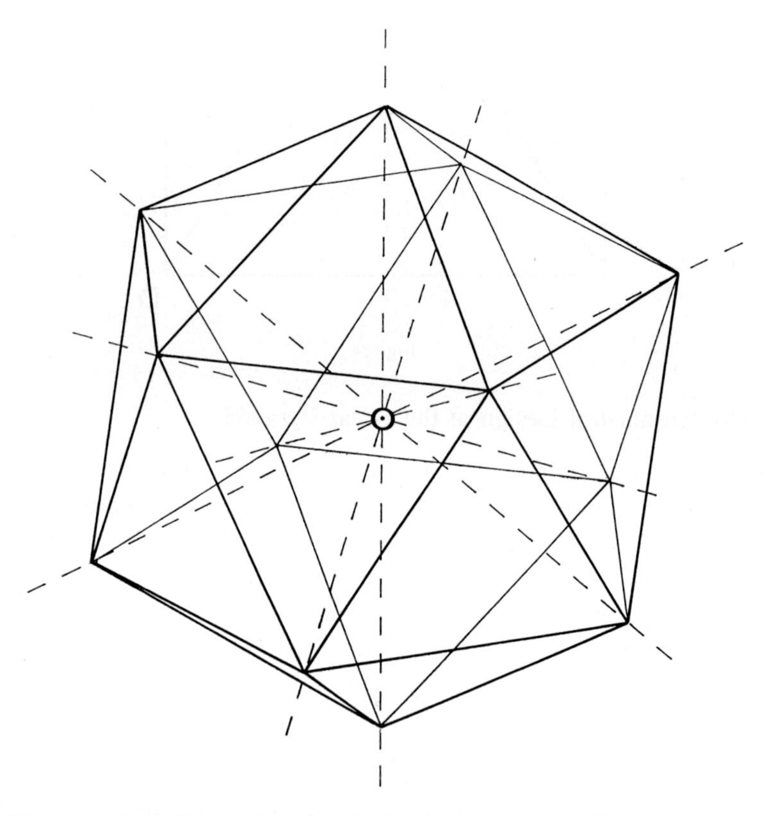

Fig. 15 The six fivefold axes of rotations in the icosahedral group of isometries go through the twelve vertices of the icosahedron, and together with its 30 edges form the 30 central triangles, all half golden rhombi.

[u] here the rhombus is a square, just an "accident".

triangular pyramids (Fig. 16a) and pentagonal pyramids (Fig. 16b), which also function as polyhedral kaleidoscopes.[1]

The key of the conceptual design of the Great Pyramid was the Egyptian Triangle 3:4:5, which is – as explained in Sec. 5 – related to the golden triangle. The central triangles of the icosahedron were distributed round a square (Fig. 17), to obtain a third type of pyramid of half golden rhombi, closing off the triangle 3:4:5.

7.2 *The Mystical–Geometrical Concept of the Great Pyramid*

The question now arises: why was the square pyramid with golden faces chosen? Many people are convinced that the Great Pyramid contains a sort of mystical message for future civilizations. In his chapter on the Great Pyramid, Martin Gardner[13] mentions a passage of John Taylor[6]: "The Pyramid was a symbol of the True Church of Christ as the upper keystone." (This symbol is otherwise very popular in Christian

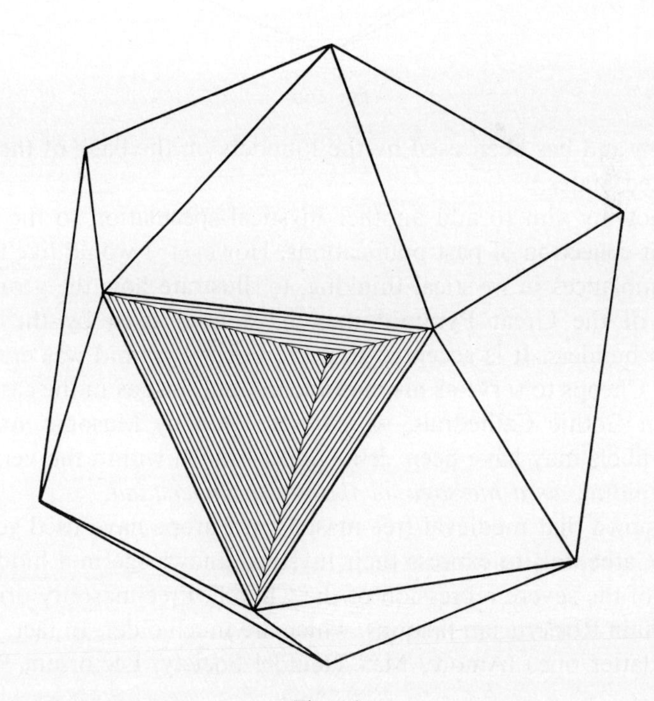

Fig. 16a

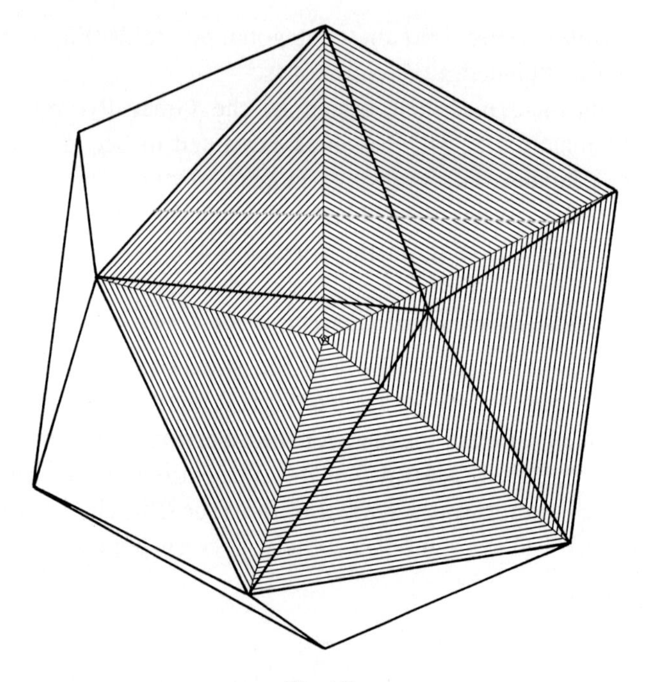

Fig. 16b

mysticism and has been used by the founders on the back of the seal of the United States.)[v]

It is not my aim to add another mystical speculation to the already abundant collection of past publications. However, I would like to point out resemblances in mystical thinking, to illustrate how the geometrical concept of the Great Pyramid may have been used by the ancient Egyptian builders. It is accepted that the Great Pyramid was erected by Pharaoh Cheops to serve as his mausoleum, but, just as in the case of the European Gothic Cathedrals, which were *built* by Masonic architects, some symbols may have been deliberately hidden within the very shape of the building, *as a message for those who understand.*

It is known that medieval free-masons in Europe have used geometry in their Cathedrals to express their mystical knowledge in a hidden way because of the severe repression of the Church. Free-masonry originated from certain Rosicrucian factions, which are much older. In fact, the still existing latter ones (Amorc, Max Heindel Society, Lectorium Rosicru-

[v] also found on the backside of a $1 note.

cianum) all claim that their knowledge goes back to Ancient Egypt, e.g., the Amorc faction, which has its headquarters in San José, California, still gathers in temples that are decorated in an Ancient Egyptian way. Hence, a clue to ancient Egyptian mystics may be found here, in order to come to an understanding of the mystical way of thinking by the conceivers of the shape of the Great Pyramid.

J. Van Rijckenborgh (Lectorium Rosicrucianum) explains in his books on the Ancient Egyptian Gnosis[19] how the mystical religion of the Egyptian is related to Christian mysticism. Basically, the central idea in mysticism is the fall of mankind and how to return to the Garden of Eden. This mysticism may have originated in Ancient Egypt, or even in civilizations before that, and may have found its way through Persia and Greece to Rome, where it became equated with Christianity.

Although they give no mystic geometrical explanation for the shape of the Great Pyramid, Free-masons and Rosicrucians all use the square as a symbol for mankind in its fallen status. From this square, the "free building" work starts, which means; the return of the microcosmos to the universe of origin. This universe is characterized by 12 cosmic Forces (to be compared with the 12 signs of the zodiac), and a thirteenth Force

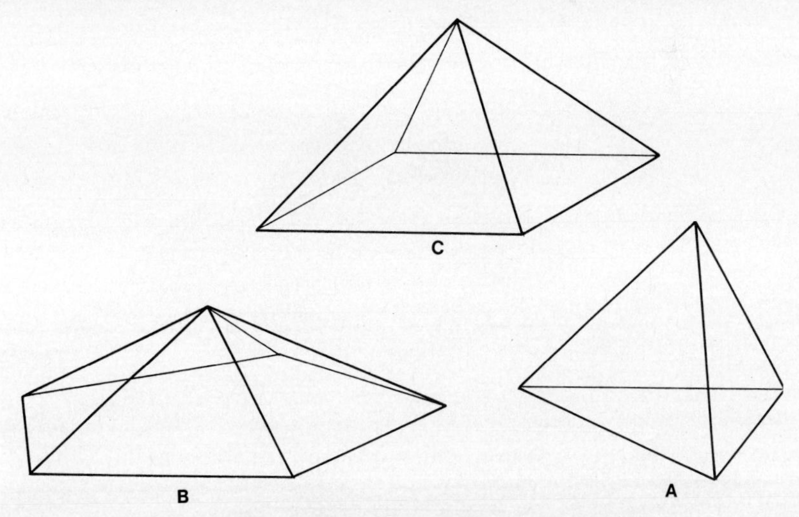

Fig. 17 Pyramids of central triangles of icosahedron:
a. triangular (in icosahedron)
b. pentagonal (in icosahedron)
c. square (the great pyramid)

whose function is to uplift fallen mankind back to its origin. This thirteenth Force is called the Christ, and appears in the middle of the cosmos, but also in the middle of each microcosmos (each individual being in its totality), where it is called the "rose of the heart" (hence, the expression "Rosicrucian"). The geometrical concept of 7.1 can be related to this philosophy, where a sphere would represent the universe of order and 12 cosmic forces appear as vertices of an icosahedron. The thirteenth force is then the center of the sphere, and triangular and pentagonal pyramids of half golden rhombi are formed. The third type of pyramid in the Egyptian triangle, namely the *square pyramid* is out of this cosmic order, but points from fallen mankind to the center of the sphere, thus giving the clue to how this thirteenth Force (Christ) can be found within the heart of each human being. As such, the symbol shows how Christ is the Keystone, as Taylor mentioned. *It represents a bridge to the universe of origin*; or as Christianity calls it: the return to the Garden of Eden.

If this way of thinking was truly applied by the building masters of the Great Pyramid, it would have been an act of great love for mankind, for they were convinced in their belief that they had to eternalize this

Fig. 18 The pyramid complex at Giza: (front to back) the three smaller pyramids, Mycerinus, Chephren, and Cheops (the Great Pyramid).

message of salvation in the monstrous proportions of a construction, which the King thought was meant only for himself; if not, the mere geometry is of the greatest beauty.

References

1. H.F. Verheyen, *The Golden Rhombus in Applied Constructive Art* (to be published).
2. I.E.S. Edwards, *The Pyramids of Egypt*. Penguin Books, London (1985).
3. J.H. Cole, The determination of the exact size and orientation of the Great Pyramid of Giza, *Survey of Egypt*, Trans. 39, Cairo (1925).
4. H. Vyse and J.S. Perring, *Operations Carried on the Pyramids of Gizeh*, I–III. London (1840–42).
5. B.T. Sporry and A.A. Tadema, *De Pyramiden van Egypte*, 2nd Edn. Fibula-Van Dishoeck, Haarlem (1977).
6. J. Taylor, *The Great Pyramid: Why Was It Built and Who Built It?* J. Taylor, London (1859).
7. C. Piazzi Smyth, *Life and Work at the Great Pyramid*, I–III. Edmonton & Douglas, London (1864–1867).
8. C. Piazzi Smyth, *Our Inheritance in the Great Pyramid*. 3rd Edn. Daldy, Isbister, London (1877).
9. C. Piazzi Smyth, *New Measurements of the Great Pyramid*. R. Banks, London (1884).
10. F. Petrie, *Pyramids and Temples of Gizeh*. London (1883).
11. H. Neikes, *Der goldene Schnitt und die Geheimnisse der Cheops-Pyramide*. Dumont-Schauberg, Cologne (1907).
12. H.A. Naber, *Meetkunde en Mystiek*. Theosofische Uitgeversmaatschappij, Amsterdam (1915).
13. M. Gardner, *Fads and Fallacies in the Name of Science*. Dover, New York (1967).
14. R.J. Gillings, *Mathematics in the Time of the Pharaohs*. Dover, New York (1972).
15. H.W. Turnbull, *The Great Mathematicians*, 4th Edn. Methuen, London (1951).
16. D'Arcy Thompson, *On Growth and Form*. Cambridge University Press, London (1951).
17. Ph.-E. Legrand, *Hérodote, Livre II Euterpe*. Les Belles Lettres, Paris (1948).
18. Plutarch, *De Iside et Osiride*.
19. J. Van Rijckenborgh, *De Egyptische Oer-Gnosis, I–IV*. Rozekruis Pers, Haarlem (1960–65).

A MYSTIC HISTORY OF FIVEFOLD SYMMETRY IN JAPAN

Koji Miyazaki

1. A Geometrical Aspect of Japanese Aesthetics

Foreigners, who visit Japan for the first time, may be informed that a regular pentagonal cherry blossom is one of the most characteristic thing for Japanese, in much the same manner as Mt. Fuji symbolizes Japan. In other words, they may think that the Japanese like a regular pentagon and the golden ratio 1 : 1.618, which is determined by the length-ratio between an edge and a diagonal of the regular pentagon. It is, however, usually said that Japanese of the past rather liked the silver ratio $1:\sqrt{2}$, which is determined by the length-ratio between an edge and a diagonal of a square.

The reason for the Japanese preference for geometric form may be based on an ancient Chinese myth dating from before Christ. It tells how the universe was created by two gods with snakelike bodies, Fu-I and Nü-Kua, the first emperor and empress in Taoism, probably man and wife.

They are portrayed in many reliefs in China. Fu-I always has a compass and Nü-Kua a carpenter's square, as tools for designing the heaven and the earth (Fig. 1). Yin-Yang theory in Taoism says that a compass as the symbol of Yang suggests a circle, male, heaven, sun, and the like, while the carpenter's square as the symbol of Yin suggests a square, female, earth, moon, and the like. The ancient Japanese, who learned much from the Chinese, seemed to be influenced by such a myth.

Fig. 1 Fu-I holding a compass (right) and Nü-Kua holding a carpenter's square (left) (copy). Stone engraving in China. 1st century.

Many gigantic ancient burial mounds built for emperors or empresses have geometrical shapes derived from a circle and a square. A circular mirror and a straight sword symbolizing a square are always set at the right and left in the sanctuary of a shrine (Fig. 2).

In addition, the ancient Japanese may have thought that the circle and square create a regular octagon, and they attached importance to it. Some typical burial mounds and other holy places in ancient times were built on regular octagonal plans. An example of this is the picture on the late Japanese 10,000 yen note which shows a place of prayer in Horyuji Temple in Nara. It was originally built by order of the then ruler Prince Shotoku-taishi, (574–622), and is the oldest wooden building in the world. From that time on, many of the temples commemorating him have a regular octagonal plan, and the number eight in ancient Japan has a most important meaning.

The regular octagon is constructed by using the silver ratio which can be easily measured by two kinds of scales marked at both sides of a Nü-Kua's carpenter's square: an ancient scale is on one side and the $\sqrt{2}$ times of it is at the other. It is usually said that the plan and elevation of Horyu-ji Temple (Fig. 3) was derived from the silver ratio. As a result, Shotoku-taishi is even now revered as a god of the carpenter in Japan.

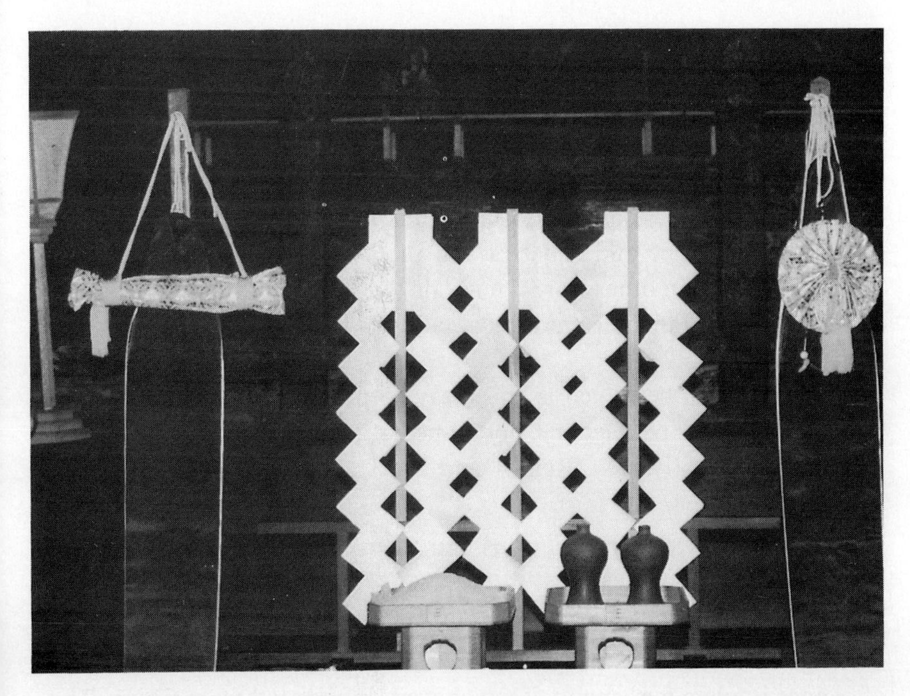

Fig. 2 A circular mirror (right) and a straight sword (left) in a shrine. Both are enclosed in red clothes.

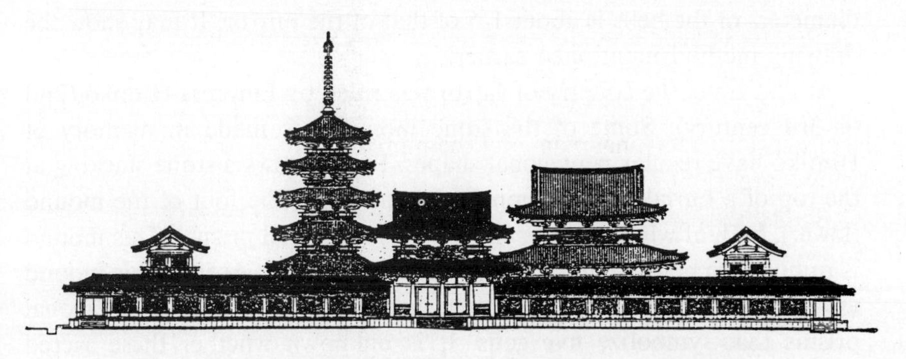

Fig. 3 The front elevation of Horyu-ji Temple. 8th century, Nara.

The asymmetric design of this temple, having a thickset hall on the right and a slender five-storied pagoda on the left, may represent Fu-I's compass (the right) and Nü-Kua's carpenter's square (the left). Today, this design is often thought to be the origin of the traditional Japanese taste which denied the symmetrical form.

Consequently, it is sometimes said that the golden ratio is the symbol of beauty in Europe and America and the silver ratio is the symbol of beauty in Japan. In other words, it seems that fivefold symmetry was not so important to Japanese traditional culture.

Is it true or not?

This riddle will be solved in this essay by presenting a brief history of regular pentagons, pentagrams, dodecahedra, and icosahedra. For further details, refer to Miyazaki, Refs. 1 and 2.

2. A History of Regular Pentagons and Pentagrams in Japan

2.1. *Ancient Pentagonal Religious Objects*

Religious life in Japan began before the time of Christ, and made use of copper implements, many of which were introduced from China. The circular mirror and a straight sword were among the most typical forms.

Curiously enough, some of the circular mirrors have pentapetalous or decapetalous patterns, each of whose curves of petals can be drawn as an epicycloid (Fig. 4). Some such circular mirrors have small spherical decorations, which are sometimes bells. Fig. 5 shows that each of the diameters of the bells is about 1/5 of that of the mirror. It may show the drawing method mentioned earlier.

At that time, the country of Japan was ruled by Empress Himiko (2nd or 3rd century). Some of the stone monuments made in memory of Himiko have regular pentagonal shapes. Fig. 6 shows a stone stacking at the top of a burial mound (upper), and a well at the foot of the mound (lower), both of whose shapes are regular pentagonal prisms. This mound is rumored to be a candidate of the tomb of Himiko. Around this mound are some stone monuments whose shapes are also regular pentagonal prisms that symbolize five gods. It is unknown whether these sacred solids truly relate to Himiko or not, but it is important to realize that the early Japanese were aware of the significance of the regular pentagon.

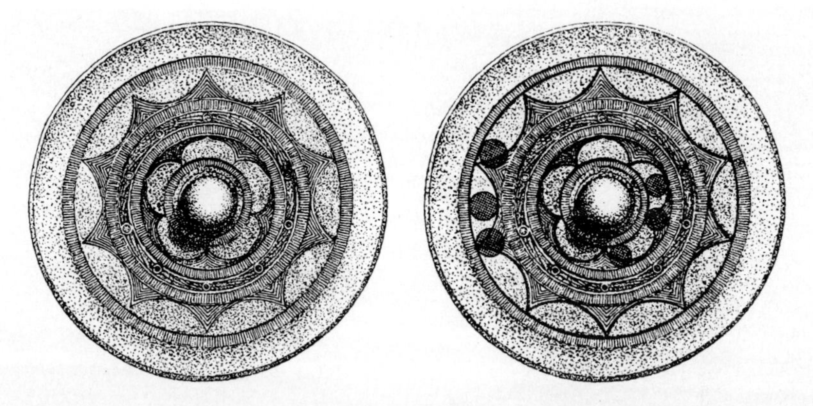

Fig. 4 A geometrical analysis of petalous patterns in a circular mirror, Kofun era, using epicycloids (right). The epicycloids can be precisely drawn according to rotation of small dark discs in the figure. The original shape of the mirror is shown at the left.

Fig. 5 A circular mirror having five small spherical bells. Kofun era.

2.2. *The Geometry of the Chinese System of Five-Elements*

According to a Chinese classic "Weishī-Wojen-zhuàn" (c.250), Empress Himiko eagerly learned Taoism and the *Yin-Yang* theory, which was introduced from China, and used its teachings as a political tool.

Taoism states that the whole universe is made of Chinese five-elements under intermediation of Fu-I (*Yang*) and Nü-Kua (*Yin*). The Chinese five-elements are wood, fire, earth, metal, and water. Even today, in Japan, many significant things and regular functions are named according to the five-elements.

Fig. 6 A stone stacking (upper) and a well (lower) whose shapes are pentagonal prisms. Period unknown. Yano-shinzan, Tokusima Prefecture.

These elements have regular pentagonal and pentagramic mutual relationships, as shown in Fig. 7. In this figure, each element is shown with the specific symbol assigned to it (refer to Fig. 23). The regular pentagon as the outer shape shows a productive relationship between two neighboring elements: water and its relationship to wood, wood to fire, fire to earth, earth to metal, and metal to water. In contrast, the regular pentagram constructed of the five diagonals shows a destructive relationship between two neighboring elements: wood and its relationship to earth, earth to water, water to fire, fire to metal, and metal to wood.

According to Yoshino,[3] traditional narratives or fairy tales whose heroes and heroines relate to water, fire, metal, etc. were born from these relationships.

2.3. *Pentagrams as Talismans*

After Prince Shotoku-taishi first organized a political system which standardized law in Japan, Yin-Yang Fortunetellers, who took charge of astrology and astronomy held important positions in the ancient government.

Abe-no-Seimei (10th or 11th century, Heian era) was the most famous one of them. According to some tales, e.g. the *Konjyaku-monogatari* Romance from the Heian era, his mother was a fox and he had supermental capabilities. He could open and close a door by his thinking,

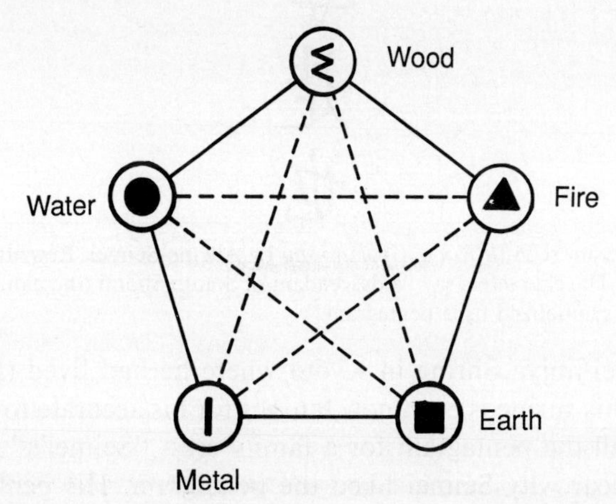

Fig. 7 Pentagonal and pentagramic mutual relationships among Chinese five-elements. Each symbol mark is also seen in Fig. 23.

as though he were using a remote control, and he could perceive movements of people without actually seeing them.

His symbol mark was a regular pentagram. There are many pentagrams in his main book *Kin'u-Gyokuto-shu* ("A magical book on the sun and moon"). Fig. 8 shows a sentence of the book. Many pentagrams can be

Fig. 8 A pentagram seen in *Kin'u'-Gyokuto-shu* by Abe-no-Seimei. Rewritten in the 14th century. The characters say "a descendant of Somin Shorai (the name of an ancient hero) is symbolized by a pentagram."

seen in Seimei-jinjya Shrine in Kyoto where he had lived (Fig. 9). The guardian of this shrine is even now famous for his accurate fortunetelling. Today, we call the pentagram for a family crest "Seimei's" bellflower.

It is not clear why Seimei liked the pentagram. His pentagram may show a relationship to the Chinese five-elements, or he might have known about the famous relationship between a pentagram and Pythagoras.

Fig. 9 Pentagrams in Seimei-jinjya Shrine. Rebuilt in the 19th century. Kyoto.

Yasuzou Okada, a modern historian, has compiled a great many such mystic meanings of the pentagram in Japan (Refs. 4 to 10).

According to Okada, there is a curious mark which is usually seen as a pair of pentagrams (Fig. 10, upper). This is called a "Kuji" ("nine letters"). A bible for esoteric Buddhists, *Uho-Senketsu* ("On a manner of steps for monks"), says that this mark means a magic square of 3 × 3 which exhibits a magical power when a monk steps in the order of numbers. In Taoism, it is said that the magic square was devised by Fu-I at 4000 B.C.

Such pentagrams and Kujis were often painted, engraved, or embroidered on various objects or places as a talisman after Abe-no-Seimei. For example, pearl divers in Toba had a pentagram named "Seimei" and a Kuji embroidered or engraved on their caps and tools until quite recently (Fig. 11). In Hana-matsuri Festival (a flower festival) held in Toei, women dress in traditional Japanese clothing, kimonos, which have pentagrams and

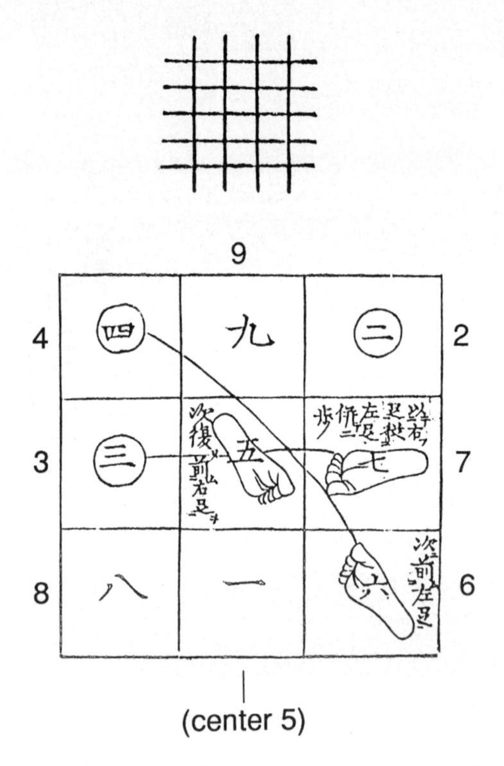

Fig. 10 *Kuji* (upper) and foot marks of a monk on a magic square of 3 × 3 (lower). From *U'ho-Senketsu* (1835) by Hokkai Aoki. The Arabic numerals are attached by the author for explanation.

Fig. 11 A pentagram and Kuji engraved on the handle of a knife of a pearl diver. Pearl Museum, Toba, Mie Prefecture.

Fig. 12 A pentagram and a simplified Kuji embroidered on Japanese clothes used at Hana-matsuri Festival. Toei, Gifu Prefecture.

rhombi embroidered on them (Fig. 12). According to Okada, the rhombus stands for a Kuji.

A pentagram can also be seen at the northeastern quarter (the ominous direction) of houses, temples, castles, and so on, as a talisman. Fig. 13 shows an example on a ridge-tile of a temple and Fig. 14 on a stone wall of a castle. There are many such examples. Also, there is a modern temple (Butsugen-ji Temple, Ito, Shizuoka Pref.) which is highly decorated by many pentagrams. The reason may be that this temple is situated at the ominous direction of the residence of the patron.

In the 18th or 19th century, soldiers on a battle-field were said to have been protected by pentagrams, which were embroidered on top of their military caps (Fig. 15). Sometimes they flew a regimental flag with a small pentagram and Kuji on it (Fig. 16).

A pentagram painted on a talisman named *Somin-Shorai* is also effective to ward off diseases and ill fortune. The word *Somin-Shorai* was originally the name of a man, Somin Shorai in the age of myths, already explained in Fig. 8. He sacrificed himself, and in doing so, helped

Fig. 13 A pentagram seen on a gargoyle of a worship-hall in Myoshin-ji Temple. 17th century. Kyoto.

Fig. 14 A pentagram seen on a corner of a wall in Kanazawa-jyo Castle. 16th century. Kanazawa, Ishikawa Prefecture.

Susa'noh-no-Mikoto, who was known as a powerful and violent god. This talisman is said to possess the magical power of God Susa'noh.

According to the oldest Japanese classic *Kojiki* (8th century), Susa'noh had an elder sister, the goddess (not the god) of the sun, i.e. the circle, and an elder brother, the god (not the goddess) of the moon, i.e. the square (the circle and square have switched places). Probably because of the myth, the shape of a Somin-Shorai Talisman is usually a regular octagonal prism (Fig. 17) which is born between a cylinder and a quadrangular prism, though it is often deformed as a regular hexagonal or pentagonal prism (Fig. 18). In many cases, a pentagram can be seen on this talisman.

Fig. 15 Pentagrams embroidered on the top of the military cap of a field marshal at the end of the 19th century (Meiji era).

Fig. 16 A pentagram and *Kuji* embroidered on loops of the regimental flag of Ogura-han Clan. 19th century (Edo era).

Fig. 17 A Somin-Shorai given at Sasano, Yamagata Prefecture, whose shape is a regular octagonal prism.

Fig. 19 shows the other curious Somin-Shorai Talisman given at Matsushita-sha Shrine which is sacred to Susa'noh himself. A Kuji can be seen here.

Another pentagram appears as one of Gozan-okuribi bonfires which are burned in the middle of each August in Kyoto. At night, five linear shapes of bonfires rise on the slopes of five mountains surrounding Kyoto. Of them, a Chinese character "Dai" which means "big" is the most typical shape (Fig. 20).

What does "Dai" mean?

There are some opinions about this riddle. One is that it symbolizes the dead to be mourned, because this character looks like a human body reclining. The other is that this character stands for the Indian five-elements, which are called "big" elements in Japan (see Sec. 3.3). This ceremony is said to have originated in Edo era by an Esoteric Buddhist who emphasized Indian five-elements. In addition, according to Wazaki,[11] there is the idea that this shape was modeled after a pentagram as a talisman for Kyoto.

Fig. 18 Two Somin-Shorais given at Kokuseki-ji Temple, Mizusawa, Iwate Prefecture, each of whose shape is a regular hexagonal and pentagonal prism.

2.4. *Pentagons and Pentagrams in Traditional Architecture*

As is well known, traditional houses in Japan were made of wood, and in most cases, their plans were polygons having even number (4, 6, or 8) of edges. In contrast, are there ruins of ancient wooden houses whose plans were regular pentagons?

It is quite within the bounds of possibility that, when the country was ruled by Empress Himiko, there were many wooden houses whose plans were the regular pentagons or polygons having odd number of edges. Regular pentagonal ruins of ancient houses of the 3rd or 4th century were recently excavated in Onaka-iseki Historic area.

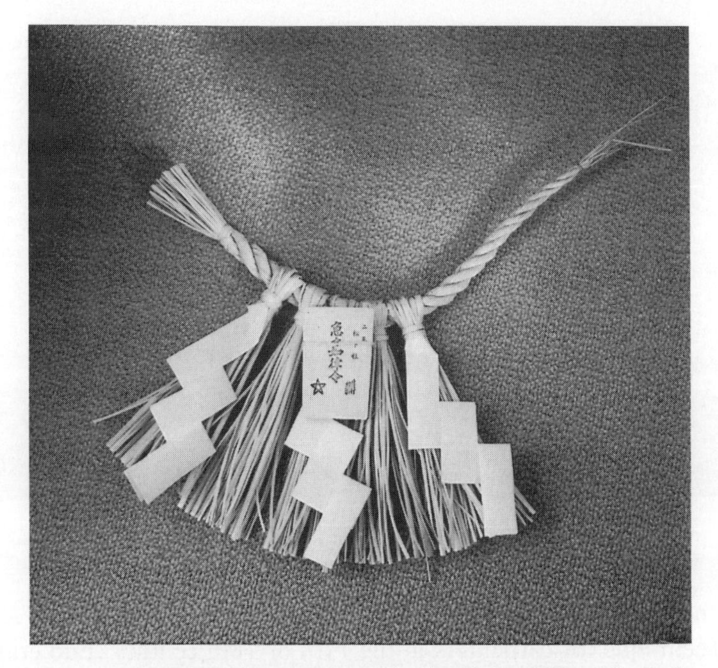

Fig. 19 A Somin-Shorai given at Matsushita-sha Shrine, Toba, Mie Prefecture.

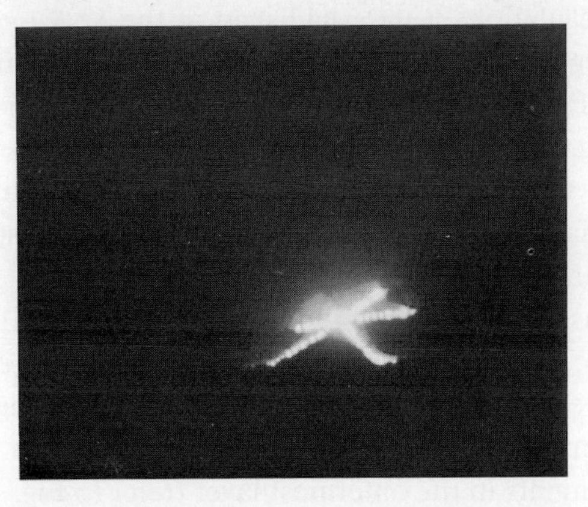

Fig. 20 A character *Dai* as one of Gozan-okuribi Bonfires. Kyoto. The length of the horizontal line is about 80 meter.

Among the ruins, we can find a regular pentagon in the company of a square and a regular hexagon (Fig. 21), though it has usually been said that the plans of houses at that time were circular.

Fig. 21 A regular pentagonal ruin of an ancient house. 2nd or 3rd century. Ohnaka-iseki Historic area, Harima, Hyogo Prefecture.

Japanese traditional building techniques were compiled into *Ki-Ku-jutsu* Technique for carpenters in the 17th or 18th century (Edo era), (*Ki* means a circle or a compass, and *Ku* a square or a carpenters's square).

Many tight regulations were established in the *Ki-Ku-jutsu.* According to one of the regulations, the design of wooden capitals under the eaves of temples (see Fig. 13) must be derived from an aperiodic tessellation of pentagrams (Fig. 22).

According to Okada,[12] carpenters in the Edo era sometimes used a fortunetelling-diagram in order to decide the lucky orientation of a house. One of the diagrams is shown on a set of a circular and square wooden plates which was used in the rebuilding of Akoh-jyo Castle (Fig. 23). On the circular disc, we can see a pentagram and Kuji at either end of a diameter. There are also the well-known famous eight marks for Yin-Yang Fortunetelling in the innermost layer and the five marks for Chinese five-elements in the outermost layer (refer to Fig. 7). At the same time, a magic square of 3×3 which was shown in Fig. 10 (lower) is engraved on the square plate.

The sites of some modern castles also show a relationship between a pentagram and traditional architecture. One of them is Goryokaku Castle (Fig. 24). This plan has rather a decagonal shape which can be subdivided

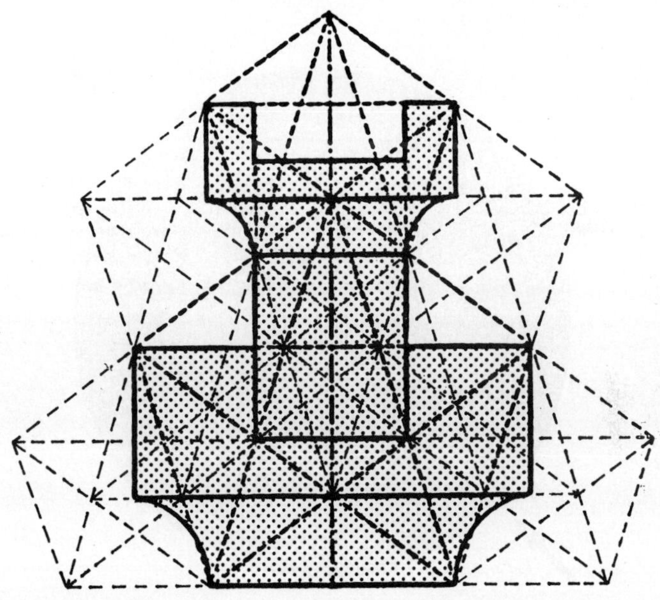

Fig. 22 Derivation using Ki-Ku-jyutsu Technique of a design of wooden capitals under the eaves of a temple (see Fig. 13).

into two kinds of rhombi seen in one of the famous Penrose's aperiodic patterns, as shown to the right of the figure.

This castle was designed by a Japanese architect based on a modern French castle, so that it could be easily defended in the event of an attack by a hostile army. It may also have had to do with the magical power of a pentagram.

3. A History of Regular Dodecahedra and Icosahedra in Japan

3.1. *Introduction of Regular Dodecahedra and Icosahedra*

In the 8th century (during the Nara era), many valuables in the Orient were gathered around the capital, Nara, as tributes to the then emperors. These are now kept in Shosoin Treasure house in Nara. In them, there is a copper incense burner from Persia which has an open-worked regular dodecahedral arabesque pattern (Fig. 25). It may have been actually made by a Persian, and has been kept with great care out of public view.

When were Japanese aware of the geometry of regular dodecahedra and icosahedra? It was probably after the appearance of *Wasan*, the tradi-

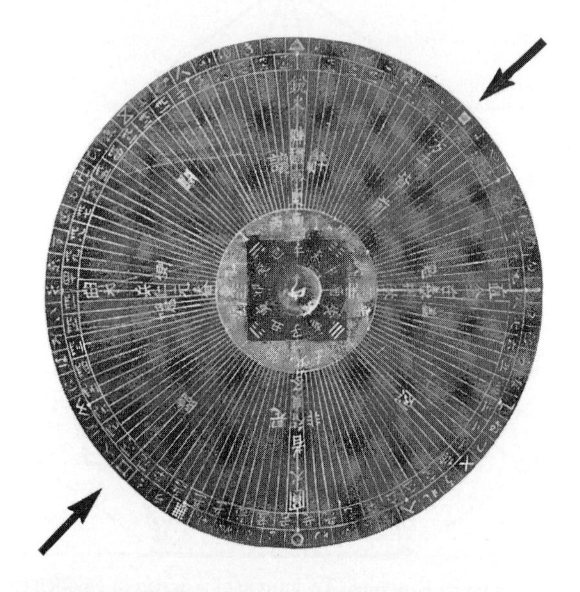

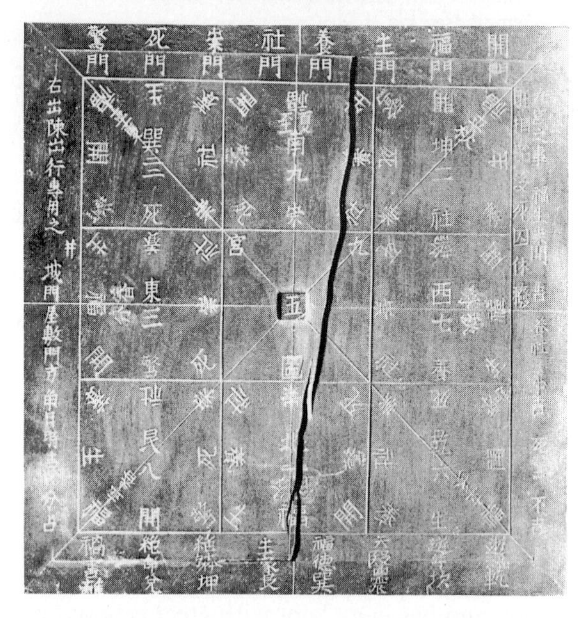

Fig. 23 A set of circular and square wooden plates for fortunetelling. 17th century. Ohishi-jinjya Shrine, Akoh, Hyogo Prefecture. The positions of pentagram and *Kuji* are shown by the arrows (see also Figs. 7 and 10).

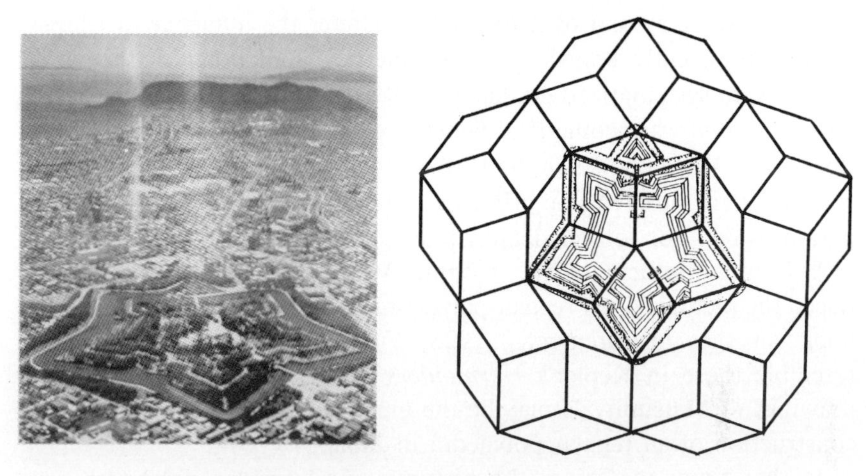

Fig. 24 The regular pentagonal site of Goryokaku Castle (left, photo courtesy: Hakodate city office) and a Penrose pattern covering over its plan (right). 19th century. Hakodate, Hokkaido.

Fig. 25 A copper incense burner having an open worked regular dodcahedral pattern. 18th century. Shosoin Treasure house, Nara.

tional Japanese system of mathematics. Under the influence of Chinese mathematics, early knowledge of geometrical and other mathematical figures were systematized gradually as *Wasan* Mathematics. The leading part of *Wasan* was geometrical figures for the sake of familiarity, and in the earliest *Wasan* books in the 17th century (during the Edo era), we can also find some problems on primitive polyhedra such as cubes, tetřahedra pyramids etc. For further details, refer to Miyazaki.[13]

A *Wasan* Mathematician, Toshino Matsumiya (1686–1780), introduced the pictures of all regular polyhedra in his book *Bundo-Yojutsu* ("A book on the measurement," 1728, Fig. 26). These pictures rather resemble those in Kepler's *Harmonices Mundi* (1619, Fig. 27). These drawings of Matsumiya's may be the first attempt to show the accurate construction of all regular polyhedra in Japan.

At that time, in China, the pictures of all regular polyhedra were introduced in *Cèliáng-Quángyi* ("A book on the measurement," 1631), *Shūlǐ-Jīnyùn* ("A full commentary on mathematics," 1722), etc. by using pictures which bear a striking resemblance to those in Pacioli's *De Divina Proportione* (1509), Dürer's *Unterweisung der Messung* (1512), etc.

A little later in Japan, Yoriyuki Arima (1714–1783), a lord of Kurume in the south of Japan and a *Wasan* Mathematician, said in his

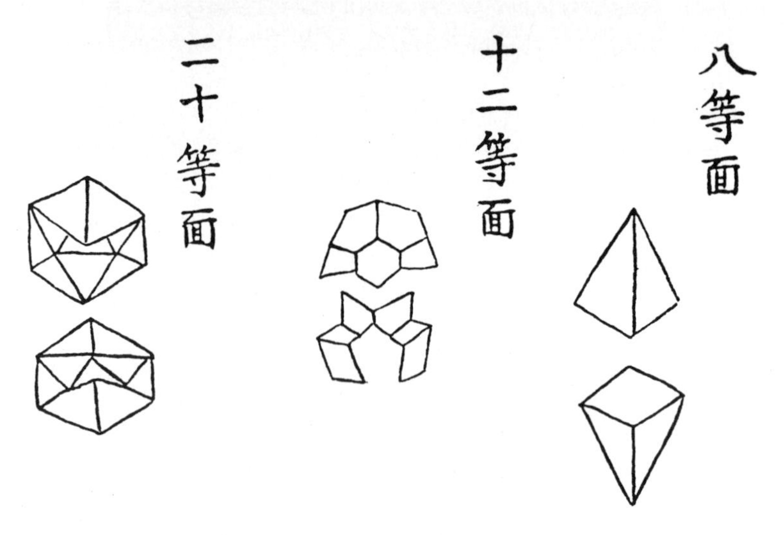

Fig. 26 From *Bundo-Yojutsu* (1728) by Toshino Matsumiya.

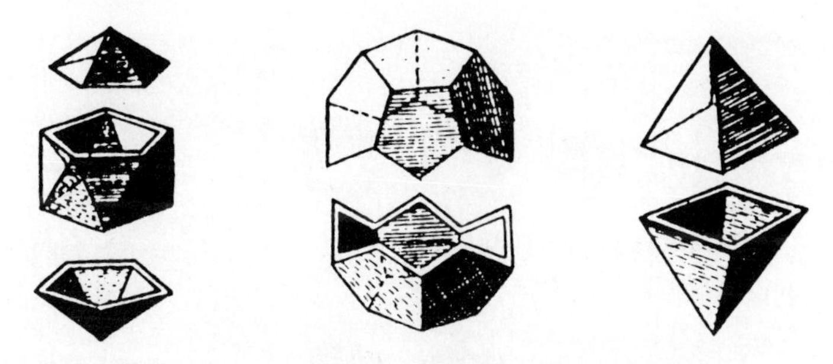

Fig. 27 From *Harmonices Mundi* (1619) by Kepler.

mathematical book *Shuki-sanpo* ("A collection of mathematical problems," 1769) that everyone knows about regular polyhedra and can easily visualize the shapes without their pictures.

To the contrary, Eken Shinpo (1657–1753), a priest in Hachinohe in the north of Japan and a *Wasan* Mathematician, took the credit for single-handedly discovering the regular dodecahedron and icosahedron. According to *Shinpo-teisanki* ("A mathematical book compiled by Shinpo's pupils," 1751), he was convinced that no book in China nor Japan described these curious polyhedra at that time, and he decided to make a *Sangaku* Tableau of these to disseminate this knowledge extensively. A *Sangaku* Tableau is a large votive tablet given to a shrine or temple. He thought that, as soon as people were aware of these, they would begin to apply their curious shapes to an incense-box, a dice for fortunetelling, a medicine-chest, etc. Shinpo differed from Pythagoras, who is said to have kept the existence of the regular dodecahedron secret.

The *Sangaku* given to the temple by Shinpo is now missing. Therefore, Hideo Kuwabara, a contemporary *Wasan* Mathematician, made a new *Sangaku* to commemorate Shinpo, and gave it to the Koryu-ji Temple at Hachinohe where Shinpo lived (Fig. 28). He posed the problem which asks for the edge length of each spherical regular polyhedron when the diameter is given. Kuwabara got a hint for the problem from a *Temari*, a traditional Japanese artifact (Fig. 29). (In the figure, dodecahedral and icosahedral patterns were presently devised by a modern geometer.)

Furthermore, Shinpo also said that the twelve faces of a regular dodecahedron can form a die for fortunetelling, using the 12 animals of the Japanese zodiac: rat, cow, tiger, rabbit, dragon, snake, horse, sheep, mon-

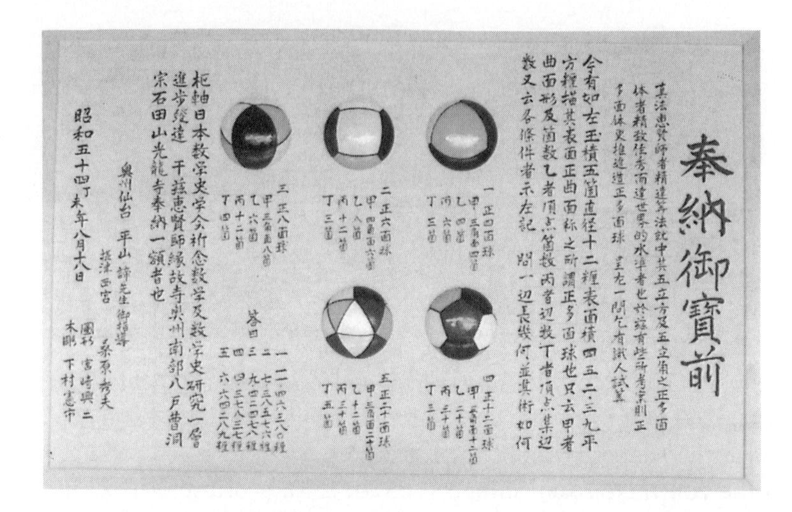

Fig. 28 A *Sangaku* Tableau, painted problem on regular polyhedra. Offered to Koryu-ji Temple by Hideo Kuwabara. 1979. 970mm × 600mm (a golden rectangle). Hachinohe, Aomori Prefecture. Designed by the author.

Fig. 29 Regular polyhedral patterns seen on Temaris. Made by Kiyoko Urata, a modern artisan.

key, cock, dog, and wild boar. Such a die is now sold as a modern artifact in Takayama (Fig. 30), but it may have no connection with Shinpo.

3.2. *Icosahedral Geodesic Patterns as Talismans*

After Shinpo and Arima, in the north and south of Japan, many *Wasan* Books on polyhedra were written. *Sanpo-Kiriko-shu* ("A mathematical

Fig. 30 Two kinds of regular dodecahedral dice made of paper. Modern artifacts of Takayama, Gifu Prefecture.

collection of polyhedra," period unknown) by Yasuaki Aida (1747–1817) is the most typical one of them. In it, he introduced many icosahedral geodesic polyhedra (Fig. 31).

Icosahedral geodesic patterns such as *Kagome*, composed of many pentagrams and hexagrams, has even now a mystic meaning in Japan. It is a powerful talisman having many pentagrams and also hexagrams.

In the 15th and 16th centuries (the age of civil wars in Japan), soldiers on a battle-field pitched camp under a *Uma-jirushi*, a solid landmark for each camp or for each horse. This mark became a *Matoi*, a solid landmark for each group of firemen in Edo era. There were many

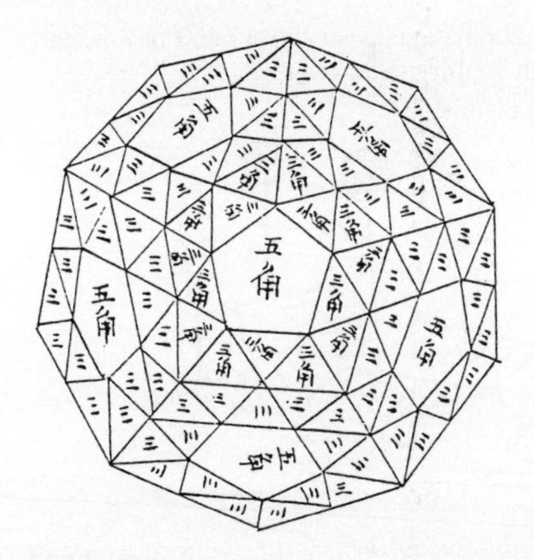

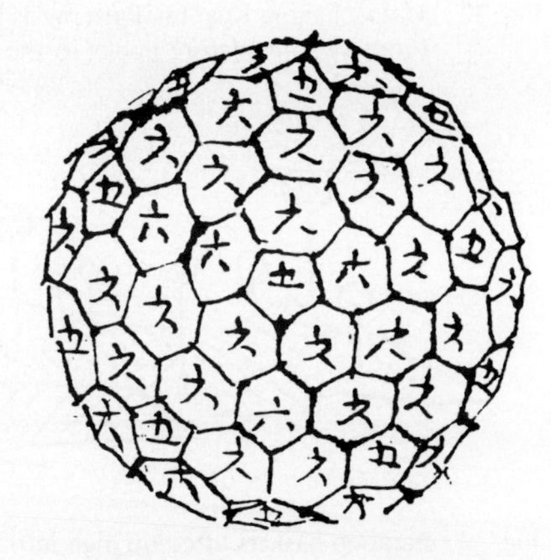

Fig. 31 From *Sanpo-Kiriko-shu* (18th or 19th century, Edo era) by Yasuaki Aida.

bamboo baskets having *Kagome* patterns as *Uma-jirushis* or *Matois* (Fig. 32). They surely had a miraculous power as talismans in war and in fire. The mystic power of such bamboo baskets are still revered in some areas since the Edo era. On February 8th and December 8th, people lift up a bamboo basket high into the sky to drive out evil spirits. (Fig. 33).

They are also used at the *Gion-matsuri* carnival in Kyoto held in the middle of July each year. It is the typical festival after Somin-Shorai Talisman. This carnival originally dated from the end of the 10th century

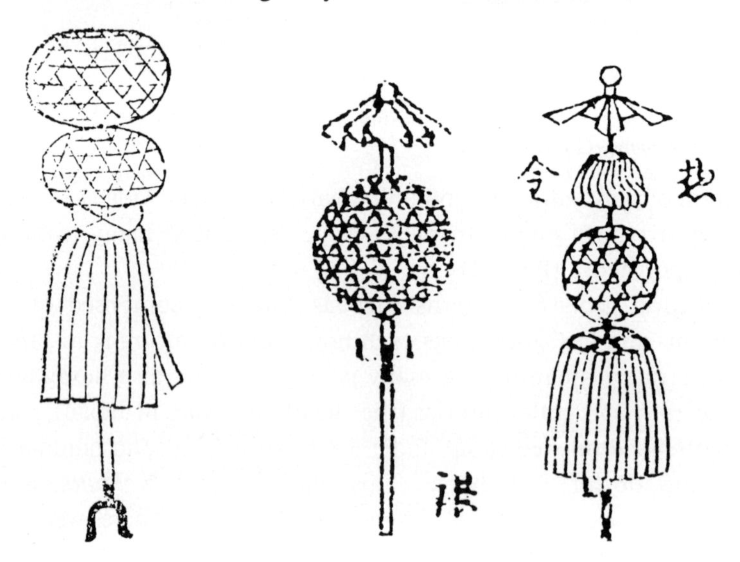

Fig. 32 *Matois* having Kagome Patterns as bamboo baskets. Usually, about 2 or 3m high. From various picture books in the 19th century (Edo era).

Fig. 33 Bamboo baskets lifted up high into the sky over Edo Town (today's Tokyo). From *Morisada-mankoh* (19th century, Edo era).

(during the Heian era) as a divine service at Yasaka-jinjya Shrine, whose principal image is Susa'noh-no-Mikoto. Somin Shorai himself was also enrolled among the gods and the descendents of Abe-no-Seimei had taken charge of fortunetelling in the shrine.

The typical large devices of the carnival is a *Yama*, which is a mountain-like hemi-sphere covered by a red cloth, and a *Hoko*, which is a halberd-likeline segment (Fig. 34). Judging from Yin-Yang Theory, *Yama* may represent a circle as Yang and *Hoko* represent a square. Both of them are retained in the elevation of Horyu-ji Temple (Fig. 3).

The main event of the carnival is the *Yama-Hoko* parade. Many big dancing carts, each of which take on a *Yama* or a *Hoko*, are drawn by the crowd along main streets in Kyoto, to scatter Somin-Shorai Talismans with no pentagrams (Fig. 35).

Where are the pentagrams? Astonishingly enough, many of them are inside each Yama covered by a red cloth. The skeleton of this hemi-sphere is always a bamboo basket of *Kagome* pattern (Fig. 36). People believe that this basket possesses mystic power as a talisman against thunder.

3.3. *A Geometry of Indian Five-Elements*

As mentioned before, it seems that the geometrical pictures of all regular polyhedra were first introduced into Japan in the 18th century. In spite of this, many Japanese may have appreciated their geometrical meaning as early as the 11th or 12th century.

Fig. 34 *Yamas* (low thickset, carts) and *Hokos* (tall slender carts) in Gion-matsuri Carnival). From *Kyoto-Gion-ezu* Tableau (1895).

Fig. 35 A Somin-Shorai Talisman given at Gion-matsuri Carnival.

Fig. 36 A *Kagome* pattern as the skelton of a *Yama*.

Fig. 37 A typical *Gorinto.*

At that time, a *Gorinto*, a five-storied small pagoda, was devised by a Japanese as a symbolic tower for Buddhists (Fig. 37). It consists of five blocks, each of whose shape is: a cube, sphere, square-pyramid (or sometimes a trigonal-pyramid), hemi-sphere, and a *Hoju* gem (chestnut-shaped), from bottom to top. On each block, generally, a Sanskrit letter or Chinese character is inscribed or carved, naming each of Indian five elements: earth, water, fire, air, and heaven, always in the same order. Sometimes, each block is colored yellow, white, red, black, and blue, always in that order.

Why was this composition adopted? There are various opinions about this. One is that it was modeled after a pagoda of the ancient India, and the other, after the human body. That is, from bottom to top: a sitting leg, abdomen, chest, face, and head.

On the other hand, some historians express another view that a *Gorinto* may represent a stacking of five regular polyhedra from bottom to top; a cube, regular icosahedron, regular tetrahedron, regular octahedron, and regular dodecahedron (Fig. 38). There is a curious correspondence between this opinion and the description by Plato in his *Timaeus*. If so, it may be that a *Hoju* gem represents a regular dodecahedron or regular pentagon (Fig. 39). According to Ohsato,[14] it is taught that the head of the portrait of the previously mentioned Kukai, who is said to have devised a *Gorinto* in the first place, must be drawn inside the shape of a regular pentagon (Fig. 40).

Fig. 38 An original *Gorinto* as a stacking of five colorful regular polyhedra. Made by the author.

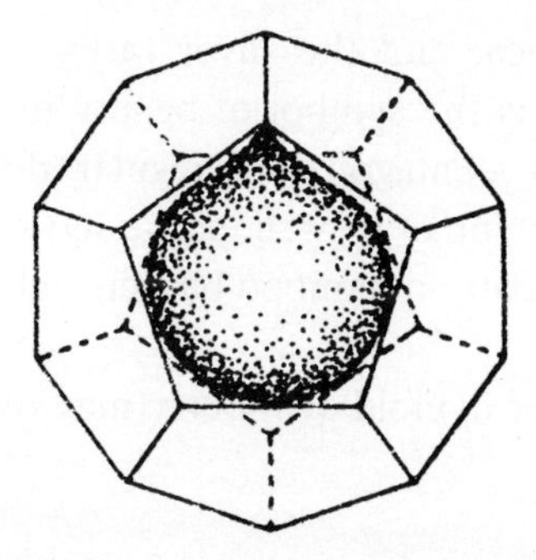

Fig. 39 A *Hoju* gem in a regular pentagon or dodecahedron.

It is often said that early Japanese were not aware of the existence of regular dodecahedra, though these were every place in shrines or temples as *Hoju* gems. The uppermost small spherical decoration of the five-storied pagoda in the Horyu-ji Temple (Fig. 3) is, of course, a *Hoju* gem.

4. Conclusion

It is sometimes said that the golden ratio, 1:1.618 (in other words, a regular pentagon or regular dodecahedron, etc.), is the symbol of beauty

Fig. 40 A portrait of Kukai, whose head is drawn inside a regular pentagon (copy).

in Europe and America, and the silver ratio, $1:\sqrt{2}$ (in other words, a square or cube, etc.), is the symbol of beauty in Japan.

However, a regular pentagon and regular dodecahedron have held significant meaning in Japan as well, as we have seen here. Both can be seen everywhere in Japan, in bamboo baskets, talismans, and *Hoju* gems, a universal shape.

The mystic power of fivefold symmetry may overwhelm everything in Japan.

Chronology

Era	Period	Key terms in this essay
(Mythological age)		(Fu-I), (Nü-Kua) (Susa'noh-no-Mikoto) (Somin Shorai)
Jyomon	– c.200B.C.	
Yayoi	c.200B.C.–c.300	Himiko
Kofun	c.300–551	
Asuka	551–644	Shotoku-taishi
Nara	644–783	Horyu-ji, Shosoin
Heian	783–1192	Kukai, Abe-no-Seimei
Kamakura	1192–1333	
Nanbokucho	1333–1392	
Muromachi	1392–1573	
Momoyama	1573–1615	
Edo	1615–1868	Eken Shinpo, Toshino Matsumiya Yoriyuki Arima, Yasuaki Aida Goryokaku
Meiji	1868–1912	
Taisho	1912–1926	
Showa	1926–1989	
Heisei	1989–	

Acknowledgments

I am grateful to Professor Yasuzo Okada, Osaka Seikei Women's Junior College, Higashi-yodogawa, Osaka 533, Japan. He has contributed to this essay by his comments and suggestions, as well as by sending me many unpublished materials. Almost all the figures in Sec. 2 and Sec. 3 (from Fig. 8 to Fig. 19), and Figs. 23, 32, and 33 were corrected by him. He was the first man to investigate the mystic history of the pentagram in Japan.

Furthermore, I must make special mention of Professor Harriet E. Brisson, Rhode Island College, who has many interests in Japanese traditional culture and kindly edited the English version of this essay.

References

1. K. Miyazaki, *An Adventure in Multidimensional Space.* Wiley, NY (1986).
2. K. Miyazaki, *Plato and a Five-Storied Pagoda* (in Japanese). Jinbunshoin, Kyoto (1987).
3. H. Yoshino, *Yin-Yang Fortunetelling and Religious Services for Children* (in Japanese). Jinbun-shoin, Kyoto (1986).
4. Y. Okada, *Mystic Crests Engraved on the Stone Wall in a Modern Castle* (in Japanese). Trans. of Osaka Seikei Women's Junior College, No. 15 (1978) 11–35.
5. Y. Okada *Traditional Crests as Talismans on Stone Walls* -1- (in Japanese). *op. cit.,* No. 21 (1984) 111–131.
6. Y. Okada, *Traditional Crests as Talismans on Stone Walls* -2- (in Japanese). *op. cit.*, No. 22 (1985) 1–13.
7. Y. Okada, *Traditional Crests as Talismans on Stone Walls* -3- (in Japanese). *op. cit.*, No. 23 (1986) 113–121.
8. Y. Okada, *Traditional Crests as Talismans on Stone Walls* -4- (in Japanese). *op. cit.*, No. 24 (1987) 137–146.
9. Y. Okada, *Traditional Crests in Myoho-in Temple* (in Japanese). *op. cit.*, Special Issue (1988) 1–17.
10. Y. Okada, *Traditional Crests as Talismans in Kyushu Area* (in Japanese). *op. cit.*, No. 26 (1989) 95–102.
11. H. Wazaki, *Ethnology on the Bonfire Shaped "Dai"* (in Japanese). Kobun-do, Tokyo (1987).
12. Y. Okada, *A Pair of Fortunetelling-Diagrams Inherited in Akoh-jyo Castle* (in Japanese). Akoh-Gishi-kai (1984).
13. K. Miyazaki, *Polyhedra and Architectures* (in Japanese). Shokoku-sha, Tokyo (1979).
14. K. Ohsato, *On the Portrait of Kukai* (in Japanese). Kikan-Butsuga, No. 2, (1983) 22–25.

FIVEFOLD SYMMETRY IN THE LITERATURE*

Nenad Trinajstić

"The hand that signed the paper felled a
city;
Five sovereign fingers taxed the breath,
Doubled the globe of death and halved a
country;
These five kings did a king to death."

Dylan Thomas (1914–1953): *The hand that signed the paper*

It appears that symmetry in general and fivefold symmetry in particular provide artists and writers with a rich source of themes, images and metaphors (see e.g., Refs. 1–12). Fivefold symmetry in the arts will be reviewed by other authors in this volume.[a] Here we will present some examples of the use of fivefold symmetry in the literature. Another contribution from this area can be found elsewhere in this volume.[b]

To detect symmetry in the literature is not an easy task, though in some cases the writer may use in a direct way a symmetric and/or asymmetric object as a descriptive means or a metaphor. The abstractness of language makes literary works difficult to analyze in terms of symmetry operations. Symmetry operations such as rotations, reflections or translations do not appear discernible in a simple way in the literature; they are, if they exist, usually hidden. Lately, however, efforts have been made[9] to associate the structure of a literary work (e.g., poem, novel, short story, play) with a diagram or a geometric object with some characteristic symmetry or antisymmetry. A diagram may be used, for example, to visualize the relationship (sometimes very complex) between the characters in a novel

* This essay is dedicated to the memory of Miroslav Krleža (1893–1980), the greatest of Croatian writers in this century, whose work is so rich with all kinds of symmetries including fivefold symmetry.

[a] See for examples, M. Fleurent.

[b] See L. A. Cummings.

or a play (see e.g., Ref. 13). A whole literary work may be associated with some geometric (spatial) form, or it may be associated with several spatial forms, tightly or loosely related (see e.g., Refs. 9,14,15). Very often in poetry we find poems in graphical shapes which are isomorphic with diagrams that possess recognizable characteristics of symmetry or asymmetry. A good example to illustrate this point is a funnel-like bilaterally symmetric shape of a poem by the German neoromantic poet Christian Morgenstern (1871–1914) entitled "Die Trichter."[16]

To associate the structure of a literary work with a 3D object is not simple because we deal with the non-spatial objects to which we ascribe spatial features. Such an analysis and interpretation is usually highly subjective, and a given non-spatial configuration may be associated with different symmetric or asymmetric figures depending on the imagination of a person carrying out such a study (see e.g., Ref. 9).

As we stated elsewhere,[17] numbers are often used by poets.[18] Here we show a poem entitled *Quinta essentia*[19] by an important modern Croatian poet Boro Pavlović (b. 1922) in which he plays "in quantized harmony," as a poet and Nobel-Prize winning scientist Roald Hoffmann (b. 1937) would say[20] with the first five numbers in five languages (English, Latin, Greek, Italian and Croatian).

QUINTA ESSENTIA

one	unus	eis	un(o)	je(da)n
one	una	mia	una	je(d)na
one	unum	en	un	je(d)no
two	duo	dyo	due	dva
two	duae	dyein	due	dvije
three	tres	treis	tre	tri
four	quattour	tessares	quattro	čet(i)ri
five	quinque	penta	cinque	pet
		PANTA		
		&		
		TUTTA		
		QUANTA		

The formal structure of this poem may be associated with inverted pentagonal pyramid whose base is the regular pentagon (see Fig. 1).

Each vertex of a pentagon represents one language, and each edge the translation process from one language into the other. The bottom of the

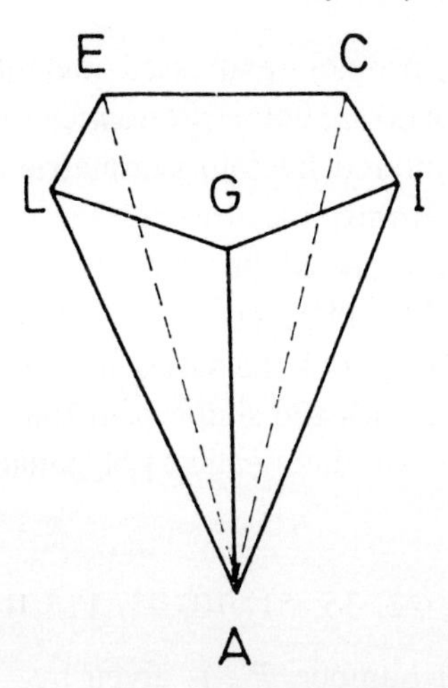

Fig. 1 The geometry of a poem "Quinta essentia" by Boro Pavlović. The labels have the following meaning: A = Arabic numerals, C = Croatian language, E = English language, I = Italian language, G = Greek language, and L = Latin language.

pyramid is labelled by Arabic numerals. Rotation by $2\pi/5$ radians always brings a different code for a given number at the bottom of the pyramid which serves as a decoding device (see below one example of this).

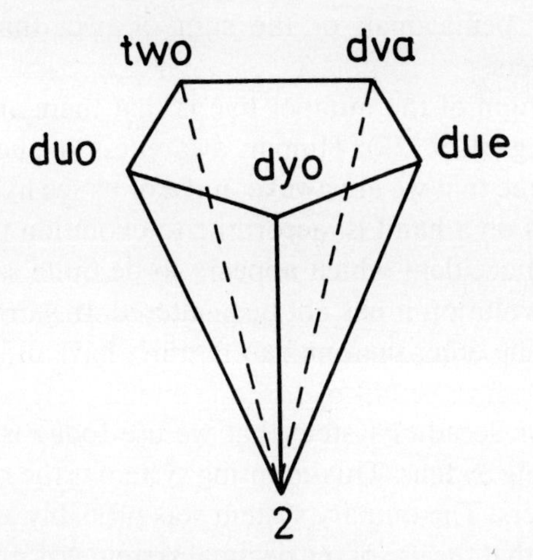

Since the meaning is always the same, all the vertices are of the same kind, but since the coding process (language) is different, the vertices are labelled by different letters. Therefore, the hidden structure of this poem possesses pentagonal symmetry. The idea of this poem is, as far as we can see, that the differences between peoples (expressed metaphorically by different linguistic terms for numbers) are not so pronounced as we are led to believe.

The most often encountered fivefold symmetric objects in the art and the literature are pentagrams, i.e., the regular pentagon and the five-pointed star.[21] At their basis is the number five.[17] For example, the number five is a pentagonal number.[22]

Pentagonal numbers belong to the series whose name is derived from association with patterns with the shape of regular pentagons. They are part of a special class of numbers called polygonal numbers.

The pentagonal number series is

$$1, 5, 12, 22, 35, 51, 70, 92, 117, 145, \ldots$$

and the n-th pentagonal number, P_n, is given by

$$P_n = \frac{3n^2 - n}{2} .$$

Pierre de Fermat (1601–1665) found that a strikingly simple relationship exists between the polygonal number series and the natural number series. This may be expressed as the following simple statement: Every number is either pentagonal, or the sum of two, three, four or five pentagonal numbers.

The possible origin of the number five is that there are five fingers on each hand (see e.g., Ref. 23). Human awareness of the number five is natural: All the time that we are awake, most of us see five fingers on each hand. Five fingers on a hand is, according to evolution theory, the result of the anatomical accident which appears to be quite satisfactory, since in the course of evolution it has not been altered. In support of the above is the fact that some other mammalian families have different number of "fingers."[24]

The decimal (or decadic) system that we use today is founded on the practice of counting by tens. This counting system is the result of counting by the use of fingers. The quinary system was probably also used by early people. It may be that the use of the decimal system was much simpler than the use of the quinary system for practical purposes, and consequently the former system has developed to be used exclusively. In addition, the development of language to cover abstractions such as numbers has been rather slow. It is probable that the difficulties in producing numerical verbal expressions were the smallest in the case of the decimal system, and this perchance also favoured the base 10 system. It is perhaps for this reason that the modern languages are built almost solely around the decimal system. A mathematician may argue that more interesting counting systems than quinary or decimal systems would emerge if Hominids had four (octal system) or six (duodecimal system) fingers on each hand.

Artists have been inspired by the number five. There is an interesting painting by Charles Demuth from 1928 entitled *I Saw the Figure 5 in Gold*. It may be found in the Alfred Stieglitz Collection of the Metropolitan Museum of Art in New York. This painting is also reproduced in the Blackwell book on geometry in architecture.[6]

The number five was (and is) used occasionally by poets as a symbol or metaphor (see e.g., Refs. 25–28):

THE SONNET ABOUT THE NUMBER FIVE

5 5 5 5 5 5 5 5 5 5 5
5 5 5 5 5 5 5 5 5 5 5
5 5 5 5 5 5 5 5 5 5 5
5 5 5 5 5 5 5 5 5 5 5

five 5 5 5 5 5 5 5 5 5
5 5 5 5 5 5 5 5 5 5 5
5 5 5 5 5 5 5 5 5 5 5
5 5 5 5 5 5 5 5 5 5 5

5 5 5 5 5 5 5 5 5 5 5
5 5 5 5 5 5 5 5 5 5 5
5 5 5 5 5 5 5 5 5 13 5

5 5 5 5 5 5 5 5 5 5 5
5 five 5 5 5 5 5 5 5 5
5 5 5 5 5 5 5 5 5 5 5

Trinajstić, 1989

We have already seen that the possible origin of the number five is the fact that there are five fingers on each hand. Hands have also been used by writers in different contexts. For example, the Croatian writer Ranko Marković (b. 1913) has produced a wonderful short story entitled *Hands*[29] in which the hands possess a life of their own. Each hand differs – the right hand is more aggressive and stronger than the left hand. However, when they work, the work is efficiently carried out only if they act together. (We should remember that two hands are chiral objects, identical in shape, but not superimposable because one is right and one is left. They represent a prototypical chiral pair.[30]) The author in this story has shown that each hand is unique (an asymmetric figure), but by acting together they complement each other (the left hand and the right hand together form a symmetric figure).

It is amusing to note that there is a horror story related to the above by William Fryer Harvey entitled *The Beast With Five Fingers*,[31] in which one severed hand possesses all the (even evil) characteristics of a human being: It walks, talks, writes, hates, etc., the same as *Hands* in the above story. The element of horror in the story is produced by the persistent persecution of the main character (Eustace Borlsover)[c] by the severed right hand of his uncle (Adrian Borlsover) because he offended it. But there is nothing unfathomable about the hand, as Eustace said to Saunders (the secretary of the late Adrian), who admitted that he was scared stiff by the hand's continuous harassment: "You've no need to be. There's nothing supernatural about that hand, Saunders. I mean, it seems to be governed by the laws of time and space. It's not the sort of thing that vanishes into thin air or slides through oaken doors. . . ." This short story is also the basis of a movie under the same name made by Warner Brothers in 1947 with two superb character actors Peter Lorre and J. Carroll Naish and directed by the horror specialist Robert Florey.

There are many other stories in which hands have a life and a purpose of their own. One that should be mentioned is a sinister tale by Thomas Burke (1886–1945) entitled: *The Hand of Mr. Ottermole*,[33] perhaps the finest short story in the genre of the detective story. At the climatic end of this story, the main character (the sergeant, Mr. Ottermole, expressed the following idea, which might be considered the common idea of all the three stories about hands mentioned here: ". . . Can't ideas live in nerve and muscle as well as in brain? Couldn't it be that parts of our bodies aren't really us, and couldn't ideas come into those parts all of a sudden, like ideas come into — into' — . . . —' into *my hands*."

An example of the direct use of a fivefold symmetric object, e.g., the pentagon, may also be found in literature. An example of this is the spectacular novel *Poland*, by James A. Michener (b.1907)[34]:

> . . . instead they came up with a chain made of woven hair from a cow's tail on which was suspended a curious pentagon-shaped medal dating back to some pre-Christian

[c] English writers seem to be very fond of the name Borlsover. For example, this name has been used, amongst others, by Edmund Crispin (1921–1978) in his amusing story entitled "Abhorred Shears."[32]

times. "What's this?' a soldier asked, and Biruta said truthfully:
"We've always had it.'
'Why was it hidden?'
"It's our good-luck charm."
This was too complex for the men, so they summoned Krumpf, and as soon as he surmised that it must be some early Germanic medallion, a souvenir of the time when Teutonic greatness began, and he snatched it from the soldier. As he stomped off with his prize, Biruta thought: How strange. A man from this village, centuries ago, took that medal from a pagan. Now the pagans have reclaimed it.

In a science-fiction classic about life in a 2D world entitled *Flatland* by a noted Shakespearean scholar Edwin A. Abbott (1838–1926), pentagons are used in two ways. First, almost all of the houses in Flatland are pentagonal.[35] The following lines explain why pentagonal houses are preferred in Flatland:

"The most common form for the construction of a house is five-sided or pentagonal, as in the annexed figure. The two Northern sides RO, OF, constitute the roof, and for the most part have no doors; on the East is a small door for the Women; on the West a much larger one for the Men; the South is usually doorless.

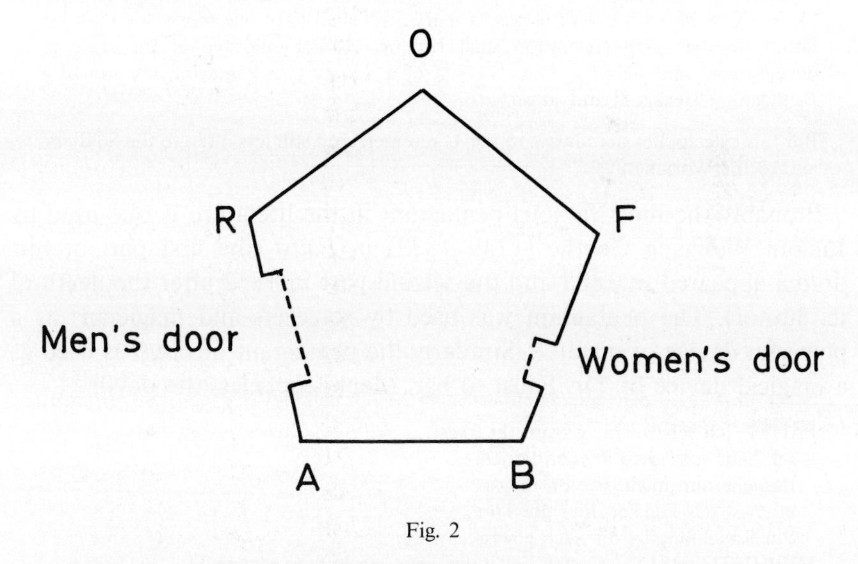

Fig. 2

Square and triangular houses are not allowed, and for this reason. The angles of a Square (and still more those of an equilateral Triangle), being much more pointed than those of a pentagon, and the lines of inanimate objects (such as houses) being dimmer than the lines of Men and Women, it follows that there is no little danger lest the points of a square or triangular house residence might do serious injury to an inconsiderate or perhaps absent-minded traveller suddenly therefore, running against them: and as early as the eleventh century of our era, triangular houses were universally forbidden by Law, the only exceptions being fortifications, powder-magazines, barracks, and other state buildings, which it is not desirable that the general public should approach without circumspection.

At this period, square houses were still everywhere permitted, though discouraged by a special tax. But, about there centuries afterwards, the Law decided that in all towns containing a population above ten thousand, the angle of a Pentagon was the smallest house-angle that could be allowed consistently with the public safety. The good sense of the community has seconded the efforts of the Legislature; and now, even in the country, the pentagonal construction has superseded every other. It is only now and then in some very remote and backward agricultural district that an antiquarian may still discover a square house."

Unfortunately even in Flatland rooms are not pentagonal, though in some houses the Hall and the Cellar have the shape of an irregular pentagon.

Not only houses, but some people are also pentagonal in Flatland:

"Our Professional Men and Gentlemen are Squares (......) and Five-Sided Figures or Pentagons." The sons have one more side than the fathers, but this rule does not apply to all classes:

"It is a Law of Nature with us that a male child shall have one more side than his father, so that each generation shall rise (as a rule) one step in the scale of development and nobility. Thus the son of a Square is a Pentagon; the son of a Pentagon, a Hexagon; and so on.

But this rule applies not always to the Tradesmen, and still less often to the Soldiers, and to the Workmen; . . ."

Probably the most famous pentagram in the literature is one used by Johann Wolfgang Goethe (1749–1832) in *Faust* (the first part of this drama appeared in 1808 and the second part in 1832 after the death of its author). The pentagram was used by sorcerers and magicians as a powerful device of magic.[21] Similarly the pentagram in *Faust* is used as a magical device by Dr. Faust to ban Mephistopheles, the devil:[36]

FAUST. Ich sehe nicht, warum du fragst.
Ich habe jetzt dich kennen lernen,
Besuche nun mich, wie du magst.
Hier ist das Fenster, hier die Türe,
Ein Rauchfang ist dir auch gewiss.
MEPHISTOPHELES. Gesteh' ich's nur! dass ich hinausspaziere,
Verbietet mir ein kleines Hindernis,
Der Drudenfuss auf Eurer Schwelle -
FAUST. Das Pentagramma macht dir Pein?
Ei sage mir, du Sohn der Hölle,
Wenn das dich bannt, wie kammst du denn herein?

Wie ward ein solcher Geist betrogen?
MEPHISTOPHELES. Beschaut es recht! Es ist nicht gut gezogen;
Der eine Winkel, der nach aussen zu,
Ist, wie du siehst, ein wenig offen.
FAUST. Das hat der Zufall gut getroffen!
Und mein Gefangner wärst den du?
Das ist von ungefähr gelungen!

More about the pentagram in Goethe's *Faust* may be found elsewhere in this volume.[d]

Several illustrative examples of the use of the number five and fivefold symmetry in the literature that we have shown indicate that these examples are not as abundant as those with other types of symmetry.[9] To conclude we may say that the use of the number five and fivefold symmetry in the art and literature is related in some way to the search for unusual beauty inherent in and induced by this particular type of symmetry. However, we should remember that the aesthetic effects resulting from symmetry of an object or a literary work lie in the psychic process of perception, and this process is highly individualistic. Therefore not everyone will necessarily be as enchanted by fivefold symmetry as we are.

Acknowledgments

The author would like to thank the following people for their discussions and correspondence on symmetry and fivefold symmetry in mathematics, science, engineering, art and literature: D. Bonchev (Burgas), W. Carter (Reading, England), S.J. Cyvin (Trondheim), L.A. Cummings (Waterloo), B.M. Gimarc (Columbia, SC), I. Hargittai (Budapest), Ž. Jeričević (Houston), L. Klasinc (Zagreb), D.J. Klein (Galveston), A. Lakhtakia (University Park, PA), S. Nikolić (Zagreb), B. Pavlović (Zagreb) and M. Randić (Ames, IA).

Early versions of the essay were examined by W. Carter, S.J. Cyvin, L.A. Cummings, I Hargittai, Ž. Jeričević, D.J. Klein, A. Lakhtakia and S. Nikolić. The author is grateful to them for their critical comments and many helpful suggestions.

References

1. L. Pacioli, *De Divine Proportione*. Venice (1509), reprinted in 1956.

[d] J. Brandmüller, The Fivefoldness in Mathematics, Physics, Chemistry and Beyond, in this volume.

2. D'Arcy W. Thompson, *On Growth and Form,* 2nd Edn., reprinted. University Press, Cambridge (1942) Vol. II, p. 726.
3. H. Weyl, *Symmetry*. University Press, Princeton, NJ (1952).
4. G. Pischel, *Storia Universale dell'Arte* (translation from Italian into Croatian). Mladost, Zagreb (1966) Vol. 2, p. 159.
5. A.V, Shubnikov and V.A. Koptsik, *Symmetry in Science and Art*. Plenum Press, New York (1974).
6. W. Blackwell, *Geometry in Architecture*. Wiley, New York (1984).
7. I. Hargittai and M. Hargittai, *Symmetry Through the Eyes of a Chemist*. VCH, Weinheim (1986).
8. Plato, *Timaeus and Critias*. Penguin, Harmondsworth, Middlesex (1986).
9. B. Pavlović and N. Trinajstić, On symmetry and asymmetry in literature, *Comput. Math. Appls.* **12B** (1986) 197–227. Reprinted in I. Hargittai (ed.), *Symmetry: Unifying Human Understanding*, Pergamon Press, New York (1986).
10. V. Molnar and F. Molnar, Symmetry-making and -breaking in visual art, *Comput. Math. Appls.* **12B** (1986) 291–301. Reprinted in I. Hargittai (ed.), *Symmetry: Unifying Human Understanding*, Pergamon Press, New York (1986).
11. L.A. Cummings, A recurring geometrical pattern in the early renaissance imagination, *Comput. Math. Appls.* **12B** (1986) 981–997. Reprinted in I. Hargittai (ed.), *Symmetry: Unifying Human Understanding*, Pergamon Press, New York (1986).
12. L. Bencze, Uncertainty principle and symmetry in metaphors, *Comput. Math. Appls.* **17** (1989) 697–798. Reprinted in I. Hargittai (ed.), *Symmetry: Unifying Human Understanding,* Pergamon Press, New York (1986).
13. S. Lasic, *The Structure of Krleža's "Banners"* (in Croatian). Liber, Zagreb (1974).
14. C.H. Whitman, *The Geometric Structure of the Illiad.* Norton, New York (1965).
15. M. Rose, *Shakespearean Design*. Belknap, Cambridge, MA (1972).
16. C. Morgenstern, Die Trichter, in: *Galgenlieder Gingganz und Horatius Travestitus*. Zbinden, Basel (1972) p. 34.
17. N. Trinajstič, The Magic of the Number Five, reported at *MATH/CHEM/COMP* (Dubrovnik, June 25–30, 1990).
18. S. Buchanan, *Poetry and Mathematics*. The University of Chicago, Chicago (1975), reprint.
19. B. Pavlović, Quinta essentia, in: *Encyclopaedia Poetica* (in Croatian). Exhibition, Zagreb (1958).
20. R. Hoffmann, *The Metamict State*. University of Central Florida, Orlando, FL (1987) p. 101.
21. J. Chevalier and A. Cheerbrant, *Dictionary of Symbols* (a translation from French into Croatian). Matica Hrvatska, Zagreb (1983) pp. 713–714.
22. H.E. Huntley, *The Divine Proportion*. Dover, New York (1970).
23. C.B. Boyer, *A History of Mathematics*. Wiley, New York, 1968, Chap. I.
24. A.E. Brehm, *Der Farbige Brehm* (a translation from German into Croatian). Prosvjeta, Zagreb (1982).
25. E.E. Cummings, *Collected Poems*. Harcourt, Brace & Co., New York (1938).

26. J. Updike, *Seventy Poems* (translation into Croatian). Prosvjeta, Zagreb (1981), p.132.
27. M. Stojevič, ed. *Chakavian Poetry in the 20th Century* (in Croatian). Izdavački centar, Rijeka (1987).
28. R. Šerbedžija, *Black-and-Red* (in Croatian). Globus, Zagreb (1989).
29. R. Marinkovič, *Ruke* (in Croatian, but there is a translation into German available: "Hände," in *Erzahlungen*. Steingruben, Stuttgart (1961)), 4th Edn. Mladost, Zagreb (1981) pp. 134–145.
30. V. Prelog, Nobel lecture: Chirality in chemistry, reprinted in (in Croation) *Croat. Chem. Acta* **48** (1976) 194 and (in English) *Science* **193** (1976) 17.
31. W.F. Harvey, The beast with five fingers, in *The Ghouls*. P. Haining, (ed.) Futura, London (1976) Book Two, pp. 31–55.
32. E. Crispin, Abhorred shears, in *Beware of Trains*. Penguin, Harmond-Sworth (1987) pp. 56–63.
33. T. Burke. The hands of Mr. Ottermole, in *Great Tales of Detection*, D.L. Sayers, (ed.), Everyman's Library, London (1976) pp. 154–170.
34. J.A. Michener, *Poland*. Corgi, London (1984) p. 685.
35. E.A. Abbott, *Flatland*. Dover, New York (1952).
36. J.W. Goethe, *Faust*. Wilhelm Goldmann Verlag, München (1928) pp. 45–46.

THE ALLEGORICAL METHOD AND SYSTEMS OF FIVES
A beginning discussion exemplified in *Beowulf*

Laurence A. Cummings

1. Introduction

In this essay, I shall first present briefly a theory of how we apprehend the world, which presentation will lead easily into a description of the allegorical mode of creating and interpreting works, especially human ones. After emerging from the troublesome terminology that has grown up around allegory, I hope to expose how it works in practice, emphasizing its tendency to produce fivefold meanings.

In discussing meanings in our experience of events in our daily lives and in human messages we are aware of their multiplicity of import. One psychic event (and everything in the world must be received through the psyche if we are to be aware of it) radiates many possible meanings: recollections of yesterday, reconstructions of history, contemplation of present objects touched by men, such as a poem or a trolley or an essay in a collection of essays.

This writing is an attempt to describe a way that men have developed to render intelligible the multifold impact of meanings on our imaginations and our understanding. In this volume, we are all concerned with fives. Multiplicities of meaning from a single event occur in twos, threes, fours, fives, and even greater sets. Thinkers have long been aware of this necessity in our mental process.[1–8] This paper will first deal with the manifoldness of meanings. Then it will attempt to describe an account of these meanings called allegory in the confusing plenitude of terms used to describe the methodology. It is easier to deal with a registered object than a transient one, which is a reason for using objects preserved in some form in an unmoving (or theoretically static) past. Rituals, literature, art works, and works of architecture are remnants from our past which are seemingly in stasis in that, to each man, they present what seems to be the same data. Using these objects we are able to see whether or not the allegorical method can reveal any patterns of recountable meanings, hopefully in fives. In the limits of an essay, only an entry can be made into possibilities, and one very old literary work will be observed as specimen of the allegorical method.

2. Our Modes of Thought

Much of our talk and more of our thought moves indirectly. Often, perhaps always, the object of our contemplation — the thing — is in actuality unspeakable and in itself incomprehensible. Our sensory apparatus presents to us an image, its fidelity being limited by our reporting apparatus. The Thing is imaged in our mind; that image is not the Thing.

In my mind, I am in charge, it seems, or at least aware of all that's out there beyond my awareness. I am agent. Therefore, this image is subject to all the tiny adjustments — conscious and more likely not conscious — which I give it to make it "clear." This word "clear" implies that it — the image — and perhaps the object as well — is really unclear. Sometimes, if I do not keep my wits about me, the image adjusts or is adjusted to suit my convenience; for instance, if it is very fearful, I make it less so, so that I can tolerate it. Or again, if it doesn't fit in with my preconceptions, and I instantaneously see that a great, complicated, awkward, lengthy, tiring readjustment in my values or attitudes will be necessary to accommodate this new image as it is seen in my mind, then I round it off, augment it, pull it in here and out a bit there to make my mental life easy, Then I convince myself that this edited image is a simulacrum of the Thing, by a vaguing out of the original image, of the memory-file of that original, and of the very process by which I adjusted it and erased its original. The end product of this cosmetic treatment is labelled a sensory fact, or a "real" bit of data. It seems reasonable to suppose that what has "actually" happened is that there have been a score or scores of minute acts of will, which masquerade as acts of judgement. We think, or try to think in parallel with the Thing. We think in analogue, as if listening at home to tapes which are not the actual music or voice in the first room, hoping that we have a high fidelity. This analogous trick of ours enables us for the most part to get by.

Ordinarily, a sequence of images occurs, and then we are in the domain of time. In eternity, there is no time, and doubtless the Thing is single, simple, direct, and our image may become identical with it. Here in time we are presented with motion, change, flux, streams, sequences. One event is accompanied by others, and all are preceded by others, and, so far as we can predict out of our past, all events will be followed by others. Our experience of life is continuous and episodic, even in the dream world, even in the mirror world. When one episode is followed by another, and preceded by yet others, we have a sequence of events. This phenomenon we call a story. A recognizable story, one that we seem to recall, is a myth. There is also another recognition that may be called iconic. If we fix on some element — some phantom of the flux of outside and internal event all clamoring for notice — and intensify our attention on it, we are creating or recognizing an icon that gives a message like a photograph radiant with importance. What an iconographic object does (when it does do something) is to establish or to solicit recognition, to recall a pose or a disposition, but when the icon moves, we see it as its ritual act; the act and its causes and its outcome are the story, the myth. A myth occurs in time; its sequence is chronological, rather than logical or even particularly spatial.

Logical ordering utilizes certain fictions: it collects images by similarities into categories or classes (*X* is an animal), and within these mental places, *topos*, it disperses them by differentiation (*X* is an animal that smiles), implying that some animals, or one, do/does not. Hence, we refer to this classification and differentiation, this labelling (smiling, animal *X*), as topical. It is a most useful sort-and-file system. Another logical ordering attempts to face up to time by proposing that we imagine event *A* as being caused by event *B* (if *B*, then *A*). A billiard ball (*B*) in motion strikes another ball (*A*), and we see *B* stop or deflect, and *A* take on motion. We fictively suppose that *B* and *A* have necessary, uniformly applied, unavoidable relationship in such arrangement that *B* "causes" *A*. Billiard ball *B* causes the motion of ball *A*. As if it were part of cosmic nature or the nature of the cosmos, we call this natural causation. This ordering has been a useful theory in the past, a sort of mechanical myth.

Spatial ordering asks for dimensionality to be assumed; my imaging of length and width and height is assumed to be an accurate description of the spatial thing, and therefore that these directions do in fact occur outside my awareness. We can then arrange images to left and right, forward and back, up and down, over and under and on, inside and outside, beside, through, and so on. This device implies shadow and light, colour, dullness and shine. This attempt at sketching the Thing is, like logical fictive constructs, a convenient way of attempting to manage the Thing so that we can play with it. It is most important, perhaps as one of the fundamental ingredients in constructing or identifying an icon, or a mythical landscape.

Just as the image — if unedited — apes the Thing, within the limits of our preceptorial apparatus, so the myth apes the serial manifestations of the Thing, the events in time. The myth tries to parallel the series. But the myth is still unordered. It has no *topos*, no topic under which to put it. It has no arrangement in space, like the icon. We have the simulacrum of the sequence of events, but as yet we do not understand it. It is mere reportage. We do not contain it in a way to relate it to other data in the mind. The mother tells the plain story to the child, and, an impudent inquirer, it turns up a troubled face, and says, "But what does it mean?" And the mother must come forth with the moral of the tale, some comforting sentence that connects this last story with all the others that the child has heard and with all the other concepts beginning to occupy its mind. It is in some such fashion that we rewrite our own personal history. What is useful to our life, we keep; the rest, we edit or erase. So that the active file of information is in constant playback to the enormous data bank of past "facts" and conclusions and generalizations deduced from these facts, which we call memory; this playback reveals when a new image does not compute, and we often bend, fold, spindle, and mutilate it until it does.

But the Thing stands unadjusted and unadjustable. When the Thing and our imaging of it in consonance with our total array of memorial images (edited and other) become sufficiently remote from each other, we become awkward to live with, and if our divorce from the Thing becomes more or less permanent, we are committed to a mental institution of one kind or another: a prison, a hospital, an asylum, or a university.

3. Allegory

Once we understand that the Thing is impossible to conceive, we can recognize that our ways of understanding are but by shadows flickering on the wall.[9] We must think, or most of us must, and so we chatter on to ourselves about the shadowy images. We try to keep some image of the actual Thing by using the presumed similitude of our apprehension of it in our image: we hope and try to maintain in the image a constancy, a congruity, a parallelism in contour to the Thing.

Why did old, impotent Abraham[10] climb Mount Moriah with wood-bearing Isaac to sacrifice his only son born of a free woman, when the Promise was that the son's seed would bring forth the inheritance and the people? He did it because he must. By virtue of the absurd, he believed that although the sacrifice would be made, the Promise would be fulfilled. That is what the story, the history said. This is the literal meaning. But at once says the interpreter, the story has other senses, other levels (higher and deeper). Abraham did it because unawares he was the prefiguration — the figure before — of another figure in the grand Divine scheme; St. Augustine calls this an archetype.[1] Abraham's rite was the ritual enactment of the story of Jesus the Man carrying the wooden cross up Golgotha Hill to be sacrificed by His people in order that the very Promise that had been made to Abraham would be fulfilled — that Promise of the inheritance of life to His people. Abraham is Mankind or even the providential Father of Heaven working out our salvation. Man must pay the debt of original sin by dying. At the crucial instant, the ram bleated, the angel pointed, and the Agnus Dei, the Lamb of God, was substituted for the Man. God was on the rood. Thus the act of prefiguration under the Old Law was figured forth in actuality in the New, when the archetype is fulfilled in the person or event itself. The prefiguration was for the *figura*, for the *sacra*, for the Holy One of Israel and Isaac and old Abraham, for the crucified God. Likewise, the prophets of the Old Testament — wild, crazy mediums of Jehovah to the Israelites — were indeed the visionaries of those murmuring, headstrong tribesmen. Literally. But their oracles were often into the future, a future which contained their fulfillment for the Christian West in the New Testament.

Their incoherent stories and lightning wisdoms contained in them the Redemption. The prefigurations, the prophecies were archetypal until the Thing itself happened. Then no more waiting, no imperfection, no incompleteness. *CONSUMMATUS EST.*

We speak of a sense or a level of meaning in a myth, then. The word *level* is capable of misleading us, as are the frequent companion words *higher* or *deeper* meanings. This visual metaphor seems to say that there was one meaning and then others with which one could trick out this basic (real?) meaning. That is not how many Ancient, all Medieval, and most Renaissance men read. The meanings were instantaneously present. The image should be the concentric layers of a pearl (or an onion) — all are there at once to make up the onion or pearl. St. Augustine said in his *De Doctrina Christiana* that we poor fallen creatures must first look for the literal meaning and stop there if it suffices, and not watch for other meanings unless compelled to by the fact that we find something in these inspired words disharmonious with Truth or right conduct as enunciated elsewhere in the Holy Writing.[2]

Over the centuries the allegorical method of reading, writing, seeing, composing, speaking, and hearing collected a number of terms to describe its operation, and there was a shifting about in the usage of words that can lead to ambiguity in attempting to understand a statement about it.[11–13] The mode was called allegory (Latin *allegoria*), consisting of the Greek *allos* meaning "other" and *agoreuein* meaning "speak in the agora," that is openly, publicly, indicating that allegory was some discourse other than open and public in meaning.[12]

It was customary to call the open meaning the literal (*littera*), or when dealing with a narrative, historical because the early use of the method in the Christian world was in the interpretations of Scripture, especially the Old Testament, wherein historical events were seen to present later occurrences in veiled form. The open meaning of the letter and history was accompanied by less open, less immediately publicly available meanings.

Critics described these latter meanings as one multifaceted meaning, or divided the facets into two or three meanings or senses. These enrichments were termed allegorical. They were also termed figurative, especially when an Old Testament passage involved a prefiguration, that is, the presentation of an historical person who stood for a later figure, as

Isaac stood for Jesus the man. In this denial of time, an event could prefigure a later event: the substitution of the ram for Isaac who was likened to the Lamb of God who was substituted for guilty Man in the redemption of Man from his fall in Eden. In much the same way, the fulfillment of an Old Testament prophecy by an act in the New was a species of figural completion; when Isaiah said, "Behold, a virgin shall conceive," the figure was fulfilled when the Virgin Mary conceived Jesus. Sometimes the prefiguration is called an archetype, so that the enriched meaning is called archetypal, to be perfected with the appearance in time of the type (Latin *typus*, "figure, image, form"; Greek *typos*, "impression, mold") and the typological or archetypological meaning is finished.

The meanings beside the literal or historical one were called other names besides allegorical, figurative, archetypal, or typological. They were also called spiritual, as being beyond the ordinary meaning that would be available in the agora and its involvement in time and the present. They were also called mystical, as if they were steps toward the divine meaning. Sometimes critics use the word *metaphorical* to describe these "hidden" meanings. When these meanings were analyzed into parts, a further terminology appeared. Various interpreters divided these meanings in various ways, but here we shall look at a widespread and traditional division into three parts, since the account has thus far been a melange of notions from various authors writing at various times and has not been presented as a correct chronology. Here the word *level* is avoided because, as has been said, it can lead to a visual image of higher and lower strata when literal and allegorical meanings are simultaneous and not separable in texts and images. The word *sense* when used is not meant to carry a vectorial suggestion but only the idea of having a meaning. While some of our discussion thus far has made reference to Christian interpretations, we must not suppose that the allegorical method was primarily an exegetical tool for the Old and New Testaments only. The polysemous mode was used not merely in hermenuetics but in other applications as well. In fact, it is a not uncommon way of composition and interpretation at the present moment.

It might be easiest to start with a little rhyme made by Nicholas of Lyra (ca. 1270–1340) as a mnemonic for hermenuetical study:

> Littera gesta docet,
> Quid credas allegoria,

Moralis quid agas,
Quo tendas anagogia.

The letter teaches history,
What you believe is allegory,
The moral is what you do (your conduct),
Whither you reach toward is anagogy.

The first-mentioned sense beyond the historical is called allegorical (the same name given to the whole approach or to all the senses), or as Nicholas suggests, the doctrinal, that is dealing with that to which one gives intellectual and/or emotional assent. In terms of the Church, this would be the creed and similar teachings of the Church out of its *magisterium.* As the word suggests, the term *doctrine* can apply to any teaching the composer intends and/or to the Horatian dictum that poetry should *teach* as well as delight. But this sense can also be called figurative (also a term for all the senses besides the literal), which for the Christian would mean that the prefigure of the literal is answered by the figure of the Church or of Christ, so that it becomes ecclesiastical or christological. Also, this sense is called typical or typological, which seems to mean the same thing as figurative, except that its use is restricted to one person, object or incident in the literal history.

The third sense (third as customarily listed) is the moral, which is that which suggests the right conduct of life. It is the soteriological meaning that directs the Christian soul as to its acts in its time in this world — the alignment of the mind and will to the will of God. To parallel this in a secular work, such a sense would refer man to actions in the way of justice, courage, or whatever virtue would be thought appropriate by the maker or the interpreter. This sense is also called the tropological. Trope (Latin *tropus* from Greek *tropos*, "a turning") apparently refers to a turn of the sense in that it associates the doctrinal with its moral application to the individual life.

In Patristic, Mediaeval, and Renaissance exegesis, the fourth sense is the anagogical, also called the mystical or the eternal. This meaning calls attention to the parallel in the eternal to the temporal situation of the individual in the moral sense. It frequently puts the meaning in an eschatological interpretation, which would find reference to the end for which things exist in God's plan. Thus the literal history of Abraham and Isaac is a typological prefiguration of the execution of Jesus for man on

Calvary. In the moral sense we are instructed to have the same firm faith in God's promises that Abraham had that the Promise would be fulfilled even if his son were killed. And finally, in God's eternal unfolding of man's destiny, we recognize our redemptions from death by the sacrifice of Jesus, the Lamb of God.

Such an exegetical mechanism as allegory encourages the exposure of polysemous meanings, but, as the confusing nomenclature that developed symptomatically indicates, also often uncovers more meanings than the four. In reading the work of allegory or of allegorical criticism, especially of earlier times, one sometimes feels that there has been an extraordinary striving to reach the four, or that there has been a squeezing down to limit the meanings to four, though more meanings are present. In more recent allegorical interpretations, the fourfold old method is not so richly endowed, perhaps because many modern men do not look into the Scriptures for guidance, and lacking a cosmic system other than that provided by nature, they are unable to hear harmonies available to the earlier Christian. The critic sometimes observes a multiplicity of meanings but does not regard four as a guide. Also he secularizes what once were the overtly religious meanings. For him the multiple meanings from a single psychic event that are available to the human mind can be added to by such senses as the philosophical meaning, the political meaning, the social meaning; these senses are used especially in composing or witnessing secular works which involve no cosmic value system such as the Christian universe. A Marxist allegory might see in history a gradual evolution toward an ideal society; thus, an historical event, such as the rise of the merchants, tradesmen, and artisans in the Middle Ages, is seen as a necessary step prophetic of the demise of the feudal aristocracy, and this would be seen as good. The allegorical mode often requires an eschatology that looks into the future; it supports the existence of an order of good and bad, desirable and undesirable, and that teleologically speaking, things and events have a final end toward which they willy-nilly move.

4. Beowulf in Outline

It remains to be shown how allegory works in practice. It was originally my purpose to apply it to several literary, artistic, and liturgical works from several epochs, but this ambitious project must be replaced by a more careful analysis of a single literary work. In my experience as a critic, I have noticed that the number of senses in a work of this nature seems contrived to exceed the conventional four meanings, often carrying the number, when toted up, to five simultaneous meanings, if not more. I have noted this in Christian liturgy, in the architecture of the mediaeval churches, in Dante and other conscious allegorists, in plays of the Middle Ages and the Renaissance, in some fiction of the Enlightenment, and in the work of the great Nineteenth-Century American novelists and short-story writers from which source it has spread widely in European story through such authors as James, Kafka, and Joyce. The method has taken a similar course in painting, architecture, and similar arts. But a fuller study of this peculiar European mode must await a book. In this essay, the reader will be given one somewhat full example of a complex story in the hope that this will establish initial credit for the probable success of a wider application of the allegorical mode with a frequency of five as often as any other number of senses. First I shall recount the plot of *Beowulf* in order to familiarize the reader quickly with the main events and a few details that aid my explanations. Then I shall show how a set of five meanings emerge from the work and relate these to the structure of the allegorical method. *Beowulf* is not ordinarily opened out in this way, probably because literary scholars are most interested in examining the language, the historical background, the archeological implications, the sources of the contents, and other very important investigations.[17–21]

Beowulf is an Old English (Anglo-Saxon) epic poem composed in the Eighth (some say earlier or later) Century by a Christian writer in the tradition of Germanic court singers. He looks back at the heroic times of the Fifth and Sixth Centuries using historic persons and events as well as legendary figures to relate the life of a monster-killer who defends the people with his courage, adroitness, and great strength. Beowulf and his fifteen retainers (a loyal group called a *comitatus* by Tacitus) travel overseas to Denmark to the famous, splendid hall of Heorot ruled by Hrothgar, King of the Scyldings, because he has heard that the hall has been under nightly attack for months by a troll monster named Grendel, who devours men. Though the effort of his *comitatus* are vigorous and loyal, they are futile, but Beowulf kills Grendel and later kills his mother when she exacts revenge. In doing so, he relied on the grace of God (ll. 669–70, 685–87), for he was also in these battles as a defender of the King of glory (665–66). Beowulf acknowledges that he could not have triumphed over Grendel without God's willing it (966–69), nor could he have defeated his dame (1657–58), as the poet in his own voice also acknowledges:

> But the Lord granted
> Sea-Geats success, gave them solace, help
> fortunate fate, so that foes of theirs
> many conquered through the might of one
> by his own power. Almighty God,
> to tell the truth, controlled ever
> the days of man.
>
> (696–702)

Grendel, as a descendant of Cain (104–14), bears God's wrath (711), is God's foe (786, 811); he is called a fiend (104, 636, 726, 737, 970), a hellfiend (731), hell's captive (787), a demon (707). His lair is called a devil's den (756). At his death, "Hell received him" (852), where he will be "controlled by fiends" (808). Beowulf and his men return to the land of the Geats whom he serves faithfully; after the death of the king and the king's son, Beowulf becomes king and rules ably for fifty years. Then a precious cup is stolen by a slave from an ancient, unknown treasure hoard; in hot anger, its guardian firedrake ravages the land with his blazing breath, killing people. He burns down the hall of the Geats. The aged Beowulf is distressed:

> The wise man wondered if the World Ruler,
> the eternal Lord, were truly angry
> at a broken law. His breast in tumult
> from clouded forebodings, an unaccustomed mood.
>
> (2329–32)

Was God displeased at the theft of the flagon? Nevertheless Beowulf decides it is his duty to fight the dragon. He takes with him eleven chosen warriors for support, and they set out for the remote cave, with the dastardly slave as guide to the spot. In the encounter Beouwulf is being overwhelmed but only one retainer comes to his aid — his kinsman Wiglaf, last of the Waymundings. Together they kill the dragon, but Beowulf has been fatally bitten. The dragon is never associated with Hell or fiendishness. In the battle, neither Beowulf nor the narrator refer to God, except that before engaging in combat, Beowulf says dubiously,

> but how the Guide of men favors our fate, we must find at the wall.
>
> (2525–26)

Dying, he asks Wiglaf to bring some of the ancient treasure for him to see. In his death speech, Beowulf gives thank to God that he has seen these ancient riches that he has won for his people, but at the barter of his life (2794–99). Wiglaf denounces the cowardly retainers for violating the warrior code which requires that the comitatus support their leader in battle even unto death.[23] A messenger tells the main body of warriors waiting in the distance of the outcome of the fight. He warns them to prepare for battle soon against their old enemies, the Franks and Frisians, previously held off by Beowulf's kingship. Also the Swedes will be foes, looking for revenge for battles. He prophesizes the fate of the Geats:

> but treasure plundered,
> aching-hearted, in alien lands
> they must wander wide, now the wise leader
> laughs no longer, leaves sport and joy.
> Therefore must spearpoints spin, speed cold at dawn,
> hands hurling them; not one harp will sound
> to waken warriors; as the wan raven
> dines on the doomed, with dinning gabble
> telling the eagle of his tempting meal,
> while with the ravening wolf he rends the corpses.
>
> (3018–3027)

Wiglaf supervises the preparation of the funeral pyre and mound under which all the treasure hoard will be buried. As the pyre burns, a woman raises a keen, fearing "that ravage and ruin, arrest and abasement" await the Geats.

5. Beowulf as Allegorical

In this story, Beowulf is at first a great victorious hero, later a good king who is defeated in battle. But his death leaves his kinsman Wiglaf as the sole survivor of the Waymundings, so that presumably the tribe will be extinguished at Wiglaf's death. The Geats also seem bound for defeat and dispersal as a people.

The death of Beowulf meant the end of his tribe and the end of the people over whom he had been king. A more universal meaning was the failure of the picked warriors to support their chief in his hour of need (in contrast to the actions of the retainers in the fights with Grendel and Grendel's dame), implying a dissolving of the traditional war code and an end to the whole heroic age. The poem thus is an elegy of loss — the death of Beowulf, the extinction of his line and the end of the Geats, and the departing of the old warrior values and the society which was supported by them. Such a gloomy conclusion would remind the poet's possibly half pagan, half Christian audience of the end of things predicted in the old Nordic religion — Ragnarok, the great cosmic battle between the evil giants and the good gods and men in which the good would go down to defeat and evil emerge victorious. So that the poet, albeit he is a Christian, sings out of the old teleology of the Germanic peoples. On a cosmic level, he composes an elegy of the defeat of what is noble in man and of the end of the world that had sustained their ancestors from time immemorial.

Thus we see that the poem has four meanings: the death of Beowulf, the end of the Waymundings and of the Geats, the end of the old warrior ethos, and the analogue of Ragnarok. All are united in a lament over unrecoverable loss and tragic waste.

However, a fifth meaning also emerges. A salient fact is the identification of Grendel (and hence his mother) with Hell and diabolical evil and his descent from Cain. Cain, the cursed first murderer, is identified by St. Augustine[1] as the first founder under Satan of the City of Man and the beginner of the opposition of that city to the City of God. He who opposes the City of Man is on the side of the City of God. In his troll fights, Beowulf is clearly a citizen of the Heavenly City or community. God by the operation of His grace gives victory to Beowulf, of which Beowulf is well aware. To the Christian, Beowulf is a pilgrim or exile in this world on his journey toward the Celestial City. The support (in the opposition of the Divinely supported hero against the devil-fiend) of this meaning for the troll battles in the cleansing of Heorot from evil is continuously reasserted in the poem so as to make unmistakable this significance. In the fight with the firedrake, Beowulf is not sure whether or not the dragon's attacks are permitted by the will of God because of the theft of and the failure to return the stolen cup. He decides to follow the heroic code: injury must be personally avenged,[23] and since a chieftain is leader of his folk, he must lead them when vengeance is necessary. The Divine sanction of the first part of the poem disappears. There is no cosmic conflict between good and evil. Beowulf retains some

Christ-like pretensions in that he is one of thirteen (counting one caitiff slave) who go to seek to fight the dragon. The dragon could easily be identified with the Old Enemy, the snake of the Garden of Eden, but the poet refrains from this meaning. Beowulf, in destroying the dragon, who was after all following the code in avenging the dishonor felt by it in the theft of the cup, is destroyed in following that same code. God has not supported him in this kind of a conflict, and though Beowulf the good thanks his heavenly Chieftain in his dying words for the barter of his life (like Christ) for his people, his adherence to the heroic code rather than a sure following of God's will has ended in his demise. The elegaic intent here seems to be that the Christian God's will is to replace the stern vendetta of the heroic age. Though Beowulf doubtless reaches the City of God on his passing into eternity, he has also become an emblem of the passing of the old heathen way in favor of the mightier new way of Jesus. Even so should the Christian direct himself to following what God wills, rather than follow the warrior code and heathen fatalism.

The allegorical sense is something like this: literally and "historically" Beowulf is the great Germanic hero who, after initial victories over trolls, dies in killing a dragon. Doctrinally, we learn that even such a hero or his people cannot survive unless God's favor assists them. Morally, man must act only in conformity to the will of God if he is to be victorious in overcoming evil. Anagogically, God's will is not on the side of straightforward vengeance against the wrong-doer. Beyond these four senses is another, which one could term eternal or eschatological, except that these terms would ordinarily belong to the anagogical sense: In the great spread of time till Doomsday, the old warrior code with its loyalties and honor and blood guilt was to be defeated by the new era of grace with its virtues, just as the end of things in the defeat of the gods—shown as now empty and without hope—was to be replaced by the new dispensation of the allwielding God.

6. Fivefoldedness

The complexity of such an explanation makes one wish that literary exigesis were as straightforward and linear as much of our analysis of other forms of knowledge is. This difficulty draws on the testy attitude of some critics who favor a kind of fundamentalism in their interpretations.

But art is exceedingly complex. A simple reading is usually a slashing simplification, that is, a falsification of what is there.

What this examination of Beowulf has tried to show is that the meaning of the poem is fivefold. As was said before, this result comes up in analysis of several other art works over all of Western culture. I do not propose for an instant that five is the magic number of meanings that an art object necessarily has any more than I would accept four as that number. Some objects have more or less meanings than these numbers. There is nothing inherent in man-made objects that generates any given number of meanings. All depends on what the maker puts into his work and on what the person witnessing the work brings with him. Nevertheless, it seems that fivefoldedness recurs often enough to suggest that much art work has a receptivity to accepting such a large number of meanings.

References

1. St Augustine, *Concerning the City of God Against the Pagans.* Translated by Henry Bettenson, Harmondsworth (1984).
2. St Augustine, *On Christian Doctrine.* Translated by D.W. Robertson Jr., Indianapolis, (1958).
3. Bede, 'De Schematibus et Tropis.' Translated by Gussie Hecht Tenenhaus. *Quarterly Journal of Speech.* XLVIII No. 3 (October 1962) pp. 237–53.
4. R.M. Grant, *A Short History of the Interpretation of the Bible.* New York (1963).
5. G.B. Phelan, *Saint Thomas and Analogy.* Milwaukee (1941).
6. Beryl Smalley, *The Study of the Bible in the Middle Ages.* Notre Dame, Indiana (1964).
7. J.S. Preus, *From Shadow to Promise: Old Testament Interpretation from Augustine to the Young Luther.* Cambridge, Mass (1969).
8. Steven Ozment. *The Age of Reform 1250–1550: An Intellectual and Religious History of Late Mediaeval and Reformation Europe.* New Haven (1980).
9. Plato, Book VII of *The Republic*, conveniently in W.H.D. Rouse, ed., *Great Dialogues of Plato.* New York (1956).
10. S. Kierkegaard, *Fear and Trembling and the Sickness unto Death.* Translated by Walter Lowrie. Princeton (1974).
11. W.F. Lynch, *Christ and Apollo: The Dimensions of the Literary Imagination.* New York (1963).
12. A. Fletcher, *Allegory: The Theory of a Symbolic Mode.* Ithaca (1964).
13. J. MacQueen, *Allegory.* London (1970).
14. C.S. Singleton, ed., *Interpretation: Theory and Practice.* Baltimore (1969).
15. M.W. Bloomfield, ed., *Allegory, Myth and Symbol.* Cambridge, Mass. (1981).
16. S.J. Greenblatt, ed., *Allegory and Representation: Selected Papers from the English Institute 1979–80.* Baltimore (1981).

17. Fr. Klaeber, ed., *Beowulf and the Fight at Finnsburg*. 3rd Edn. Boston (1950).
18. C.L. Wrenn, ed., *Beowulf with the Finnesburg Fragment*. Boston (1953).
19. R.W. Chambers, *Beowulf: An Introduction to the Study of the Poem with a Discussion of the Stories of Offa and Finn*. Supplement by C.L. Wrenn. 3rd Edn. Cambridge (1963).
20. G.N. Garmonsway et al., *Beowulf and its Analogues*. London (1969).
21. J.F. Tuso, ed., *Beowulf: The Donaldson Translation, Backgrounds and Sources, Criticism*. New York (1975).
22. *Beowulf*. Translated by Ruth P.M. Lehmann. Austin (1988). I use this translation throughout.
23. C. Tacitus, Germania, in *The Complete Works*, translated by A.E. Church and W.J. Brodribb, edited by Moses Hadas. New York (1942).

CERTAIN QUINARY ASPECTS OF THE HINDU CIVILIZATION

Akhlesh Lakhtakia

Fivefold, verily, is this whole world. With the [knowledge of the] fivefold, indeed, one wins the fivefold.

Taittreya Upanishad 1.7

1. Introduction

The number five does not seem to have been appreciated by most cultures in terms of enumeration; the only known exception is a South American Arawakan language named Saraveca in which counting is exclusively with base five. Some other Central American cultures do give some significance to 5 by naming the numbers 6 through 9 as 5 + 1, 5 + 2, etc.[1] Like most Indo-European civilizations, the Hindus had become firmly decimal long before they invented the symbol for zero and gave the world the positional decimal system we now commonly employ.[2] The number five by itself and to the exclusion of all other small integers,[a] was of little symbolic consequence for the Hindus, particularly in a civilization that conceived of time periods as large as 4.29×10^9 years,[3] lengths as small as 1.3×10^{-7} inch (the diameter of a molecule), and numbers as large as $(8{,}400{,}000)^{28}$.[2]

This is not a work in the manner of Martin Gardner's Dr. Matrix. For instance, the *Rig Veda*, the first religious book of the Hindus, contains 1,028 hymns with a total of 10,562 lines, which are distributed over 10 books. No attempt will be made to find the significance of the number five in the following fashion: the digital sum of 10,562 is 5, the number

[a] For an interesting discussion of the significance of the numbers three and seven in the *Rig Veda*, the reader is referred to E. W. Hopkins, "The holy numbers of the Rig-Veda," in *Oriental Studies* (Oriental Club of Pennsylvania). Ginn, Boston, 1984.

of books is 2 times 5, and so on. Nor will any attempt be made to enumerate all occurrences of the number five or groups of five in Hindu religious and secular literature, art, society, etc.; given the antiquity of Hinduism, that task could well occupy several lifetimes. In particular, the occurrences of the number five have been taken from religious–philosophical and religious–historical works, so that examples, if any, from purely secular and scientific literature may not be found here. Instead, the thrust of this work is to acquaint the reader with some of the important facets of the Hindu cultures that come in groups of five. However, a comprehensive discussion of the facets themselves is not envisaged; nor is any special significance, in the overall framework of the Hindu civilization, of the number five deduced. Furthermore, by no means does this essay imply that the number five dominates, or has ever dominated, all other small numbers in the Hindu civilization.

Introducing a novice to Hinduism is a task of no small magnitude. For this reason, some familiarity with Hinduism has been assumed here, the focus being on the number five; the level of familiarity necessary is the one obtainable from popular books on the major religions of the world (e.g., Ref. 4). It is also to be noted that Jainism and Buddhism, which are heterodox offshoots of Hinduism, have not been considered here. The Sanskrit and Tamil words mentioned here have not been spelled using diacritical marks, given the obscurity of such representation in other than philological and philosophical circles. Finally, many of the situations described may no longer be found today, although the Hindu civilization has been extremely continuous in time; this feature is responsible for the mixture of (grammatical) tenses in the sequel!

2. Salient Features of Hinduism

Hinduism, as is now quite well known in the West, is not a cult, a sect, or a church; it is a conglomerate of very diverse philosophies, doctrines, dogmas, and ways of life. It integrates a vast number of heterogeneous elements of a largely continuous whole, and embraces all aspects of the lives of its adherents. In principle, it incorporates all manner of discourses on the human experience, ranging from polytheism to monotheism to atheism to an absolute monism. What is divine is to be revered in all its manifestations, the form and the nature of reverence dependent on the capacity of the individual. Simple persons believe in what may be almost a folk religion, having faith in one or more local gods, in many instances of primitive aspects. More advanced individuals may revere a Trinity — the Creator, the Preserver, and the Destroyer. The cult of the Creator (*Brahma*) was never prominent and is virtually extinct nowadays, while most urban Hindus set store by either the Preserver (*Vishnu*) or the Destroyer (*Shiva*) without actually disparaging the other two elements of the Trinity. A Hindu of a philosophical bent of mind will believe in a single Unity — the Unity (*Brahman*)[b] that is All, — or may choose to believe in the absence of any divinity at all, celebrating merely the relationship between the sentient creature (*purusha*) and the inanimate nature (*prakriti*). Even so, few ideas in Hinduism are considered to be finally irreconciliable in this mosaic: whatever be the beliefs of a particular individual, the beliefs of other individuals are equally important parts of his or her religion. It is not difficult to find a philosophical Hindu observing the rituals typical of comparatively primitive elements because of domestic and social considerations, while a simpler minded adherent respects the more intellectual outlook of another Hindu.

Hinduism is a complex faith. It is also the oldest extant faith. And it is extremely syncretic and all encompassing, being, as stated above, more a way of life than a mere church or a theocracy. Any student of the religion (I will use this word, though it is very imprecise for Hinduism[5]) must learn to emulate taxonomists and be prepared to count and classify. Thus, the proto-religion of the Indo-Aryans consisted of a pantheon of 32 gods and one goddess, all of whom personified natural phenomena such as air, water, and fire. As a result of assimilating the prevalent religions of the pre-Aryan Indians and subsequent scholarly analyses, the pantheon eventually expanded to 330,000,000 elements of varying prowess and sophistication.[c] The chief members, however, are the constituents of the Trinity mentioned above. Later rationalization explained these myriads of elements to be the multitudinous manifestations of the one God. Almost simultaneously, vigorous enquiries were initiated into the relationship of Man with Nature, the concept of a God, the nature of Matter; all such enquiries required enumeration of various parameters and gave

[b] *Brahman* is the Spirit, the ultimate entity of the absolute monism; *Brahma* is the Creator in the Trinity; and *Brahmins* are members of the priestly caste.

[c] To this day new deities keep on being created. An example is the deity *Satya Narayaana*, who has no foundations in the ancient scriptures and was not known at all till about two centuries ago, but gained a considerable following after the 1940s; see Ref.39.

rise to six systems of philosophy, which are the same time secular and religious. Finally, what has come to be the essence of Hindu philosophy is this: know thy Self, for thou art All. Prodigious feats of analysis and synthesis were conducted during the evolution of the Hindu philosophical traditions, which must be differentiated from the ritualism of the common practices.

3. The Nature of *Brahman*

What is the shape of God, the Spirit,[d] the ultimate Unity, the *Brahman*? A quote from the *Brihadaranyaka Upanishad*,[6] which records famous debates held in the forest (*aranya*), suggests that the definition of God must be given in the negative, but with the positive connotation:

> They describe Spirit as "Not this; not that."
> The first means: "There is nothing except Spirit;" the second means: "There is nothing beyond Spirit."

Thus the Spirit must be described as enabling, yet transcending, all senses. This has given rise to the following fivefold definition of the Spirit in the *Kena Upanishad*[6]:

> That which makes the tongue speak, but needs no tongue to explain, that alone is Spirit; . . .
> That which makes the mind think, but needs no mind to think, that alone is Spirit; . . .
> That which makes the eye see, but need no eye to see, that alone is Spirit; . . .
> That which makes the ear hear, but needs no ear to hear, that alone is Spirit;. . .
> That which makes the life live, but needs no life to live, that alone is Spirit; . . .

But the acme of the Hindu philosophy is the realization that the Spirit is the Self; for instance, the act of creation is thus described in the *Aiteraya Upanishad*[6]:

> There was in the beginning one sole Self: no eye winked. He thought: "Shall I create territories?"

Ergo, the attainment of wisdom is the realization that each territory is the Self, and nothing but the Self; indeed, a sublime passage in the *Chhandogya Upanishad*[6] concludes with:

> A wise man, leaving his body, . . . is one with His own nature. That nature is Self, fearless immortal Spirit.

[d] This Spirit should not be confused with the Holy Spirit of Christianity.

In the present context, a definition of the Self is also fivefold, as given in the same *Upanishad*:

> Who sees through the eye, knowing that He sees, is Self, the eye an instrument whereby He sees; who smells through the nose, knowing that He smells, is Self, the nose an instrument whereby He smells; who speaks through the tongue, knowing that He speaks, is Self, the tongue an instrument whereby He speaks; who hears through the ear, knowing that He hears, is Self, the ear an instrument whereby He hears; who thinks through the mind, knowing that He thinks, is Self, the mind an instrument whereby He thinks.

Furthermore,

> He who knows this . . . remains one, though multiplied threefold, fivefold, sevenfold, elevenfold. . .

The definitions of the Self and the Spirit given above use five attributes, but not the same ones. In this connection, the *Mandookya Upanishad*[6] instructs us that the Self enjoys coarse matter using 19 different agencies, which are classified as: five organs of sense (for hearing, touching, seeing, tasting, and smelling); five organs of action (for speaking, handling, walking, generating, and excreting); five vital airs (*praana*, *apaana*, *wyaana*, *udaana*, and *samaana*); and four mental states (discursive mind, discriminating mind, mind-material, and personality). From these 19, five have been taken to define the Self and five to define the Spirit. Consider the exposition of the Spirit in the *Bhagavad Gita*[7] as well:

> Those know Me *Brahma*; know Me Soul of Souls,
> The *Adhyatman*; know Karma, my work;
> Know I am *Adhibuta*, Lord of Life,
> And *Adhidaiva*, Lord of all the Gods,
> And *Adhiyajna*, Lord of Sacrifice;
> Worship Me well, . . .

where too five appellations of the Spirit are made known.

Why five and not some other number? The use of five qualities or attributes becomes quite clear on examining the etymological evolution of the Sanskrit work *pankti*; it originally meant a "set of five." But by the tenth and the last book of the *Rig Veda*, the first sacred work of the Hindus, it came to denote a "series" generally.[8] That is, it was deemed sufficient to delineate a series by five terms, just as mathematicians commonly denote a series by its first few terms.

The relationship of the worldly being to the Self has also been explored in the *Upanishads*. The *Taittreya Upanishad*, for example, describes the enlightenment of the ego through a five-stage process[9]: first, the being is *annamaya* (immersed in food), which corresponds to the corporeal body; then, it becomes *praanamaya*, or endowed with the vital airs; then, it becomes *manomaya* and acts consciously; next, it becomes *vigyaanamaya*, endowed with knowledge; and finally, it becomes *aanandamaya* (joyous), upon realizing that it is the Self. These five body sheathes (*panchakosha*) have formed part of the subsequent schools of thought, and particularly of the post-Vedic Shaivism.[10]

4. *Eeshvara* and Shaivism

The *Brahman* (or the Spirit) is *nirguna*; that is, it is devoid of form and incorporates all attributes, good or bad.[e] Hence, it cannot be revered; it can only be understood. The concept of *saguna* God follows: since this conception has a form and personifies the good (*sattva*) it can be installed in temples and revered. Therefore the *saguna* God is also called *Eeshvara*, the personal God.

There are three great traditions of *saguna* God, all being non-Aryan in origin.[11] Salvation, defined by the Hindus as the acquisition of the true knowledge, can come from worshipping the Mother Goddess (*Devi*). The worship of *Devi*, in her manifold forms, can be traced to the Indus Valley civilizations as well as to the Mediterranean fertility cults. Another important current in the following of *Devi* comes from indigenous village cults. The early Indo-Aryans were positively hostile to Mother Goddess, and it took a long time for female deities to come to be ranked as the equals of the male gods of Hinduism. In the *Shaakta* tradition, *Devi* is the Energy (*shakti*) that fuels the Universe, and salvation comes when the *Devi* finally reveals her true identity to the devotee.

Salvation can also come from worshipping *Vishnu* or *Shiva*. *Vishnu* is partly the solar god of the early Aryans and partly an early Dravidian sky-god (Tamil *vin* = sky) whose color was blue. Indeed some modern commentators (see Ref. 11) think of Vaishnavism, the following of Vishnu, as the accumulation of a number of non-Aryan deities conjoined

[e] It is significant that the Hindu deities (*devas*) never seek to eliminate the demons (*asuras*); they merely rout them from time to time, and the demons regain their strength after each encounter with the deities.

with some proto-Aryan mythology. In fact, *Vishnu* can claim but five whole hymns among the 1,028 of the *Rig Veda.*[9] Modern Vaishnavites, however, insist that their religion is Vedic and hold that the *nirguna Brahman* manifests itself in the *saguna* form as *Vishnu.*

The number five is of special significance for the Shaivites, the followers of the cult of *Shiva*, the Destroyer. The Shaivite sects within the Hindu fold developed between the second century B.C. and the second century A.D. It appears that the Vedic apotropäic destroyer-god *Rudra* was amalgamated with a Dravidian god *Shiva* of destruction/creation during this period, after a somewhat acrimonious struggle; indeed, during these times some Shaivites were denounced as non-Vedic (*avaidik*), and the period literature attests to the discord between the Shaivite and the Vedic pantheon.[12] Eventually, the differences were reconciled, and in the true spirit of Hinduism, a compromise seems to have been reached around the second century A.D.: both *Shiva* and the Vedic god *Rudra* were identified as *Agni*, the proto-Aryan fire god; in addition, temples dedicated to *Vishnu* began to have icons of *Shiva* as well, and vice versa. To this day, the Shaivite sects are particularly strong in south India, though Shaivism finds adherents elsewhere too, notably in Kashmir.

There are several schools of Shaivite thought, which range from pluralistic realism to absolute monism. All Shaivite schools postulate three distinct entities — God (*Pati*), Soul (*Pashu*), and Nature (*Mal*) — all three coexistent and eternal, with the Soul and the Nature energized by God identified as *Shiva.* The iconography of these sects imparts *Shiva* with five faces, each face representing an active energy (*kriya shakti*). These five energies are: (a) *srishti*, which creates the world; (b) *sthiti*, which preserves the world; (c) *sanhaara*, which destroys the world; (d) *tirobhaava*, which conceals the true characteristics of the world from the Soul; and (e) *anugraha*, which reveals the truth of the world to the Soul. Thus, the Soul cannot know Itself without the intercession of *Shiva.* It should be noted here that in this scheme of things, *Shiva* as God has preempted the roles of the other two members of the Hindu Trinity.

The very philosophic *Shaiva-Siddhaanta*,[13] the most developed form of Shaivism, was propounded in Tamil Nadu during the thirteenth and fourteenth centuries A.D. It is a highly logical system in which each contention is either proved or disproved through the usual Hindu fivefold process of reasoning: proposition, reason, example, application, and conclusion. The number five also figures in the *pancha-akshara* (five letters) dear to the Shaivites: *shi*, *va*, *ya*, *na*, and *ma*: respectively, these letters signify God, *shakti*, soul, *tirobhaava*, and nature.

The iconography of *Shiva* is as multifaceted as is his following. In abstract form, *Shiva* is represented by the *lingam* (phallus), in coitus with the *yoni* (vagina), the symbolism expressing reverence for the origin of life. In many instances, the top portion of the *lingam* shows a five-headed *Sadaashiva* figure.[11] Anthropomorpic images of Shiva and his various incarnations abound[12]: of these, the *Sadaashiva* images have five faces and ten arms, while the *Mahaasadaashiva* images have 25 faces, possibly corresponding to the 25 principles of the nontheistic *Saankhya* philosophy. Arguably, the most famous anthropomorphic icon of *Shiva* is the *Nataraaja* (master of dancing). The dance *aanandam taandavam* of *Nataraaja* has a special relevance to the five activities (*panchakritya*); to cite from *Tirukuttu Darshana*, a work of the Tamil commentator Tirumular,[14]

> His fivefold dances are in *sakala* (corporeal) and *nisakala* (non-corporeal) form
> His fivefold dances are His *panchakritya*:
> With His grace He performs the five acts.

The Tamil text *Unmai ulakham* thus explains the symbolism[11]:

> Creation arises from the drum [held in one hand]; protection from the hand of hope; from [the encircling ring of] fire proceeds destruction; from the foot planted upon [the demon] Mooyalahan proceeds the destruction of evil; the foot held aloft gives salvation. . . .

while the *tiruvasi* [*tiras* = *awry, vasi* = apparel] round *Nataraaja* symbolizes the act of obscuration (*tirobhaava*). *Shiva's* dance, thus, is a continual saga of creation and destruction involving the universe. As Capra[15] puts it, it is the "dance of subatomic matter;" to wit, the creation and the annihilation operators of quantum mechanics.

5. Logic and Atomism

Shaiva-Siddhaanta was developed quite specifically to counter the six systems of Hindu philosophies in general, and the *Sannkhya*[16] in particular. The word *sankhya* means *number* or *statistic* in Sanskrit, and this philosophy is also based on the number five, or rather on the square of five; finer ontological analyses lead to other schemes of enumeration, but 25 is the most commonly accepted number of elements.[17] Although it too has a non-Vedic origin,[18] in contradistinction to *Shaiva-Siddhaanta*, *Saankhya* proposes a consistent dualism of nature (*prakriti*) and soul (*purusha*): nature and soul suffice to explain the universe without the need and soul to hypothesize God.

Nature and soul are essentially separate entities in *Saankhya*, but the soul mistakenly identifies itself with some aspects of nature; right knowledge consists, therefore, of the ability to resolve the correct identity of the soul by the soul. This misidentification comes about because the soul obtains an ego-consciousness (*ahankaara*) upon being encircled by nature. The treatise *Saankhya-pravachana-sootra*, dating from the fourteenth century A.D., states that the universe is the result of various arrangements of 25 principles, which are thus stated[5]:

> *Prakriti* is the state of equilibrium of *sattva*, *rajas*, and *tamas*. From nature evolves *mahat* (intellect); from *mahat*, *ahankaara* (I-consciousness); from *ahankaara*, the five *tanmaatras* (subtle elements) and the two sets of *indriyas* (senses or instruments); from the five *tanmaatras*, the gross elements. Then there is the self. Such is the group of the twenty-five principles.

The five gross elements in this scheme are space, air, water, fire, earth; the five subtle elements are sound, touch, sight, taste, smell; the five organs of perception (*gyaana-indriyas*) are with which to hear, touch, see, taste, smell; and the five organs of activity (*karma-indriyas*) are with which to speak, grasp, move, procreate, evacuate. *Saankhya's* doctrine of the three attributes of *prakriti*, which cause goodness (*sattva*), passion (*rajas*), and lethargy (*tamas*) in beings, is widely echoed in many other schools of Hindu thought.

To a great extent, *Saankhya* is the distillate of, and represents advancement on, two other schools: *Nyaaya* and *Vaishesika*. Both these schools are highly analytical. The word *nyaaya* means that by which the mind is led to a conclusion; since reasoning is required to conclude, the special feature of *Nyaaya* is the development of the canons of logical proof (*prameya*). *Vaishesika*, on the other hand, concerns itself with the particularity (*vishesa*) of the entities being examined; it is, therefore, a system of physics and metaphysics, and it is especially known in the West for its theory of atomism. Neither system concerns itself with any divinity, except to state that a prime mover is required to explain the existence of the universe because neither atoms nor the casual postulation of *karma* (activity) suffice for that purpose.

The earlier versions of *Nyaaya* admitted 11 divisions of reasoning (*tarka*), but more recent versions admit only five[19]: (a) *reductio ad absurdum* (*pramaanabaadhithaarthaprasanga*), (b) *regressus ad infinitum* (*anavasthaa*), (c) *ignoratio elenchi* (*aatmaashraya*), (d) dilemma (*anyonyaashraya*), and (e) circular reasoning (*chakra*). The process of reasoning is syllogistic, the members (*avayava*) of a syllogism being five in number: the proposition (*pratigya*), the reason (*hetu*), the example (*udaaharana*), the application (*upanaya*), and the conclusion (*nigamana*). In earlier versions, the syllogisms had ten members, but the seer Gautama applied the principle of parsimony to pare them down. The classes of fallacious reasoning are also five: the discrepant (*savyabhichaara*) or the indeterminate (*anaikaantika*), the contrary (*viruddha*), the counter-balanced (*satpratipakshi*), the unreal (*asiddha*), and the contradicted (*baadhita*). It was, of course, understood that no debate between two logicians can take place without protocols for eliminating insincerity and frivolousness, as well as for agreement on basic definitions.

The early *Vaishesika* syncretism recognized that all things that can be known and named fall under six categories[19]: substance (*dravya*), quality (*guna*), activity (*karma*), generality (*saamanya*), particularity (*vishesa*), and inherence (*samavaaya*). More modern treatments add the seventh category of nonexistence (*abhaava*), which is possessed, for instance, by darkness. *Vaishesika* recognizes nine substances: the self (*purusha*) and the mind; time and space; ether; and the four atomic substances (earth, water, fire, and air). The orthodox list has 24 properties which a substance can possess.[5,19]

All substances possess the five general qualities: number, dimension, individuality, conjunction, and disjuction. Space and ether are to be differentiated by the fact that ether has sound while space is the distance between two noncongruent objects, a distinction of some importance to modern physics. The four atomic substances have the five general properties, as well as priority plus posteriority (i.e., causality); air also has temperature and velocity; fire has the qualities of air, plus color and fluidity; water has the qualities of taste, gravity, and viscidity, in addition to the fire's qualities; and earth has the same qualities as fire, but it also has the properties of smell, taste, and gravity. The nine substances are eternal, but the products of their aggregations are transient. In particular, dimension is primarily of two types: minute (*anutva*), as in the case of the atom, and large (*mahattva*), as the ether is.[19]

The *Vaishesika* atoms are of four kinds — earth, water, fire, and air — and at least three atoms of a kind are necessary for the formation of an aggregate that is perceptible. Atomicity here has been related to minuteness, and not to the "un-cut-ability" of Demomcritus; fractal geometers might have cause to rejoice in *Vaishesika*. The various properties are eternal in an atom, but transient in the aggregate, as mentioned above for the nine substances. Motion is defined as a continuum of conjunctions and disjunctions of the object from the place; it is necessarily evanescent, since only aggregates can have it, and it has been analyzed into five kinds: throwing up, throwing down, expansion, contradiction, and going (which covers any other form not covered by the first four). Time is defined as the knowledge of causality and simultaneity, while space is that which gives rise to the recognition of the distance between two (noncongruent) objects.[19] Ergo, both space and time are constructs that only the mind can perceive, a notion of some importance in the understanding of the theory of relativity.[15]

For a very detailed exposition of the various Hindu views of atomism, including the *pancha-bhautika* (*bhoota* = elementary matter type, quark?; *bhautikee* = physics) theory, reference is made to Gaur and Gupta[20] as well as to Seal.[21] The *pancha-bhootas* are variously described in different commentaries, but a full discussion is beyond the scope of this essay.

6. The Caste System

Aditi is the five kinds of all men who are.

Rig Veda 1:89;10

So says the first book of the *Rig Veda*, the first book of the Hindu civilization. These five kinds (*panchajana*) are the members of the four castes, plus the non-Aryans,[5] though there are other interpretations possible.[22] The caste system, the one hallmark of the Hindu culture known the world over, had very early beginnings. So pervasive has its influence in India been that even Muslims, who believe in the universal brotherhood of Islam, have a caste system in India.[23] At the apex were the *Brahmins*, the priestly class, who regulated the spiritual and temporal life of all Hindus. Next came the *Kshatriyas* whose function was to protect and rule. The *Vaishyas*, farmers and tradespeople, were followed by *Shoodras*, the menials. The fifth initially comprised of non-Aryan

Indian and other peoples who became subservient to the Aryans but were never fully absorbed by the Hindu culture; for instance, *Garuda*, the mighty humanoid avian carrier of *Vishnu*, was enjoined to devour the non-Aryan *Nishaada* people of the Chhota Nagpur plateau to satisfy its appetite.[24] Also included in this fifth set were the foreigners (*mlechhas*) with whom the Aryans may have had some contacts. This division of the Aryan society is idealistic, what the priests in all probability wished for. Mixed castes arose at rather an early date, otherwise the ancient Hindu lawgiver Manu would not have explicitly approved of only endogamous marriages.[3] Outcastes from the three upper castes also formed castes of their own, such as the *Kaayasthas* of Uttar Pradesh and the surrounding regions. Furthermore, the incorporation of the local non-Aryan populations gave rise to regional castes and subcastes.[25] It needs be added that the caste-to-profession correlation is weakening by the day; however, the caste system still retains almost its full force in personal and political life[26] and appears to have successfully adapted to the democratic polity of the modern Republic of India.

Theories about the origin of the Hindu caste system abound, and even the ancient lawgivers seldom agreed on the issues. The one aspect that the ancient seers were unanimous about is the *dharma* appropriate to each caste. *Dharma* is too nebulous a concept for accurate translation into a Western language; loosely stated, it is a code for conduct, in private as well as in public, for the individual to grow in stature toward the achievement of salvation, but with due respect for the sustenance of the community as a whole. We need not concern ourselves with the various facets of *dharma*: for the present purposes it will be sufficient to give the five types of what is not *dharma*, as enumerated in the *Shaanti Parvana* of the epic *Mahaabhaarata*.[27] These are: (a) *vi-dharma*, which contradicts one's own *dharma*; (b) *para-dharma*, which is the *dharma* of somebody else (unlike oneself); (c)*upa-dharma*, which opposes established morals and is hypocritical; (d)*chhala-dharma*, which deceitfully masquerades as *dharma*; and (e) *dharmaabhasa*, which is what one does to satisfy one's own will but is not *dharma*. There is, of course, *adharma*, which is totally and unambiguously immoral for everybody. Furthermore, from the same source, knowledge, bravery, alertness, strength, and courage have been identified as a man's five natural helpmates in living up to his *dharma*.

In addition to *varna-dharma*, conduct peculiar to the members of a caste for the sustenance of the community at large, there is the *aashrama-dharma* for the individual. A man passes through four stages in life, namely, childhood and youth, married life, as a forest-dweller, and finally, that of an ascetic who has renounced all material pleasures. *Aashrama-dharma* enables the man to successfully negotiate these four stages; similar but not identical *dharma* is prescribed for women too. In the present context, a householder was enjoined to perform five great sacrifices (*pancha-mahaayagyas*) every day. These are (a) *brahma-yagya*, eulogy to the learned sages; (b) *pitri-yagya*, offerings to ancestors; (c) *deva-yagya*, oblations to the gods; (d) *bhoota-yagya*, offerings to propitiate potentially harmful spirits; and (e) *manushya-yagya*, hospitality to all guests. By means of these five sacrifices, the householder expiates sins that are unavoidably committed at the five slaughterhouses (*panchasoona*) that exist in every home, namely, the hearth, the grinding stone, the broom, the pestle and mortar, and the water-vessel. While using any of these implements, it was reckoned that the householder, unwittingly or otherwise, causes the destruction of little creatures.[9,27]

Complex rules have evolved for the interaction between castes and subcastes on the one hand, and between members of the same caste or subcaste on the other. One may generally call them rules of conduct, which also form part of the individual's *dharma*. Infractions are, of course, punishable: one favorite manner prescribed is to give *pancha-daana* (five offerings) to suitable *Brahmins*, many times five in number. These offerings consist of land, grain, cows, gold, and clothes; such an offering may also be freely given to a *Brahmin* as alms to earn merit. Very severe infractions have to be punished very severely: a common enough procedure in olden days for the offender was to partake of a mixture of the five secretions of a cow: milk, yogurt, clarified butter, urine, and dung.[28] This mixture, called *pancha-gavya*, derived its sanctity from the cow, the most sacred animal for Hindus, a representation of the Mother Earth. Another mixture of five substances is the *panchaamrita*, a sweet liquid made by blending milk, curds, liquefied butter, honey, and sugar. It is offered as the blessing of the deity at Hindu temples to worshippers and is particularly tempting to youngsters.

The *panchaamrita* may have another claim to fame. The *Agni Puraana* declares that by bathing in this mixture for a year, and by donating a cow to a *Brahmin* at the close of that period, a man becomes a king in his next life. Since a ruler has "purchased" his subjects thus, he is their lord and must be obeyed[29]; a Sanskrit word for king is *naresha*, the lord of men. This is, however, no argument for the "divine right of the king to do wrong"; it merely suggests the divinity of the kingship.[3]

The criminal and civil codes prevalent in ancient India reflected the influence of the caste system. In general, the higher the caste of the person, the fewer the types of criminal charges that could be brought against him[28]; on the other hand, a person of a higher caste was liable to be punished more severely than one of a lower caste on the same charge, the severity of the punishment being proportional to the knowledge of the Vedic laws the indicted person was supposed to possess.[3,29]

The most grievous sin in Hinduism is *bhroonahatyaa*, the abortion of an embryo; this sin is inexpiable.[8] Besides this, the four major sins a *Brahmin* could be held guilty of are drinking alcoholic beverages, killing a *Brahmin*, sleeping with the wife of his teacher, and stealing gold.[8,29f] It has been held[30] that these five heinous sins (*pancha-mahaapaatakas*) cannot be remediated by any means, but these were certainly punishable by courts of law. In his *Arthashaastra* , the statesman Kautilya (fourth to third centuries B.C.) prescribes branding the sinner's face with a flag of vintners, a human torso, female private parts, and a dog, respectively, for the latter four sins.[29] Inducing abortion was held to be punishable by death in the *Manu Dharmashaastra*.[3]

Any account of the caste system would be incomplete without mention of the five semidivine or superhuman entities.[28] These were (a) the *naagas* and *naagins*, serpents, generally cobras, capable of assuming human forms; (b) the *yakshas* and *yakshinis*, guardian spirits of trees; (c) the *vidyaadharas*, andromorphic magicians fond of seducing young nubile women; (d) the *gandharvas*, celestial singers and musicians connected with the erotic arts; and (e) the *apsaraas*, water nymphs who delighted in arousing lust in men, and particularly in ascetics. *Naagas* are worshipped to this day on *Naagapanchami*, which falls on the fifth day of the bright half of the lunar month *Shraavana* (July-August). Snakes, particularly cobras, were special. For one injured by a snake, a specified

[f] Consumption of flesh — especially that of a cow — considered taboo by the orthodox Hindus of today, was permissible in former times; see Keith[9] and Dey.[31] The influence of Buddhism and Jainism gradually led to the abandonment of animal sacrifices; Mahatma Gandhi's advocacy of *ahimsa* (nonviolence) was thus derived from post-Vedic influences.

ritual was recommended; it commenced in the lunar month *Bhaadrapada* (August-September) and consisted of revering a five-hooded snake of wood or clay for a whole year.[9] There were also ghouls and ghosts, variously called *pretas*, *pisaachas*, and *brahma-raakhshas*, who had to be propitiated by the householder everyday by the *bhoota-yagya*.

7. Literature

Religion pervaded all aspects of Hindu life till quite recently, as was evident in Mahatma Gandhi's evocation of *Raam-raajya* during the struggle for Indian independence. Thus, motifs for literature in nonreligious contexts were commonly derived from the Hindu pantheon. The oral tradition of the later Vedic age had five distinct forms: (a) *gaatha*, which initially meant simply a song but gradually specialized to celebrate heroic deeds; (b) *naraashamsi*, which was an eulogy to the (human) ancestors; (c) *aakhyana*, or a dramatic dialogue; (d) *itihaasa*, or narration of historical events; and (e) *puraana*, meaning ancient lore. From the early Vedic age, indisputable evidence exists only for *gaathas* and *naraashamsis*. Eventually, however, the first three forms were absorbed in the *itihaasas* and the *puraanas*; indeed, some commentators consider the giant epic *Mahabhaarata* to be a giant *itihaasa* made up of *aakhyanas*, etc.

The 18 *Puraanas* and their derivates, the 18 *Upapuraanas*, are probably the most comprehensive works and have come down to us generally intact.[31] The chief characteristic of a *Puraana* is its treatment of five different subjects (*pancha-lakshana*). These are (a) *sarga*, the creation of the universe; (b) *pratisarga*, its destruction and recreation; (c) *vanshas*, genealogies of royal houses; (d) *manvantaras*, reigns of the different Manus,[3] and (e) *vanshaanucharitas*, genealogies of celebrated beings, both gods and men. Accounts of astronomy, geography, anatomy, and medicine abound, the range of the *Puraanas* being quite encyclopedic. For some two millennia, these works have been the vehicle for instructing the common Hindu folk about their history, as well as religious and social duties, and have conveyed to them a sense of continuity and rootedness; a considerable number of the *Puraanas* are particularly relevant to the Vaishnavites. The *Aagamic* literature generally performs a similar function for Shaivites but is of later origin, around the eighth century A.D. Still another, the *Taantrik* literature is even more recent, some *tantras* having been written not three centuries ago.[31] The chief elements of Taantrism are the five m's (*pancha-makaara*): liquor (*madya*), meat (*maansa*), fish (*matsya*), lucre (*mudra*), and sexual intercourse (*maithuna*). To some, the *Taantrik* literature appears to possess a distinctly pornographic flavor, and the *Taantrik* practices do involve vulgar sensualism. It may be said, however, that these practices are supposedly redemptive and meant for the lowest type of Hindu, the one overly possessed with *tamas*.

Derivatively, the work *itihaasa* means "truly thus it happened." The earlier interpretation, "ancient events," however, was broadened later to include all forms of historical composition. These were arranged into stories to illustrate moral, aesthetic, spiritual, and material truths. In the medieval age, the *itihaasas* were narrowed down to accounts of events resulting in the achievement of royal fortune by kings.[32] Composed by court bards to glorify their masters and their royal houses, and written in heavily ornate styles, the medieval historical works cast the royal glory in the form of a lovely princess symbolizing the goddess of royal fortune, *Raajyashri*. The obtaining of the *Raajyashri* was invariably developed in five stages of the beginning (*praarambha*), the efforts (*prayatna*), the hope of achieving the end (*praapti-aasha*), the certainty of achievement (*niyataapti*), and finally, the achievement (*phalaagama*). These stages supplied a sequence of ordered actions for the development of the story. Incidentally, the use of ordered sequences of achievements, rather than of frameworks of dates and years, is a characteristic of all premodern Hindu literature.

Apart from religious–philosophical and religious–historical works, other forms of literature fluorished[g]: lyrical poetry, drama, as well as treatises on medicine, botany, mathematics, the sciences, and law.[33] But perhaps the best known Hindu work of literature is the collection of stories assembled as the *Panchatantra*. Assuming varied forms in their native India, these stories have traveled across Asia and Arabia to Europe, and, for more than two millennia, they have delighted countless individuals. An early nineteenth-century French priest, Abbé Dubois,[30] who worked in India and is generally known for his excessive disparagement of the contemporary Hindu society, even suggested that Aesop's

[g] There is no reason for the number five not to occur in literary critiques and technical works, but my limited survey failed to yield any instances of particular significance.

fables were copies of the *Panchatantra*. The word *Panchatantra* means five *books*, though the Abbe, ever distrustful of the Hindus, would rather translate it as five *tricks*. Each of the five books is independent, consisting of a framing story into which numerous other stories have been inserted as narrated by one or another character of the framing story. This style, known to Western readers familiar with the *Arabian Nights*, is a common enough Indian style: the megaepic *Mahaabhaarata* has been told in much the same manner.

Supposedly the stories were told by a Pandit Vishnusharman to educate the three sons of a certain king in south India. The three princes, supreme blockheads that they were, had been found incapable of receiving education in the customary manner, for which reason their teacher resorted to imparting them wordly savvy by telling them stories. Indeed,[34]

> Whoever learns the work by heart,
> Or through the story-teller's art
> Becomes acquainted,
> His life by sad defeat—although
> The king of heaven be his foe—
> Is never tainted.

Considering that the *Panchatantra* stories have been used in India for centuries in lieu of formal education,[35] per Hindu tradition, could street-smarts be taught through only five, albeit long, stories?

8. Geography and Polity

The early Vedic society in India being expansionist in character, many geographical accounts have come down to us. The traditional division of India into five regions is quite commonly found in literature and dates back at least to the *Atharva Veda*, which is recognized as the fourth and the last *Veda*, after the *Rig*, *Yajur*, and *Saam Vedas*. In the *Aiteraya Brahmana*, we find India to be divided into five *diks* (directions): *Praachya* (Eastern), *Dakshina* (Southern), *Prateechee* (Western), *Udeechee* (Northern), and *Dhruvaa Madhyamaa* (Central). Some of the *Puranaas* add two more: *Vindhya*, which corresponds to the mountainous range separating the southern part of India from the northern part, and *Himavanta*, the region of the Himalayas. The tenth-century geographer Raajashekhara gave the boundaries of the five regions very explicitly, although other older as well as contemporaneous works let us know that there was little consensus on the boundaries of the central zone.[36]

These divisions recognized India as a landmass, not as either a country or a nation. The subcontinent was broken up into a carpet of states, big and small, some sovereign and the others in vassaldom. Occasionally, the ruler of a small state expanded his overlordship over a large part of India and then styled himself as a *Chakravarti Samraat* (Universal Emperor).

Though republican constitutions were not unknown to ancient India, Hindu political theories have predominantly been concerned about the governance of the monarchic state.[h] Hindu thinkers did not indulge in utopian dreams: either they were teachers of the sacred law, which was binding upon the monarch, or, as diplomatists and economists, they laid down rules of policy for princes and ministers to emulate.[37] Kings were usually *Kshatriyas*, although *Brahmin* and "low-born" kings were not unheard of; women, however, were generally considered unfit to be monarchs.[3,37] Various theories of the evolution of the powers of the Hindu monarch exist but are not germane to the present issue.

Two extremes of royal duties were prescribed in ancient works. The first requires the king to be the *Dharmaraaja*, the one who upholds the Vedic traditions and the sacred laws by creating appropriate conditions for each of his subjects to practise his or her own *dharma*. The second definition is purely secular: the king is to be concerned only with *artha* (wealth), and he is to conduct the affairs of the state to secure economic prosperity for all his subjects. In practice, however, scholars noted that three other functions were also to be followed by the monarch. *Vyavahaara* seems to have stood for contractual law, which the king was bound to uphold and enforce. *Charitra* was usage or custom, which the king was supposed not to oppose. Finally, *shaasana* was rule by decree, which was the royal prerogative during exigent circumstances.[38] Altogether, thus, there were five royal duties suggested.

Kings came and kings went but, apart from occasional disruptions due to warfare or natural calamities, the life of the average Indian went on. This pattern persists to this day, and is, perhaps, single-handedly responsible for the continuance of the Hindu traditions over several thousands of years. During this long period of time, the lowest unit of

[h] The presence of a monarch at the helm of the affairs appears to be a must in the early political literature. Nevertheless, a wayward king could be easily discarded by his subjects, who would then proceed to enthrone another without delay.

local government has remained the village (*graama*), led by a headman (*graamika*). In earlier times, this functionary used to be assisted by the *talvaathaka* (accountant), the *doota* (messenger), the *seema-karmakaara* (land assessor), and various *karnis* (clerks). The village has always been largely autonomous and was treated as a fiscal unit as late as during the British colonial times. Justice was administered by village elders assembled in a council called *pratishthaa* by the lawgiver Brihaspati; there could be two, three or five members. These individuals were powerful figures in the rural life: indeed, they judged according to the custom of the people, not that of the king. The king was enjoined to accede to the council unless the elders were at variance with one another.[39] Similar constitutions also prevailed in the guilds of various types of artisans.

In later times, the council came to be called *panchaayat* (council-of-five) and is now the most ubiquitious secular system in India. Munshi Prem Chand, a twentieth-century Hindi litterateur of Dickensian charm, has woven a popular sentiment in his short story *Pancha Parmeshwara*: the *panchaayat* is emblematic of the Supreme Deity meting out justice. The civil and criminal codes of the Republic of India recognize this ancient institution and have invested the *panchaayats* with modest judicial powers. At the time of this writing, a bill had been introduced in the Indian Parliament to further strengthen the village *panchaayats* as well as similarly modeled urban bodies. There is, however, an interesting twist. Nowadays in India, decrees imposed to quell riotous situations frequently ban assemblage in public areas of five or more persons, not related to each other by blood!

To a large extent, the theoretical Hindu monarchic state survives today in the Himalayan kingdom of Nepal. This country is currently ruled by an almost absolute king,[i] whose powers are somewhat tempered by a council of ministers as well as by a four-tier system of *panchaayats*. *Panchaayats* at the village level are elected by a show of hands. These are followed by district and zonal *panchaayats*, with the *Raashtriya Panchaayat* at the apex being the national parliament. Political parties are not permitted, the *panchaayats* being supported by nonpolitical groups of

[i]From the beginning of the nineteenth century till 1950, the de factro rulers of Nepal were the hereditary mayors of the palaces. First, it was the Thapa family, and then the Rama family, which held sway over the affairs of Nepal and held the monarch captive.

workers and farmers. The king, however, is not bound to follow the advice of these bodies.[j]

9. Conclusion

The foregoing material was culled from a survey of intellectual works and commentaries written by intellectuals for other intellectuals, regardless of the period of composition. This poses the greatest challenge to any student of the Hindu civilization: almost all of the Hindu literature — ancient as well as premodern — and almost all of the latter-day Western-style scholarship of Indian or Western origin, have concentrated on the ideal, not on the real. With the possible exception of a school of history championed by Kosambi,[25,40] scholars have not examined in depth, or may not have sufficient materials to so examine, everyday life in the various parts of the earlier India (where undoubtedly more fivefold facets are waiting to be discovered!). It must be borne in mind that the rapid growth of modern means of communications as well as the import of a largely nonindigenous system of administration have gravely complicated the efforts to reconstruct the life of the ordinary Hindu in earlier times from observations of the current society.

The diaskeuasts of the epic *Mahaabhaarata* exclaimed that[25]

> Whatever is here might be elsewhere,
> But what was not here could hardly be found. [1.56.33]

On the contrary, despite the comparatively rare occurrence of the number five in the Hindu writings of earlier times, my survey of a large number of commentaries, expositions, translations, and transliterations forces me to be very modest about the exhaustiveness of the present work. Nevertheless, it is hoped that the various fivefold facets of the Hindu civilization identified here will be of some use to the readers in organizing their own ideas on fivefold symmetries in the cultural context. I will now take my reader's leave with transliterations of two pieces of verse from the *Taittreya Samhita*[22] on the Creation:

> The creator did it with the five that he created five-and-five sisters to them [each].
> Their five courses assuming various forms, move on in combination. [IV.3, 11; 5]

[j]At the time of going to press, it appears that Nepal is well on its way to becoming a constitutional monarchy.

Five milkings answer to the five dawns;
the five seasons to the five-named cow (i.e., the earth).
The five sky-regions, made by the fifteen,
have a common head, directed to one world. [IV.3, 11; 11]

Acknowledgments

This essay is dedicated to the affectionate memory of my father, Servesh Kumar Lakhtakia, who instructed me in the *Vedaanta.*

Several interesting discussions with my colleagues, Professors V. K. Varadan and V. V. Varadan, are acknowledged. My colleagues Professors L. N. Mulay, K. Vedam, and M. G. Sharma have rendered me yeoman service by their keen critiques of this essay.

To my wife Mercedes, thanks are due for reading and proofing the manuscript several times during its preparation; all remaining errors I claim as mine, however, and are not to be deemed as community property.

References

1. D. Wells, *The Penguin Dictionary of Curious and Interesting Numbers.* Penguin, London (1987).
2. B. Datta and A.N. Singh, *History of Hindu Mathematics.* Asia Publishing House, Bombay (1962).
3. K. Motwani, *Manu Dharma Shaastra.* Ganesh, Madras (1958).
4. G. Parrinder, *World Religions.* Facts on File, New York (1984).
5. S. Radhakrishnan and C.A. Moore, *A Source Book in Indian Philosophy.* Princeton University Press, Princeton, N.J., (1957).
6. S.P. Swami and W.B. Yeats (Transl.), *The Ten Principal Upanishads.* Faber & Faber, London (1937).
7. Sir Edwin Arnold (Transl.), *The Song Celestial: Shrimad Bhagavad Gita.* Interprint, New Delhi (1978).
8. A.A. MacDonnell and A.B. Keith, *Vedic Index of Names and Subjects.* Motilal Banarsidass, Delhi (1967).
9. A.B. Keith, *Religion and Philosophy of the Vedas and Upanishads.* Harvard University Press, Cambridge, Mass. (1925).
10. E.V. Daniel, *Fluid Signs: Being a Person the Tamil Way.* University of California Press, Berkeley (1984).
11. K.K. Klostermaier, *Mythologies and Philosophies of Salvation in the Theistic Traditions of India.* Wilfred Laurier University Press, Waterloo, Ontario (1984).
12. T.A. Gopinath Rao, *Elements of Hindu Iconography, Vol. II.* Paragon, New York (1968).
13. V. Paranjoti, *Shaiva Sidddhaanta.* Luzac, London (1938).
14. A.K. Coomaraswamy, The dance of Shiva, *Siddhanta-Dipika*, XIII (July 1912).

15. F. Capra, *The Tao of Physics*. Shambala, Boulder (1975).
16. S.S.S. Sastri (Transl.), *Saankhya Karika*. University of Madras, Madras (1935).
17. A.B. Keith, *The Saamkhya System*. Oxford University Press, London (1918).
18. A. Sen Gupta, *Saamkhya and Advaita Vedaanta: A Comparative Study*. M. Sen, Lucknow (1973).
19. A.B. Keith, *Hindu Logic and Atomism*. Greenwood, New York (1968).
20. D.S. Gaur and L.P. Gupta, The theory of *panchamahaabhoota* with special reference to *Aayurveda*, *Indian J. Hist. Sci.* 5:1 (1970) 51–67.
21. B. Seal, *The Positive Sciences of the Ancient Hindus*. Motilal Banarsidass, Delhi (1985).
22. B.G. Tilak, *The Arctic Home in the Vedas*. Tilak Bros., Poona (1925).
23. A.C. Mayer, Municipal elections: A Central Indian case study, in C.H. Phillips, ed. *Politics and Society in India*. Praeger, New York (1962).
24. T.A. Gopinath Rao, *Elements of Hindu Iconography, Vol. I*. Paragon, New York (1968).
25. D.D. Kosambi, *Ancient India*. Pantheon, New York (1965).
26. C. von Fürer-Haimendorf, Caste and politics in South Asia, in C.H. Phillips, ed. *Politics and Society in India*. Praeger, New York (1962).
27. P.H. Prabhu, *Hindu Social Organization*. Popular Prakashan, Bombay (1963).
28. J. Auboyer, *Daily Life in Ancient India*. Macmillan, New York (1965).
29. J.W. Spellman, *Political Theory of Ancient India*. Clarendon, Oxford (1964).
30. J.A. Dubois, *Hindu Manners, Customs and Ceremonies*, H.K. Beauchamp (Transl.). Clarendon, Oxford, 1906. [The original manuscript in French was purchased by the East India Company in 1807. The Abbé is better known in France for publishing a translation of the *Panchatantra* in 1826].
31. N. Dey, *Civilization in Ancient India*. Indological Book House, Varanasi, 1972. [This work was serialized in the *Bengal Magazine* during 1877–1878.]
32. V.S. Pathak, *Ancient Historians of India*. Asia Publishing House, New York (1966).
33. G. Sastri, *A Concise History of Classical Sanskrit Literature*. Oxford University Press, Calcutta (1960).
34. A.W. Ryder (Transl.), *The Panchatantra*. University of Chicago Press, Chicago (1956).
35. L. Sternbach, Legal interpretation of the Pañcatantra, in *Proceedings of the XVI Indian Oriental Conference*, Vol. 2 (1951), pp. 78–94.
36. R.N. Mehta, *Pre-Buddhist India*. Examiner Press, Bombay (1939).
37. U.N. Ghoshal, *Hindu Political Theories*. Oxford University Press, Madras (1927).
38. P. Masson-Oursel, H de Willman-Grabowska and P. Stern, *Ancient India and Indian Civilization*. Barnes & Noble, New York (1967).
39. H.N. Sinha, *The Development of Indian Polity*. Asia Publishing House, New York (1963).
40. D.D. Kosambi, *Myth and Reality*. Popular Prakashan, Bombay (1962).

ON THE SHAPE OF FIVE IN EARLY HINDU THOUGHT

Haresh Lalvani

1. Introduction

This article complements A. Lakhtakia's article *Certain Quinary Aspects of the Hindu Civilization*[1] and focuses on the geometry and symmetry associated with the number five during the Vedic period (1300 B.C.–500 B.C.[2]). Lakhtakia has provided a fairly large catalogue of the appearance of five in various aspects of Hindu philosophies. This is preceded by the earlier work of Hopkins[3] on the use of numbers in the *Vedas*, the written texts of the Vedic period. In addition, Hopkins[4] focused on the "holy number" in *Rig-Veda*,[5] the earliest of Vedic texts which dealt with sacred hymns and were written "prior to 1000 B.C."[6] However, the numeric works are catalogs and do not address the reason or basis for the use of specific numbers. This author has focused on the morphologic basis of five, and has drawn from a few texts of the Vedic period to demonstrate the *structural* reason for five as a significant number in early Hindu metaphysical taxonomy. This structure of five was determined by the geometry of space known then. Literary evidence of geometric and architectural concepts associated with five in the ancient texts also indicates that some meta-modelling techniques were known and used during this time, i.e., spatial and geometric concepts influenced and perhaps even *shaped* thinking during the Vedic period. Such techniques deal with modelling of concepts and, when supported by the knowledge of geometry, imply familiarity with proto-representation schemes.

The examples presented here are by no means exhaustive, and better examples by Vedic scholars may be found. Some of the references are taken from the *Rig-Veda*, and some others from *Taittireya Samhita*,[7] a *Yajur-Vedic* text on sacrificial rituals, written around 600 B.C.[8] Most of the literary examples are taken from Muller's English translation of the *Upanishads*,[9] the philosophic treatises "composed from about 700 B.C. onwards"[10] during the latter part of the Vedic period. This literary evidence is supported by *Sulva-Sutras*[11–13] (literally, "*Sutras* of the cord"), the earliest Hindu texts on geometry which describe the precise measurements, design and construction of the sacrificial altars (*Vedis*) used during the Vedic period. The *Sulva-Sutras* of Baudhayana, written around 500 B.C., preceded the *Sulva-Sutras* of Apasthamba written around 350 B.C.[14] Both are practical manuals describing geometric construction and establish the intimate correlation that existed between geometry and religious practice of that time. The priests (*yagnikas*) performing the sacrificial rituals were expected to be familiar with a body of geometrical knowledge and recited mantras while the bricks for the altars were being laid. From the *Sulva-Sutras* it is clear that the knowledge of geometry was advanced during the Vedic period.

2. Number, a Device for Cataloging

From a survey of these texts it is clear that number played an early role as a literary device, possibly a mnemonic one. The *Rig-Veda* describes the heavenly deities (*Aswins*) in terms of three:

> "Aswins, thrice bestow upon us riches: thrice approach the divine rite:
> Thrice preserve our intellects: thrice grant us prosperity: thrice food.
> The daugther of the sun has ascended your three-wheeled car."
> (l, 3, 4, 5).

In another verse, the number seven is used similarly:

> "The seven who preside over this seven-wheeled chariot (are) the seven horses who draw it, seven sisters ride in it together, and in it are deposited the seven forms of utterance."
> (II, 3, 8, 3)

Hopkins describes three and seven as the "most revered cardinals of the *Rig-Veda*,"[15] and the use of five is far less frequent. It is clear, and even more evident from *Upanishads*, that some numbers acquired significance as a means to catalog the physical and metaphysical worlds, and their component parts. This catalogic use of number has continued through the subsequent development of Hindu thought where the deviations from this idea seem quantitative (e.g., five becomes six). Following Lakhtakia, the examples shown in this paper have focused on the use of five, though some relationship to other numbers is described. Five provides an interesting case study from a morphologic viewpoint.

3. The Number Five

Many of Lakhtakia's fives[1] are morphologically different and each appearance of five must be studied on its own. However, a few classes of fives can be identified. For example, the "five kinds of men" include the four castes and one other for non-Aryans or foreigners. This is a special decomposition of five into four + one. Consider another example of the "five royal duties" which consist of two major duties and three minor ones. This is another decomposition of five, into three + two. The five kinds of motion consisting of two (up and down), two (expand and contract), and one (going) is a decomposition into two + two + one. These three classes of fives are clearly different from five physical senses, see-hear-speak-touch-smell, or the five elements, earth-air-fire-water-space (or ether), to name two examples which represent a different class, that of five *equivalent* things.

There are other instances of five mentioned by Lakhtakia which are arrived at in successive periods by adding and subtracting, or by simply changing the number to another in later periods. For example, the eleven divisions of reasoning of earlier versions of the *Nyaya* system of philosophy became five in later versions. Other types of five from Lakhtakia include the five-stage enlightenment of the ego, the five stages of a story, or the five books of Panchatantra. From these and other examples, it is clear that the number five is a dynamic one, which like Shiva, has many manifestations, some of which change in time, others change in space, some may be internally symmetric, others just happened to add up to five, and still others which became five at one stage or another in the philosophical-religious evolution.

The present paper focuses on the groupings of five equal or equivalent things for which a consistent internal structure can be found and supported by the cited texts. The evolution of five is not addressed, though some of it is implied. An extension to a few higher numbers is discussed, though not chronologically nor from any historical standpoint. The emphasis is on the *morphology* of Vedic thought. The occurrence of five is presented first, its geometric derivation is then described, its representation is inferred and supported by architectural geometry known during that time, and extension to a few higher numbers is touched upon, followed by some remarks on the relation to fivefold symmetry.

The ubiquitous occurrence of five is declared in the *Taittireya Upanishad*:

> " 'The earth, the sky, heaven, the four quarters, and the intermediate quarters,'—
> '*Agni* (fire), *Vayu* (air), *Aditya* (sun), *Chandramas* (moon), and the stars,'—
> 'Water, herbs, trees, ether, the universal Self (*virag*),'— so much with reference to material objects (*bhuta*).
> Now with reference to the self (the body):
> '*Prana* (up-breathing), *Apana* (out-breathing), *Vyana* (back-breathing), *Udana* (out-breathing), and *Samana* (on-breathing),'—
> 'The eye, the ear, mind, speech, and touch,'—
> 'The skin, flesh, muscle, bone and marrow.'
> Having dwelt on this (fivefold arrangement of the worlds, the gods, beings, breathings, senses, and elements of the body), a *Rishi* said:
> 'Whatever exists is fivefold (*pankta*)."
> (I, 7, 1)

There could be many different reasons for the choice of five, some of which may have acted concurrently. A purely functional reason, an anthropomorphic one, for selecting five as an important organising number may rest in our five fingers which provide a portable module for enumeration and counting, and hence a device for classification. This reason has little to do with Hinduism intrinsically and any civilization could have come up with a five-based system. Lakhtakia[1] mentions the South American Arawakan language Saraveca which counts with a base five, and other Central American cultures which give "some significance to 5." It would be interesting to find out if there exists any literary evidence for a five-based counting system in the Vedas. The *Rig-Veda* mentions "a hundred and a thousandfold" (II, 1, 4, 3) and the

Aitareya-Aranyaka Upanishad mentions "Ten tens are a hundred, ten hundreds are a thousand....." (II, 3, 5, 5), suggesting the knowledge of a ten-based or decimal system.

4. The Spatial Five

The following examples suggest that there may be an alternative reason, a spatial one, for the choice of five. This relies solely on the geometry of space. It is based on the decomposition of space into four parts, and *not* five. These were termed "the four quarters" in the Vedic texts, representing the fourfold division of space into East, South, West and North (Fig. 1), an idea not unknown to other cultures. These "quarters" were commonly enumerated in that cyclic sequence, always beginning with the East. *Rig-Veda*, already describes *Ushas*, the dawn, ". . . . is beheld in the east" (II, 1, 4, 3), and "She goes to the west. . ." (II, 1, 4, 7). The "four quarters" were already mentioned in the first line of the earlier quote from the *Taittireya Upanishad.*

4.1 *The Fourfold Division of Space*

This division of space into four quarters produces a conceptual reference system that provides at least one firm structural organisation for Vedic theology and philosophy. For example, a foot of *Brahman*, the "Supreme Principle"[16] is described in the *Chandogya Upanishad* as:

> "The eastern region is one quarter, the western region is one quarter, the southern region is one quarter, the northern region is one quarter. This is a foot of Brahman, consisting of the four quarters....."
> (IV, 5, 2)

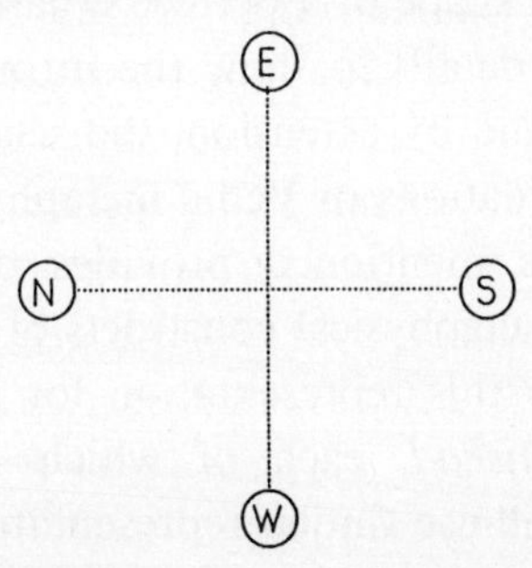

Fig. 1 The "four quarters" of space in Vedic texts, referring to the division of space into a fourfold symmetry.

This fourfold division of space was used as a meta-modelling scheme, albeit a simple one, to structure the worlds, the luminous objects, the human senses, and so on. In the sequel aphorisms, three other feet of *Brahman* are described thus:

> "The earth is one quarter, the sky is one quarter, the heaven is one quarter, the ocean is one quarter. This is a foot of Brahman, consisting of the four quarters....." (IV, 6 3).
> "Fire is one quarter, the sun is one quarter, the moon is one quarter, lightning is one quarter. This is a foot of Brahman, consisting of the four quarters....." (IV 7, 3)
> "Breath is one quarter, the eye is one quarter, the ear is one quarter, the mind is one quarter. This is a foot of Brahman, consisting of the four quarters....." (IV, 8, 3).

The *Sulva-Sutras* provide the precise geometric procedure for constructing this fourfold division of space. This was part of building the sacrificial fire altars which were aligned along the East-West axis, with the priests facing the East, the direction of the rising sun. Fundamental to his building procedure was the construction of a square using a string, equivalent to the use of a compass.[17] The East-West line (*prachi*) of a fixed length is drawn first, it is then bisected at right angles to derive the North-South line of equal length thereby producing the orthogonal cross. The center of the cross marks the center of the square which is then constructed from four equal intersecting circles. From *Sulva-Sutras* it is clear that this procedure and other elaborate geometric details were known to the Vedic priests performing the ceremonies. According to Thibaut, the geometric precision was important: ".....care had to be taken that the sides really stood at right angles on each other; for would the *ahavaniya* fire have carried up the offerings of the sacrificer to the gods if its hearth had not the shape of a perfect square?"[18] It is important to mention some of these details to show the intimacy between geometry and religious practice, and by extension, the relation between geometic knowledge and the foundations of Vedic metaphysics and epistemology.

The orthogonal cross mentioned provides one possible geometric representation for the metaphysical constructs of the Vedic philosopher-priests. Figs. 2a–c show this representation for the above three quotes from *Chandogya Upanishad*, each of which describes one foot of Brahman. The author will use similar representation schemes throughout this paper to accompany the literary quotes. The Vedic knowledge of representation schemes can only be inferred. Their aim was not to

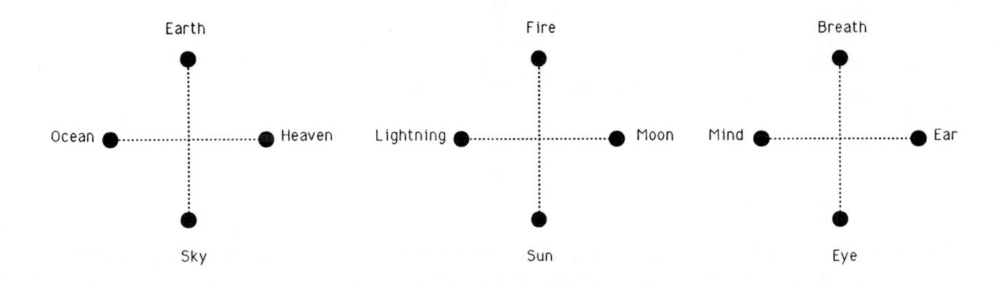

Fig. 2 The use of "four quarters" to represent four physical or human elements.

represent *per se*, but the architectural-geometry of their altars did provide a physical diagram which embedded their metaphysical constructs. This aspect will be discussed further later in the paper.

The simple model of an orthogonal cross is two-dimensional, lying flat on the *horizontal plane*. It was extended at various times to include more spatial elements, but the orthogonal bias continued. It is important to realise that the four directions of space, from a symmetry standpoint, is a point group of fourfold symmetry, and is only one of the infinite ways to cover *entire* space around a point and along the horizontal plane. However, four was the relevant case and covered *all* horizontal space. Therefore, the invention of four words, four senses, etc. was conceptually *complete* and hence sufficient. This idea of completion of a model is important, and its significance is described later in Sec. 5.

4.2 *Five with Fourfold Symmetry*

Restricting to five, there are two ways of converting four to five. One case remains two-dimensional, the other acquires the third dimension, but both require the addition of a fifth point. The former is obtained by adding the center (Fig. 3) to produce the familiar cross arrangement, and is called the quincunx in Christian thought. The latter is obtained by a vertical dimension above this center (Fig. 4). This point is the zenith, and covers all vertical space. Both models retain the fourfold symmetry, though in each case five things, each represented by a point in the figures, are being organised. Both models, in their own way, "complete" the earlier four. The former covers all positions on the horizontal plane, the latter all directions in three-dimensional space (above the ground), thus

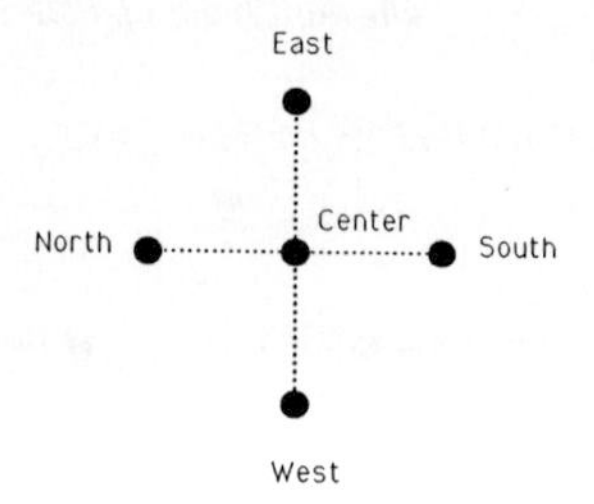

Fig. 3 The derivation of five from four by adding a fifth position in the center. A two-dimensional five is produced but retains its fourfold symmetry.

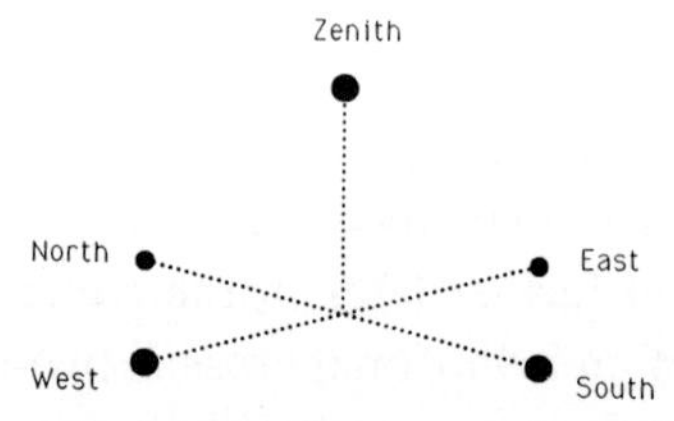

Fig. 4 An alternative derivation of five from four by adding a fifth position in the center. A three-dimensional five is produced but retains its fourfold symmetry.

completing the four in two different ways. Examples are described for the two cases, and suggest that both models were used.

4.2a The Two-Dimensional Case

4.2a(i) The Anthropomorphic Five

The two-dimensional model of five produced a simple one-to-one correspondence with the shape of a human being. Thus the fivefold decomposition of a human being into a head, the left and right side, the center and the seat or support (Fig. 5). The "side" is replaced by "wing" in a bird, and the "seat" by "tail" in both bird and animal. Consider a example from the *Taittireya Upanishad* (II, 1–5) where five components

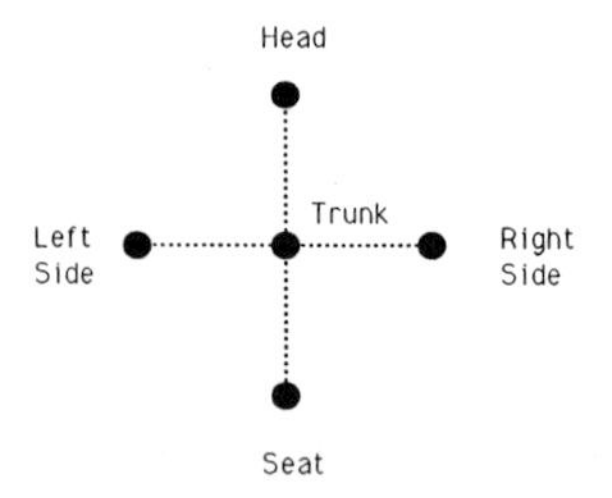

Fig. 5 The decomposition of the human body into five parts and its representation.

of bliss — food, breath, mind, understanding, bliss — are described in terms of the shape of man (Figs. 6a–e):

> "Man thus consists of the essence of food. This is his head, this his right arm, this is his left arm, this is his trunk, this the seat (the support)"
> (II, 1).

The shape of breath is described as:

> "It also has the shape of man.....*Prana* (up-breathing) is its head. *Vyana* (back-breathing) is its right arm. *Apana* (down-breathing) is its left arm. Ether is its trunk. The earth the seat (the support)"
> (II, 2).

The shape of mind is described in terms of five sacred texts:

> "It also has the shape of man.....*Yagus* is its head. *Rik* is its arm. *Saman* is its left arm. The doctrine (...the *Brahmana*) is its trunk. The *Atharvangiras* (*Atharva*-hymns) the seat (the support)"
> (II, 3).

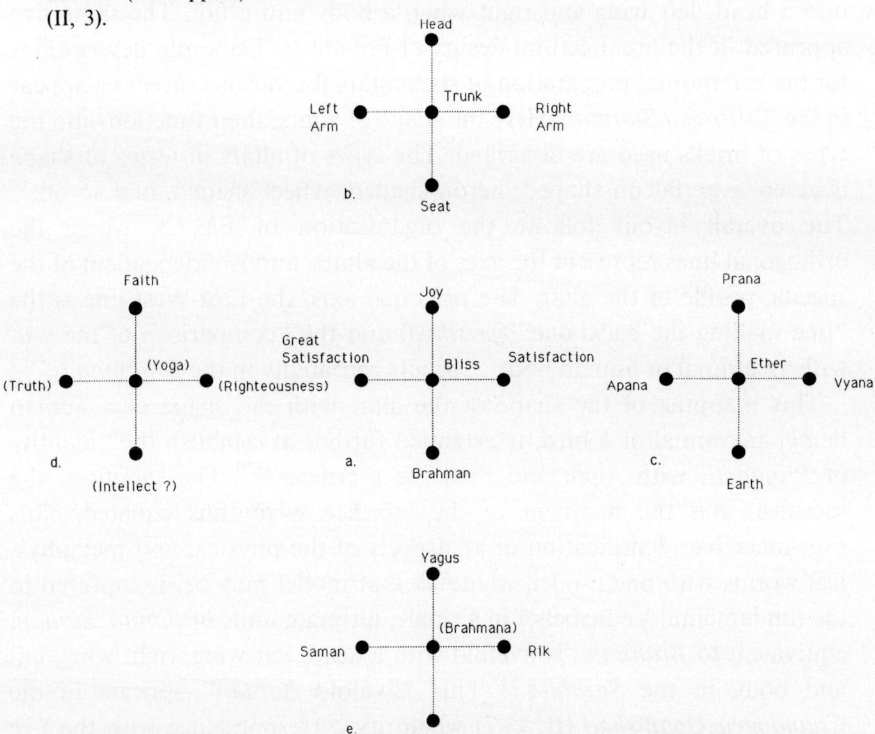

Fig. 6 The anthropomorphic representation of five fives for the human body, various human attributes and the sacred texts (based on Fig. 4).

The shape of understanding is described as:

"It also has the shape of man.....Faith is its head. What is right is its right arm. What is true is its left arm. Absorption (*yoga*) is its trunk. The great (intellect?) the seat (the support)"
(II, 4).

The shape of bliss is described as:

"It also has the shape of man.....Joy is its head. Satisfaction is its right arm. Great satisfaction is its left arm. Bliss is trunk. *Brahman* the seat (the support)"
(II, 2).

4.2a (ii) Five in Design of Altars

The same two-dimensional model of five also produced a correspondence with the shape of bird or animal. The fivefold bird decomposed into a head, left wing and right wing, a body and a tail. The same five appeared in the architectural design of fire altars. Elaborate descriptions for the ceremonial preparation of such altars for various sacrifices appear in the *Taittireya Samhita.* Here the shape of altars, their function, and the types of bricks used are described. The types of altars in terms of shape is given, e.g., falcon shaped, heron shaped, wheel shaped, and so on.[19] The overall layout follows the organisation of Fig. 3, where the orthogonal lines represent the axes of the altars, and is independent of the specific profile of the altar. The principal axis, the East-West line is the "line making the backbone" (*pristhya*) and the "comparison of the *vedi* with an animal or human body ... occurs repeatedly in the *Brahmanas.*"[17]

This mapping of the shape of the altar with the shape of a human being, an animal or a bird, is extended further to establish the "identity of *Prajapati*, with *Agni*, and with the sacrificer."[20] The sacrificer, the sacrifice, and the recipient of the sacrifice were thus equated. This ever-inclusive identification of all aspects of the physical and metaphysical worlds with one model, whatever that model may be, is captured in the fundamental Vedic belief in a single, ultimate unit, *Brahman. Saman*, equivalent to *Brahman*, is fivefold with a head, left wing, right wing, tail and body in the *Samhita.*[21] This "fivefold *Saman*" appears in the *Chandogya Upanishad* (II, 2–7) where its correspondence with the five worlds, (the five stages of) rain, all (five) waters, the (five) seasons, (five) animals, the (five) *pranas* or senses is described.

To take a more precise architectural example, four square or rectangular bricks can be laid around a central one to produce an arrangement of five bricks (Figs. 7a,b). The *Sulva-Sutras* describes the use of such square or rectangular bricks.[22] This organisation matches the two-dimensional model shown earlier in Fig. 3; in geometric terms. Figs. 3 and 7a are duals of each other. Such an organisation using five squares as shown is an alternative representation scheme for five things. A reference to "five bricks" appears in the *Maitrayana-Brahmana Upanishad* where the relation with the shape of an animal or bird is more specific (Figs. 8a–c):

> "This fire (the *Garhapatya*-fire) with five bricks is the year. And its five bricks are spring, summer, rainy season, autumn, winter; and by them the fire has a head, two sides, a centre, and a tail."...
> "*Prana* is Agni (the *Dakshinagni*-fire), and its bricks are the five vital breaths. *Prana, Vyana, Apana, Samana, Udana*; and by them the fire has a head, two sides, a centre, and a tail."...
> "That (*Indra*) is the *Agni* (the *Ahavaniya*-fire), and its bricks are the *Rik*, the *Yagus*,

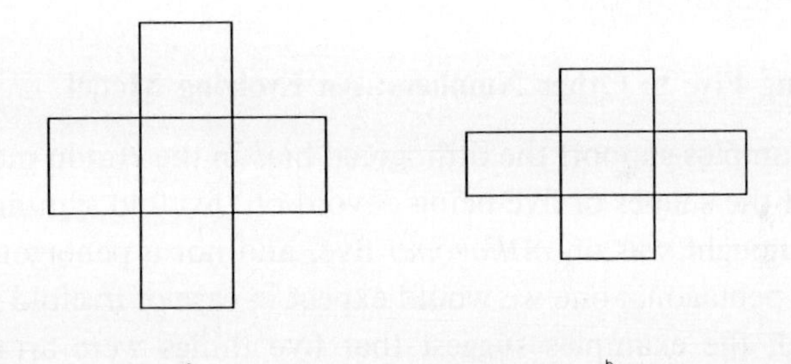

Fig. 7 The arrangement of five square or rectangular bricks used for fire altars. This arrangement retains the organisation of earlier Figs. 3 or 5, but now has a physical basis in brick construction. (Incidentally, Fig. 7a is the geometric dual of Fig. 3.)

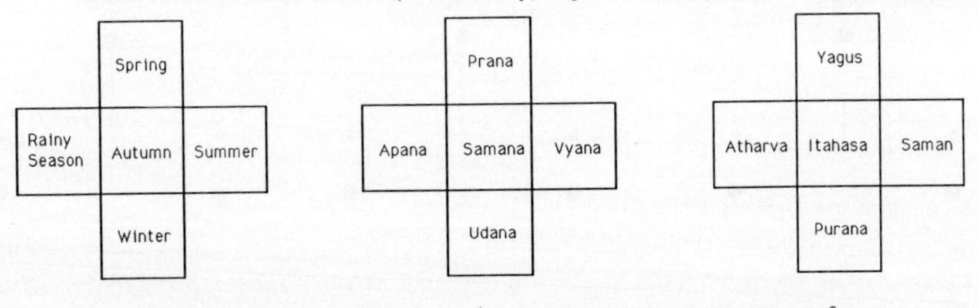

Fig. 8 The arrangement of five bricks and their mapping with the five seasons, the five breaths and the five sacred texts.

the *Saman*, the *Atharvangirasas*, the *Itihasa*, and the *Purana*; and by them the fire has a head, two sides, a centre, and a tail."...
(VI, 33)

From this example it can be inferred that this representation using squares (or rectangles) was known to the Vedic architect-priests.[23]

4.3 *The Three-Dimensional Five*

The three-dimensional model of five was used for organising the five breaths, the five senses, and the five heavenly elements as described in the *Chandogya Upanishad*:

"For that heart there are five gates belonging to the *Devas* (the senses).
The eastern gate is the *Prana* (up-breathing), that is the eye, that is *Aditya* (the sun) ...
The southern gate is the *Vyana* (back-breathing), that is the ear, that is the moon ...
The western gate is the *Apana* (down-breathing), that is the speech, that is *Agni* (fire) ...
The northern gate is the *Samana* (on-breathing), that is the mind, that is *Parganya* (rain) ...
The upper gate is the *Udana* (out-breathing), that is the air, that is ether ..."
(III, 13, 1–5) (Figs. 9a–c)

5. Extending Five to Other Numbers: An Evolving Model

These examples support the orthogonal bias in the Hindu metaphysics, and suggest the shapes of five being devoid of "fivefold symmetry." Five in Hindu thought was an *orthogonal* five, and not a pentagonal one, or the regular pentagonal one we would expect in case of fivefold symmetry. But instead, the examples suggest that five things were arranged in a fourfold symmetry. The fourfold division of space into East, West, North

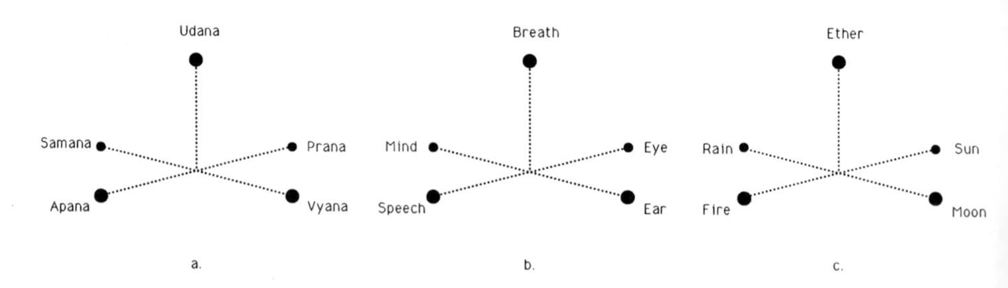

Fig. 9 A three-dimensional five for the five breaths, the five senses, and the five heavenly elements.

and South produces this basis for four, which with the addition of the center, or of the vertical axis (zenith), produces the basis for five. This model will persist till such moment when additional positions or directions in space are identified and the model will undergo a *transformation*, just as four was transformed to five.

At each stage of this transformation process, the model provides a relative level of completion. Each level is determined by the morphology of space and provides a way to fix a number which forms the basis for cataloging the number of physical or metaphysical elements. It is easy to imagine as the model changes, the associated number acquires significance and the process provides a way to discover or invent more physical and metaphysical units, including new gods. It seems possible that such a transforming model (or models) was at work during the Vedic period which spanned several centuries, was composed by several independent groups of writers, and evolved during the period.

Just as four was extended to five, a few extensions of five into higher numbers is shown by transforming its two- or three-dimensional spatial model. Six is produced by adding the nadir, and seven by adding the center to the six.[24] The eight comes from the further division of four into the North-East, North-West, South-East and South-West (Fig. 10); nine adds the center (Fig. 11); ten adds the zenith and nadir to the eight (Fig. 12), and eleven adds the center to ten. Two examples of three-dimensional extensions of five are described. The first model shows how five can become six, and the second converts it into eleven. The conversion to seven, eight, nine or ten is embedded in the latter example.

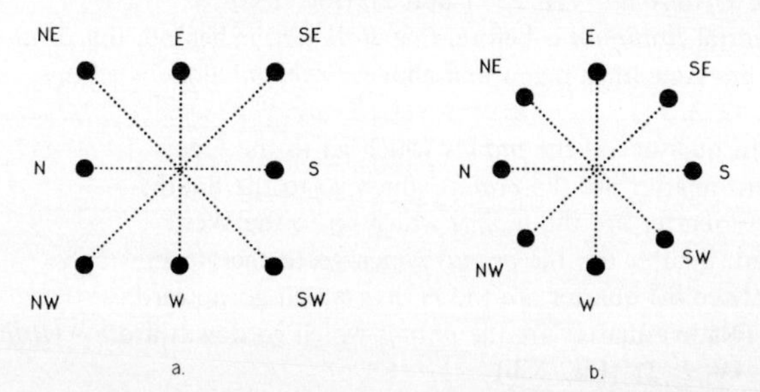

Fig. 10 Extension of four to eight by introducing the four intermediates quarters to Fig. 1. Two alternatives are shown; one retains the square shape, the other is a regular octagon turned on its point.

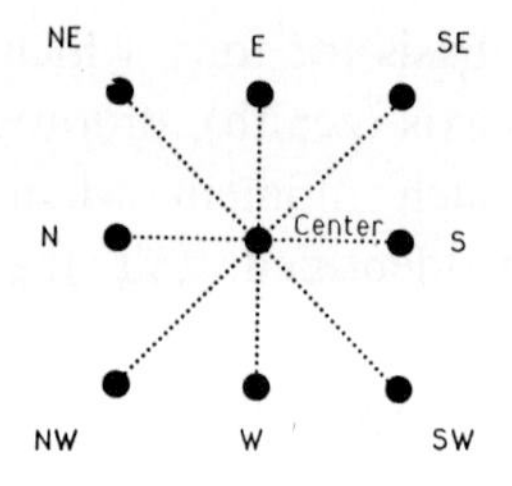

Fig. 11 Extension of five to nine by introducing the four intermediates quarters to Fig. 3.

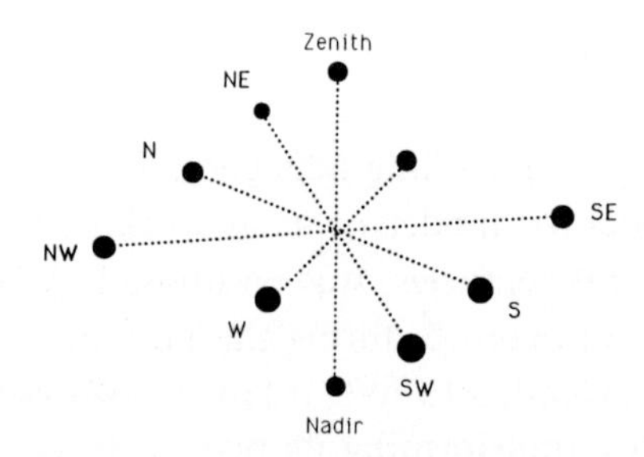

Fig. 12 Extension of eight to ten by adding the zenith and the nadir to Fig. 10. The two-dimensional model now becomes a three-dimensional one.

The five are a subset of six which is obtained by extending the vertical axis below the horizontal plane (Figs. 13a and b). This completes the *x-y-z* axis, the basis of the Cartesian co-ordinate system. Consider following excerpts from four different *Upanishads*:

The Infinite (as the I, and also the Self) is "below, above, behind, before, right and left — it is indeed all this".
(*Chandogya Upanishad*, VII, 25, 1 and 2). (Fig. 13a)
"That immortal *Brahman* is before, that *Brahman* is behind, that *Brahman* is right and left. It has gone forth below and above; *Brahman* alone is all this;..." (*Mundaka Upanishad*, II, 2, 11)
"His Eastern quarter are the *pranas* which go to the East;
His Southern quarter are the *pranas* which go to the South;
His Western quarter are the *pranas* which go to the West;
His Northern quarter are the *pranas* which go to the North;
His Upper (Zenith) quarter are the *pranas* which go upward;
His Lower (Nadir) quarter are the *pranas* which go downward;..." (*Brihadaranyaka Upanishad*, IV, 2, 3). (Fig. 13b)
"He (*Brahman*) was one, and infinite; infinite in the East, infinite in the South, infinite in the West, infinite in the North, above and below and everywhere infinite." (*Maitrayana Upanishad*, VI, 17)

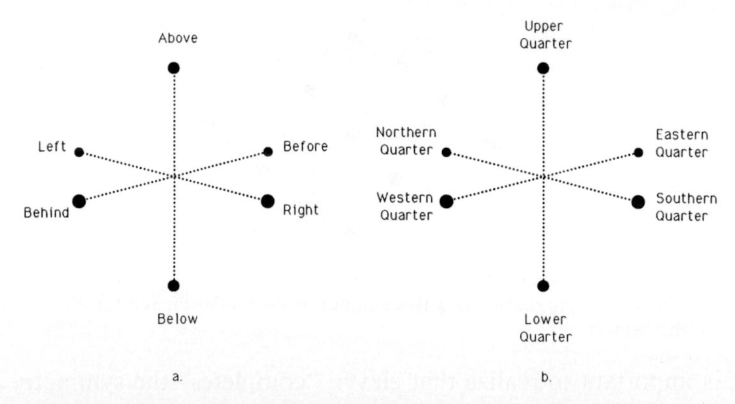

Fig. 13 Extension of four or five to six by completing the three orthogonal axes and adding the "above" (or "upper") and "below" (or "lower") to the four quarters.

Incidentally, the outer points of the six axes define the vertices of an octahedron, a regular polyhedron (Fig. 14). In some sense, the Vedic thinkers anticipated the octahedron,[25] or at least all of its fourfold axes of symmetry.

In another extension of the model, four intermediate quarters are added to the existing four. This along with the center, the zenith and the nadir, make a model of eleven or (ten without the center), the outer points of which describe the vertices of a square bipyramid (Fig. 15) or an octagonal bipyramid (Fig. 16).

> "The head was the Eastern quarter, and the arms this and that quarter (i.e. the N.E. and S.E., on the left and right sides). Then the tail was the Western quarter, and the two legs this and that quarter (i.e. the N.W. and S.W.) The sides were the Southern and Northern quarters, the back heaven, the belly and sky, the dust the earth." (*Brihadaranyaka Upanishad*, 1, 2, 3).

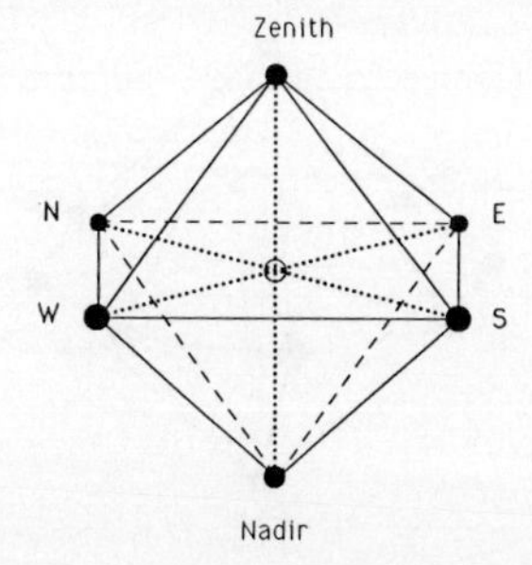

Fig. 14 The regular octahedron as a three-dimensional model for six (or seven, if the center is included).

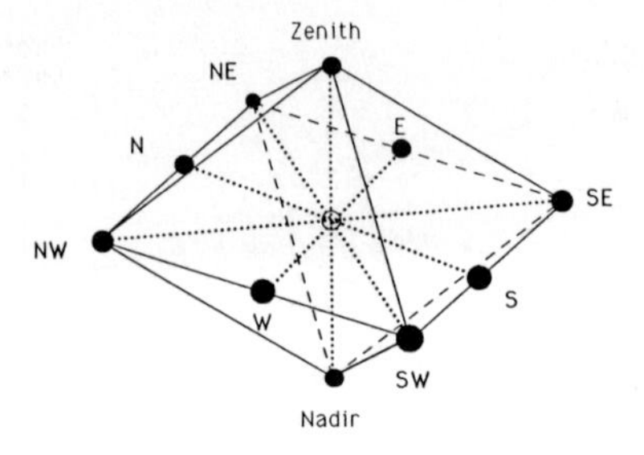

Fig. 15 The square bipyramid as a three-dimensional model for ten (or eleven including the center).

It is important to realize that eleven "completes" the symmetry of the square bipyramid (or the square prism).[26] In symmetry terms, the four quarters define one set of twofold axes, the four secondary quarters define another set of twofold axes, both the nadir and the zenith define the fourfold axis of symmetry, and the eleventh point lies at the center of symmetry. These eleven are further embedded in an octahedral (or cubic) space where the six vertices of an octahedron (or faces of a cube) are occupied, four lie on the mid-edges of the horizontal mid-plane, and one is at the center. It would be interesting to know how much of the complete cubic symmetry was covered in this manner, since the cubic symmetry represents the next structural level of completion. The full cubic symmetry has 27 points and can be thought of as three parallel layers of nine, where each layer is identical to Fig. 11. The threefold division of the vertical axis into three layers is already present in the *Rig-Veda* where vertical space is divided into earth (*prithvi*), sky (*antariksha*) and the heaven (*div*). One of the vertical planes was also

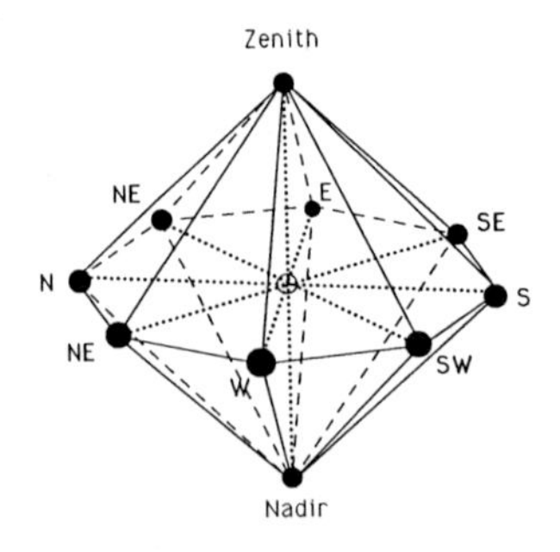

Fig. 16 The octagonal bipyramid as an alternative three-dimensional model for ten or eleven.

known since the movements of the sun were described in seven positions along this plane,[27] and so on.

6. Fivefold Symmetry

It would also be interesting to find out if the regular pentagon or pentagram was known during the Vedic period. The *Sulva-Sutras* describe two types of pentagons, both *irregular*. The first one (Fig. 17) was called the *hansamukhi*, the swan-faced, it has bilateral symmetry and was derived from an orthogonal grid in which the diagonals are included.[28] The second one (Fig. 18) has no symmetry and has a more complex geometry.[29] Both were used as bricks in the design of the bird-shaped altars. The *Shilpa-Sastras* of Manasara, a later architectural treatise (estimated approximately between 500–700 A.D.[30]) does illustrate a *regular* pentagonal column (the *Siva-kanta*) as one of the "five" orders of columns (Fig. 19). The association between five and Shiva is clearly made; the five attributes of Shiva are well-known to the Shaivites.[31] This architectural example points to an indirect correlation between Siva and fivefold symmetry. Unless an earlier evidence is found, any association between five and pentagonal symmetry in Hindu civilisation cannot be pushed much further back. It would be interesting to find out how far this indirect relation between fivefold symmetry and

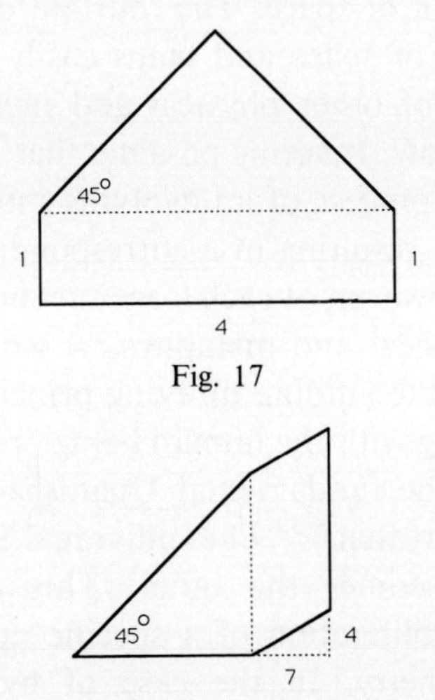

Figs. 17 and 18 Two shapes of nonregular pentagonal bricks described in *Sulva-Sutras*.

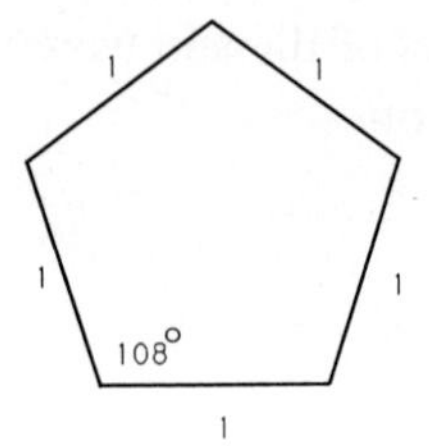

Fig. 19 The regular pentagon, the basis of the pentagonal column (*Siva-Kanta*) which appears in the *Shilpa-Shastras*.

Siva has trickled through in subsequent periods into the Hindu arts and iconography. The iconographic appearance of five abounds.[32]

7. Metaphysical Symmetry

Whether pentagonal symmetry was known during the Vedic period or not, the *Sulva-sutras* suggest that the geometric knowledge during that period was sufficiently developed. What is more interesting is that the concepts of space helped shape the morphology of Vedic knowledge. The occurrence of five as a significant number for the number of senses, the elements, the sacred texts, the worlds, the presiding deities, etc. provide an interesting case to explore this concept. The number five occurs as a by-product of fourfold symmetry, and is derived by adding a fifth element to the fourfold division of space. The fourfold division is based on the four primary quarters of space and maps easily onto the human body. Conformal mappings of other physical and metaphysical entities, and birds and animals follow. It seems possible that five occurred as a stage in the transformation process of an evolving morphologic model which changed through time, resulting in a corresponding shift of emphasis to a different number. However, at each stage the model maps one-to-one in all aspects of the physical and metaphysical worlds as required by the concept of *Brahman*, the guiding unifying principle.

Further, the mapping with the human being provides an anthropomorphic basis to realise the fundamental Upanishadic equation *Atman* = *Brahman*, i.e., "thou art that."[33] The individual Self equals the collective Self, and from one comes the other. This equation provides the mechanism for the proliferation of a specific number through a metaphysical mirror symmetry. In the case of five such a reflection is supported by the following quote:

"By means of the one fivefold set (that referring to the body) he completes the one other fivefold set."
(*Taittriya Upanishad*, I, 7, 1).

Multiple reflections,[34] or iterations, produce the result which recapitulated in the words of a Rishi:

"Whatever exits is fivefold (*pankta*)."
(*Taittiriya Upanishad*, I, 7, 1).

References

1. A. Lakhtakia, Certain quinary aspects of the Hindu civilization (preceeding paper in this book)
2. A.A. McDonnell, *History of Sanskrit Literature*. Motilal Banarsidass, New Delhi (1962) p. 9.
3. E.W. Hopkins, Numerical formulae in the Vedas and their bearing on Vedic criticism, *Amer. Orient. Soc. Jour.* **16** (1896) 275–281.
4. E.W. Hopkins, The Holy Numbers of the Rig-Veda, *Oriental Club of Philadelphia Oriental Studies...*, Boston (1894) p. 141–159.
5. H.H. Wilson, trans., *Rig-Veda Sanhita*, E.B. Cowell and W.F. Webster, eds., Vols. 1-6, N. Trubner and Co., London (1864–68).
6. R. Thapar, *History of India 1*, Penguin, Middlesex (1966) p. 31.
7. A.B. Keith, trans., Veda of the Black Yagus School (entitled) *Taittiriya Sanhita*, Books 1–7. *Harvard Oriental Series*, Vols.18 and 19, Cambridge (1914).
8. Keith in Ref. 7 mentions "at least 600 B.C." for the date of the *Brahmana* protions of the *Samhita*, p. clxxii.
9. M.F. Muller, trans., *The Upanishads*, Parts 1 and 2, Dover, New York (1962); reprint of *The Sacred Books of the East*, Vols. I (1879) & XV (1884), Clarendon Press, Oxford.
10. Thapar, Ref. 6, p. 33.
11. G. Thibaut, The Sulva-Sutras of Baudhayana, *The Pandit*, Old Series, Vols. IX and X (1875–6).
12. G. Thibaut, On the Sulva-Sutras, *Journal of the Asiatic Society of Bengal*, part III, Vol. XLIV, 227–275 (1875).
13. A. Burk, Das Apasthamba-Sulba-Sutras, *Zeitschrift Deutschen Morgenlandische Gesellschaft* **56** (1902) 327–391.
14. Keith, Ref. 7, p. xlvi.
15. Hopkins, Ref., 4 p. 141.
16. Variously translated as the "Highest Self", the "Universal Soul", the "Supreme Principle", etc., and generally understood as the one principle unifying all (see also Lakhtakia, Ref. 1).
17. Thibaut, Ref. 11, Vol. IX, 292–8 (May 1875).
18. Thibaut, Ref. 12, p. 232.
19. Keith, Ref. 7, V, 4, 11.
20. Keith, Ref. 7, p. cxxxi.
21. Keith, Ref. 7. p. cxxxi.
22. Thibaut, Ref. 11, Vol. X (Dec. 1875) p. 169.

23. Later in the *Shilpa-shastras*,[30] this repesentation is seen in an advanced stage and used for the layout of temple plans; for example, 25 deities are allocated over 25 squares in a 5×5 square grid in *Uda-Pitha* plan of 25 square plots (Chap. VII, Sheet V).
24. No example for the concurrence of seven with this organisation has been found by this author so far.
25. The regular octahedron is one of the five Platonic "solids" attributed to Plato, contemporaneously or possibly slightly later. K. Critchlow in *Times Stands Still*, (G. Fraser, 1980), has already shown examples of stone-cut objects that demonstrate the knowledge of the five Platonic solids over a 1000 years before Plato.
26. Hopkins, Ref. 3, p. 280, mentions 33 Vedic gods, and that "all gods are included in this number." Further, in Ref. 4, p.150, these "three elevens" are distributed as "eleven in heaven, eleven on earth, eleven in the waters." Each eleven may correspond to the square bipyramid model, though this remains to be verified.
27. *Chandogya Upanished*, II, 9, 1–8.
28. Thibaut, Ref. 11, Vol. X (Jan. 1876) p. 192.
29. Burk, Ref. 13, p. 380.
30. P.K. Acharya, transl. *Architecture of Manasara Silpashastra*, Oxford Univ., London (1933) p. 152; for an illustration see the companion volume I *Illustrations of Architectural and Sculptural Objects*, Sheet No. XLIV.
31. See for example, J. Singh, *Siva Sutras*, Motilai Banarsidass, New Delhi (1979) p. xiv.
32. G. Liebert, *Iconographic Dictionary of the Indian Religions*, E.J. Brill, Leiden (1976).
33. *Chandogya Upanishad* (VI, 8–16), last verse in each section where Muller translates the same as "Thou art it", and elsewhere as "Thou art He" (p. xxxv, Part I of Ref. 7).
34. It is tempting to think of this self-similarity of five (or other numbers) as a recursive notion similar to fractals (see Fig. 6, as a suggestive example). Such an idea would need additional support from literary sources or the *Sulva-Sutras*.

ALBRECHT DÜRER AND THE REGULAR PENTAGON

Donald W. Crowe

1. Introduction

The author's intention, in this note, is to give a summary of Dürer's geometric writings connected with the regular pentagon and other regular polygons. All this information is contained in, or suggested by, one or more of the five main references: the translation,[1] by W.L. Strauss, of Dürer's chief mathematical work, the *Underweysung der messung/mit dem zirckel und richtscheyt/in Linien ebnen unnd gantzen corporen* (referred to hereafter as the *Underweysung*), the 19th Century analyses of the *Underweysung* by S. Günther[2] and H. Staigmüller,[3] the later mathematical study by M. Steck,[4] and E. Panovsky's standard life of Dürer.[5] Important supplementary information comes from Cantor's history,[6] Vahlen[7] and Jaeggli.[8]

For those readers who do not read 16th Century German, the Strauss edition of the *Underweysung*, consisting of photocopies of Dürer's original edition of 1525 (including the misspellings and broken type), together with an English translation and notes on facing pages, is invaluable. This translation provides ready access not only to the mathematical content of the *Underweysung*, but also (for the reader of modern German) to Dürer's own charming mathematical and literary style by making intelligible the more obscure words, phrases and abbreviations on the facing 16th Century page.

For the detailed analysis of the mathematical content of the *Underweysung* all subsequent writers, including Steck, are particularly indebted to the detailed studies, and calculations, of Günther and Staigmüller. Indeed, if it were not for the relative inaccessibility of these works, the present note would hardly be needed. Staigmüller's paper[3] is the first part of a school program for the end of the school year 1890–91 of the Königliche Realgymnasium of Stuttgart. (The second part contains a description of the courses taught during the year, together with the names of their teachers, etc.) Günther's paper[2] has a similar origin, as a supplement to the annual report of the school year 1885–86 in Anspach. The approximately 100th anniversary of these two papers is an appropriate time to call attention to them once more.

The book[4] by Max Steck is a comprehensive study of Dürer's mathematical work, complete with a 448-item bibliography of commentaries up to 1948. In keeping with this thorough scholarship, Steck reproduces a handwritten copy of *Geometria Deutsch.*[9] This is a 1484(?) version of a ten-page workshop manual which contains much of what we know, aside from what can be deduced from the *Underweysung*, of craftsmen's geometry of the times. From the *Geometria Deutsch* we learn that various constructions, including the approximate construction of the regular pentagon once thought to be Dürer's own invention, were taken over by Dürer from standard workshop practice. Conversely, Dürer's inclusion of Ptolemaic and Euclidean constructions not given in the "Geometria Deutsch" provides evidence that he had access to the mathematics of antiquity only beginning to become available to contemporary craftsmen.

Panofsky's study[5] is the modern standard life of Dürer, which provides the reader with the background of Dürer's life and work. Its final chapter, "Dürer as a Theorist of Art," is a convenient short introduction to Dürer's writings.

The fact that most references to Dürer's mathematical work are in German stems from the fact that his *Underweysung* was not immediately translated into any modern language. His early fame as a theoretician was spread through a Latin translation (Paris 1532, reprinted many times). Although not generally thought of today as a universal thinker of the style of his near contemporary Leonardo da Vinci, in the sixteenth century, "Dürer was acclaimed not only as a painter and engraver, but also as a civil engineer and mathematician. To this extent, Dürer probably enjoyed similar celebrity as a theoretician, just as Leonardo does today, the latter's thoughts having been conveyed to us by the modern techniques of reproduction which made his manuscripts known." (A. Jaeggli, in Ref. 8, p. 215).

2. *Underweysung der Messung*

Dürer's *Underweysung* was published in 1525, only three years before his death in 1528. A posthumous edition, published in 1538, contains certain additions, probably due to Dürer himself, including nets for two more Archimedean polyhedra and some additional mechanical devices for making perspective drawings. Unless otherwise stated, the present discussion applies to the first edition, as reproduced by Strauss.[1]

In passing, among the interesting bibliographic details supplied by Jaeggli in Ref. 8, pp. 210–217, is his opinion that the printed *Underweysung* is "one of the most beautiful examples of German Renaissance printing." The 1538 edition carries the name of Hieronymus Formschneider (originally Andreä, one of the inventors of the German "fraktur" typeface used in its perfected form for the first time in the *Underweysung*) as printer. He is also thought to be the printer of the 1525 edition, as well as various works of Luther, including the little prayer book of 1527. (This same Formschneider was reprimanded, as a printer, by the Nuremberg town council for having publicized Hans Sachs, of *Die Meistersinger* fame, after the latter had been ordered to refrain from his literary pursuits.) However, Jaeggli is convinced that "The perfect form of the layout and the art with which the illustrations have been so harmoniously inserted into the text make it obvious that a highly talented artist supervised the work. This was none other than Dürer himself . . ."

The *Underweysung* was written as a text in practical geometry for the instruction of young craftsmen and artists, "not only painters, but goldsmiths, sculptors, stonemasons, and cabinetmakers," as Dürer says in the letter to his friend Willibald Pirckheimer, which serves as his Introduction. In the foreword to another work, *Von Menschlicher Proportion*, this purpose is described in more detail:

"It is obvious that German painters have no little ability in drawing and the use of color, although up to now they may have had some failings in the use of geometry and perspective and the like. Therefore it is to be hoped that when they obtain these skills, and thereby combine their theoretical and practical abilities, they will in time yield first place to no other nation."

In bold letters at the top of the first page of the main text is the disclaimer, "The learned Euclid has put together the foundations of geometry, and whoever knows Euclid already will have no need to read further." However, it is clear from Dürer's words that he does not imagine that many young craftsmen will have any acquaintance with Euclid.

In fact, most of the information contained in the *Underweysung* is not contained in Euclid. Some of the practical methods Dürer describes certainly come directly from workshop practice. Thus, historically, the book is of interest both from the point of view of the mathematics it presents to the craftsman, as well as the common workshop knowledge it presents to the mathematician. As a case in point, it shows the craftsman the (theoretically) exact construction of the regular pentagon as given by Ptolemy. This is followed immediately by an approximate construction apparently taken directly from the *Geometria Deutsch*, which has two advantages for the craftsman: First, it is closely related to the familiar construction of a regular hexagon and has an attractive symmetry lacking in Ptolemy's construction; second, like the regular hexagon construction, which can be obtained from the same figure, it involves only a fixed ("unverruckt") compass opening. This restriction to the use of a "rusty compass," which Dürer particularly stresses, would perhaps not have come to the attention of professional mathematicians except for the emphasis Dürer placed on it, both here and elsewhere.

The *Underweysung* consists of four "books." The first deals with one-dimensional figures. Beginning with definitions much like Euclid's, it continues with various plane curves. According to Panofsky, this is the first discussion, in German, of conic sections. Even here, however, Dürer's treatment is one particularly oriented toward the workshop. In the discussion of the ellipse, for example, he constructs an ellipse, as a plane section of a cone, by (to quote from Ref. 5, p. 255) "the ingenious application of a method familiar to every architect and carpenter but never before applied to the solution of a purely mathematical problem, let alone the ultra-modern problem of the conic sections: the method of parallel projection." Dürer's incorrect drawing of the ellipse, based on this theoretically correct construction, has attracted considerable discussion, including Pedoe's comments in Ref. 11.

The second book deals with two-dimensional figures, beginning with curvilinear triangles and quadrilaterals. Then come the constructions of regular polygons, which we will discuss in the following sections. An arc is trisected. The polygons are applied in the design of various decorative stars and tilings. Then there are ways of constructing figures similar to given figures, but with greater area, such as a square having seven times the area of a given square. There are also constructions of new figures of areas equal to given figures, including two ways, one exact and one approximate, of constructing a square of the same area as a given equilateral triangle. Included here is a construction of a square having the same area as a circle (implicitly assuming $\pi = 3\ 1/8$), which is supplemented in the 1538 edition by another such construction in which $\pi = 3\ 1/7$. (In connection with squaring the circle, Dürer notes explicitly that the scholars have not yet been able to do this, but that the described method will do as a temporary measure or for small areas.) The Pythagorean theorem is given in a general form where arbitrary "equilateral" polygons are drawn on the three sides of the right triangle. Since this statement is verbally incorrect (except for triangles, where "equilateral" implies "equiangular"), we may wonder here, as elsewhere, whether Dürer knew the correct theorem.

The third and fourth books deal primarily with three-dimensional objects. An exception to this is the long section on geometrical construction of Roman and Gothic letters of the alphabet for the use of stonemasons and book illustrators. The third book is especially concerned with specific projects for architects and engineers, including the appropriate content of monuments to military victories over rebellious peasants, as well as over more orthodox opponents. Since buildings must sometimes be equipped with sundials, Dürer describes their construction, including a method for dividing a right angle into 90 equal parts (two trisections and a bisection, followed by a division of each of the eighteenths into five equal parts).

The fourth book is the mathematical counterpart of the three-dimensional architectural projects of the third book. Dürer describes the five regular "Platonic" polyhedra, as well as seven semi-regular ("Archimedean") polyhedra. (Two others are added in the 1538 edition.)

He represents each of these by "nets," with detailed instructions to enlarge them and draw them onto two layers of paper pasted together, so that when a sharp knife is used to cut through one layer along the edges the net can be folded up to make a solid model. Although such nets are familiar in our own time, and the regular dodecahedron had been drawn in this way by Leonardo in Pacioli's *Divina Proportione* (1509) (according to Ref. 4, plate 18), apparently Dürer was the first to use this method to represent the Archimedean polyhedra. In Staigmüller's opinion, this is one of Dürer's most important achievements as a mathematician. Dürer also draws nets for some polyhedra of his own invention. After giving three ways of solving the classical problem of duplicating the cube, he concludes with a long section on perspective and other methods of systematically transforming one figure into another.

3. The First Regular Polygons

Since Dürer's constructions of the regular pentagon appear in the context of constructions of other regular polygons, it is appropriate to describe this book (II), which deals with plane figures, in more detail. A plane figure, says Dürer, is "ein ding das durch die linien geendert und abgesondert wird und noch kein Corpus schleust" ("a thing which ends in, and is bounded by, lines and does not enclose a solid"). Among these plane figures are the regular polygons, for which Dürer gives the construction of those with n sides, for $3 \leq n \leq 16$, and $n = 28$. The inclusion of the regular 28-gon (derived from the approximate construction of the regular heptagon) can perhaps by explained by the fact that the St. Lorenz church in Nuremberg contained, over its main portal, a huge rosette window whose outer ring of stonework is based on a regular 28-gon. (See the photograph in Ref. 1, p. 447.) It would hardly be suitable for Dürer to write a Nuremberg manual for stonemasons that did not include the prerequisites for the spectacular achievement of earlier masters which they could see each day with their own eyes!

The order of presentation of these constructions illustrates the pedagogical nature of Dürer's manual, since he begins, not with the one with the fewest sides, but with the simplest. This is the regular hexagon, inscribed in a circle. Dürer's construction is the well-known one:

> "Take a compass. Place it with one leg on center point a and with the other leg draw a circle, large or small, as you like. Then place one leg of the compass on the

periphery and step around the circle with both legs of the compass. This gives 6 points which you mark with the numerals 1, 2, 3, etc. Then join points 1.2 and 2.3 and 3.4, etc. with straight lines."

This familiar construction is made without changing the setting of the compass. This feature of many of Dürer's constructions (including one of the regular pentagon constructions) led Panofsky to suggest that Dürer's work was the main influence leading to the "obsession" of Italian geometers of the sixteenth century with construction problems using a compass with only a fixed opening. (Interest in such problems seems to have disappeared with the discovery by Poncelet and Steiner that every construction possible with straightedge and ordinary compass is possible not only with straightedge and compass with a fixed opening, but even with straightedge and a single circle with a given center.)

From the regular hexagon inscribed in a circle Dürer gives a natural construction of an equilateral triangle in the same circle ("ein dryangel in ein zirckel schlyss"): just join points 1 and 3, 3 and 5, 5 and 1 of the hexagon.

The next construction is the regular heptagon, placed here by Dürer because his (approximate) construction is the particularly simple one of half the side of the equilateral triangle inscribed in the same circle. According to Günther,[2] "The proposition '*Die Siebenecksseite ist gleich der halben Dreiecksseite*' is a venerable one, and has kept a tenacious hold on life throughout the centuries." Kepler called this "Dürer's Rule," but Dürer himself makes no claim to originality. Among the specific earlier sources cited by Günther is the *Geometria Deutsch*. However, the latter makes no mention of the equilateral triangle in its construction of the regular heptagon. In fact the procedure, as written there, is an inadequate description if we are restricted to using straightedge and compass in the usual way. The result is the same as Dürer's, but the method is different from either of Dürer's two methods. The fact that Dürer's method is simpler than that found in *Geometria Deutsch* illustrates that Dürer did not simply copy all the available regular polygon constructions from that source.

This regular heptagon construction is remarkably accurate, since the half-side of an equilateral triangle subtends an angle of $51° \ 19' \ 4''$ at the center of the circle, compared with the correct value of $(360°/7) = 51° \ 25' \ 43''$. That is, the error is an essentially imperceptible 1/10 of a degree.

In the course of his next constructions, of the regular pentagon, Dürer obtains a third construction of the (approximate) side of a regular heptagon. Although the construction itself is different, an easy calculation shows that the length obtained is the same as the half side of the equilateral triangle just described, namely $\sqrt{3}/2$ in a circle of radius 1.

From the regular heptagon, Dürer shows (with both words and picture) how to construct a regular 14-gon by bisecting the arcs of the circumscribing circle, and "in the same way one can divide the circle into 28 parts and double it." Dürer's own words, "*nim das trum des zirckels im 7 ecke zwischen 1.2 und teyl das in zwey gleiche felt/und trit mit der leng im zirckelryss herumb so werden dir 14 punckten die zeuch mit geraden linien zusamen/so wirdet ein 14 ecket figur darauss*," are ambiguous at this point, and Strauss has (mis?) translated them as "divide one of the sides [*sic*] in half." However, since in the immediately following doubling of a square to get a regular octagon, and the regular octagon to get a regular 16-gon, Dürer's words are unambiguously correct, it seems reasonable to credit him with the correct construction here also.

4. The Regular Pentagon

Dürer next gives two constructions of the regular pentagon, the first exact, the second approximate. We will read into these two short pages a good deal of the approach and method of the whole *Underweysung*.

The first construction is the one given by Ptolemy, The Almagest, Book 1, Ch. 9 (Ref. 3, p. 23). The description is accompanied by a drawing (see Fig. 1) which makes a convenient visual mnemonic since it shows the sides of the regular pentagon, regular decagon, and (approximate) regular heptagon, labelled 5, 10, 7 respectively, all on the same concise figure. The full description of this construction reads:

> "Draw a circle with center *a* and draw a horizontal line through the center *a* and where it cuts the circle mark *b* and *c*. Then through the center *a* draw a perpendicular line with equal angles and where it cuts the circle put a *d* [missing description of *e* as midpoint of *ac*]. Then draw a straight line *ed* and take a compass and place it with one foot in point *e* and the other in *d* and draw from there around to the horizontal line *bc*. Where it meets, place an *f* and join *fd* by a line. This length *fd* is one side of a pentagon. Then *fa* is one side of a decagon. After this divide *ac* into two equal parts by point *e*. Then you can draw a perpendicular line from *e* up to the circle. In this way you have a seventh of the circle, approximately ('*mechanice*')." (Fig. 1)

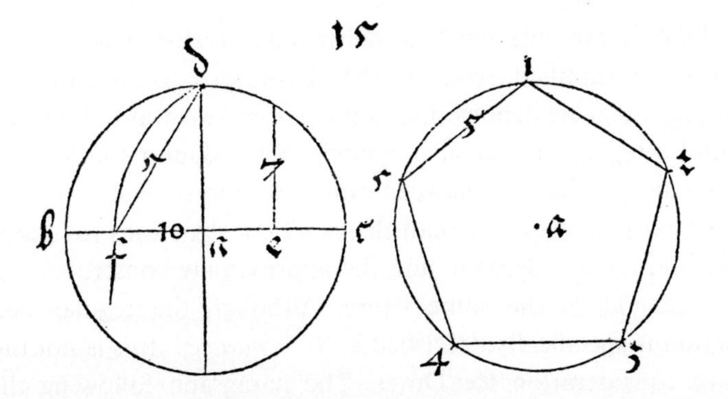

Fig. 1 Dürer's exact construction for the regular pentagon, after Ptolemy.

Even in this short passage, much of Dürer's style and method comes through. First, the informal, almost chatty style of his presentation is in sharp contrast with academic mathematical writing today. Second, the pedagogically motivated duplication in the phrase "a perpendicular line with equal angles" assists the reader who may not quite be aware that if the line at *a* is perpendicular to *bc* then necessarily the angles it makes with *bc* will be equal, and vice versa. (This same phrase is used two paragraphs earlier in constructing a square inscribed in a circle. After drawing one diameter we are instructed to "draw a perpendicular at the center, both above and below, at equal angles.") Third, the absence of any attempt at a proof that the construction is correct is characteristic of the *Underweysung*. It is enough that the reader should understand *what* to do, not *why*. Although Dürer does not specifically describe this pentagon construction as exact ("*demonstrative*"), the fact that the heptagon edge in the same figure is called approximate ("*mechanice*") suggests that he believed the pentagon (and decagon) construction to be exact. Fourth, the missing sentence which should describe the location of *e* suggests, with other available evidence, that Dürer copied his proof directly from some available version of Ptolemy. In other places the original source can be traced because the lettering of his illustration and the wording of the procedure are essentially identical to an original known to be available to Dürer through Pirckheimer or the mathematician Johannes Werner (Ref. 3, pp. 40–42; Ref. 13, p. 363.)

Fifth, Dürer's presentation of a multiplicity of ways of solving a single problem is exemplified here in the appearance of a third way of constructing a regular heptagon, and the first of two ways of constructing the edge of a regular pentagon in a circle. Sixth, Dürer's lack of concern for the practical difference between exact and approximate constructions is shown by his inclusion of both the exact constructions for the regular pentagon and regular decagon and the approximate construction for the regular heptagon in the same figure. Although the regular heptagon construction is specifically described as "*mechanice*," this is not the most important consideration for Dürer. The paragraph following this one begins "*Aber ein fünfeck auss unverruckten zirckel zu machenn* . . ." and continues to give an approximate construction of the regular pentagon, using a compass with fixed opening. The fact that the paragraph begins with this distinction indicates that Dürer placed more importance on the fixed compass opening than on its approximate nature. Indeed, for this construction, and the subsequent approximate construction of the regular nine-gon, there is no explicit mention that they are not exact. (In contrast, the constructions of the regular eleven-gon and thirteen-gon are again described as "*mechanice*.")

Finally, this passage raises the question about Dürer's dependence on Euclid, and his relation to Pacioli and his *Divina Proportione*. The latter relation, especially, has been the subject of much speculation, some of which will be touched on here. It has been suggested that Dürer deliberately avoided Euclid's construction (Book IV, Prop. 11) of the regular pentagon. This construction depends on constructing the isosceles triangle formed by the two diagonals at one vertex of the regular pentagon, together with the side opposite this vertex. Each base angle of this triangle is twice the vertex angle (Euclid, IV. 10). The construction of such a triangle in turn depends directly on the construction (in Euclid, II. 11) of the "golden ratio," Pacioli's "divine proportion." The golden ratio, $(1 + \sqrt{5})/2$, occurs here as the ratio of the long side of this triangle to its base, i.e. the ratio of a diagonal of the regular pentagon to its side.

The background for the sustained interest shown in this apparently insignificant omission on Dürer's part goes back to Dürer's second trip to Italy. During his first trip to Italy, made in 1494 while he was still a young man, he was very much a spectator of the sights of Venice as a world port and student of the distinguished artists and the influence of the renaissance excitement on their work. In his second trip, lasting a year and a half beginning in 1505, he went as an established artist, the co-worker and rival of artists in Italy.

Whether he met Pacioli or saw a manuscript of the *Divina Proportione* (first actually published in 1509) has been much debated, though most agree that he was particularly influenced by Pacioli. Now the *Divina Proportione* is a paean to the glories of the golden ratio: "*del primo proposto effecto, del secondo essentiale effecto de questa proportione, del terzo suo singolare effecto, del quarto eneffabile effecto*," and so on up to the thirteenth "*suo dignissimo effecto*" where he stops in honor of Jesus and his twelve disciples. It is remarkable that in his theoretical writings Dürer makes no explicit reference to the golden ratio whatsoever, and even deliberately avoids the implicit reference that would be made by using Euclid's regular pentagon construction. Three "explanations" of this omission, beginning with the most grandiose, are described in the next paragraphs.

Christine Papesch (pp. 201–211 in Ref. 8) is of the opinion that in Dürer's systematic study of perspective and of the Italian theoretical proportions of the human figure, he eventually balked at the rigid application of the golden ratio as a suitable guiding principle dependent neither on "subjective taste nor on metaphysical principles." Dürer abandoned the idea of an ideal beauty that based the drawing of the human body on the geometry of the golden ratio, and he undertook direct organic and anthropometrical studies of the human figure. In his engravings, this change is manifested in the contrast between the ideal conception of *St. Jerome in his Study* and the frustration with scientific possibilities expressed in *Melencolia I.* Having abandoned his faith in the golden ratio as an anatomical model, he rigidly excluded it from his geometrical guide for young artists.

Staigmüller's[3] explanation for Dürer's lack of reference to the golden ratio is more prosaic. He agrees neither with those who conclude that Dürer had never seen the *Divina Proportione*, nor with Papesch's theory of its deliberate exclusion. He thinks it quite likely that Dürer saw a manuscript version of Pacioli's work in Venice. There are general similarities between his book and Pacioli's that support other evidence for this. These similarities include the statement of purpose in Dürer's introductory letter to Pirckheimer, as well as the general contents of the two books, such as the constructions of letters of the alphabet and the treatment of regular and semi-regular polyhedra. On the other hand, suggests Staigmüller (reflecting, no doubt, his own experience, as well as that of all writers before the photocopy age), Dürer no longer had a copy of the book in front of him when writing the *Underweysung*, and his mathematical abilities were certainly not such that he could reproduce the many arithmetical and geometrical associations of the golden ratio for himself.

The present writer is inclined to extend Staigmüller's unromantic analysis to the particular choice of the Ptolemaic construction of the regular pentagon instead of Euclid's. We know that Dürer had considerable assistance with the purely mathematical part of the *Underweysung*, both from his scholarly friend Pirckheimer and the prominent Nuremberg mathematician Johannes Werner. It is easy to imagine that Dürer did not want to digress from his main theme to describe Euclid's somewhat complicated and unmotivated construction of the special isosceles triangle, since his other constructions do not depend heavily on anything but simple lemmas, such as bisecting angles and constructing perpendiculars. We can imagine his reaction when Pirckheimer or Werner showed him Ptolemy's simple direct proof. "This will do perfectly well, and it has the advantage of not requiring a lengthy discussion of the golden section, about whose virtues I am a bit suspicious just now anyway."

The second construction of the regular pentagon given by Dürer is the same approximate construction which appears in the *Geometria Deutsch*. For everyday workshop use, its advantages have already been mentioned: it uses only a fixed compass opening, it is closely related to the simple, well-known, construction of a regular hexagon, and it is symmetric. His description is:

> "To make a pentagon with a compass having a fixed opening, do it as follows. Draw two circles, one through the center of the other, and join the two centers *a, b* with a straight line. This will be the length of one side of the pentagon. Where the circles cross, put *c* above and *d* below, and draw line *cd*. Now take the unchanged compass and place one leg at *d* and with the other draw an arc through the two centers *a, b*. Where it meets the two circles again, put *e, f*. Where it meets the upright *cd*, put *g*. Then draw line *eg* clear through to the circle and mark this point *h*. Then draw another line *fg* through to the circle and mark this *i*. Then draw the lines *ia* and *hb*. Now there are three sides of the pentagon. From this, draw two equal sides up from *i, h*, to get a pentagon, as I have drawn below."

(Note that in Fig. 2, to which Dürer refers in this passage, he has not labelled points *c* through *i*.)

This construction, except for the very last step, is copied almost verbatim from the *Geometria Deutsch*. In this last step, Dürer has made the construction more symmetric by finding the fifth vertex as the intersection of two arcs from the symmetrically placed points *i, h*. The *Geometria Deutsch* constructs this fifth vertex as the intersection of one of these arcs with the line *cd*. Except that all the relevant points are labelled, and slightly different letters are used, the figure given in the *Geometria Deutsch* is identical to Dürer's.

Dürer does not tell us whether this construction is exact or approximate. Perhaps he did not know. A hundred years later, Cataldi[10] devoted some ten pages, most of which are elaborate arithmetic calculations, to showing that this construction, which he attributes to Dürer, is incorrect. Staigmüller also mentions two writers before Cataldi who showed that Dürer's construction was inaccurate. Then, in a brief footnote, he calculates the angles at *a* and *b* to be $108^\circ\ 21'\ 58''$ instead of the 108° needed for an equiangular pentagon. (The equal angles at *i* and *h* are too small, and the upper angle is too large by more than a degree.) Dürer's pentagon is easily seen to be equilateral, because of the fixed compass opening.

It is interesting that, after giving these two regular pentagon constructions that carefully avoid Euclid, Dürer now gives a construction of the

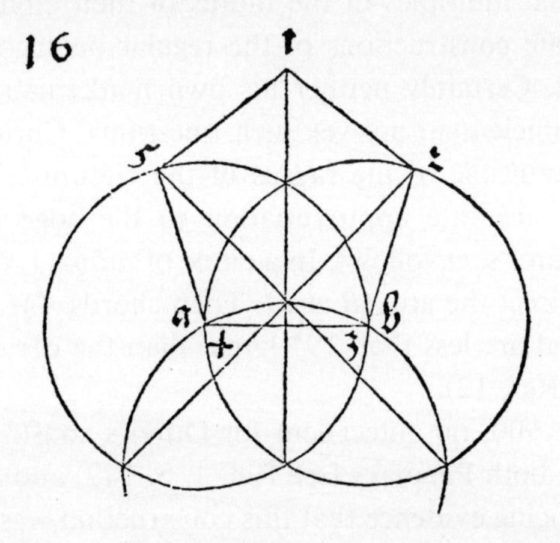

Fig. 2 Dürer's approximate construction of the regular pentagon, from *Geometria Deutsch*.

regular fifteen-gon that follows that of Euclid IV. 16 exactly. An equilateral triangle and a regular pentagon are inscribed in a circle so that they have one common vertex. The vertices thus divide the circle into arcs of lengths 3/15, 2/15, and 1/15 of its circumference. Euclid (somewhat awkwardly, it would seem) finds an arc corresponding to the side of a regular fifteen-gon by bisecting one of the 2/15 arcs. Dürer also, instead of simply choosing the 1/15 arc, repeats Euclid's unnecessary bisection.

5. Dürer's Construction of Regular Nine-, Eleven-, and Thirteen-gons

Dürer's construction of the (approximate) regular nine-gon is ingenious and elegant, though, in contrast to his other regular polygon constructions (except the approximate regular pentagon), it does not directly yield a nine-gon inscribed in a previously prescribed circle. He emphasizes that the beginning of the construction requires only one compass opening. However, he makes no comment as to whether the construction is approximate or exact. The construction has some general resemblance to his approximate construction of the regular pentagon, both using a segment determined by a "fish bladder." Perhaps because of their comparative accuracy he did not know whether either was exact. His instincts may have told him that his constructions of the regular eleven-gon and thirteen-gon could not be correct, since their side lengths are simple rational multiples of the radius of their circumcircles, while the more elaborate constructions of the regular pentagon and nine-gon might be correct. Certainly neither his own mathematical abilities nor interests were sufficient to answer such questions. Curiously enough, a simple rational multiple of the radius of the circumscribed circle does give a far more accurate approximation to the edge of an inscribed nine-gon than Dürer's, as follows: In a circle of radius r, draw a chord AB of length $9r/7$. Bisect the arc AB at M. Then chords $AM = MB$ subtend central angles that are less than 19″ larger than the correct value of 40° (Ref. 7, p. 257; Ref. 12).

Until at least 1900, no antecedent for Dürer's construction had been found. However, both Panofsky (see Ref. 1, p. 149) and Steck (Ref. 4, p. 49) found convincing evidence that this construction was taken by Dürer from standard craftsmen's practice.

Dürer's description, as it comes through to modern ears, is translated in full. Fig. 3 is his:

> "A nine-gon can be found from a triangle. Draw a large circle with center *a*. Then draw, with unchanged opening, three fish bladders. Call the upper end on the circle *b*, the other ends on the sides *c, d*. After that, draw a vertical line *ab* in the upper fish bladder. Divide this line with two points 1, 2 in three equal parts, so that 2 is the nearest to *a*, and draw a straight cross line through point 2 at the same angles [with] *a, b*. And where it cuts through the fish bladder on both sides put *e, f*. After that take a compass, put one foot at the center *a* and the other at the point *e* and draw a circle in a ring through *f*. Then the length *ef* goes nine times around this circle, as I have drawn below." (see Fig. 3)

In manuscript (Ref. 13, p. 337), Dürer improves this in the obvious way by pointing out that three such edges give 3/9 of the nine-gon and bisecting the three remaining long arcs gives all the vertices of the nine-gon. The same source[13] reproduces drawings to show that Dürer made very careful original drawings, and that he did indeed bisect the remaining arcs to get the nine-gon, thus distributing the error rather than copying the single edge around the circle nine times according to his instructions in the *Underweysung*. Moreover, another drawing reproduced in Ref. 13 extends the lines from the circle center to the nine points on the small circle to meet the outer, original circle, thereby giving the vertices of a nine-gon in that original circle.

Not much needs to be said about Dürer's straightforward (approximate) construction of the regular eleven-gon. In fact, he only devotes three lines to it himself:

> "If I now want to draw an eleven-gon in a circle, I take a quarter of the circle's diameter and extend it by an eighth of itself, then go around the circle with this length. Incidentally, this is approximate, not exact."

This yields a central angle of $32° \ 40' \ 12''$ instead of the correct $360/11 = 32° \ 43' \ 38''$. Remarkably, this simple procedure is in this case more accurate than either of the standard general approximation methods quoted by Günther and Staigmüller, and described in Vahlen[7] in "Renaldini's Rule" and the construction by Karl Bernhard zu Sachsen-Weimar-Eisenach.

Dürer's intended construction of the regular thirteen-gon is not completely clear, though Hunrath's reading[12] is by far the most persuasive. The version of some commentators (including Strauss and the usually careful Staigmüller) is that Dürer takes half the edge of a regular

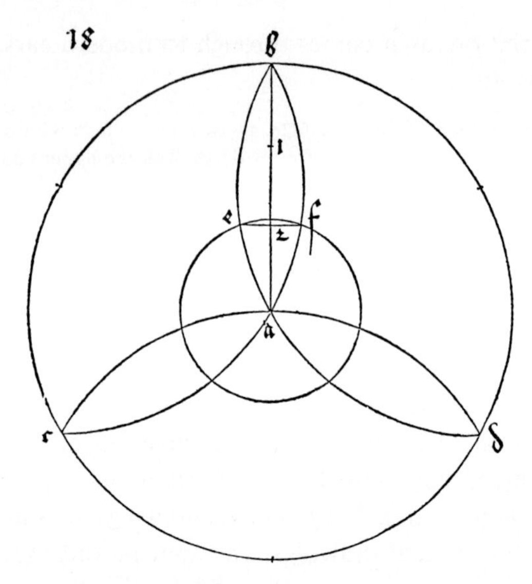

Fig. 3 Dürer's approximate construction of the regular nine-gon.

hexagon to be the edge of a regular thirteen-gon in the same circle. (Does this construction seem plausible because it is analogous to the very accurate construction of the edge of a regular heptagon as half the edge of a regular triangle in the same circle?) In fact, an easy calculation shows that in a unit circle this method yields a side length of .5000, which is noticeably closer to the side length .5176 of a regular twelve-gon than to .4786 ($= 2 \sin (180°/13)$), that of a regular thirteen-gon.

Hunrath notes that Dürer's own words are, "Draw a circle with center *a*. Then draw a radius *ab* and cut it in the middle by a point *d*. Then use the length *cd* 13 times around the circle." (Note that the point *c* is not specified, and that the apparently erroneous interpretation replaces *c* by *b*.) Hunrath's simple explanation is that Dürer intended the sentence "From *bd* subtract one twenty-fourth of *bd* and mark this point with *c*" to appear between the last two of his sentences. We can quickly calculate the accuracy of this construction. In a unit circle, let A be the angle subtended by this side, $1/2 - (1/24)(1/2)$. Then $\sin (A/2) = 23/96$, so that $A = 27° 43' 26''$. This is remarkably close to the correct value $A = 360°/13 = 27° 41' 32''$. Dürer's drawing appears in our Fig. 4. It seems possible that the illegible symbols below the 1 at the top of the figure are the "24" referred to by Hunrath.

Staigmüller commented that "Even in Dürer's figure the mistake is disturbingly apparent at first glance." On the contrary, what is disturbing is that Dürer's figure at first glance looks correct. It is only when the reader tries to reconstruct the figure from Dürer's (apparent) instructions that it is discovered that the result has little resemblance to Dürer's figure. Is it reasonable to suppose that Staigmüller's explanation for Dürer's non-use of Pacioli's golden ratio can be applied here to Staigmüller himself? Having copied Dürer's verbal description in the library, but not having facilities to make an exact copy of the figure, he reconstructed the latter from Dürer's words and indeed found a disturbingly apparent error.

6. Regular Pentagons in Stars, Tilings, Rings, and Polyhedra

We conclude with brief mention of other occurrences of regular pentagons in the *Underweysung*. Two of Dürer's mathematical errors occur in connection with these later appearances of regular pentagons on his stage.

His section on construction of regular polygons is interrupted by an approximate trisection of an arc, and an impatient remark to those of us who continue to ask whether constructions are exact or only approximate: "*Wer es will geneuer haben/der such es demonstrative*" ("Whoever wants it more exact is welcome to look for a proof"). This is immediately followed by several pages on the use of polygons for decorative purposes. He shows a variety of stars derived from polygons inscribed in circles, including the pentagonal star of Fig. 5.

He suggests that tiles of polygonal shape could either be placed to cover a floor, or arranged to make attractive designs in the floor. In particular, he devotes two pages, including four drawings, to tilings involving congruent regular pentagons. Two of these, whose gaps can be filled with congruent rhombuses, are shown in Fig. 6. Grünbaum and Shephard (Ref. 14, pp. 52–53) have noted in passing that these are among the uncountably many ways to tile the plane using tiles congruent to these two, but that if only finitely many ($n \neq 0$) pentagons are permitted then there are exactly three possible such tilings, for $n = 1, 2$, and 6.

Dürer's strangest mathematical error occurs in arranging regular pentagons, and regular heptagons, in the decorative rings shown in Fig. 7. Not only are both rings incorrect in his figure, but Dürer explicitly states that "if you put seven-gons one after another . . . they go around in a

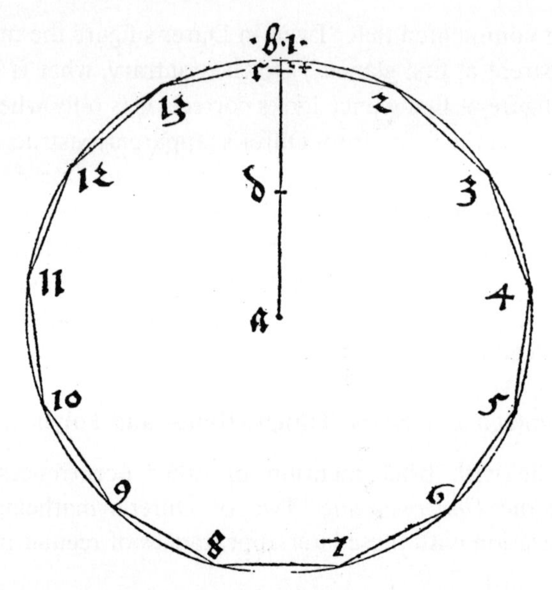

Fig. 4 Dürer's misunderstood construction of the regular thirteen-gon.

circle. However, they don't close up. The same holds for the pentagon."

In fact, in each ring in Fig. 7 there is room for exactly one more regular polygon than Dürer has drawn. This is easy to verify: In a ring of regular

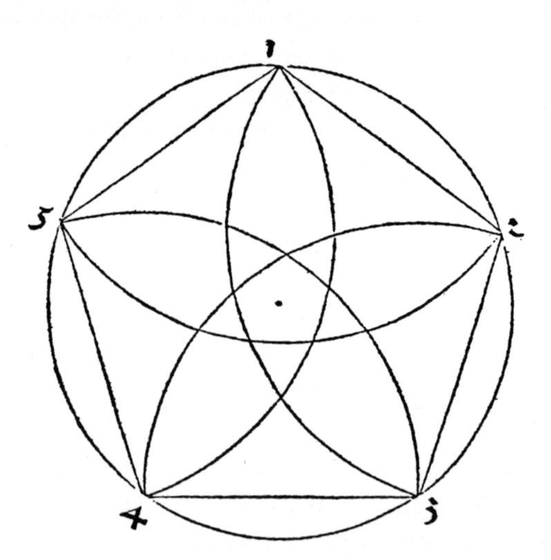

Fig. 5 A pentagonal star, for decorative purposes.

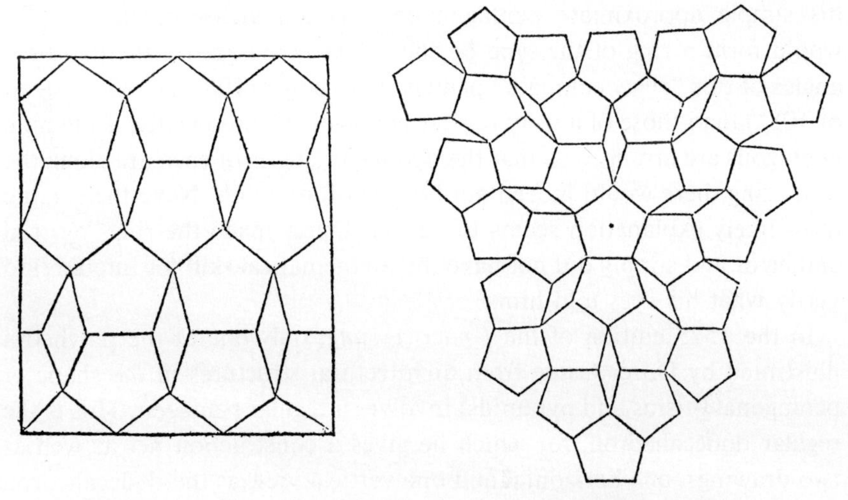

Fig. 6 Two tilings of the plane by rhombuses and regular pentagons.

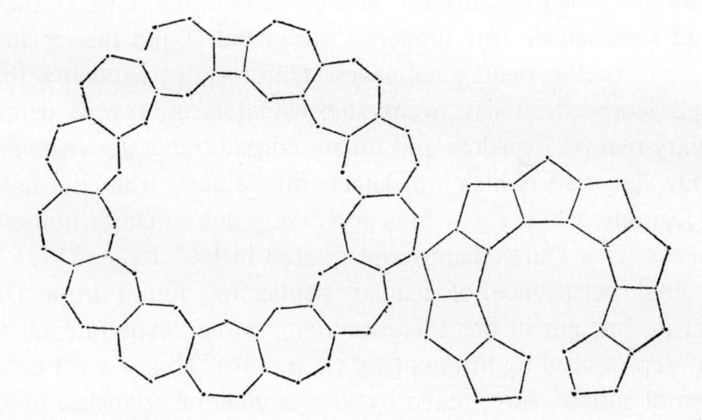

Fig. 7 Dürer's strangest mathematical error: non-closing rings of regular heptagons and regular pentagons.

pentagons, each angle between two adjacent pentagons is $360° - 2\ 108° = 144°$. This is exactly the vertex angle of the regular decagon around which they could be arranged. A similar calculation shows that Dürer's ring of regular heptagons would close up exactly around a central (non-convex) 28-gon. Dürer's error is especially strange in view of his (implicit) earlier emphasis on the construction of the regular 28-gon.

It might be asked whether Dürer's error here stems from the fact that his simple approximate pentagon construction yields pentagons that would form a ring of the type he shows. As noted above, the two base angles of this "rusty compass" pentagon are larger (108° 21′ 58″ instead of 108°) than those of a truly regular pentagon. Consequently, if the nine pentagons are arranged so that the two larger angles of each are along the inner ring there would indeed not be room for a tenth. Nevertheless, the most likely explanation seems to be that Dürer made the rings by trial and error and simply did not have the mathematical skill (or interest?) to verify what his eyes told him.

In the 1525 edition of the *Underweysung*, only one of the polyhedra illustrated by Dürer (aside from architectural structures in the shape of pentagonal prisms and pyramids) involves a regular pentagon. This is the regular dodecahedron, for which he gives a construction net as well as two drawings, one horizontal and one vertical view as the dodecahedron rests on one face. For the 1538 (posthumous) edition, two Archimedean polyhedra are added to the seven in the 1525 edition. Each of these, the truncated icosahedron (the modern "soccer ball") and the icosidodecahedron, has twelve pentagonal faces. Dürer states explicitly that the truncated icosahedron has twenty hexagonal faces, twelve pentagonal faces, sixty-two [*sic*] vertices and ninety edges ("*scharpfer seitten*"). The erroneous sixty-two (which we detect immediately from the failure of Euler's formula, $v - e + f = 2$) is apparently due to Dürer himself, as it also appears in a Dürer manuscript quoted in Ref. 13, p. 351.

One final occurrence of regular pentagons, found in a Dresden manuscript, but not in the *Underweysung*, is the "experimental folding pattern" reproduced by Strauss (Ref. 1, p. 456). This is a net consisting of a central square, surrounded by four equilateral triangles, to each of which is attached a regular pentagon.

Appendix

Some other constructions

Two general approximate constructions for the side of a regular n-gon inscribed in a unit circle are referred to by Günther and Staigmüller as if they were well-known one hundred years ago. These are Renaldini's Rule, which apparently dates to 1628, and the 19th Century (?)

construction of Karl Bernhard zu Sachsen-Weimar-Eisenach. These are described in Vahlen (Ref. 7, pp. 302–305), as follows.

To apply Renaldini's Rule, first draw an equilateral triangle ABC on one diameter AB of the unit circle. Take D on diameter AB so that $AD = (2/n)AB$. Let P be the point where line CD meets the circle. Then AP is the (approximate) side of a regular n-gon inscribed in the circle.

The case $n = 5$ is illustrated in Fig. 8. For the pentagon, the edge obtained in this way has length 1.1749, only slightly shorter than the correct value 1.1756. The corresponding central angle is less than 3′ too small. For $n = 7$, the error doubles, to more than 5′, almost as much too large as Durer's value is too small.

This method, although no longer well-known to mathematicians, has survived, for the special case of the pentagon only, in a modern workshop manual, *Nomadic Furniture.*[15]

The second construction is described by Vahlen as "very exact," and appears to be so for larger values of n. For $n = 5$, it is less exact than the preceding method. Its description is (see Fig. 9):

In a unit circle with diameter AB and center O, extend both the diameter and a perpendicular radius OR by $(1/n)\ AB$, to E and F respectively. Take D on diameter AB so that $AD = (3/n)AB$. Let G be the point nearest the diameter where line EF meets the circle. Then DG is the (approximate) side of a regular n-gon inscribed in the circle.

For $n = 5$, $DG = 1.1662$, nearly 1% less than the correct side 1.1756. For $n = 6$, the construction is exact, and for $n = 7$, the resulting central angle is barely 1′ too small, considerably more accurate than either Dürer's or Renaldini's constructions.

Finally, in a recent paper,[16] Gleason has discussed exact constructions with straightedge, compass *and an angle trisector.* In that connection, he describes a construction of the regular heptagon, due to Plemelj, which shows how the construction given by Dürer ("half the side of the equilateral triangle") can be modified, using an angle trisection, to obtain an exact construction. The theory developed by Gleason shows that the regular thirteen-gon (but not the regular eleven-gon) is also exactly constructible by straightedge, compass and angle trisector. He devises such a construction, and derives from it two further approximate constructions for the regular thirteen-gon.

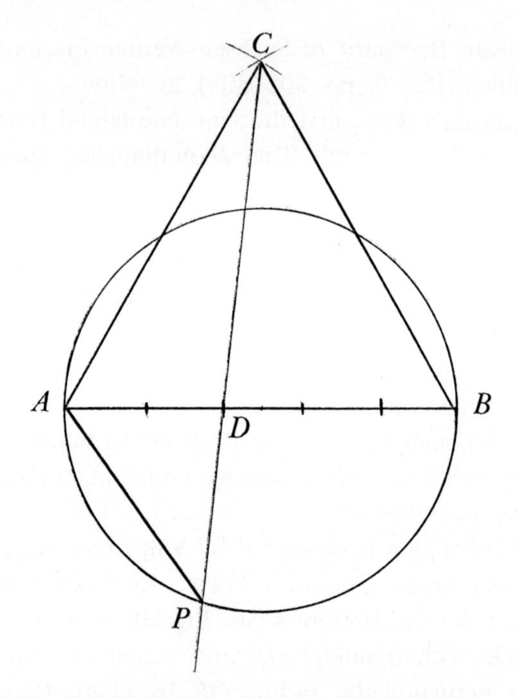

Fig. 8 Renaldini's Rule, illustrated for the regular pentagon.

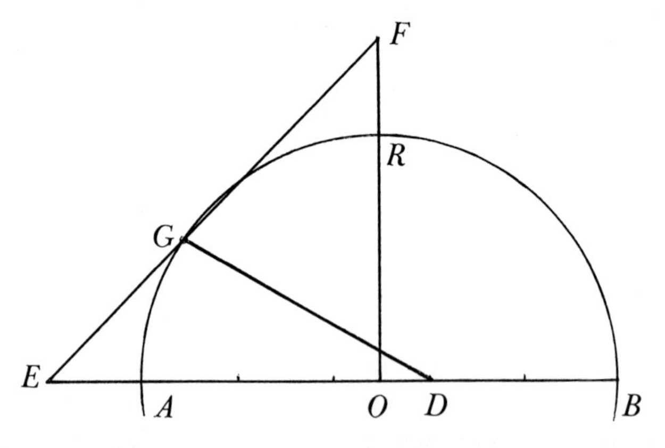

Fig. 9 The construction by Karl Bernhard zu Sachsen-Weimar-Eisenach, illustrated for the regular pentagon.

Acknowledgment

The author is indebted to Wolfgang Wasow for assistance in the translation of several passages from Dürer and Pacioli, as well as continuing discussions of Dürer as geometer. However, the final versions of the translations from Dürer are the author's own.

References

1. W.L. Strauss, Albrecht Dürer. *The Painter's Manual 1525*. Abaris Books, New York (1977).
2. S. Günther, *Die Geometrischen Näherungskonstruktionen Albrecht Dürers, Beilage zum Jahresbericht der Kgl. Studienanstalt Ansbach für* 1885/86. Ansbach (1886).
3. H. Staigmüller, *Dürer als Mathematiker, Programm des Königlichen Realgymnasiums in Stuttgart am Schluss des Schuljahrs 1890/91*. Stuttgart (1891), 1–59.
4. M. Steck, *Dürers Gestaltlehre der Mathematik und der Bildenden Künste*. Max Niemeyer Verlag, Halle (Saale) (1948).
5. E. Panofsky, *The Life and Art of Albrecht Dürer*, 4th Edn. Princeton University Press, Princeton (1955).
6. M. Cantor, *Vorlesungen über Geschichte der Mathematik*, Vols. I (2nd Edn.), II. B.G. Teubner Verlag, Leipzig (1894, 1893).
7. Th. Vahlen, *Konstruktionen und Approximationen*. B.G. Teubner Verlag, Leipzig (1911).
8. A. Jaeggli, ed., *Unterweisung der Messung mit dem Zirckel und Richtscheit* (facsimile edition). Joseph Stocker-Schmid Verlag, Dietikon-Zurich (1966).
9. *Geometria Deutsch* (1484?). Reproduced in Ref. 4.
10. P. Cataldi, *Trattato geometrico, dove si esamina il modo di formare il pentagono sopra ad una linea retta, descritto da Alberto Durero*. Sebastiano Bononi, Bologna (1620).
11. D. Pedoe, Albrecht Dürer and the ellipse, *The Australian Mathematics Teacher* **30(6)** (1974) 222–226.
12. K. Hunrath, Zu Albrecht Dürers Näherungskonstruktionen regelmässiger Vielecke, *Bibliotheca Mathematica*. Dritte Folge, **6** (1905) 249–251.
13. H. Rupprich, ed., *Dürer, Schriftlicher Nachlass*, Vol. 3. Deutscher Verlag für Kunstwissenschaft, Berlin (1969).
14. B. Grünbaum and G. Shephard, *Tilings and Patterns*. Freeman, San Francisco (1987).
15. J. Hennessey and V. Papanek, *Nomadic Furniture*. Pantheon, New York (1973).
16. A. Gleason, Angle trisection, the heptagon and the triskaidecagon, *Amer. Math. Monthly* **95(3)** (1988) 185–194.

FIVEFOLD SYMMETRY IN THE GRAPHIC ART OF M.C. ESCHER

R. A. Dunlap

1. Introduction

M.C. Escher is known almost exclusively for his graphic works. Although these clearly display considerable technical ability, Escher's work was not held in high esteem for its artistic content by most other artists. It was, rather, among scientists that Escher found his most ardent admirers. Escher himself commented,[1]

> "The ideas that are basic to them (Escher's prints) often bear witness to my amazement and wonder at the laws of nature which operate in the world around us. He who wonders discovers that this is in itself a wonder. By merely confronting the enigmas that surround us, and by considering and analyzing the observations that are made, I ended up in the domain of mathematics. Although I am absolutely without training or knowledge in the exact sciences, I often seem to have more in common with mathematicians than with my fellows artists."

Certainly more often, critical studies of Escher's works have dealt with their scientific content rather than with their artistic content. While Escher's early prints were based on reality, usually Italian architecture, his later prints were based, as he explained, on "inner visions."[1] These later works exhibit symmetry in a variety of manners. Symmetry aspects of Escher's prints involving the regular division of a plane have been considered in detail by MacGillavry[2] and are most apparent in his "Circle Limit" prints whiach show well defined threefold symmetry (e.g. *Circle Limit I*, 1958, *Circle Limit IV*, 1960) or fourfold symmetry (e.g. *Circle Limit II*, 1959,

Circle Limit III, 1958). In fact, these woodcuts were produced by printing the same block three or four times to take advantage of the symmetry of the pattern. Many of Escher's prints of three-dimensional objects indicate his interest in the structure of crystals, both from a macroscopic (e.g. *Crystal*, 1947, and *Dragon*, 1952), as well as a microscopic point of view (e.g. *Cubic Space Division*, 1952, and *Depth*, 1955). Escher kept close contact with the scientific community, in particular with mathematicians and crystallographers, and incorporated their ideas into many of his prints. In fact, Escher did not have to look for his scientific inspiration, as his father and three of his brothers were trained as scientists or engineers. Escher's fascination for crystals was emphasized by his comment[3]:

> "Long before there were people on the earth, crystals were alreay growing in the earth's crust. On one day or another, a human being first came across such a sparkling morsel of regularity lying on the ground, or hit one with his stone tool and it broke off and fell at his feet, and he picked it up and regarded it in his open hand, and he was amazed."

Fivefold symmetry can be seen in a number of Escher's prints. Although he did not experiment with quasiperiodic tilings of the Penrose type,[4,5] some examples of nonperiodic two-dimensional tilings which exhibit fivefold symmetry are known (see e.g. Ref. 6). It is, however, in his representation of three-dimensional objects that fivefold symmetry most notably appears, usually in the form of the icosahedron and the dodecahedron. Escher expressed his own interest in the regular geometric solids in the following way[7]:

> "They emphasize man's longing for harmony and order, but at the same time their perfection awes us with a sense of our helplessness. Regular polyhedra are not inventions of the human mind, for they existed long before mankind appeared on the scene."

In the present review, we discuss fivefold symmetry aspects of Escher's graphic works. We begin with a biographical introduction to help relate Escher's ideas to his life. We then proceed with an analysis of prints which make use of fivefold symmetry and we relate the ideas used in these prints to general aspects of Escher's works.

2. Historical Background

From an early age, M.C. Escher showed an interest in art. Although his father had encouraged him to study architecture at the School of Architectural and Decorative Arts in Haarlem, Escher's interests soon turned to graphic arts. During the period 1919 to 1922, Escher eagerly undertook the study of graphic technique with Jessurun de Mesquita whom he had met in Haarlem during his architecture studies. Following this, Escher travelled extensively in Spain and Italy, and finally settled in Rome in 1923, where he lived for 12 years. During this period, Escher's prints were based primarily on landscape and architectural topics. Even in these early prints, Escher's interest in symmetry was apparent (e.g. *Palm*, 1933), as well as his interest in regular three-dimensional objects (e.g. *Genzazzano, Abrozzi*, 1929, *Town in Southern Italy*, 1930). The early experimental print *Eight Heads*, 1922, even foreshadowed his later fascination for plane-filling motifs and regular tilings.

Escher's life in Italy became less bearable with the rise of the Fascist regime in the 1930s and finally, in 1935, he moved to Chateau d'Oex, Switzerland. There are few prints from the period of his short stay in Switzerland, and these tend to indicate a temporary interest in nature (e.g. *Snow in Switzerland*, 1936, *Prickly Flower*, 1936, *Libellula*, 1936). In May and June of 1936, Escher travelled by ship along the coast of Italy and Spain. This trip produced the woodcuts *Freighter*, 1936, and *Porthole*, 1937, but, more importantly, allowed Escher to make a detailed study of Moorish mosaics.

In 1937, Escher moved to Ukkel, Belgium, and then to Baarn, The Netherlands, in 1941. In 1970, at the age of 72, Escher moved to Laven, The Netherlands, where he died on 27 March 1972.

Escher's early work was largely influenced by his surroundings and, to a large extent, he chose his surroundings, i.e., immersed in the architectural heritage of Italy, to be compatible with his interests. His later work stemmed almost entirely from his inner thoughts and was influenced primarily by his interest, albeit from an amateur standpoint, in the exact sciences. This transition occurred around 1937, and Escher's extensive study of Moorish mosaics at this time may have played an important role in the increasing use of symmetry in his art.

3. Escher's Use of Fivefold Symmetry

In this section, the major prints of M.C. Escher which exhibit fivefold symmetry are considered. The aspects of fivefold symmetry in each print, and the relationship of the print to Escher's work in general, are discussed. Prints are considered in chronological order.

3.1 *St. Bavo's, Haarlem*

This india ink drawing (see Fig. 1), dating from 1920, is one of Escher's earliest known graphic works. It was made at the time Escher was in Haarlem, studying under Jessurun de Mesquita. The chandelier has a tenfold rotation axis and shows Escher's interest in symmetry, even at this early date. The drawing shows a number of features which played an important role in Escher's later prints. The vaulted ceiling appears

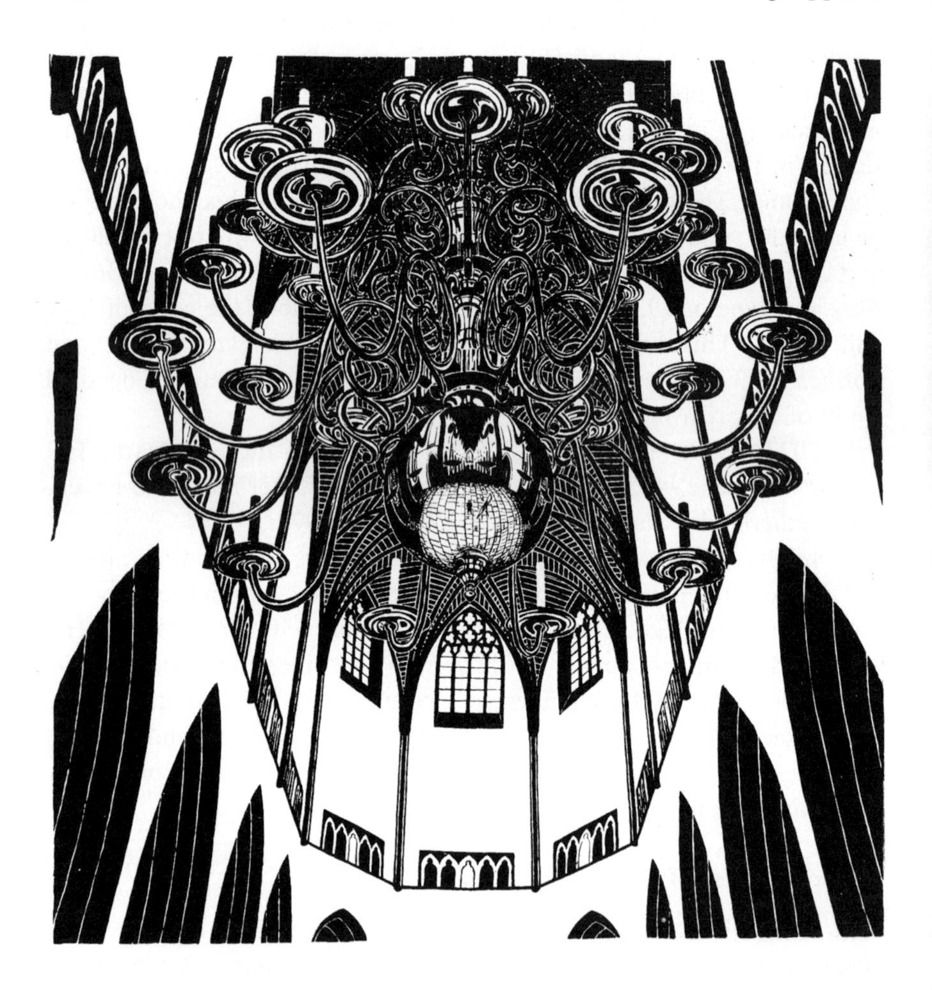

Fig. 1 St. Bavo's, Haarlem (1920). India ink, 117.0 cm high × 99.0 cm wide. © 1989 M. C. Escher Heirs/Cordon Art-Baarn-Holland.

frequently in his later work, and is, perhaps, most apparent in *Dream*, 1935. The feeling of perspective results from the highly convergent lines. Escher used this same technique, although from the viewpoint of looking down rather than looking up, in *Tower of Babel*, 1928, and *St. Peter's, Rome*, 1935. Perhaps the most intriguing aspect of this drawing is the presence of the highly distorted reflection of the artist himself. Mirror images are common in Escher's work and he has used the image of the artist in a spherical reflector on a number of other occasions. After *St. Bavo's, Haarlem*, one of the earliest examples of this is *The Sphere*, 1921, from illustrations for A. P. van Stolk's *Flor de Pascua*. Better known examples of this concept are *Still Life with Reflecting Sphere*, 1934, *Hand with Reflecting Sphere*, 1935, and *Three Spheres II*, 1946.

3.2. *Prickly Flower*

This wood engraving is illustrated in Fig. 2. It is dated "2–36" and was made during Escher's brief stay in Switzerland. It illustrates a cluster of flowers with fivefold symmetry and represents one of Escher's final subjects from nature.

3.3 *Reptiles*

One of Escher's best known prints, *Reptiles*, dates from 1943 and exhibits fivefold symmetry in the form of a regular dodecahedron. This lithograph, which is illustrated in Fig. 3, combines a number of concepts which played a major role in Escher's later graphic work. The reptile, or "alligator" as Escher refers to the creature,[7] exists initially as a regular division of a plane. This particular division of the plane has obvious sixfold symmetry and dates from a 1939 notebook. It was used previously in *Development II*, 1939, and *Metamorphosis II*, 1939–1940. Escher's interest in this concept was apparent in his early woodcut *Eight Heads*, 1922, although this may be viewed more as contour-sharing rather than the tiling concept of Escher's later works. Beginning in 1926, Escher kept a series of notebooks containing ideas for regular divisions. However, it was not until after his trip to Spain in 1936 to study Moorish mosaics, that regular divisions formed a large component of Escher's work. In *Reptiles*, the little alligator frees itself from the two-dimensional tiling and, as Escher explains it,[1] "Launches out into real life." The emergence of three-dimensional beings from two dimensions appears in a number of

Fig. 2 Prickly Flower (1936). Wood engraving, 27.7 cm high × 20.8 cm wide. © 1989 M. C. Escher Heirs/Cordon Art-Baarn-Holland.

Escher's later works. Some particular examples are *Drawing Hands*, 1948, and *Liberation*, 1955. Specifically, *Reptiles* represents not only the emergence of three dimensions from two dimensions but also a cyclic transformation. This closed cycle has allowed Escher to illustrate an infinite progression within a finite space. This cyclic nature has been used in conjunction with the two- to three-dimensional transition in *Encounter*, 1944, and *Magic Mirror*, 1946. *Reptiles* is also a print which displays contrasts, the highly ordered shape of the dodecahedron contrasting the natural structure of the succulent plant. In addition, the collection of ordinary objects contrasts with the presence of the fantastic little reptile. Both of these elements appear as well in *Order and Chaos*, 1950. Finally, *Reptile* is a good example of Escher's ever-present sense of humour. The book across which the reptiles walk is a text book on zoology. The small

Fig. 3 Reptiles (1943). Lithograph, 33.4 cm high × 38.6 cm wide. © 1989 M. C. Escher Heirs/Cordon Art-Baarn-Holland.

book in the lower left hand corner of the print is labelled "Job." Locher[2] has pointed out that Escher enjoyed the fact that this book was commonly assumed to be of biblical origin. In fact, Escher explained the truth of the matter in an introduction to a later edition of his prints as: "The booklet Job has nothing to do with the Bible, but contains Belgian cigarette papers."[8]

3.4 *Study for the Wood Engraving "Stars"*

This 1948 woodcut, illustrated in Fig. 4, served as a study for the wood engraving *Stars*, 1948, and depicts ten geometric solids. The solids are illustrated as frames and are either single, double, or triple forms. While Escher's fascination for regular solids is conspicuously apparent in many of his later works (e.g. *Tetrahedral Planetoid*, 1954, and *Reptiles*, 1943), this motif appeared less conspicuously in a number of his pre-1937 works as well. The "architectural" works based on Italian villages (e.g. *Goriano Sicoli, Abruzzi*, 1929, and *Genazzaro, Abruzzi*, 1929), are good examples of Escher's early interest in geometrical shapes. The idea of interpenetrating structures was first apparent in his print *Crystal*, 1947. *Study for the Wood Engraving "Stars"* displays fivefold symmetry in the form of a regular icosahedron and a regular dodecahedron. The figures represented in this print are (from left to right) first row: superposition of a cube and an octahedron, tetrahedron and triple octahedron; second row: a type of rhombohedron, icosahedron, and cube, and third row: superposition of two tetrahedra, dodecahedron and superposition of two cubes.

3.5 *Stars*

This 1948 wood engraving, as illustrated in Fig. 5, is based on the study produced as a woodcut during the same year. In *Stars*, the triple octahedron has become the central figure. The icosahedron and dodecahedron are still present but have become solid figures. The icosahedron appears to the left of centre in the top of the print and the dodecahedron appears to the right of centre in the bottom of the print. A second example of each of these two figures appears on the left hand edge of the print. The contrast between high symmetry and low symmetry, or analogously, between order and chaos, as has been seen previously in *Reptiles*, 1943, and was used later in *Order and Chaos*, 1950, appears

Fig. 4 Study for the wood engraving "Stars" (1948). Woodcut, 37.0 cm high × 37.5 cm wide.

here between the regular octahedra and the lizards which inhabit it. Escher himself said of this print[8]: "Two chameleons have been chosen as denizens of this framework because they are able to cling by their legs and tails to the beams of their cage as it swirls through space."

3.6 *Order and Chaos*

Fig. 6 shows the lithograph *Order and Chaos*, which was completed in 1950. The "order" is manifest in the form of a regular geometric solid, which Escher[8] calls a "stellar dodecahedron" encased in a transparent

Fig. 5 Stars (1948). Wood engraving, 31.7 cm high × 25.8 cm wide. © 1989 M. C. Escher Heirs/Cordon Art-Baarn-Holland.

sphere. The stellar dodecahedron clearly exhibits dodecahedral symmetry, and can be thought of as a dodecahedron with each of its faces extended to form one face on each of five pentagonal pyramids. Thus the figure consists

Fig. 6 Order and Chaos (1950). Lithograph, 28.0 cm high × 28.0 cm wide. © 1989 M. C. Escher Heirs/Cordon Art-Baarn-Holland.

of twelve inter-connected, star-shaped surfaces. Analogously, one may construct an icosahedron by creating twenty faces between three adjacent stellar vertices. This is the first appearance of this structure in Escher's work, but it reappeared two years later in the print *Gravity*, 1952. "Chaos" appears in the form of "a heterogeneous collection of all sorts of useless broken and corrupted objects."[8] This print, like so many of Escher's, has brought together a number of basic concepts — the contrast between order and disorder, as we have seen already in *Reptiles*, 1943, and *Stars*, 1948, and between the ordinary and the fanciful. Also, the concept of the symmetric regular solid appeared frequently in Escher's prints from the late 1940s and early 1950s (e.g. *Crystal*, 1947, *Stars*, 1948, *Double Planetoid*,

1949, *Gravity*, 1952, and *Tetrahedral Planetoid*, 1954), and developed out of an interest in regular solids expressed in Escher's early "architectural" works. Finally, we see once more the idea of a sphere reflecting its surroundings. In this case, the sphere not only shows the reflection of the objects around it, relating the ordered world to the chaotic world, but shows the reflection of the window of the artist's room and relates the subject of the picture to the world outside the picture. This latter concept, as we have seen, was present in Escher's very early work, *St. Bavo's, Haarlem*, 1922.

3.7 *Gravity*

This 1952 lithograph (see Fig. 7) uses the stellar dodecahedron that appeared previously in *Order and Chaos*, 1952. Each face of the pentagonal pyramids is partially open, forming a five-sided cage, the floor of which is the face of the inner dodecahedron. Each of these cages is inhabited by what Escher calls[8] "a tail-less monster with a long neck and four legs."

The creature stands on the face of the dodecahedron and sticks its head and four legs through the five openings in the pentagonal pyramid which encases it. The creatures have been water coloured by hand by the artist in six different colours: red, orange, yellow, green, blue and violet. The twelve-faced dodecahedron houses twelve creatures, presumably, two of each colour, all of which are visible (at least in part) except for three which live on the far side of the dodecahedron. Pairs of like-coloured creatures inhabit opposite faces of the dodecahedron. In addition to the geometrical aspects exhibited by this print, Escher has also used a concept that appeared in other prints from this period of his work, that of a small, self-contained world with its own centre of mass, and a relative sense of down at each point on its surface. This concept is displayed also in *Double Planetoid*, 1949, and *Tetrahedral Planetoid*, 1954.

3.8 *Stereometric Figure*

This 1961 wood cut is seen in Fig. 8. It is not one of Escher's better known prints, nor is it one that particularly demonstrates his graphic ability. It is, however, the one that is the most interesting in the context of fivefold symmetry. The "stereometric" figure is in fact the superposition of an icosahedron and a dodecahedron, the two regular platonic solids that exhibit fivefold symmetry. The icosahedron has twenty faces and twelve vertices, while the dodecahedron has twelve faces and twenty

Fig. 7 Gravity (1953). Lithograph with watercolor in red, orange, yellow, green, blue, and violet, 30.0 cm high × 30.0 cm wide. © 1989 M. C. Escher Heirs/Cordon Art-Baarn-Holland.

vertices. Both, however, have thirty edges, and it is an edge relationship between the icosahedron and the dodecahedron that Escher has illustrated here. By beginning with an icosahedron and constructing perpendicular bisectors to all edges, a dodecahedron is formed. Inside the figure, a cube (eight vertices and six faces) and an octahedron (six vertices and eight faces), both with twelve edges, are also related by coincidence of their edge centres. This cube-octahedron relationship was used previously in *Crystal*, 1947, and *Stars*, 1948. The size relationship of the icosahedron and dodecahedron is, therefore, constrained by the fact that the

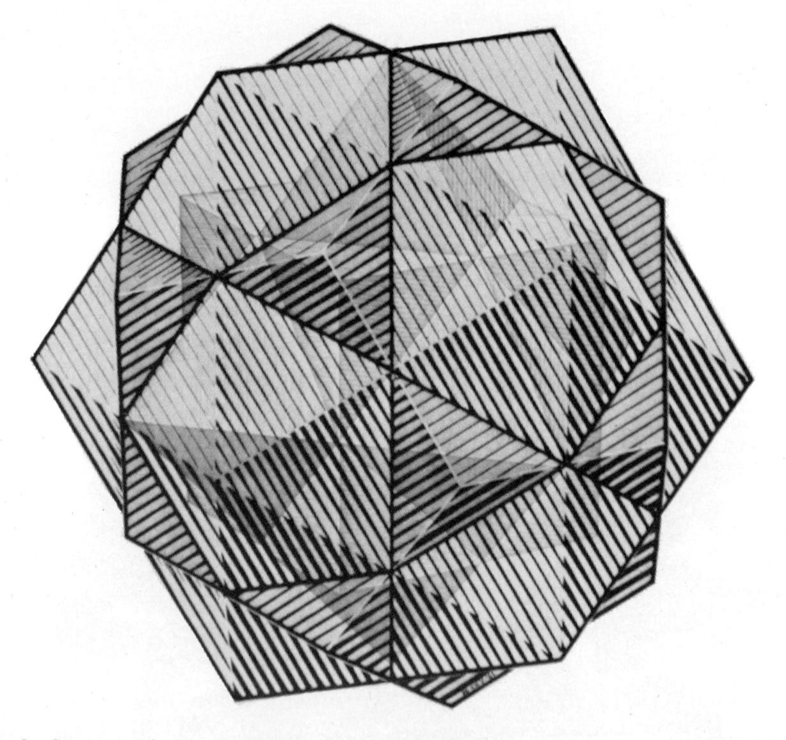

Fig. 8 Stereometric Figure (1961). Woodcut in three colours 39.0 cm high × 33.0 cm wide.

distance from the centre to the edge centres for each figure must be the same. A simple derivation relates the centre-to-edge centre distance for the icosahedron, r_i, to its edge length a_i as

$$r_i = a_i cos(\pi/5). \tag{1}$$

Similarly, for the dodecahedron, we obtain

$$r_d = 2a_d cos^2(\pi/5). \tag{2}$$

From the requirement that $r_i = r_d$, Eqs. (1) and (2) give

$$a_i = 2a_d cos(\pi/5). \tag{3}$$

The cosine term may alternately be expressed as

$$cos(\pi/5) = (1 + \sqrt{5})/4, \tag{4}$$

and Eq. (3) becomes

$$a_i = (1 = \sqrt{5})a_d/2 = \tau a_d, \tag{5}$$

where τ = 1.61803... is the golden ratio. It is this ratio that plays such an important role in the details of quasicrystal structure; e.g. the ratio of prolate to oblate rhombohedra in the three-dimensional Penrose tiling.[9] Clearly, the appearance of the golden ratio in no way indicates a premonition on Escher's part of the importance of this quantity a quarter of a century later, but stems from the basic properties of the Platonic solids and Escher's almost Keplerian fascination for their interrelationship. In *Stereometric Figure*, we see a concept that has been used before in *Crystal*, 1947, and in *Stars*, 1948 — that is, combinations of two or more of the five regular Platonic solids.

4. Conclusions

We have seen the use of fivefold symmetry in the graphic work of M.C. Escher. This appears particularly in his later work, and is commonly manifest in the form of the two Platonic solids which exhibit fivefold symmetry, the icosahedron and the dodecahedron, and in variations and combinations of these solids. We have related the use of fivefold symmetry in these prints to the general themes which occur throughout the graphic art of M.C. Escher.

Acknowledgements

The author is most grateful for the comments and encouragement provided by Drs. E. Dunlap and I. Hargittai.

The Escher illustrations in this article are copyright 1989 by M. C. Escher Heirs/Cordon Art, Baarn, Holland. The author gratefully acknowledges their permission to reproduce these prints.

References

1. M.C. Escher, "Introduction" to *The Graphic Work of M.C. Escher*. Ballantine, New York (1972).
2. C.H. MacGillavry, *Symmetry Aspects of M.C. Escher's Periodic Drawings*. Internat. Union Crystallog., Utrecht (1965).
3. M.C. Escher, Approaches to infinity, in *The World of M.C. Escher*. Abradale, New York (1988).
4. R. Penrose, The role of aesthetics in pure and applied mathematical research, *Bull. Inst. Math. App.* **10** (1974) 266–271.

5. M. Gardner, Mathematical games: Extraordinary nonperiodic tiling enriches the theory of tiles, *Sci. Am.* no. 1, **236** (1977) 110–121.
6. I. Hargittai, Real turned ideal through symmetry, in *Symmetrie in Geistes und Naturwissenschaft*, R. Wille, ed. Springer-Verlag, Berlin (1988) pp. 131–161.
7. H.S.M. Coxeter, The mathematical implications of Escher's prints, in *The World of M.C. Escher*. Abradale, New York (1988).
8. M.C. Escher, Classification and description of the numbered reproductions, in *The Graphic Work of M. C. Escher*. Ballantine, New York, (1972).
9. D.R. Nelson, Quasicrystals, *Sci. Am.* no. 2, **255** (1986) 42–51.

NOMOTHETICAL MODELLING OF SPIRAL SYMMETRY IN BIOLOGY

Roger V. Jean

1. Introduction

This chapter deals with one of the first biological subjects to be mathematized, before genetics. It concerns the structure and morphogenesis of the patterns on plants and inside their buds, e.g. on pineapples, pine cones and sunflowers, the object of the discipline called phyllotaxis. Crystallographers recently underlined the quasi-crystalline nature of the transition zones observed on the daisy, for example, where one pair of consecutive Fibonacci numbers (m, n), $m>n$, expressing the spirality of the inner part of the capitulum, is replaced by the pair $(m+n, m)$ in the outer part. Together with mathematicians they have shown the importance of packing efficiency and self-similarity in this phenomenon of growth. Self-similarity generates noble numbers, that is numbers built on $\nabla = (\sqrt{5}+1)/2$ the golden ratio, among which are the divergence angles found in phyllotaxis. A very simple algorithm is proposed that generates those numbers and orders them. This order is compared to those proposed by the two entropy models in phyllotaxis, and is seen to be in good agreement. Given that the structures analyzed here are also found in other areas of research, such as in the study of proteins, tobacco mosaic viruses, and jellyfishes, the chapter stresses the interdisciplinarity of the subject of phyllotaxis, and the importance of a systemic and holistic approach, able to transcend the strict botanical substratum of the problem of phyllotaxis.

2. Examples of Spiral Symmetry

Spiral patterns can be observed in biology, as this section shows. It also introduces basic notions useful for their descriptions. The following diagrams represent various structurally similar biological systems. Fig. 1 represents a part of a Medusa showing a pattern of tentacles. In the area of botany known as phyllotaxis (the study of spiral patterns on plants and inside their buds), this pattern would be described as representing a (2, 3) *visible opposed spiral pair* (in botany the term parastichy is used to mean spiral). This means that (i) there are two families of spirals containing respectively 2 and 3 spirals, (ii) each family goes through all the tentacles, (iii) the families wind in opposite directions with respect to the centre of symmetry, (iv) the pair is visible in the sense that at every intersection of two spirals there is a tentacle, and (v) considering any spiral, the tentacles

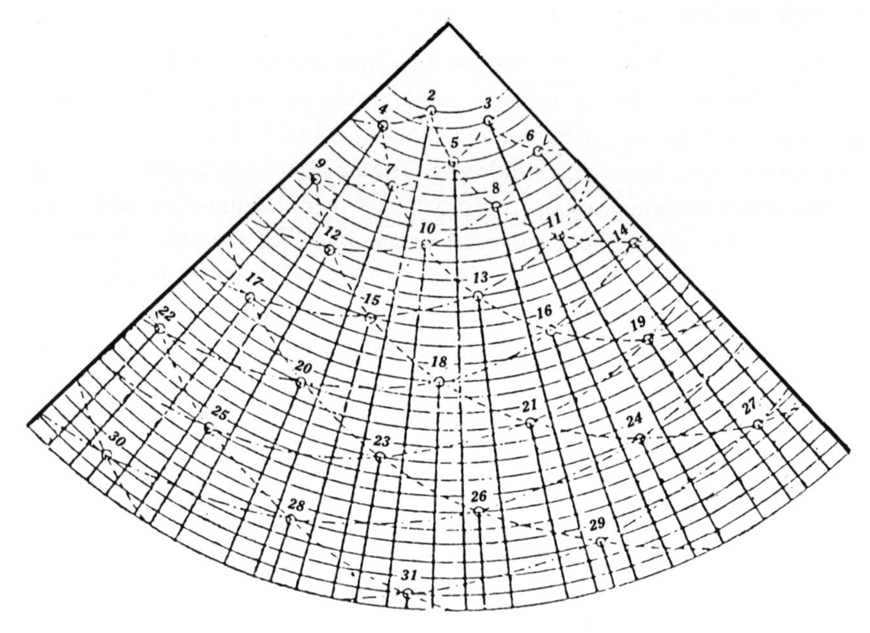

Fig. 1 Schematic representation of the arrangement of ex-umbrellar tentacles in a quadrant of *Olindias formosa* (from Ref. 1; reproduced with permission from The Zoological Society of Japan). The order of development of tentacles in *Goniomemus* also conforms with this scheme. The diagram shows a (2, 3) visible opposed parastichy pair.

are linked from 2 to 2 or from 3 to 3. The pair (3, 5), not drawn in the diagram, is also a visible opposed pair.

Fig. 2 illustrates a system of primordia found in flax. A *primordium* is a very early stage of leaf or other organ, and by extension it represents a more mature organ such as seed, scale, and floret. The black dots represent the centre of the system and the mid-points of the primordia. As in the Medusa, the spiral pairs $(4 - 1,\ 3 - 1) = (3, 2)$ and

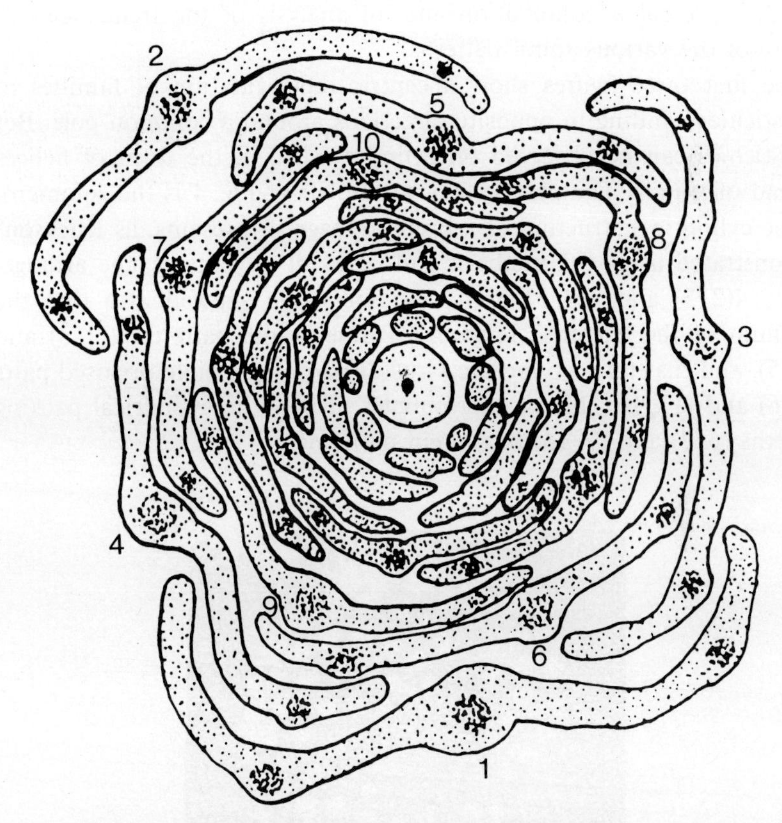

Fig. 2 Drawing of a microscopic cross-section through a growing tip of *Linum usitatissimum*. The outer ten primordia have been numbered according to their ages. Then we can draw for example three spirals defined respectively by the centres of primordia 1, 4, 7, 10, 13, . . ./ 2, 5, 8, 11, 14, . . ./ 3, 6, 9, 12, 15, ... The pairs (3, 2) and (3, 5) (not drawn) are visible opposed pairs (from Ref. 2). The pattern shows a mean divergence angle of 138.77°, and a mean plastochrone ratio (*R* in the text) of 1.0792, with standard deviations of 8% in both cases.

$(4-1, 6-1) = (3, 5)$ are visible opposed pairs. The pair (8, 5) is also visible.

In phyllotaxis, most of the time the numbers found are consecutive terms of the Fibonacci sequence 1, 1, 2, 3, 5, 8, 13, . . . , such as in the pattern of thorns on the cactus of Fig. 3 showing a (13, 21) visible opposed spiral pair. Generally speaking, those spiral patterns are made of consecutive terms of a Fibonacci-like sequence such as a, b, $a+b$, $a+2b$, $2a+3b$, $3a+5b$, . . . where each term is the sum of the preceding two (in Ref. 3, I present a compilation and an analysis of the frequencies in nature of the various spiral patterns).

The first three figures show a centric representation of families of parastichies winding in opposite directions around a common pole. But parastichies can be observed on cylinders also, in the form of helices instead of spirals. The photograph on the left of Fig. 4 is the submicroscopic cylindrical structure of a bacteriophage. It presents, as Erickson[6] demonstrated using the methods and terms of phyllotaxis, the arrangements 4(2, 3) and 4(3, 5) of its hexamers. He showed also that the flagellum of the bacteria *Salmonella* presents the patterns 2(2, 3) and 2(3, 5), and that tobacco mosaic virus displays the visible opposed pairs (1, 16) and (17, 16). Many microtubules display the cylindrical packing patterns (6, 7) and (7, 13) of protein monomers.

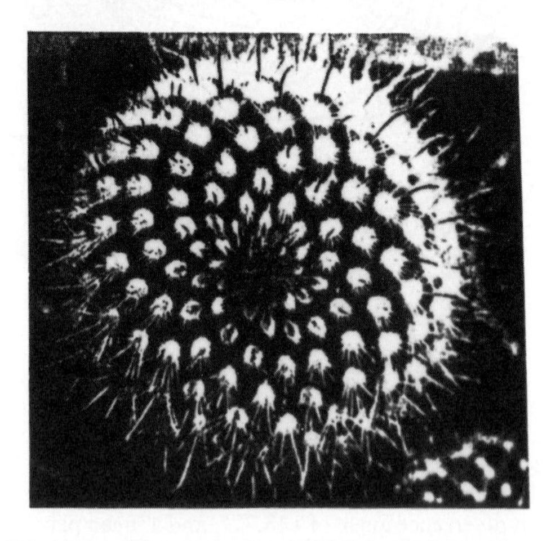

Fig. 3 A (21, 13) visible opposed pair of thorns observed on a cactus.

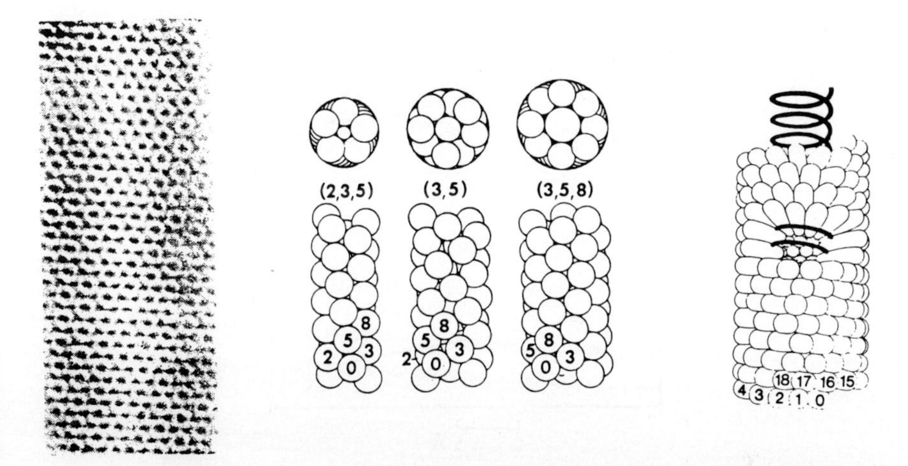

Fig. 4 On the right is a model of the protein coat of the tobacco mosaic virus showing the pattern (1, 16, 17) (from Ref. 4). On the left is the optically filtered image of the polyhead of a bacteriophage showing a 4(2, 3, 5) arrangement of its hexamers (from Ref. 5). In the middle are packings of spheres on cylinders, with which Erickson[6] determined the "phyllotaxis" of such patterns using published high resolution electron micrographs, and optical diffraction studies.

When such a cylinder is unfolded in the plane, the helices become straight lines, and we have a figure such as Fig. 5 representing the (3, 4) conspicuous opposed parastichy pair made by amino-acid residues in a polypeptide chain. By following the points along the direction we marked in the figure, we can go through all the residues along a single helix called in phyllotaxis the *generative spiral*. Denoting by r the *rise*, that is, the vertical distance between any two consecutive points along the generative spiral, it is possible to transform this cylindrical representation into a centric representation of the residues, by a transformation of the type $R = \exp(2\pi r)$. Then R is the ratio of the distances to the common pole of the spirals, of any two consecutive residues on the generative helix (I propose in Ref. 8 a "phyllotactic" analysis of these chains).

Assuming that the basis of the cylinder has a length equal to one, the *divergence angle*, a number smaller than $\frac{1}{2}$, is the horizontal distance between two consecutive points along the generative helix. In the centric representation, the divergence angle is the angle at the centre made by two consecutive points, such as the angle smaller than 180° made by points 1 and 2 in Fig. 2 (a thorough introductory analysis of phyllotaxis

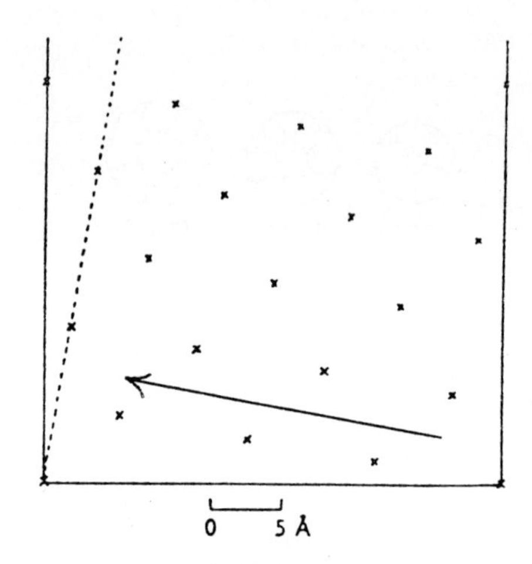

Fig. 5 The pattern formed by the side-chains of an alpha helix (from Ref. 7). Numbering the points from right to left and upward, starting with 0 on the right hand corner, we can see that points 3 and 4 are the closer to 0, so that the pair (3, 4) is a visible opposed pair of families of parastichies.

in the centric representation is given by Jean and Schwabe in Ref. 2, and the first author introduced in Ref. 9 an efficient, easy and accurate method for pattern recognition).

3. Nomothetics

Meyen[10] used the word nomothetics to describe this orientation of research in which one looks for general principles, common structures and universality among various sets of organisms or objects. "Morphological laws cannot be reduced to history, function and adaptation" (Ref. 10, p. 253). There exist some general structural principles, manifested within various groups of organisms. The phenomenon of morphogenetic parallelism allows us to extend the methods of phyllotaxis to other fields, such as the study of micro-organisms, viruses, and polypeptide chains. Urmantsev's laws of correspondence and polymorphism (see Ref. 10) have great heuristic value and can bring understanding by showing the parallelism and convergences between structures and processes in one field, and those in other areas. This point of view has an important consequence on the problem of phyllotaxis, as we underline in Sec. 9.

According to Meyen,[10] "from the point of view of the spatial position of elements, the flower having juxtaposed sepals and petals will be closer to carbohydrate molecules with the same arrangement of hydrogen atoms, than to other flowers bearing alternating sepals and petals." Stevens[11] abundantly illustrated the geometrical relationship between the arrangements of scales on fish, tortoise and snake skins (Fig. 6), on the scaly anteater, and on the seed of asclepiad (Fig. 7). He put these arrangements in relation to the geometry of soap films and bubbles, of turtle-shells and of crystal grains, stating that underneath all these phenomena there is a matter of minimizing work or surfaces of contact, or of decreasing potential energy, by means of triple junction points. Such triple points are clearly found, for example, at the junctions of the hexagonal scales on the surface of pineapples.

Fig. 6 Scales on the skin of a snake (from Ref. 11; reproduced with permission from Little, Brown & Co. Publisher).

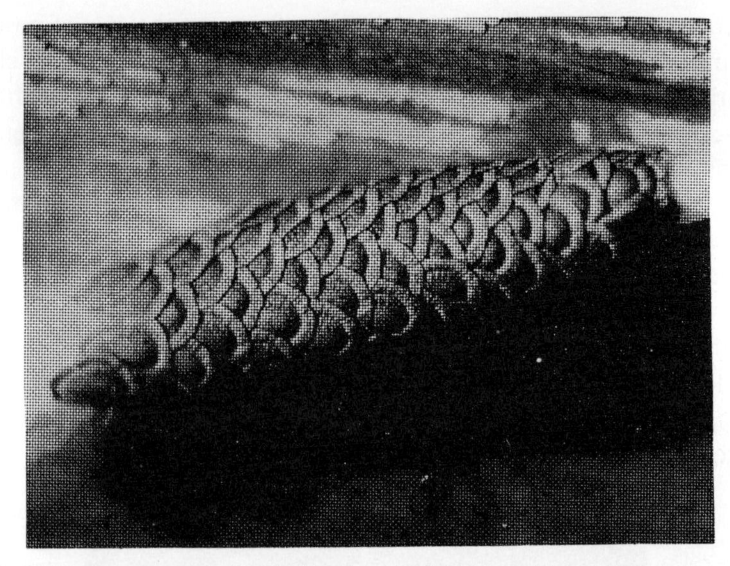

Fig. 7 The scales on a seed of asclepiad (from Ref. 11; reproduced with permission from Little, Brown & Co. Publisher). This plant is also called dompte-venin (venom), because it was formerly regarded as an antidote to poison.

Trying to explain the same structure by referring to only one context, botanical or not, seems to be impossible. But "for the overwhelming majority of biologists, the comparison of flowers with molecules, and shells of foramininifers with Roman pottery, seems either senseless or improper (or both). The overcoming of this psychological barrier, and the spreading of the application of system laws to structural investigations, will be a revolution in biology comparable to the introduction into biology of statistical thought" (Ref. 10, p. 248). Given that the black spots on the tail of a leopard or of a jaguar can be generated by a diffusion-reaction equation controlling an inhibitor and an activator,[12] then by analogy, could not the primordia (the spots) on the cylindrical surface of a plant be generated in the same way (see Ref. 13, where many models using diffusion equations are reported; see also Sec. 7 of this chapter)?

Sec. 5 considers the structural parallelism found between phyllotactic patterns and quasi-crystals. But first let us underline the overwhelming presence of ϕ in the spiral patterns mentioned above.

4. Presence of the Golden Number ϕ

The golden ratio $\phi = (\sqrt{5} + 1)/2$ looks like a constant of nature, expressing the symmetrical proportions involved in the generation of spiral patterns. This does not come as a surprise given that the ratio of consecutive Fibonacci numbers is close to ϕ, and is ϕ at the limit. This number generates the rhythmic alternation of the neighbors (5, 8, 13, 21, 34, 55, . . .) of the vertical axis in the cylindrical lattice of Fig. 8, as explained below.

Spiral phyllotaxis is a consequence of the occurrence of ϕ, regardless of the growth of the plant in diameter and in length. Growth only determines the conspicuous parastichy pair; generally, growth contributes to stress the presence of ϕ by the phenomenon of *rising phyllotaxis*, the process by which the conspicuous parastichy pair (m, n) of m and n

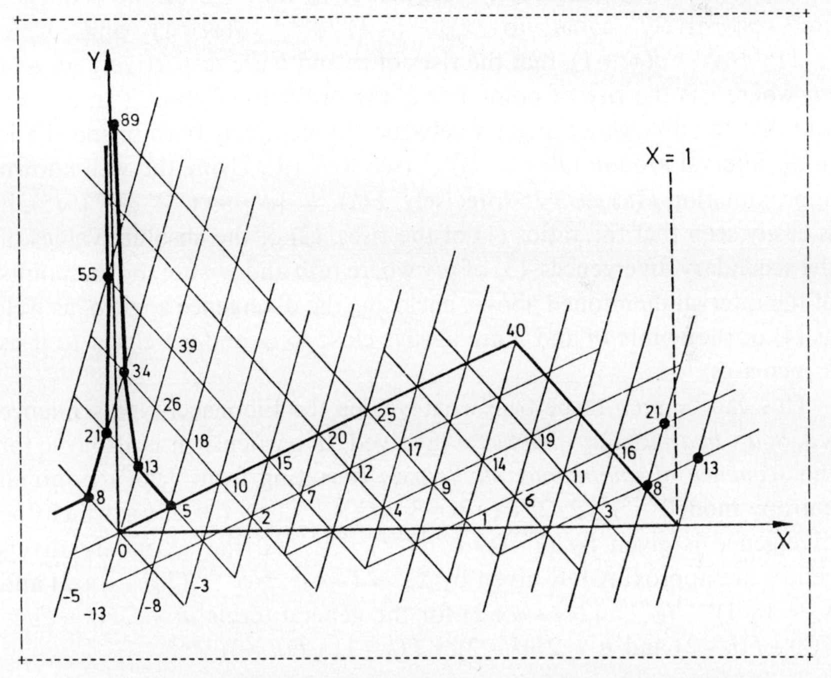

Fig. 8 The normalized cylindrical representation of a pattern with divergence $d = \phi^{-2}$ where $\phi = (\sqrt{5} + 1)/2$. The length of the base of the stripe, that is the segment between two consecutive representations of the same point p, has length 1. The Fibonacci numbers, on the polygonal lines, alternate on each side of the vertical axis due to the absence of intermediate convergents in the continued fraction of ϕ^{-2}.

spirals, $n<m$, is replaced by the conspicuous pair $(m+n, m)$, which is replaced by the conspicuous pair $(2m+n, m+n)$, ... as the rise r decreases and the apical dome enlarges relatively to the size of the primordia. This phenomenon does that as the ratio of the two terms in the consecutive pairs gets closer to ϕ.

Let us consider the regular cylindrical representation of a plant growing according to the *normal phyllotaxis sequence* $J<1, t, t+1, 2t+1, 3t+2, \ldots>$, where $J \geq 1$ and $t \geq 2$ are integers, and where each term is the sum of the preceding two. For example, Fig. 8 represents a lattice where $J=1$ and $t=2$, giving a divergence angle (the horizontal distance between consecutive points in the lattice) of ϕ^{-2}. Call $F(k)$ the kth term of the Fibonacci sequence. It can be shown for the sequence of normal phyllotaxis that the abscissae (called secondary divergences) of points $m = J(F(k)t + F(k-1))$ and $n = J(F(k-1)t + F(k-2))$ of the sequence are respectively equal to $x_m = (-1)^k/\phi^{k-1}\ J(\phi t+1)$ and $x_n = (-1)^{k-1}/\phi^{k-2}\ J(\phi t+1)$, that the rises of m and n are respectively mr and nr, where r is the rise of point 1 (e.g. the ordinate of point 1 in Fig. 8), and that the divergence angle d between consecutively born primordia is in an interval around $(J(t+\phi^{-1}))^{-1}$ (see Ref. 14). Using the well-known approximation $F(k) \simeq \phi^k/\sqrt{5}$ (precisely, $F(k) = [\phi^k - (-\phi)^{-k}]/\sqrt{5}$), it is easily seen that the ratios (1) of the rises, (2) of the absolute values of the secondary divergences, (3) of u/v where u/m and v/n are the endpoints of the interval mentioned above, enclosing the divergence angle d, as well as (4) of the points m and n are always close to ϕ, and get closer to it as k increases.

This can be seen to be true whatever be the Fibonacci-type sequence $<a, b, a+b, a+2b, 2a+3b, \ldots>$ involved. In particular it is also true for the *sequence of anomalous phyllotaxis* (resulting from Jean's minimal entropy model[15,16]) $<2, 2t+1, 2t+3, 4t+4, \ldots>$, $t \geq 2$, for which the divergence is given by $(2+(t+\phi^{-1})^{-1})^{-1}$,[14] and the secondary divergences are approximately given by $x_m = (-1)^{k-2}/\phi^{k-2}\ (2\phi t+\phi+2)$ and $x_n = (-1)^{k-3}/\phi^{k-3}\ (2\phi t+\phi+2)$ for the general terms $m = 2F(k-1)t + F(k) + F(k-2)$ and $n = 2F(k-2)t + F(k-1) + F(k-3)$.[17,18]

5. Fivefold Symmetry and Quasicrystals

Let us come back to the idea of morphogenetical parallelism and underline the fruitfulness of the dialogue between crystallography and phyllotaxis. It is interesting to recall that brothers L. and A. Bravais, the pioneers who put forward the cylindrical representation of phyllotaxis 160 years ago, were respectively a botanist and a crystallographer.

In spiral phyllotaxis, three families of n, m, $m+n$ parastichies, (determined for example by points 5, 8, and 13 in Fig. 8), giving triple contact points $(n, m, m+n)$, and two visible opposed parastichy pairs (n, m) and $(m+n, m)$, corresponds in Rivier's crystallographic approach[19] to the lattice planes of conventional crystallography. Phyllotaxis, then, appears to be a special case of two-dimensional crystallography. Comparing Fig. 9 with Figs. 4 and 8 already suggests relationships between the two subjects. Rivier,[21] as well as Ridley[22] and Marzec[23] stress that phyllotaxis is the solution of a structural variational principle giving maximal uniformity to the patterns and requiring divergence angles, whose continued fractions terminate as $1/\phi = [0; 1, 1, 1, \ldots]$. Let me explain a bit.

As a generative spiral Rivier uses Vogel's cyclotron parabolic spiral (see Ref. 22) defined by $r = a\sqrt{n}$ and $\theta = 2\pi nd$, $n = 1, 2, 3, \ldots$, where d is the divergence angle. This spiral is also used by Ridley to show that $d = \phi^{-2}$ gives the best packing efficiency of the fruits in the sunflower head. Crystallographic requirements deliver what Rivier and other crystallographers call *noble numbers* for possible values of the divergence angle d, precisely those numbers whose continued fractions terminate as $1/\phi$ (among which is ϕ^{-2}, observed in most of the cases in plants). The cyclotron spiral gives uniform floret density and the noble numbers optimize the local homogeneity and the self-similarity of the patterns. The set of noble numbers is precisely what Marzec[23] obtained. It is his set ϕ_G of divergence angles that distribute the leaves around the stem of a plant more uniformly than the neighboring angles can do. Outside this set the pattern becomes very strongly spiralling, because it is dominated by a set of very long parastichies, and the structures generated are not self-similar.

The crystallographers Rothen and Koch[24,25] clearly illustrated what this uniformity and efficiency is about, by considering the Voronoi-polygon (that is, the Dirichlet domain, or Wigner-Seitz cell) of the cylindrical lattice. Given a point p of the lattice (see Fig. 10), such a polygon encloses by definition all points of the plane nearest to p than to any other point of the lattice. The scales on a pineapple define such

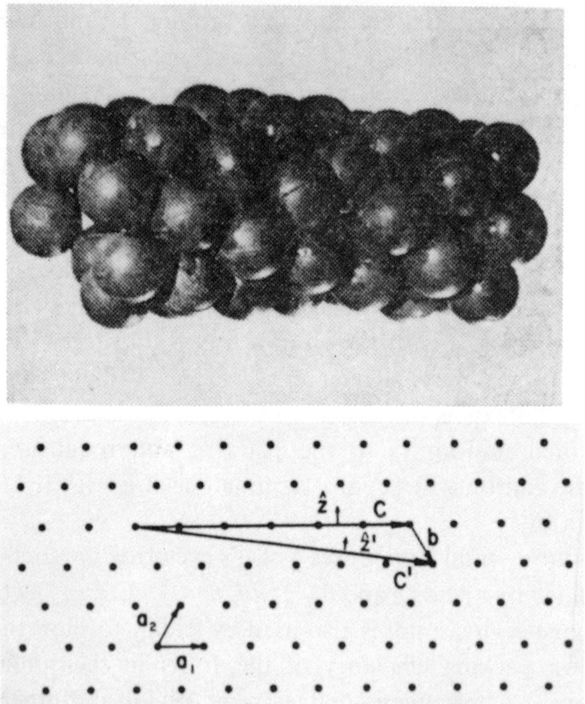

Fig. 9 A dislocated cylindrical crystal made from rubber balls, and its planar representation (from Ref. 20; reproduced with permission from Gordon and Breach Science Publisher).

polygons with respect to their centres. These authors showed that when the value of the internode distance or rise r changes in the cylindrical lattice (the divergence angle being constant), corresponding to the operation of inflation–deflation in quasi-crystals, the Voronoi-polygon does not change in form, exclusively when the divergence d is a noble number. In other words, these numbers give shape invariance or self-similarity of the Voronoi-polygons around the points of the lattice. When the divergence angle is not a noble number, the shape of the polygon changes with the rise.

It is particularly important that this polygon does not change when one considers the natural transitions observed in the daisies or sunflowers showing a system (m, n) in the centre of the capitulum, $(m, m+n)$ in the middle portion of the capitulum, and $(2m+n, m+n)$ in the outer portion, corresponding to changes in r (rising phyllotaxis). In such

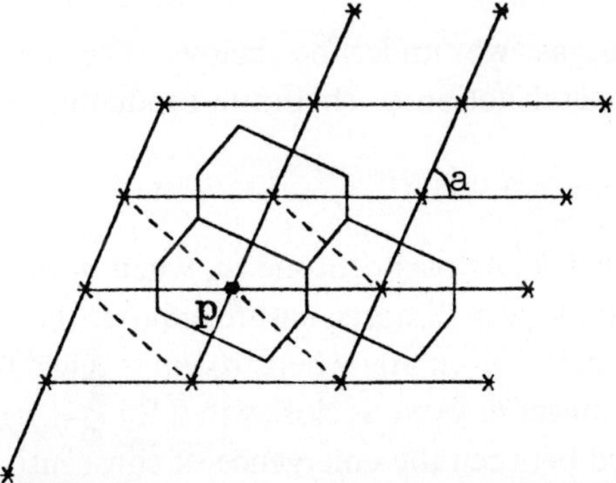

Fig. 10 A visible opposed pair in a lattice, determines a set of congruent parallelograms filling the space completely and without overlapping. Each point of the lattice, such as point p, belongs to six triangles, such that six sides meet at point p. Drawing the perpendiculars in the middle of these sides determines the Voronoi-polygon. When the parallelogram is a rectangle, the polygon is a rectangle instead of an hexagon. Three hexagons meet at triple points.

capituli there are two circular defect lines having a quasi-crystalline structure. Self-similarity means that the same plant needs only one algorithm, that is the same code, to produce both small and large capituli, and to change the phyllotaxis inside a capitulum.

Why packing efficiency and self-similarity is so important for nature is the question that brings us deeper into the causes. In botany it has to do with Zimmermann's telome (primary cylindrical axis) theory (see Ref. 14) according to which higher plants would come from propteridophytes of the *Rhynia* type via six elementary morphogenetic principles (resting on empirical material). In biology it has to do with Lyell's principle of evolutionary continuity according to which a large part of the cellular organization is very remote and is common to all organisms. In physics, it has to do with the principle of optimal design, operationally formulated by Rosen,[26] a form of which has been used by Jean[15,16] under the name of principle of minimal entropy production.

The preference of nature for self-similar structures gives predominance to the golden number ϕ in natural constructions. Let us not inquire further here on the reasons for this choice. Let us rather build on it, and introduce a very simple minimality principle, an algorithm which at first sight does not seem to have much to do with morphogenesis, that will allow us to generate the noble numbers and to order them. This order has to do with the decreasing relative frequencies of the various divergence angles in nature, as we underline below. The principle implicitly incorporates a general notion of rhythmic production of primordia.

6. The ϕ-model

Minimality principle: At plastochrone m, when m primordia exist, and a rhythm of gnomonic growth starts, nature requires the primordia to show the conspicuous pair $(m-n, n)$, where $n<m$ is such that the function $\log(m/k\phi)$ of the integer k, $k<m$, is closest to 0 for $k = n$ (the plastochrone is the time elapsed between the emergence of consecutive primordia).

The point n is unique, and $M_m = \log(m/n\phi)$ is called the minimum of the function. If n satisfies the minimality principle for m, then m satisfies the minimality principle for $m+n$, and $m+n$ satisfies the minimality principle for $2m+n$, and so forth. I showed[27] that each one of the two extensions $(a, a+b)$ and $(a+b, b)$ of a visible opposed pair (a, b) is a visible pair, and that one of them is opposed. Given that the consecutive minima are decreasing towards zero while alternating in signs, the consecutive alternate extensions of the conspicuous pair $(m-n, n)$, that is, (m, n), $(m, m+n)$, $(2m+n, m+n)$, ... are visible and opposed (this rhythmic alternation is a basic requirement in the Bravais and Bravais treatment of phyllotaxis, as mentioned below). This means that if the system starts to experience the minimality principle at plastochrone m, the system enters into a loop for the first time, that is a rhythm of growth is created, generating conspicuous pairs along the Fibonacci-type sequence $<m-n, n, m, m+n, \ldots>$. To such a sequence corresponds a unique divergence angle which is a function of ϕ (see Sec. 4 and Ref. 14).

The present model does not explain why nature packs primordia with the efficiency underlined above, but assuming it does, it makes use of ϕ which sets a natural rhythm of growth. The golden number is responsible for the alternation of the extensions mentioned above, due to the absence of intermediate convergents in the continued fraction of the divergence angle, starting from a certain term of the fraction (see Ref. 28). The minimality principle orders *all* the noble numbers (ϕ_G set). This ordering is based on the necessity to establish a rhythm, on ϕ, as soon as possible. If it is established at plastochrone m, then a member of ϕ_G is selected which maximizes the packing efficiency of the pattern for this value. Depending on which plastochrone the rhythm of growth starts, we then have the following Table.

Table 1

Values of n satisfying the minimality principle for given values of $m>n$; the divergence angle and the sequence derived (from Ref. 29). The numbers in parentheses at the beginning of the sequence are those for which the minimality principle is not satisfied by that sequence. The # corresponds to a value of m for which the model and Jean's minimal entropy model give the same value of n. The case $m = 4$ is not possible in the latter model. The consecutive + corresponds to the ordered values of m for which the divergence angles obtained from these two models are the consecutive divergence angles obtained from Marzec's entropy model.

		m	n	conspicuous pair $(n, m-n)$	divergence d	sequence
	#	3	2	(2, 1)	137.51°	<1, 2, 3, 5, 8, ...>
		4	3	(3, 1)	99.5°	<(1), 3, 4, 7, 11, ...>
+	#	5	3	(3, 2)	137.51°	<1, 2, 3, 5, 8, ...>
	#	6	4	2(2, 1)	68.75°	2<(1), 2, 3, 5, 8, ...>
+	#	7	4	(4, 3)	99.5°	<(1), 3, 4, 7, 11, ...>
	#	8	5	(5, 3)	137.51°	<1, 2, 3, 5, 8, ...>
	#	9	6	3(2, 1)	45.84°	3<(1), 2, 3, 5, 8, ...>
	#	10	6	2(3, 2)	68.75°	2<(1), 2, 3, 5, 8, ...>
+	#	11	7	(7, 4)	99.5°	<(1), 3, 4, 7, 11, ...>
		12	7	(7, 5)	151.1°	<(2, 5), 7, 12, 19, ...>
	#	13	8	(8, 5)	137.51°	<1, 2, 3, 5, 8, ...>
+	#	14	9	(9, 5)	77.96°	<(1, 4, 5), 9, 14, ...>
		15	9	3(3, 2)	45.84°	3<(1), 2, 3, 5, 8, ...>
	#	16	10	2(5, 3)	68.75°	2<(1), 2, 3, 5, 8, ...>
+	#	17	11	(11, 6)	64.08°	<(1, 5, 6), 11, 17, ...>
		18	11	(11, 7)	99.5°	<(1), 3, 4, 7, 11, ...>
+	#	19	12	(12, 7)	151.1°	<(2, 5), 7, 12, 19, ...>
		20	12	4(3, 2)	34.38°	4<(1, 2), 3, 5, 8, ...>
		21	13	(13, 8)	137.51°	<1, 2, 3, 5, 8, ...>
	#	22	14	2(7, 4)	49.75°	2<(1, 3, 4), 7, 11, ...>
		23	14	(14, 9)	77.96°	<(1, 4, 5), 9, 14, ...>
		24	15	3(5, 3)	45.84°	3<(1), 2, 3, 5, 8, ...>
		25	15	5(3, 2)	27.5°	5<(1, 2), 3, 5, 8, ...>
		26	16	2(8, 5)	68.75°	2<(1), 2, 3, 5, 8, ...>
		27	17	(17, 10)	106°	<(3, 7, 10), 17, 27, ...>

The criterion is universal: for each m there is a unique n giving the minimum. But for the numbers at the beginning of a Fibonacci-type sequence, the minima can be given by numbers in other sequences. Consider for example the sequence <1, 4, 5, 9, 14, 23, ...>. For consecutive numbers greater than 5 in the sequence, the minimality criterion is satisfied, but for the first four terms it is not. Indeed, for $m = 9$, $n = 6$ gives the minimality criterion, for $m = 5$ we have that $n = 3$, and for $m = 4$, $n = 3$ (for each of these cases the sequence then obtained can be read in the table); thus, the parentheses in the sequence on line 14 of Table 1. Of course, these parentheses do not mean that the visible opposed parastichy pairs (1, 4) and (5, 4) are excluded by the algorithm, given that when the sequence $<1, 4, 5, 9, 14, \ldots>$ arises in a plant, at whatever plastochrone m it may be, say $m = 14$ for example, then all consecutive alternate extensions of (5, 9) are visible opposed pairs as we have seen, but also (see Ref. 27 or Ref. 30) the contractions of (5, 9), which are (1, 4) and (5, 4), are visible opposed pairs, and are observable on the plant.

Of course, the ordering of the noble numbers corresponding to the angles given in Table 1 (divide these angles by 360 to obtain approximations of these numbers, whose exact values are given in Sec. 4) has to do with the presence of elements different from 1 at the beginning of the continued fraction of those numbers. For example the continued fraction corresponding to the sequence $<1, 4, 5, 9, 14, 23, \ldots>$ is $[0; 4, 1, 1, 1, \ldots]$. To this sequence corresponds the divergence $360(4 + \phi^{-1})^{-1} \simeq 78°$, and before the appearance of $1/\phi = [0; 1, 1, 1, \ldots]$ in the continued fraction, that is before the rhythm starts, the minimality principle is not given by a member of this sequence.

7. Dynamic vs. Static Approaches

Marzec's study for primordial generation in phyllotaxis uses a one-dimensional model first developed by Thornley[31] (see Ref. 13 pp. 129–135), where the rise r is purged from the problem. In this model, the leaves on the stalk are abstractly treated as points on a circle, each point acting as a morphogen source. Phyllotaxis becomes a dissipative structure, that is, a non-equilibrium system sustained by a flow of matter and energy through its boundaries, by the creation, diffusion and decay of the morphogen, according to a diffusion-reaction equation. Thornley's model is dynamic, setting a new leaf at the position of the absolute minimum of the concentration field established by the existing leaves, leaf by leaf, until a constant divergence $\Delta\mu$ emerges. The strength of each leaf as a morphogen source is decreased by a factor $\beta < 1$ after each plastochrone (unit of time between the emergence of consecutive leaves). Then, setting a constant divergence angle for about one hundred primordia, and using an idea of perturbation, the model tries to reproduce that angle. Thornley found bands of angles (78–85°, 97–101°, 105–110°, 132–134°, 136–143°, and 146–180°), in which some observed divergence angles are, and some others are not. He obtained that when $\beta > \frac{1}{2}$, only the angle ϕ^{-2} is possible, this angle being approached for values of β near 1.

Marzec's treatment is, first of all, static. It establishes a set ϕ_G of angles whose continued fractions do not have intermediate convergents after a finite number of terms. It is shown that this criterion is sufficient for uniform spacing of leaves. Any angle in the set possesses a purely geometrical minimization property (see Ref. 32, p. 216, th. 9) producing a more even distribution of points on the circle than any neighboring angle, each leaf being set so that all the leaves have the maximum space. The results of this minimum principle are then applied directly to the minimization of a static concentration function. He uses an informational entropy function defined by the formula $S = -2\pi\langle C\rangle \int_0^{2\pi} p(\mu) \ln p(\mu) d\mu$, where $\langle C\rangle$ is the mean concentration, and $p(\mu)$ is a morphogen concentration field established by the leaves, depending on the constant divergence $\Delta\mu$. The entropy of the morphogen distribution is considered to measure the spacing of the leaves. Considered as a function of $\Delta\mu$, it is found that S is a maximum and the rates of entropy production are minimal at a $\Delta\mu$, which asymptotically approaches, as β approaches unity, one of the members of ϕ_G.

Both the dynamic and the static approaches have been considered in Jean's minimal entropy model. The search for a suitable bio-entropy formula, that is, $E = -\sum_{\tau=1}^{w} [\log S(t)/X(t)]$, where X is the complexity, S the stability, w the rhythm, t the level of the hierarchy representing the spiral pattern (see Ref. 16, p. 255), was preceded by the consideration of the formula $E = -\log[S(t)/X(t)]$ (not incorporating a notion of rhythm). When we choose among the hierarchies, the one minimizing the latter function for each consecutive level $t = 1, 2, 3, \ldots$, corresponding to a dynamic approach, unfortunately only the pattern defined by the Fibonacci sequence arises (as in Thornley's model for $\beta > \frac{1}{2}$). Similarly, for low values of β the only solution of the dynamic problem in Marzec's model is also ϕ^{-2}.

Why is that? Marzec conjectures that this robustness of ϕ^{-2} underlies its predominance in plants. It appears that those dynamic approaches do not incorporate any notion of rhythm, so fundamental in growth processes. In Marzec's static approach the rhythm is settled by the presence of $1/\phi$ in the continued fraction of every member of ϕ_G. It is ϕ that brings the rhythmic alternation of the neighbors of the vertical axis that we observed in Fig. 8. In Jean's minimal entropy model, the formula for entropy incorporating a notion of rhythm expresses a dynamic approach up to the time $m = T_r$ (when T_r primordia exist and the rhythm starts), and a static thereafter, the system having become rhythmic and deterministic, by the presence of ϕ from a certain term in the continued fraction of the divergence. Adler's model[30] is also dynamic up to the time T_C when contact pressure begins, and static afterwards.

8. Similarities Between Models

Table 1 suggests that there are convergences and resemblances between the models, even when the approaches to the problems are very dissimilar. The very simple mechanism in the ϕ-model presented here as extending models dealing with packing efficiency, which generates noble numbers, shows a good correspondence with the minimal entropy model.

In the minimal entropy model the problem of pattern generation in phyllotaxis is treated according to an interpretative approach, in contrast to a mechanistic approach where a mechanism is assumed to be responsible for the generation of the patterns. An entropy-like function is defined, and it is shown that when the principle of minimal entropy production is used to discriminate among the phyllotactic lattices it succeeds to find the angles and sequences of normal and anomalous phyllotaxis. The model offers an abstract understanding, and a physical basis I called the hierarchical structure of phyllotaxis,[14] in which the fact of neguentropy in living organisms is given meaning. In this model, phyllotaxis is interpreted in terms of its ultimate effect, which is assumed to be the minimization of the entropy of plants. The formula for bio-entropy mentioned in Sec. 7 of this chapter acts as a "functional cost" that allocates a cost to every phyllotactic pattern. Then we look in the set of costs for the minimal cost. Ordering the patterns according to increasing entropy costs leads, for example, to the predictions that the pattern expressed by the sequence <1, 2, 3, 5, 8, . . .> is more frequent than the pattern expressed by the sequence 2<1, 2, 3, 5,

8, . . .>, which is more frequent than the pattern expressed by the sequence <1, 3, 4, 7, 11, . . .>.[15] This is in good agreement with the available observations (see Ref. 3).

Marzec's entropy model delivers some of the elements of ϕ_G, in the order presented in Table 1 with the consecutive +, as the parameter β increases. "In rough order of their appearance, the most prominent members of ϕ_G are 137.5°, 99.5°, 78.0°, 64.1°, 151.1°, 54.4°, 158.1°, and 106.4°." This is akin to what happens in Jean's models. Indeed, the angles obtained consecutively in Marzec's model correspond respectively to the increasing values $m = T_r = 3$ or 5, 7 and 11, 14, 17, 19, 20, 25 respectively under the minimal entropy condition, and to the value $m = 27$ under the minimality condition in the ϕ-model. The three models show the same order of generation for the values 5, 7, 11, 14, 17, and 19 (the + in Table 1). The + in Table 1 appears where there are #.

Marzec's[23] and Adler's[30] models incorporate a notion of rhythm, though these authors do not speak about such a notion. Church[34] stresses the importance of rhythm in phyllotaxis. The rhythms in the models considered in this paper generate the usual types of patterns. When at $m = T_r$ the rhythm starts, the ϕ-model and the minimal entropy model give the same value of n, the same sequence of integers and the same divergence angle for m = 3, 5, 6, 7, 8, 9, 10, 11, 13, 14, 16, 17, 19, 22, . . . (see the # in Table 1). Among these divergence angles are the usual values 137.5°, 99.5°, 68.75° and 151.1°, corresponding to well-known types of phyllotaxis. Under Adler's minimax principle, Fibonacci phyllotaxis is unavoidable if contact pressure begins before $T_C \geq 6$. The following propositions, resulting from Table 1, show a similar result, and other similarities between the models.

Propositions: (i) Under the minimal entropy hypothesis a sufficient condition for Fibonacci phyllotaxis is that $T_r < 6$. (ii) If $120° < d < 180°$, then a sufficient condition for the Fibonacci pattern to arise is that $T_r = m < 12$, either when the minimality condition of the ϕ-model applies, or when the minimal entropy condition applies. (iii) If $90° < d < 120°$ then a sufficient condition for the pattern governed by the sequence <1, 3, 4, 7, 11, . . .> to arise under the minimality condition is that $m < 27$; under the minimal entropy condition this pattern always arises, whatever be the value of m.

Jean's minimal entropy model may appear to be more formal than the contact pressure model or Marzec's entropy model, but it may not be. Indeed, the latter models are more or less biologically grounded. They have varying amounts of appeal given such "physical" principles as contact pressure and diffusion of an inhibitor, and these postulates can presumably undergo scrutiny in the laboratory, but recent experimental researches (e.g. Refs. 35 and 36) show that such a scrutiny brings discredit to them. Even if the growth, form and position of young leaves depend on physical and chemical constraints, we must be careful not to identify these constraints with the postulates above.

Fujita's[37] a priori sequences, and thus divergence angles, in phyllotaxis, permit the existence of a wide variety of patterns. Fujita's divergence angles form a proper subset of Marzec's set ϕ_G, which makes possible the existence of all Fibonacci-type sequences. The ϕ-model may produce all noble numbers, but we obviously do not need that many angles, and thus, sequences in phyllotaxis. This possibility of so many angles might be an inherent limit for a model. For Popper,[38] "every 'good' scientific theory is a prohibition: it forbids certain things to happen." But a model which does not reproduce patterns known to exist shows another type of limitation. Out of *Bravais and Bravais' biologically plausible mathematical assumptions*, Coxeter[28] deduced the sets of angles and sequences expressing normal phyllotaxis. These assumptions are simply that the divergence d must be an irrational number and that by going upward along the vertical axis, the neighbors of that axis on the two polygonal lines (as in Fig. 8) must alternate on each side of it. Of course, if the sequence of neighbors does not start to alternate from the beginning, then other noble numbers can be obtained. The angles permitted by the minimal entropy model, that is, normal and anomalous divergence angles, form a proper subset of the set of Fujita's angles. From my recent survey of the spiral patterns observed over 160 years,[3] it can be said that the types of spiral patterns permitted to exist by the minimal entropy model are sufficient to cover all observable cases.

9. A Problem Redefined

The crux of the botanical problem of phyllotaxis is the factors determining the position of plant primordia. But this problem is a part of a more general problem concerning arrangements of primordia in general. Indeed, there are primordia in such living systems as *Hydra* and

Botryllus "where the over-all process of column growth and bud formation is strictly comparable to that of leaf primordia" (see Berrill, Ref. 39, p. 228 and p. 423). Tentacles, canals and zooids in some medusae conform exactly to (2, 3) and (3, 5) phyllotaxis. Furthermore, one must not forget that phyllotaxis-like patterns arise at other levels of organization, such as in microorganisms showing 2(2, 3) and 2(3, 5) patterns, and proteins with (7, 11) phyllotaxis and 99.5° divergence.

For Rivier,[19,21] who used the language of crystallography to describe the phyllotactic pattern of the daisy, it is clear that "phyllotaxis does indeed belong to crystallography." Rothen and Koch[24,25] consider that phyllotaxis belongs to exotic crystallography. With the discovery of quasi-crystals in 1984, the discipline of phyllotaxis has become a playground for physicists and crystallographers. They use terms such as parastichy (spiral) and rising phyllotaxis to describe their lattices, and their own terms, such as cylindrical crystal, dislocation and disclination, scale-invariance, inflation and deflation, quasi-periodic lattices are beginning to invade the field. For Hermant[40] (see Ref. 14), phyllotaxis is a purely formal, architectural problem. There is certainly a fundamental principle applicable to all the patterns reported here. Beyond the tremendously complex processes involved in the growth of these organisms, general and simple laws apply, such as the principle of optimal design, and the principle of minimal entropy production, the simple laws of packing efficiency and self-similarity, and those of fractal geometry generating structures, such as the hierarchical organization of the vascular systems (see Refs. 14, 33). The simple minimality principle proposed here as a basis for the ϕ-model and the minimal entropy principle in the systemic model[15,16,41] are applicable in other areas than phyllotaxis.

If the problem of phyllotaxis transcends the domain of botany, its solution can be given without reference to the strictly botanical substratum.[42] This bears upon the future of experimental, biochemical and physiological researches in phyllotaxis. A first important consequence is that the models based on diffusion of phytohormones, which have always been favored in botanical circles (despite the fact that not only are they not supported by any data, but also that there are contradictory data), have to be reevaluated, if not discarded. And if Fibonacci patterns are more frequent in the plant kingdom, then there must be special biological conditions favoring their emergence.

These statements also stress the importance of mathematics in the resolution of the problems. The place of mathematics in phyllotaxis, which too often is considered only as possibly useful for a posteriori descriptions, that is considered as not being able to reach the causes and generating principles, has also to be redefined. For example, the question "why is maximal uniformity so important in plant growth?" does not only belong to the realm of biology, but to that of mathematics as well. Phyllotaxis is indeed the simple, yet not elementary, mathematical solution of an optimality problem, caused by complex environmental and genetical constraints which can be modelled as a whole via systemic concepts, such as entropy. Not many models in phyllotaxis offer predictions to be confronted with the data available. And given that data collecting in phyllotaxis is rarely organized around the idea of verifying models, there is a lack of data that would allow us to check the available predictions, despite the abundance of data in general. For example, we do not know if the divergence angle of 78°, corresponding to the sequence <1, 4, 5, 9, 14, . . .>, is more frequent than the divergence angle of 151.1°, corresponding to the sequence <2, 5, 7, 12, 19, . . .> (see Ref. 3), while a model predicts that the former is more frequent.

10. Epilogue

This paper shows a variety of areas in biology where spiral lattices are found and where the methods of phyllotaxis are applicable. It shows the importance of number ϕ in spiral patterns and proposed an algorithm called the ϕ-model, which gives a very simple way to obtain and order the noble numbers (whose continued fractions end with $1/\phi = [0; 1, 1, 1, 1, \ldots]$), among which are well-known divergence angles. The importance of these numbers springs from studies on packing efficiency of plant primordia. The ϕ-model offers a formal link between Marzec's and Jean's entropy models in phyllotaxis. The author stresses the fact that the application of the methods of other fields of research in phyllotaxis is bound to deeply and irreversibly affect the way to look at the problems and at the approaches to the problems of phyllotaxis. Future research in this field of pattern recognition and pattern generation will have to give a strong emphasis on holistic and phylogenetic concepts.

Acknowledgement

To the CRSNG of Canada, for the three-year grant A6240.

References

1. T. Komai and I. Yamazi, Order found in the arrangement of organs and zoöids in some medusae, *Annotationes Zoologicae Japonensis* **26** #1 (1945) 1–7.
2. R.V. Jean and W.W. Schwabe, Shoot development of plants module UMAP unit 702, in *UMAP Modules Tools for Teaching*, COMAP, Lexington, MA (1990) pp. 169–208.
3. R.V. Jean, Predictions confronted with observations in phyllotaxis, *J. Inf. Deduct. Biol.* **3** (1986b) 1–19.
4. M.A. Lauffer and C.L. Stevens, Structure of the tobacco mosaic virus particle; polymerisation of the tobacco mosaic virus protein, *Advan. Virus Res.* **13** (1968) 1–63.
5. D.J. DeRosier and A. Klug, Structure of the tubular variants of the head of bacteriophage T4 (polyheads) I. arrangement of subunits in some classes of polyheads, *J. Mol. Biol.* **65** (1972) 469–488.
6. R.O. Erickson, Tubular packing of spheres in biological fine structures, *Science* **181** #4101 (1973) 705–716.
7. F.H.C. Crick, The packing of α-helices: simple coiled coils, *Acta Cryst.* **6** (1953) 689–697.
8. R.V. Jean, A geometric representation of α-polypeptide chains revisited, *J. Theor. Biol.* **112** (1985) 819–825.
9. R.V. Jean, A mathematical model and a method for the practical assessment of the phyllotactic patterns, *J. Theor. Biol.* **129** (1987) 69–90.
10. S.V. Meyen, Plant physiology in its homogenetical aspects, *Bot. Rev.* **39** #3 (1973) 205–260.
11. P.S. Stevens, *Patterns in Nature.* Little, Brown & Co, Boston (1974).
12. J. Murray, Les taches du léopard, *Pour la Science.* May (1988) 78–87.
13. R.V. Jean, *Mathematical Approach to Pattern and Form in Plant Growth.* Wiley-Interscience, New York (1984).
14. R.V. Jean, A basic theorem on, and a fundamental approach to pattern formation on Plant, *Math. Biosci.* **79** #2 (1986a) 127–154.
15. R.V. Jean, A systemic model of growth in botanometry, *J. Theor. Biol.* **87** (1980) 569–584.
16. R.V. Jean, Model of pattern generation on plants based on the principle of minimal entropy production, in *Thermodynamics and Pattern Formation in Biology*, I. Lamprecht anf A.I. Zotin, eds. W. de Gruyter pub., West Berlin, (1988b) 249–264.
17. R.V. Jean, Allometric relations in plant growth, *J. Math. Biol.* **18** (1983a) 189–200.
18. R.V. Jean, A fundamental problem in plant morphogenesis, from the standpoint of differential growth, *in Mathematical Modelling in Science and Technology*, X.J.R. Avula, R.E. Kalman, A.I. Liapis and E.Y. Rodin, eds. Pergamon Press, New York (1983b) 774–777.
19. N. Rivier, Crystallography of spiral lattices, *Mod. Phys. Letters* **B2** (1986) 953–960.
20. W.F. Harris and L.E. Scriven, Moving dislocations in cylindrical crystals cause waves of bending, *J. Mechanochem. Cell Motility* **1** #1 (1971) 33–40.

21. N. Rivier, A botanical quasicrystal, *Journal de Physique*, Colloque C3, Supplément au **47** #7 (1988) 299–309.
22. J.N. Ridley, Packing efficiency in sunflower heads, *Math. Biosci.* **58** (1982) 129–139.
23. C. Marzec, Phyllotaxis as a dissipative structure, in *Mathematical Modelling in Science and Technology* (Proc. Vth ICMM, Symposium on Phyllotaxis, July 1985, U. of California at Berkeley), X.J.R. Avula, G. Leitmann, C.D. Mote, E.Y. Rodin, eds. Pergamon Press, New York (1986) 740–745.
24. F. Rothen and A.J. Koch, Phyllotaxis, or the properties of spiral lattices, I: Shape invariance under compression, *Journal de Physique*, in press (1990).
25. F. Rothen and A.J. Koch, Phyllotaxis or the properties of spiral lattices, II: Packing of circles along logarithmic spirals, *Journal de Physique*, in press (1990).
26. R. Rosen, *Optimality Principles in Biology*. Butterworth, London (1967).
27. R.V. Jean, Number-theoretic properties of two-dimensional lattices, *J. Number Theory* **29** (1988c) 206–223.
28. H.S.M. Coxeter, The role of intermediate convergents in Tait's explanation of phyllotaxis, *J. Alg.* **20** (1972) 167–175.
29. R.V. Jean, A synergetic approach to plant pattern generation, *Math. Biosci.* **98** (1990) 13–47.
30. I. Adler, The consequences of contact pressure in phyllotaxis, *J. Theor. Biol.* **65** (1977) 29–77.
31. J.H.M. Thornley, Phyllotaxis I: A mechanistic model, *Ann. Bot.* **39** (1975) 491–507.
32. C. Marzec and J. Kappraff, Properties of maximal spacing on a circle related to phyllotaxis and to the golden mean, *J. Theor. Biol.* **103** (1983) 201–226.
33. R.V. Jean, The hierarchical control of phyllotaxis, *Annals of Botany* (London) **49** (1982) 747–760.
34. A.H. Church, *On the Interpretation of Phenomena of Phyllotaxis*. Hafner Pub., New York (1920).
35. R.D. Meicenheimer, *Growth characteristics of Epilobium hirsutum shoots exhibiting bijugate and spiral phyllotaxy*. Ph.D Dissertation, Department of Botany, Washington State University (1980).
36. L.F. Hernandez, *Organ morphogenesis and phyllotaxis in the capitulum of sunflower* (*Helianthus annuus* L.). Ph. D. Dissertation, The University of New South Wales, Australia (1988).
37. T. Fujita, Über die reihe 2, 5, 7, 12, . . . in der schraubigen blattstellung und die mathematische betrachtung verschiedener zahlenreihensysteme, *Bot, Mag. Tokyo* **51** (1937) 480–489.
38. K. Popper, *The Logic of Scientific Discovery*. Harper-Row, New York (1968).
39. N.J. Berrill, *Growth, Development, and Pattern*. W.H. Freeman Co, San Francisco (1961).
40. A. Hermant, Structures et formes naturelles, géométrie et architecture des plantes, *Techniques et Architecture* **VI** (1946) 9–10, 421–431.
41. R.V. Jean, Phyllotactic pattern generation: a conceptual model, *Annals of Botany* (London) **61** (1988a) 293–303.
42. R.V. Jean, Phyllotaxis: a reappraisal, *Can. J. Bot.* **67** (1989a) 3103–3107.

HAWAIIAN FLOWERS WITH FIVEFOLD SYMMETRY

Magdolna Hargittai

It is no wonder that Hawaii is a tourist paradise. It has everything one can expect: sunny beaches, deep blue sea under bright blue sky, mysterious unhikable mountains extending into dark clouds and, most of all, the exotic air of the Pacific. You can feel the past, the touch of the primitive but pleasant life of the ancient Polynesians. And, of course, it is not only history, it is also the exotic fragrance of the tropics that you can feel in the air. The flowers are truly astonishing, especially for someone coming from Central Europe or other places of temperate climate.

One cannot help wondering how there can be so many kinds of flowers and trees on these islands which were once nothing but bare volcanic lava. It is believed[1] that when the first Polynesian travellers, Hawaii's first inhabitants, arrived about a thousand years ago, they found here about 400 different plants. These plants must have reached the islands floating on ocean currents, stuck to the feathers and feet of birds, or attached to logs floating on the water and washed upon the shore. The Polynesians themselves brought about 24 plants from their homeland; they were useful as food, building materials, or medicines.

Most of Hawaii's unique and picturesque flora was brought here by later travellers after Captain Cook discovered the islands in 1778. Most of them came during the last century. They originate from the world's tropical areas, in Asia, Africa, Central and South America besides the

Pacific Islands. Thousands of flowering plants make Hawaii a huge tropical garden, unique in the world.

The flowers with fivefold symmetry are not the most spectacular ones of Hawaii's flora. All the orchids, for example, have only bilateral symmetry. The striking heliconias have only translational symmetry along their stem. Nonetheless, these rather common flowers are also beautiful and give a glimpse into nature's endless variations on the same theme.

Plants generally belonging to monocots and dicots differ characteristically in their roots, stems, leaves, flowers, and fruits. Three and its multiples characterize the inner structure of monocot flowers, while four and five and their multiples characterize the inner structure of dicot flowers. Thus, all flowers displaying fivefold symmetry belong to the dicots. Fivefoldedness may appear as rotational symmetry only (C_5) or as fivefold reflection (C_{5v}). In the following compilation, the flowers with only rotational symmetry are followed by those with reflections. Needless to say, the rigor of geometrical symmetry is absent and we observe approximate symmetry which is sometimes also called material symmetry.

The first flower shown is the hibiscus, the state flower of Hawaii (Fig. 1).[a] It is not a demanding plant, and since cross-pollination is easy, it comes in many varieties. Hawaii has the largest known number of hybrid hibiscuses in the world,[2] more than five thousand. There are all sorts of colors, from white and yellow to orange and red. The flowers bloom throughout the year and are seen everywhere. They are popular for flower arrangements, because they do not wilt, but they last for only a day, even on the shrub where they close and fade at nightfall.

A relative of the hibiscus is the hau. It was among the few plants that the early travellers brought with them in their canoes. The wood, the bark, the fiber, and even the flower had many uses. The hau tree has bright yellow flowers at its branch tips (Fig. 2). These flowers turn to orange later in the day and they change to red before they fall at night.

Large velvety golden-yellow flowers characterize the allamanda plant (Fig. 3). The brownish buds are arranged into clusters and only a few of them open at a time. Its Hawaiian name means heavenly chief.

[a] In figure captions besides the common name (if any) of the flower, the Hawaiian name (if found), the botanical name in italics, the family in parenthesis, and the places of origin are given in this order. Refs. 2 and 3 have been consulted for identification.

Fig. 1 Hibiscus, Aloalo, *Hibiscus rosa-sinensis albus*, (Malvaceae), Tropical Asia.

Fig. 2 Hau, *Hibiscus tiliaceus*, (Malvaceae), Polynesia.

Fig. 3 Yellow allamanda, Lani-alii, *Allamanda cathartica* var. *hendersonii*, (Apocynaceae), Brazil.

Fig. 4 shows the flower of the oleander. This tall evergreen ornamental shrub can be found frequently as hedges and on the sides of highways. Their flowers are usually white or pink but there are some red varieties as well. All oleanders are poisonous in all their parts, and their fragrance

Fig. 4 Oleander, Oliwa, oleana, *Nerium indicum*, (Apocynaceae), East Asia.

can cause sickness in a closed space. Even insects are afraid to attack them.

Different varieties of plumeria are shown in Figs. 5–8. The flowers of the common plumeria (Fig. 5) appear at the tips of tree branches by themselves or in clusters in spring and continue to bloom until winter. The individual flowers are yellowish with creamy white edges and backs. Another variety is the Plumeria obtusa (Fig. 6), which has white flowers with a yellow center and more oblong leaves than the common plumeria. It is evergreen and blooms continuously. Plumeria rubra (Fig. 7) is a variety with smaller and more pointed red or rose-colored flowers. The clusters of flowers appear on long thick stalks. Hybrids of the red, white, and yellow plumerias give variations ranging from white and pink to red and scarlet. One example is shown in Fig. 8. Plumeria is the most popular and most common flower for leis, which are the famous flower garlands of the Islands. The flower is durable, does not stain clothes and has a

Fig. 5 Plumeria or Frangipani, Pua melia, *Plumeria acutifolia* or *Plumeria acuminata*, (Apocynaceae), Tropical America.

Fig. 6 Singapore plumeria, *Plumeria obtusa*, (Apocynaceae), West Indies.

Fig. 7 Red plumeria, *Plumeria rubra*, (Apocynaceae), Tropical America.

Fig. 8 Hybrid plumeria.

sweet characteristic fragrance. It is probably the most abundant flowering tree on the Islands. It was named after the French botanist, Plumier, but they made a mistake in the official spelling and hence Plumeria and not Plumiera. The other name of the flower, Frangipani, comes from "frangipanier," French for coagulated milk, referring to its milky sap which is poisonous.

Madagascar periwinkle is shown in Fig. 9, a popular everblooming shrub used for edges or as a ground cover. The flowers are usually white, sometimes pink, with or without a red throat; they grow in small clusters at the branch tips. The plant is said to have medicinal properties, but is poisonous to cattle.

Fig. 9 Madagascar periwinkle, Kihapai, *Catharanthus roseus* syn. *Vinca rosea*, (Apocynaceae), Tropical America.

The so-called "flower of love" is seen in Fig. 10. This is an evergreen shrub, which has glossy green leaves, sweet-scented creamy-white flowers with five pinwheel-like petals. The flowers are fragrant at night.

Fig. 10 Flower of love, *Tabernaemontana (Ervatamia) corymbosa*, (Apocynaceae), India to Malaysia.

The rose-flowered jatropha (Fig. 11) is an evergreen shrub growing to about one to two meters high with small pink or scarlet flowers in branched clusters.

Fig. 11 Rosa-flowered jatropha, *Jatropha pandurifolia*, (Euphorbiaceae), West Indies.

The yellow ilima in Fig. 12 is the flower of Honolulu. Different forms of ilima plants can be found in Hawaii, and besides the yellow one there are some orange and red varieties. In earlier times, the yellow form was used for leis which were supposed to be worn only by the *alii* (chiefs and noble people) because yellow was the color of royalty.

Fig. 12 Sida fallax, Ilima, *Sida fallax*, (Malvaceae), Pacific islands.

Bougainvilleas (Fig. 13) are woody vines, they have a long flowering time, and give a striking effect of bright vivid color. Commonly they are purple, but there are some brick-red, pink, and white varieties as well. Interestingly, the color is not due to the flowers themselves but to colored leaves which form a three-part bract around the true flower. The flower is small, yellow, tubular, and might be mistaken for the ordinary yellow center of a flower.

The mussaenda flower is also very small (Fig. 14) and is surrounded by hairy reddish leaflike calyx lobes. Another variety of mussaenda (Fig. 15) has the same flower, but the surrounding calyx lobes are white.

Fig. 13 Bougainvillea, Pukanawila, *Bougainvillea spectabilis, (Nyctaginaceae), Brazil.*

Fig. 14 Mussaenda, *Mussaenda erythrophilla "Rosea",* (Rubiaceae), Java.

Fig. 15 Mussaenda, *Mussaenda philippica*, (Rubiaceae), Philippines.

Fig. 16 shows the flower of the natal plum. The shrub has long sharp thorns and bright red fruits. It is used for hedges, because the thorns make the shrub impenetrable. The flowers are very fragrant; the fruit makes tasty jelly.

Fig. 16 Natal plum, *Carissa grandiflora*, (Apocynaceae), South Africa.

The flower in Fig. 17 is called the "Star of Bethlehem," because of its attractive, erect, white flowers. It is a nice low plant with long, narrow, tooth-margined leaves and flowers with spreading petals and long slender tubes. It is a weed in Hawaii and its juice is very poisonous.

Fig. 17 Star of Bethlehem, Pua hoku, *Hippobroma longiflora*, (Sapindaceae), Tropical America.

Fig. 18 shows the plumbago flower. The plant is a climbing shrub with clusters of delicate, baby blue flowers which are, nevertheless, strong and therefore used for leis. The plant is popular for hedges and as ground cover.

The ruellia (Fig. 19) is frequently used as a ground cover in tropical and subtropical countries.

Fig. 18 Plumbago, *Plumbago capensis*, (Plumbaginaceae), South Africa.

Fig. 19 Ruellia, *Ruellia ciliosa*, (Acanthaceae), Subtropical regions.

Fig. 20 shows the crown flower. Its Hawaiian name refers to its crown-like central part. They were Queen Liliuokalani's favourites, as the emblem of royalty. There are both white and lavender varieties. The flower is used for leis; either the whole flower or just the center crowns

Fig. 20 Crown flower or giant indian milkweed, Pua kalaunu, *Calotropis gigantea*, (Asclepiadaceae), India.

broken into two. In its native India the flower is sacred to Shiva. Among the Vellalas of India a fourth marriage is considered unlucky, therefore no father would give his daughter to be a fourth wife. The difficulty is solved with the help of the crown flower; the bridegroom "marries" a crown-flower shrub and after that he is free to marry a fifth wife.

The stephanotis flower (Fig. 21) is fragrant, waxy, and trumpet-shaped. It grows in clusters on the vine and is popular for leis. Brides like to wear it, hence its Hawaiian name which means "marriage flower."

Fig. 21 Stephanotis, Pua male, *Stephanotis floribunda*, (Asclepiadaceae), Madagascar.

Pentas lanceolata is an ornamental tropical plant with bright green, hairy leaves and tubular flowers. They appear in showy clusters on branch tips and can be white (Fig. 22), pink, or red.

Fig. 22 *Pentas lanceolata*, (Rubiaceae), Tropical.

The carrion is a small, leafless plant with four-angled cactus-like erect stems. Already the flower buds display pentagonal symmetry (see Fig. 23). The flowers resemble a starfish in their color, shape, and size (see Fig. 24).

Fig. 23 Carrion flower, *Stapelia gigantea pallida*, (Asclepiadaceae), Africa.

Looking around the islands' gardens it seems surprising at first that more plants from the temperate climate zone cannot be found here. One would think that they should thrive in the mild climate of Hawaii. Apparently, however, they don't. Temperate zone plants are conditioned to changing seasons; they need not only the warm but also the cold weather, when they can rest. Therefore they do not grow well in these parts of the world, if they do at all.

Plants imported from other tropical or subtropical areas, however, love the Hawaiian climate and adjust easily to the living conditions here. They have come from jungles and deserts, from monsoon areas, and from other ocean islands. Probably no other tropical area has such a large and spreading collection of brilliant flowering trees, lush tropicals and exotic blossoms as Hawaii. They paint these Islands with beautiful, bright colors all year round, living happily together — just as the people who have come here from distant lands and brought with them their favorite trees and flowers.

Fig. 24 Carrion flower, *Stapelia gigantea pallida*, (Asclepiadaceae), Africa.

All photographs in this paper were taken by István and Magdolna Hargittai.

References

1. L.E. Kuck and R.C. Tongg, *Hawaiian Flowers & Flowering Trees*. Charles E. Tuttle Co., Rutland, VT (1958).
2. M.C. Neal, *In Gardens of Hawaii*. Bishop Museum Press, Honolulu, HI (1965).
3. A.B. Graf, *Tropica*. Horowitz & Sons, Fairfield, NJ (1978).

INDEX

A

B

C

D

E

F

G

H

I

J

K

L

M

O

P

Q

R

T

U

V

W

Y

Z

Artist's rendition of fivefold symmetry: "Pentagonal Interference II," by Ferenc Lantos, Pécs, 1991.